NUMBER WORDS AND NUMBER SYMBOLS

A Cultural History of Numbers

Karl Menninger

Translated by Paul Broneer
from the revised German edition

DOVER PUBLICATIONS, INC., New York

Tolle numerum omnibus rebus
et omnia pereunt.
Take from all things their number,
and all shall perish.
ISIDORE OF SEVILLE (CA. 600)

Translated from the German with permission of the Vandenhoeck & Ruprecht Publishing Company, Göttingen, Germany.

English translation copyright © 1969 by The Massachusetts Institute of Technology.

All rights reserved under Pan American and International Copyright Conventions.

Published in Canada by General Publishing Company, Ltd., 30 Lesmill Road, Don Mills, Toronto, Ontario.

This Dover edition, first published in 1992, is an unabridged, unaltered republication of the English translation (published by The MIT Press, Cambridge, Mass., 1969) of the revised German edition of *Zahlwort und Ziffer: Eine Kulturgeschichte der Zahlen*, published by the Vandenhoeck & Ruprecht Publishing Company, Göttingen, Germany, 1957–58.

The Dover edition is published by special arrangement with Vandenhoeck & Ruprecht GmbH & Co. KG, Theaterstrasse 13, Postfach 3753, D-3400 Göttingen, Germany, and with The MIT Press, Massachusetts Institute of Technology, 55 Hayward Street, Cambridge, Massachusetts 02142.

Manufactured in the United States of America
Dover Publications, Inc , 31 East 2nd Street, Mineola, N.Y. 11501

Library of Congress Cataloging-in-Publication Data

Menninger, Karl, 1898–
    [Zahlwort und Ziffer. English]
    Number words and number symbols : a cultural history of numbers / Karl Menninger ; translated by Paul Broneer from the revised German edition.
        p.   cm.
    Translation of: Zahlwort und Ziffer.
    Originally published: Cambridge, Mass. : M.I.T. Press, 1969.
    Includes index.
    ISBN 0-486-27096-3 (pbk.)
    1. Numerals—History.   2. Numeration—History.   3. Abacus—History.
I. Title.
QA141.2.M4513   1992
513.5—dc20
                                                                91-42709
                                                                CIP

Preface

The concept of number exerts a dual enchantment. When it uncovers the rich relationships between pure magnitudes to the scientist, it fills him with the satisfaction of intellectual insight; for the sake of these relationships the sovereign structure of mathematics rests upon the concept of number. For others it exerts a fascination by its deep interconnection with the daily life of the people. Is it not true that every tribe spoke and noted down numbers, that every tribe had to calculate whenever it faced life on this planet? Did man's relationship to his environment not also necessitate his relationship with numbers?

This book tries to trace this interweaving of numbers and human life. It strives to demonstrate how the clear concept of number adopts a multiple image when it grows with a people. Together the history of knowledge and of culture create a single picture: the first establishes the boundaries, the second fills the spaces with color. And all along the wealth of detail, no matter how much it takes up the reader's attention, permits the music to be heard of the eternal stream at whose banks the fate of humanity is fulfilled.

Number takes shape both in spoken and in written language, in "number word and number symbol." Just as a child learns to count one, two, three long before it learns to write or calculate with the numerals 1, 2, 3, the spoken number language precedes the written language; at least the two are so essentially separated in character that it seemed justified to separate the cultural history of number into two series of chapters, the first covering Number Sequence and Number Language and the second covering Written Numerals and Computation.

This book is written for the lover of intellectual and cultural history, but the professional historian will discern many things in it not previously expressed. Of course the plain knowledge of the number series and number symbols of practically all cultures, the Greeks, Romans, Egyptians, Babylonians, Chinese, Indians, and so on, has to be cited first, but ethnology and ethnography contribute a most colorful addition: the history of language, of culture, and of politics. There are few things of this world in which these branches of research meet each other in such an exciting and fertile manner as the concept of number. The area of its symbolic and mythical interpretation is not even included.

With this overwhelming wealth of detail it became difficult to pursue the great art of following the threads in this closely woven fabric, to separate them without destroying the fabric itself. The many illustrations and the form of presentation are intended to permit the reader to arrive at his own judgment so that he can participate intimately in the stimulating spectacle in which number has found form both in spoken and written language.

It is not often that the lover of numbers becomes acquainted with the intrinsic connection of his special area with cultural history; it is equally rare that the friend of the history of culture becomes aware of the relationship between his field and the life of numbers. I hope that the present work will serve both groups to gain the insight and the joy which comes from all knowledge of creative intellectuality in the diversity of men and peoples.

It is a very personal need for me to express my thanks for the support which I found wherever I turned. In the course of the frequently discouraging detail work of collecting, it was a happy feeling of intellectual interplay when each bit of information obtained was backed up by friendly readiness.

Above all I am grateful to the Hesse State Library in Darmstadt for laboriously securing sources for me, and also to the University Library in Heidelberg, the manuscript collection of the Bavarian State Library, the Municipal Archives in Augsburg, and the Hessian State Archives in Darmstadt. Among state collections I am particularly obligated to the Deutsche Museum in Munich, the Ethnographic Museums in Frankfurt, Stuttgart, and Munich, the Museum for Ethnology in Berlin, the numismatic collections in Munich, Karlsruhe, and Darmstadt, the Historic Museums in Basel and Vienna, the British Museum in London, the State Museums in Athens, Dublin, Helsinki, Naples, Ostia, and Rome, the Cabinet des Medailles of the Bibliothèque Nationale in Paris, the Musée de l'Oeuvre Notre Dame in Strasbourg, and the former German Institute of Archeology in Cairo.

For personal support I want to express my sincere thanks to G. Beaujouan of Paris, H. F. Deininger of Augsburg, E. Delp of Bad Nauheim, W. Eilers of Marburg, P. Hammerich of Dinkelsbühl, H. Horst of Tehran, A. König of Frankfurt, Nobuko Yokota of Tokyo, O. Neugebauer of Providence, Rhode Island, J. Rak-Sell of Singapore, the late J. Ruska of Berlin, R. Schlösser of Hannover, and K. Vogel of Munich. I am particularly beholden to Mr. Wolfram Müller of Frankfurt and Tokyo.

Heppenheim an der Bergstrasse   KARL MENNINGER
Spring 1958

List of Phonetic Symbols

ś represents a sound between s and sh

š is pronounced like sh in English "shoe"

č is pronounced like ch in English "church"

ə represents a suppressed or murmured e, like the last syllable in "America"

þ in Gothic and Old Nordic is pronounced like th in English "three"

ě corresponds to Russian ye

l, m, r represent syllabic sounds, without vowels

q is pronounced like k

u̯ is a semivowel, between u and w

Contents

NUMBER SEQUENCE AND
NUMBER LANGUAGE

Introduction

What image of numerical values is evoked by the curious succession of the words "one," "two," "three," and so on, which we call the number sequence? Going back deep into history, we find that the number sequence did not spring into existence fully formed, but rather that it evolved stepwise from one numerical boundary to the next. Such "rudimentary" first stages explain a series of otherwise unintelligible peculiarities inherent in full-fledged, "mature" number sequences.

These early difficulties have been overcome by an analysis of the number sequence. In the groupings that precede the sequence and in the gradation of numbers we discern the two basic laws governing both the number sequence and the written number symbols. The question of how these rules of succession are observed by the individual languages opens up a wide range of possibilities — such as succession by size, overcounting, ciphering, specification, and many others — possibilities that bear witness to the astonishing inventiveness of primitive man but also to the conceptual difficulties with number sequences encountered by him.

The key to all such investigations lies in our own words for numbers. How did *our* number sequence originate, what was its origin in the dawn of intellectual and linguistic growth, which continues to its present form in the full daylight of recorded history? For an answer we have the vast panorama of the whole family of related Indo-European languages. For our own number system does not stand alone; it is closely interwoven with those of many other European and Asiatic languages. The latter thus shed light on ours and illuminate much that would otherwise be impenetrably obscure, while in turn our number words throw light on those of others.

One question still remains unanswered here: Have the Babylonian numbers in any way influenced our own? The "great hundred," the frequent use of the number 12, and the peculiar, characteristic break that in many Germanic number sequences follows 60, all call to mind the Roman duodecimal fractions and suggest the sexagesimal order that dominates Babylonian spoken numbers and written numerals.

At the close we let the hidden number words — that is, words whose root is a number word although so distorted that it is scarcely discernible — once more reveal the richness and color of the world of number language. And once again we are conscious of how a primordial sense of order evolved, shrouded in the mystery of early dawn, and continues to influence our every-day life.

A more specific idea of the subjects covered may be gleaned from the Table of Contents.

The reader is advised to consult the complete number sequences (pp. 92 ff.), even when this is not specifically referred to, whenever he encounters a foreign number word; this will allow him to become better acquainted with the language.

Foreign number words had to be transliterated whenever the language could not be written with our own characters (as in the case

[3]

of Greek, Russian, Arabic, and Chinese) or when the alphabetic symbols have phonetic values differing from those of our own (as in the case of Czech or Polish). Linguistics employs for each individual language a separate system of transcription designed for that language alone, except when it writes in the original script. But since the present book is addressed to a wider circle of readers who may become confused by an excess of phonetic symbols, I have followed a compromise method, using a transcription that reproduces approximately the sounds and symbols of *all* languages (see the list of phonetic symbols on p. vii). Where, however, some special pronunciation is called for, the reader will find it in the complete number sequence of the respective language (as, for example, in the case of Gothic, on p. 92). Symbols for long and short vowels are omitted. The reader is assumed to be familiar with English, French, Italian, and German pronunciations.

In this book it was not possible to use a pure and rigorous system of transliteration (such as the scientifically accepted *θǝ:'ti:n* for *thirteen*), because this would completely distort the orthographic picture; even a modified system, such as the use of the Indo-European *kwetwores* for *quetu̯ores* (four), for example, would throw the important connection of this word with the Latin *quattuor* out of focus.

The Number Sequence

The Abstract Number Sequence

" ... *decem* ...
Hic numerus magno tunc in honore
fuit.
Seu quia tot digiti, per quos numerare
solemus."

" ... ten ...
This number was of old held high in
honor,
for such is the number of fingers by
which we count."

Ovid, *Fasti* III

How do we count today?

Before we go into the historical development of our words for numbers, let us first determine *how* and *what* we actually count and what "counting" really is.

Before us lies a heap of peas, which we wish to count. How do we go about this? We arrange the peas in a row, physically or mentally, touch the first one and say "one," then touch the second and say "two," touch the next and say "three," ... touch the last and say "twenty-two"; there are 22 peas in all. What have we actually done? We have *assigned* a word to each individual pea. Counting thus constitutes *assigning* words to things.

To what are these words assigned? To the things we are counting — in this case to peas. At other times we may be counting houses, trees, people, or fingers. Can we also count things of different natures: a pen, a desk, and a cat, for example? Yes, these are three "objects." Can intangible things be counted as well, such as the conclusions of a proof, or the thoughts embodied in a thesis? Yes. Even a person's traits: intelligent, slender, lively, generous, and so on, can be *enumerated*. In short, any distinguishable entity, tangible or intangible, identical or different, can be counted. These distinguishable things, considered together, constitute a *set* and are themselves the *elements* of this set.

Thus, we now say: A set can always be counted by assigning number words to its elements.

But the number words themselves form a set, the elements of which are the words "one," "two," "three," and so on. In the process of counting, the elements of the set of number words or of the number sequence, whichever we prefer to call it, are assigned uniquely to the elements of the set of things to be counted: uniquely, because only a single number word is attached to each pea.

If we think of the separate terms of the number sequence as small boxes, labeled 1, 2, 3, and so on, we can conceive of the counting process as follows: In each box, starting with the first, we place a single pea, the first in box 1 and the last in box 22. Then 22 boxes of our number sequence are full, while all the remaining boxes from 23 on remain empty.

Now we can explain the heading, "The Abstract (Empty) Number Sequence." So long as there is no counting, it is merely there, detached from all concrete objects, unused but ready. But as soon as we count, then according to our first image the number words become assigned to the objects, and according to the second one the objects are placed in the empty boxes of the number sequence. The last number word (or the last box) indicates the *cardinality*, or *number*, of the set.

This insight is as important as it is simple. We shall see, in fact, that the "abstraction" of the number sequence from the things counted created great difficulties for the human mind. We need only ask ourselves: How would we count if we did not possess this sequence of remarkable words, "one," "two," "three," and so on? Yet there was a time when it did not exist!

Thus one achievement of our number sequence is its independence of the things themselves. It can be used to count *anything*.

But can it also be used to count arbitrarily large sets, even the sands of the sea? Yes, even these sets "without number" can be counted by using our number sequence — this is its other achievement. To each successive grain of sand it assigns a number word, tirelessly and inexhaustibly. And when the last particle of sand has been counted, it still has "infinitely many" number words with which to go on counting.

The number sequence could go on counting, even though we cannot. Yet we know for certain that it would do so correctly and in the proper sequence. We hear of three million inhabitants of a city: does anyone count them one after another, 1, 2, 3? Still we are quite sure that if this were to be done, we would eventually come to inhabitant No. 2,999,974, No. 2,999,975, No. 2,999,976 ..., and finally to No. 2,999,999 and No. 3,000,000.

Whence this certainty, which we never gained from experience? We know that our number sequence embodies the law of infinite progression; we know that every number has a successor; and we also know how that successor is formed from its predecessor.

Hence, our number sequence is not a motley collection of words arbitrarily gathered together, but an ordered creation of the mind. It incorporates the law of infinite progression, by force of which we acknowledge the countability of sets even though we ourselves cannot actually do the counting.

A finite and amazingly small quantity of number words is enough for this purpose, for the number sequence uses these words over and over again, in their proper order and context. And it is completely independent of the objects it counts: it is abstract. Therefore it can count anything.

This is our modern number sequence, the number sequence in its highest state of development. And now that we are familiar with it, the question becomes especially absorbing: was it not always so?

The Number Sequence
Used Concretely

Haven't people always counted as we do today?

We shall find the answer to this question if we descend the ladder of culture down to the very lowest steps, scarcely above the level where mind could not rise above its environment. Early man counted, too, whenever he merely gathered fruits or hunted, whether he grew his own food by more or less primitive methods of cultivation or drove his herds from pasture to pasture or whether, like many tribes living near the coast, he sought to earn his living by trade. His way of life taught him to count, the nature of his economy determining the extent of his number sequence. Why should a pygmy people, living in isolation in the primeval forest, need to count beyond 2? Anything over that is considered "many." But the cattlebreeder must count his herd, head by head, up to 100 or even more. For him "many" is something far greater, something that no longer has economic meaning to him. Thus early man's environment determined his thinking and actions, and also his counting.

So that we can understand his number sequences, which we shall now consider, let us dwell for a moment on the way primitive man perceives the world around him. It still impinges on him directly, in all its myriads of colors and forms. Things have not yet been "cooled off" for him by his intellect, which sifts them and orders them and separates them, filing their elements away in the gray, colorless pigeon-holes of concepts. On the contrary, in their immediate, hot-blooded, many-colored uniqueness they touch his innermost heart. Thus they are not objects to him, things which are alien to him and stand "outside" himself — *here* am I and *there* is the world — rather they are completely absorbed in his own life. He is a part of them, as they are of him. He is woven into the very fabric of the universe by powerful strands of religion; he does not, like "modern" man, like ourselves, stand before it in wonderment, in calculation, or indifference.

Yet some remaining fragments of that early perception of the world still loom up in our own. Many a superstition, many an oddity, survives unrecognized in the midst of the intense consciousness of our own culture. Who today knows or cares much about the number 7? Though for us it has lost its supernatural content, though it offers not the slightest advantage in measuring and reckoning time, the seven-day week still governs our whole external life. From that early interpenetration of man with the world arises the infusion of objects and numbers with mystic significance, and hence the "holiness" of the numbers 3 and 7 and the auspiciousness or bad luck of the number 13. It is the task of mythology to uncover the concepts that led early man to impart supernatural significance to certain numbers.

Our own purpose, however, is to understand how early man gave expression to things and events, with all their profusion of kaleidoscopic detail, in his own primitive language. "A man has killed a rabbit" — an American Indian would never say this in such a colorless way. His statement, broken down into its verbal component, says: "The man, he, one, animate, standing, has purposely killed by shooting an arrow at the rabbit, animate, him, one, sitting." The Indian does not put it thus because he wants to express the event in a specially picturesque manner, as our highly developed speech could also do by adding words and phrases; he *cannot* say it any other way, because that is how he has experienced the event, and he cannot free himself from its uniqueness. Generalization into pale concepts is completely foreign to him. His language proves this, since it achieves its colorful expression not as does ours, by using auxiliary words and phrases, but rather through the inflection of its words and through particles prefixed, suffixed, or incorporated in them. Just as we (in German) can indicate only the tense and the mood of a verb (*gibt, gab, gäbe* — "gives," "gave," "would give") by inflection and phonetic shift (*i—t, ä*), so can the Indian express the gender, number, intention, and detailed manner of the act of killing in one single, inflected word. Whereas we, especially in scientific language, stress only the essential aspects of an observation, shedding all incidentals and compressing the main point into a general concept, primitive man puts as many of the details observed by him as he can into his speech. We would have to reassemble a large number of general concepts in order to express the "death of the rabbit" as the Indian does. The abundant inflectional potentialities of early language and its completely different vocabulary testify to early man's keener observation and to his more intimate involvement with the world. The Lapplander has twenty different words for "ice" and twice as many again for "snow"; he can also describe thawing and freezing by a single word in almost as many various ways. How dull, by contrast, are the word forms of English, for example! The four grammatical cases *Mann, Mannes, Manne, Mann* which we express in German by inflecting the word, are all simply "man" in English and must be specified by the auxiliary prepositions "of" or "to," as in Chinese. Chinese has no inflections at all and therefore expresses the relationships among words almost exclusively by their position in the sentence. But Chinese is by its very nature an uninflected language, whereas English over the course of time abandoned its inflections, just as inflected languages generally lose their inclination and eventually even their ability to inflect.

It is also worth noting, however, that while Lithuanian, for instance, has different words for the "gray" of geese and of horses, or wool, of human hair, and so forth, it has no separate word for the generic concept of "gray," which is abstract, or "empty," and must be embodied or "filled" by actual, concrete objects. So powerfully does the idea of the unique, the real, thing persist in the mind of early men.

After this brief linguistic discussion, let us now consider the early number sequences.

NUMBERS WITHOUT WORDS

It was related by a missionary to the Abipones, a tribe of South American Indians compelled by a shortage of food to migrate (in the 18th century): "The long train of mounted women was surrounded in front, in the rear, and on both sides by countless numbers of dogs.

From their saddles the Indians would look around and inspect them. If so much as a single dog was missing from the huge pack, they would keep calling until all were collected together again. I have often since wondered how they, without knowing how to count, would tell at once, in spite of the confused throng, that one dog was missing." Yet they had only three number words and showed the strongest resistance to learning the number sequence from white men. They would indicate the size of a herd of horses by stating how much space the horses occupied when standing next to each other.

We can understand both these phenomena if we remember the far closer relationship of these people with the world around them: the keen observation that unhesitatingly notes the absence of a single animal and can say which one is missing, and the translation of a number that cannot be visualized into a clearly perceived spatial form.

The term *number sense* may be applied to the first of these manifestations. Animals show it when they immediately detect the absence of one of their young. Men also have this latent sense and can develop it. Many a teacher distinctly senses the absence of a pupil when he faces a class doing calisthenics.

NUMBERS AS ATTRIBUTES

Here I must ask the reader to focus his attention on a more subtle distinction expressed in language and offering a wealth of insight into the origin of number words.

Numbers as attributes — is a number not, indeed, an attribute, a trait? "Two cows" — "two" precedes the word "cows" just like, for example, the adjective "beautiful." But we must not let ourselves be deceived by this. "Two" is not a characteristic of the cows themselves, for one cow cannot be two; the Two could be at best an attribute of the entity "two-cows." If, however, we regard "two-cows" as a single unit, we of course no longer need to feel the "two-ness" as a particular attribute, since it is part of the essence of the concept "two-cows." Thus we see that Two is not an attribute in the same sense as "beautiful." Hence, Two is not an adjective. What is it, then? It is a special kind of word — a number word.

Nevertheless, primitive man at first always felt the number to be an adjective. We shall now demonstrate that this is so.

The Number Word in the Noun

Some primitive peoples have completely fused the number and the object into a single entity. The Fiji Islanders, for example, call 10 boats *bola*, 10 coconuts *koro* and 1000 coconuts *saloro*. Naturally this does not hold for any arbitrary number (such as 5 nuts or 23 nuts); yet in contrast to other authorities I see in these words a designation of quantity, although admittedly, tied to the object counted. In German it would be like saying *ein Malter*, a "bushel," or *eine Mandel* (15), thereby implying potatoes in the first instance and eggs in the second. We have a parallel example in the German word *Faden*, "thread." One would think that this was something in the nature of a filament, but it is not: It is a measure — "full fathoms five ..." — applied to yarn. It is as much as can be encompassed by a man's outstretched arms. Today a *Faden* has come to mean a yarn of about this length, and hence is a measure coupled to a specific object.

The examples given show that the primitive people of the Fiji Islands have no number sequence, at least not an extensive one, that has been consciously and clearly detached from objects and thus become abstract.

The Grammatical Double Number (the Dual)

The absorption of the number into the object led to remarkable word forms for the double (the dual), the triple (the trial) and, in some languages of the South Pacific islands, even the quadruple (the quaternal) number. Besides the singular, the German language has only the indefinite plural (the multiple number): *der Mann — die Männer.* If, to make up a hypothetical example, *Manna* (in German) were to mean "two men," then this would be a grammatical dual, an inflected word form indicating the specific number two. This embodiment of the number word into the noun itself is reminiscent of the primitive incorporation of all the details of the "rabbit death" (see p. 10) into a single word. Specific grammatical number words, such as "dual," thus belong to an early stage of civilization.

The ancestral Indo-European language had a form of dual which gradually disappeared from the individual languages of the Indo-European family, surviving here and there only in vestigial form. Classical Greek, for example, had a very ancient but rare dual form:

ho phílos –	*tò philō –*	*hoi phíloi –*
"the friend"	"both friends"	"the friends"
hē cheír –	*tò cheîre –*	*hai cheîres –*
"the hand"	"both hands"	"the hands."

The ancient Semitic languages, such as the Hebrew of the Bible and also Arabic (until around A.D. 700)* had a very pronounced dual, e.g.:

Arab. *radjulun –* "man"   *radjulani –* "two men"   *ridjalun –* "men."

Surviving forms like these are very suggestive evidence of man's earliest steps beyond the number One. Two has a special place among numbers. Primitive man is intensely impressed by the geminate, or paired, condition he observes in his own body or his immediate environment: their 2 (both) eyes, hands, arms, and legs. He transfers this duality to fixed couples, such as a team of horses or a yoke of oxen (Greek *hippō, bóe*), but also to siblings, friends, and deities whom he sees or wants to see together, like the pair of goddesses (*tò theó*) Demeter and Persephone. The Sanskrit *ahani,* "the day," is grammatically a dual because it includes the night; the Turkish *valid* means "the parent," whereas the dual *valid-eijn* are "the parents"; the Indo-European *nasō* means literally "both noses (nostrils)." Sometimes a pair (like the eyes) is so strongly felt to be a single whole that one of the two (one eye) is counted, as in Chinese, by a special number word (*chih* instead of *i*), or else is indicated as a "half-eye," as in the Irish *súil,* "eye," *di súil,* "the (two) eyes," *leth-súil,* "half (= one) eye." This transference of the geminate condition of the

---

* *Editor's note:* Classical written Arabic maintains a dual form to this day.

body to any pair is likewise beautifully exemplified in Chinese, where the ideogram for "pair" in the general sense is a picture of two hands.

Two trees, two people, formed not with the dual but with the number word Two, denotes merely a fortuitous and not an intrinsic or willed duality.

I — Thou. The first step beyond the One, however, was taken at a still lower level of thought. To the awakening consciousness the world is confronted with himself; the I is opposed to and distinct from what is not I, the thou, the other. Linguistically, too, it is not unlikely that the Indo-European number word *duŭo* is in some way related to the German *du* and the English *thou*. In the Sumerian number sequence, "one" and "two" have the meaning "man" and "woman," respectively.

In this primeval dichotomy of the mind, what was One before now breaks apart into One and Two. To man the Two is at first another man, a living You with whom he becomes involved in address and response, and thus in spite of the severance still feels a bond. This is echoed in the fact that the grammatical dual survived much longer in personal pronouns than in other classes of words.

"*Habt's a Geld?*" ("Do you have any money?"), asks the Bavarian peasant — here the Middle High German dual *ez* (Old High German *iz*), "both of you," still lingers on, abbreviated, in the *'s*; the genitive case is *enker*, dative and accusative *enk* (*Wir bitten enk* — "we beg you"). *Es Vogerln tragt's mein Gruss zu ihr*, "You little birds, bear her my greetings," sings the Tyrolean, too, but not the Swabian or the Swiss. Once this dual existed alongside the universal plural *ihr* ("you"); today the plural represents it. But Icelandic still distinguishes between *vid* and *þid*, "we two" and "you two," as opposed to *vjer* and *þjer*, "we all" and "you all." The Old-Frisian dual form for all three persons still survives on the island of Sylt (*wat*, "both of us"; *at*, "both of you"; *jat*, "both of them"), along with the other cases *unk*, *junk*, and *jam*. In Gothic there was *ik*, "I"; *wit*, "both of us"; *weis*, "we"; and *þu*, "thou"; *jut*, "both of you"; *jus*, "you." *Ugkara*, "of us both," and *ugkis*, "us both" or "to us both," exist alongside *unsara* and *uns* (the Gothic *ugk-* is pronounced "unk-"). The Gothic verb has a singular, a dual and a plural: *baíra*, "I bear"; *baíros*, "we both bear"; *baíram*, "we all bear". There are also corresponding forms for the second person. In Old Norse *wit Hrafn* means "we two Hrafn," that is, "Hrafn and I." Thus the personal dual lived on and still survives in languages and dialects, while the inanimate dual has long since ceased to exist.

Two as Unity. In the Two we experience the very essence of number more intensely than in other numbers, that essence being to bind many together into one, to equate plurality and unity. Our mind divides the world into heaven and earth, day and night, light and darkness, right and left, man and woman, I and you — and the more strongly we sense the separation between these poles, whatever they may be, the more powerfully do we also sense their unity. To divide the unified, to unify the divided — this opposition between separation and unification has been fixed by languages, time and again, in compounds formed from the word Two, as in the German words *Zwist*, "discord," and *Zwirn*, "twine," in *diploma* and *dispute*, in the English *twin*, which also means "to separate" (see p. 172), in the Turkish *ikiz*, "twins," and *ikilik*, "a dispute" (from *iki*, "two," see p. 113). Old Armenian has an especially appealing example: from the number word *erku*, "two," are derived the opposed pair *erkin*, "heaven," and *erkir*, "earth."

The fact that the Two may also contain what is evil and despised is rooted in that primeval antithesis: twilight is not an auspicious light, the *Doppelgänger*, the "supernatural double," is a sinister companion, nor is Faust content with the two souls that haunt his breast. To impart a pejorative connotation to the meaning inherent in Two, French sometimes uses a phonetic change, as from *bis*, "twice," to *bes-* (*ba-* or *ber*). For example, *bis-sac* is the double sack which the mendicant monk carries over his back and his chest, whence *besace*, "beggar's satchel," and *besacier*, "beggar" (see p. 174).

Number word and Dual. Quite naturally, many languages express their number 2 in the grammatical dual, thus supplementing its meaning by the grammatical form (as in the Indo-European *dṵo*). Thus the Greek *dýo*, "two," and *ámpho*, "both," as well as the corresponding Latin words *duo* and *ambo* have the dual ending -*o*, so that *duo* expresses not so much the abstract number word Two as the inclusive Both, meaning "the one as well as the other." The sense of the word Both excludes counting beyond, whereas Two implies it. The Indo-European *ambhō* gave rise to the Sanskrit *ubháu*, the Greek *ámpho*, and the Latin *ambo*; dropping the first syllable yields the Gothic *ba*, *bai*, expanded to *bajoþs*, whence are derived the English word *both* and Old High German *be-de*.

Words for numbers higher than 2, such as those for 8, 20, 200 and 2000, which retained their dual form carry us deep into early man's numerical conceptions. The Indo-European *okto(u)*, Latin *octo*, is discussed elsewhere (see pp. 23 and 147). The Latin 20, *viginti* < *dṵi-viginti*, 2 × 10, is an old dual ending in -*i*, whereas the higher tens, like *triginta*, 30, end in -*a* (see p. 150). The Greek *eí-kosi*, 20, is also clearly distinct from the following tens such as *triákonta*, 30, etc. What does all this signify? Here, too, we see the first step beyond unity, but now the unity is no longer 1 but 10, the first cardinal number reached by early man in his counting.

How strongly the Ten was felt to be a new unit is engagingly documented by cases in which 20 is expressed simply by an indeterminate plural of "ten" ("many tens") rather than by "two tens," just as though there were only ten and then twenty but no more tens beyond that. This is what happens in Semitic languages: Hebrew *'eser*, 10, by the suffixing of the plural ending -*im* becomes *esrim*, 20; similarly in Arabic *'asrun*, 10, becomes *isruną*, 20. Subsequent tens are formed as plurals of the single numbers, as in Hebrew *šaloš-im*, 30, "the many threes," from *šaloš*, 3 (see p. 115). The counterpart in our family of languages occurs in Danish, where the plural of *ti*, 10, which is *tyve*, means 20; The subsequent tens are formed from *tyve* and not *ti*; for example *fyrretyve*, 40, is literally 4 × 20 and hence 80 (see p. 65).

Not only the Ten, but also the Hundred is thus felt to be a new unit. From 100 as the new One the first step is made toward Two:

100 in Slavic is *sto*, whose indefinite plural is *sta*, but 200 is called *dve ste* (Russian *dvě sti*, Czech *dvě stě*; see p. 98). *Ste* is the dual "double hundred," so that *dve ste* is actually a redundancy. In just the same way the Lithuanian *du šimtu*, 200, is contrasted with the form *šimtai* in the subsequent hundreds (p. 93). Sanskrit uses the old dual form *dve śate* along with the common form *dviśatam*.

This principle is repeated in exactly the same manner with the next order of magnitude, a thousand: in Russian *týsjača*, 1000, *dvé týsjače*, 2000 (dual) and, oddly enough, this form of thousand is used through 4000 but then followed by *pjatj týsjač*, "five thousands" and so on (see p. 23 for the cause of this peculiar Russian word formation).

As was shown for "ten," Hebrew again forms from *elef*, 1000, the indefinite plural *alpayim*,* "thousands," for 2000, but which at the same time (literally) also serves as a round number for "thousands" and "many thousand." This ambiguity reflects the primordial opposition in thought between One and Not-one or between One and Two. The very same ancient notion is also the reason for the Latin *amb-*, "toward *both* sides," taking on the additional meaning, "at, around," that is, "toward *all* sides" — as in *ambire*, "to walk around," and for the doubled character "man" in Chinese to be read as "every man," and for "east" ("sun behind tree") as "everywhere" (Fig. 1).

Fig. 1 Chinese "everywhere," written "east-east" — that is, "sun behind tree."

Two as the Limit of Counting. All this evidence shows that Two has a special status and is not just a number like any other in the number sequence, but instead is that extraordinary number attained by early man's first hesitant step toward counting. A hesitant step, indeed — for it is not as though man had taken all the following steps at once, running through and building up the whole number sequence. On the contrary, he stopped to catch his breath, as it were. The number 2 is a frontier in counting, the first and oldest of the many we shall encounter.

This is proved not only by the grammatical dual form, or dual number, which we have just discussed, but by other fascinating pieces of evidence as well. In Arabic, for example, the numbers 1 and 2 are adjectives that modify the object counted and testify to the early stage at which number was regarded as an attribute of the thing counted, equivalent to "beautiful" or "big," and not yet as an abstract general concept independent of the object counted. But the numbers following 3, 4, 5, and so on, are nouns (p. 81).

As further evidence we can summon the peculiar formations for 11 and 12, which in many languages are constructed differently from

* *Editor's note: Alpayim* is grammatically a dual form in Hebrew; the plural is *alaphīm*.

the following numbers 13 ... 19 (see p. 84); or the formation of number words by backward counting, as in the Latin words for 18 and 19, "two and one, respectively, from twenty" (see p. 74); or the Finnish and Ainu 8 and 9, "two and one, respectively, from ten" (see pp. 75 & 69). It is always the two numbers 1 and 2 that enjoy this special status.

This becomes quite obvious in the case of the ordinals "first" and "second." In many languages these do not have the same relationship to the number words for 1 and 2 as do the subsequent ordinals "third," "fourth," etc., to "three," "four," and so on. The Indo-European preposition *pro*, "before," in the superlative degree "foremost" or in the comparative "in front of," became the Greek *prōtos*, "first"; similarly, Old-Latin *pri-* for *prae*, "before," by way of *pri-or*, "the one in front," led to *pri-mus*, "foremost, first"; whence, naturally, French *premier* and Italian *primo*. Through sound shifts from *pr > fr* and *pro >* , "before, for," arose such Germanic forms as the Gothic *frum-ists*, Anglo-Saxon *form-est*, English *fore-st > first* and German *Fürst*, "lord, prince." From the superlative degree, Gothic *air*, "early" — *airiza*, Old High German *eriro*, "sooner, earlier" — Old High German *eristo*, comes the German *der Erste*, "the first," with its original meaning of "early morning," for this word is derived from the Indo-European root *ai*, "burn, shine."

Excellent examples from non-Indo-European languages are the Hebrew *rišon*, "the first" < *reš*, "head," and the Egyptian "that which is on the head"; in modern German the word *Haupt*, like the Middle-Latin *capitaneus* > French *chef*, "captain," is restricted in meaning to the leader of a team or group.

"Second" now occasionally appears with the meaning of "the other," Old High German *andar*, Gothic *anþar*, English *other*, Latin *alter*, Lithuanian *antras*, all from the Indo-European root *anteros*, "the one of two"; and sometimes as "the following," as in the Latin *secundus* < *sequi*, "to follow," Greek *deúteros* < *deúomai*, "to lag, stay behind," and similarly in Sanskrit and Anglo-Saxon.* Among the non-Indo-European languages I may mention Finnish, in which *yksi* is one and *kaksi* is two, but *ensimäinen* means "the first" and *toinen* "the second" (the other); in Egyptian, "the second" is called "brother." Whether these formations originate from counting on the fingers or from some other concepts is less important for our purposes than the fact that 1 and 2 clearly occupy a special place in the number sequence and that there was hence a pause after 2, early man's first step in counting. We recognize, moreover, that One itself assumed, and still holds, a unique position relative to the other numbers; we shall speak of this later (p. 19).

The Step to Three. With Three a new element appears in the concept of numbers. I — You: The I is still in a state of juxtaposition toward the You, but what lies beyond them, the It, is the Third, the Many, the Universe. This statement, in which psychological, linguistic, and numerical elements come together, may perhaps roughly paraphrase early man's thinking about numbers. "One — two — many": a

* *Translator's note:* In Swedish the common word for "the second" (of or more) is *andra*.

curious counting pattern, but it is exactly mirrored in the grammatical number forms of the noun, singular — dual — plural, as in the Greek *phílos, phílō, phíloi*, where the third number form is thus the plural. An old Sakai in Malacca, on being asked his age, replied, "Sir, I am three years old." To him 2 was the You, the near and familiar with which he lives, to which he feels related and with which he interacts, but this is no longer true of the It, the 3; for him that is the Many, the Alien, the Unknowable. A magnificent confirmation of this is the ancient Sumerian number sequence, which begins: "Man, woman, many, . . . " (p. 165).

Many investigators suspect, with good reason, that in the number words "one, two, and three" are latent the roots of the personal pronouns "this, the, and that," or even that the primordial forms of these pronouns became the first three number words.

Three is the "beyond," the "trans-." It has been thought that the Latin *tres*, 3, is related to the Latin *trans*, "across, beyond" (the root of *trare*, "to penetrate"; cf. *intrare*, "to enter by force"); and correspondingly the French *trois*, 3, to *très*, "very," the English *three* to *through* and thus the Indo-European *trejes* to *tre-*. Even though this theory cannot be proved with certainty, it does have in its favor the striking linguistic resemblance and the possible interpretation of Three as the number transcending the old numerical barrier after number 2.

The writing of the Egyptians and the Chinese, however, has perpetuated the early conceptual stage of three = many in a remarkable fashion (Figs. 2 and 3). To express the "many," the plural, of an idea, they write the character for it three times (just as the Babylonian number word *eš*, "three," became the plural ending). An ancient Egyptian inscription, for example, reads: "To the King thousands

were sacrificed and hundreds offered up (see Fig. 8, p. 42)."

The number sequence as such actually begins with three: "three, four, five, . . ., etc." When a tribe of South Sea Islanders counts by twos, *urapun, okasa, okasa urapun, okasa okasa, okasa okasa urapun* (that is, 1, 2, 2′1, 2′2, 2′2′1), we distinctly feel that they have not yet taken the step from two to three. And we realize with astonishment that these people can count beyond two without being able to count to three.

The step to Three is the decisive one, which introduces the infinite progression into the number sequence. We recognize it from its action in reverse; Two is stripped of its unique position and is recognized as a number just like any other; the grammatical dual of the original Indo-European language disappears. Furthermore, from the cardinal numbers 3, 4, 5 are formed the ordinals "third, fourth, fifth," and then by analogy, going backwards, from "the other" is developed the "second." The world of numbers has entered the personal world (two = You) through the back door. Only "the first" still holds out: There is no such thing as a "firstest." Whenever such a form does occur, as in the Turkish *bir-inzi* < *bir*, "one," it

Fig. 2 Three as the plural in Egyptian: (1) flood = heaven with 3 water jugs; (2) water = 3 × wave; (3) "many" plants = 3 × plant; (4) hair = 3 hairs; (5) weep = eye with "many" (= 3) tears; (6) fear = dead goose with 3 vertical strokes, the general plural sign, next to the ideogram.

Fig. 3 Three as the plural in Chinese: (1) forest = 3 × tree; (2) fur = 3 × hair; (3) all = 3 × man; (4) speak endlessly ("much") = 3 × speak (mouth from which words emerge); (5) rape = 3 × woman; (6) gallop (ride "much") = 3 × horse.

indicates a number sequence that arose not out of a people's own early levels of conception but was constructed by analogy with an already developed neighboring number sequence (see p. 112).

The step to Three is a step across the threshold of darkness, before which the concept "number" was still deeply rooted in the life of the soul, out into the prosaic but clear and bright light of practical life. If this step means a waning of the power of detachment, of the ability to impart to each number its own distinctive features obtained from the object itself, it is compensated by the growing power to build up the number sequence as a useful structure with applications of undreamed-of scope. This was not accomplished at a single stroke, of course, but was advanced time and again, from one numerical boundary to the next, at each of which the sequence pauses to catch its breath, as it were, and waits until it is overtaken by the reality of life and pushed on to new and higher numbers.

Our analysis of the grammatical dual has yielded an abundance of insights. In addition to the grammatical dual, some primitive peoples have a triple grammatical number, and here and there we even find a quadruple. Beyond that, however, human speech has never formed a definite plural noun form. This suggests that a special position is also occupied by the number four which we shall now approach from a different tack.

NUMBERS AS ADJECTIVES

Now that we have seen how strongly the number ("four") of something that is counted ("four large houses") was felt at a very early stage to be one of its attributes like any other ("large"), we should not be surprised to find it appearing in grammatical form as well, that is, as an adjective. Two other aspects, however, may be more surprising: that only the first 4 number words take this grammatical form, and that over the millennia quite a number of languages of our own general culture have faithfully preserved this early state.

The first four number words. These words in the primitive original Indo-European language and in many languages derived from it, such as Sanskrit, Celtic, Greek, and Old Norse, not only had three genders but were also inflected, just like the true adjectives "beautiful" or "large"; in Gothic and Latin, Four has already been frozen into an nondeclinable number word.

"One" has still kept its ability to agree with the subject in gender and case in some modern languages, including German (*ein-en Baum, ein-er Frau*, "against a tree," "of, to a woman"); hence, I shall cite here only the less familiar forms for "two," "three," and "four":

Greek	*Latin*	*Gothic*
dŷo	duo, -ae, -o	twai, twōs, twa
dyoín	duorum, -arum	twaddje
dyoín	duobus, -abus	twaiþ
dŷo	duos, -as, -o	twans, twōs, twa

Greek	Latin	Gothic
treîs, tría	*tres, tria*	*þreis, þrija*
triôn	*trium*	*þrije*
trisí	*tribus*	*þrium*
treîs tría	*tres, tria*	*þrins, þrija*
téttares, téttara		
tettárōn	*quattuor*	*fidwōr*
téttarsi		
téttaras, téttara	*(unchanged)*	*(unchanged)*

Accordingly, the genitive "of the two men (women)" is: Greek, *dyoîn androîn* (gynaikoîn); Latin, *duorum virorum, duarum feminarum*; Gothic (Luke 9:16), "Then he took the five loaves and the two fishes," *fimf hlaibans jah twans fiskans*, and (Luke 9:33), "Master, it is good for us to be here: and let us make three tabernacles," *hleiþros þrins* (see also Fig. 27, p. 129). In Old Norse the inflected cases of four are *fiorer, fiogorra, fiorom, fioran*.

One in German. In German, even to this day, One is still declined and has three genders; it is thus an adjective in form. Its meaning as a number, however, has been thereby gradually attenuated, as we can see clearly in English, where the number word *one* (< Latin *unus*) has been eroded down to an article, e.g., *a tree*; only when the number is distinctly emphasized does it appear in its old and full form of *"one,"* as in *one tree*. Indonesian provides an additional example: the number word *satu* has shriveled up into the article *se:* *se orang*, "a man," but *orang satu*, "one man."

High German — that is, the literate German language of today — can convey the numerical sense of *ein* ("one") only by inflection in speech. The same is true of other languages, such as French and Italian, while Dutch uses accents even in writing: *een boom* and *één boom*, "a tree" and "one tree." Only a few German dialects still make a clear-cut distinction. In Darmstadt, for instance, one might ask, *Habt ihr en Baum im Gadde?* ("Do you have a tree in the garden?") and be answered, *Ja, awwer nur aan!* ("Yes, but only one"). Languages that have no article, such as Russian and Latin (*homo* = "man," "a man," "the man," or "one man"), and those of most primitive peoples have naturally preserved the number word from external and internal erosion: "man" means "one man."

This attenuation of meaning is connected with the fact that One is never really felt as a number. One, as the antithesis of Many, had already taken a special position as far back as the Table of Ordinal Concepts established by the Pythagoreans. Plato constantly emphasized this: Like the Now in time and the Point in space, the One among the numbers cannot be further subdivided. Hence it conceals within itself no plurality which it collects together into unity, and since it is in this that the essence of number lies, One is not a number. Since, according to Euclid, "a number is an aggregate composed of units," One is itself not a number, though it is the source and the origin (*fons et origo*) of all numbers. Throughout the Middle Ages no one thought differently. One was designated sometimes as *genetrix pluralitatis*, the "mother of plurality," sometimes as *principium quantitatis,* "origin of plurality," and then again as *radix universi numeri et extra numerum,* "the root of every number and yet itself no number." The Salem Codex, a 12th-century manuscript highly significant for the history of our numerals, which we shall often encounter, states: "Every number can be doubled and halved, except for unity; this can, it is true, be doubled, but not halved — *in quo magnum latet sacramentum,* wherein lyeth concealed a great Mysterium [God]." As late as 1537 the German arithmetician Köbel wrote in his manual of computation:

Darauss verstehstu das I. kein zal ist | sonder es ist ein gebererin | anfang | vnd fundament aller anderer zalen.

"Wherefrom thou understandest that 1 is no number / but it is a generatrix / beginning / and foundation of all other numbers."

We are quite well able to understand this view, but precisely because we do understand it, we also recognize that One was regarded as a number only after the very concept of number had been clarified relative to the higher numbers. From these the number concept extended back to One, drained it of its philosophical content and made of it a number like the others. This is a process we have already described in the case of Two (see p. 17) and which we shall meet again later, and far more strikingly, in connection with Zero. Now we can understand why a French writer of the mid-16th century, in listing the digits, wrote:

les huicts figures 2, 3, ... 9 — "the *eight* figures 2, 3, ... 9."

Michael Stevin, the man who introduced the algorism of decimal fractions, was probably the first mathematician expressly to assert (in 1585) the numerical nature of One. He proved it somewhat as follows: If from a number 3 I subtract a non-number, then 3 remains; but since $3 - 1 = 2$, then 1 is not a non-number and must therefore be a number. Yet it is for this very reason that the old view persists, as Schiller has documented (*Piccolomini* Act II, Scene 1):

"*Fünf ist
des Menschen Seele. Wie der Mensch aus Gutem
und Bösem ist gemischt, so ist die Fünfe
die erste Zahl aus Grad' und Ungerade.*"

"Five is
The soul of Man. As Man is composed
Of good and evil mixed, so is the Five
The first number holding odd and even."

Thus One is not a number, otherwise 3 ($= 1 + 2$) would be the first odd number.

Even today we still often hear the question, "What is a prime number (such as 7)?" The answer generally given is: "A number divisible only by itself." This definition forgets that it is also divisible by 1. One feels that 1 is, after all, not the same kind of number as the others. Nor does it "act" on the number a as does every other number. For example, $a \times 1 = a$, an argument likewise formerly adduced against its numerical nature.

Language, however, has fortunately preserved this special status of One. We have already spoken in some detail of the peculiar quality of the first ordinal number (see p. 16); perhaps we should add here that, for example, when the Frenchman counts the days of the month, *le premier,* "the first," *le deux, le trois,...*, *le trente,* "the two, the three,...*, the thirty," he again gives special emphasis to One as the only ordinal number. For "one o'clock" the Italian says *il tocco.* The Germans, like the French, call the number One in card games and dice the *As* [ace], just as the Greeks once specially designated it as the *oiné.*

All this evidence supports the view that unity was first recognized as a number from the direction of plurality; hence, it is wrong to suppose that the number sequence must have been mere child's play to devise once the idea of unity was at hand. The idea of One did exist, to be sure, but embodied in the object as such and not as an independent idea of number, to say nothing of being a detached, abstract number word.

Two and Three in German. Although One is still inflected, Two has already lost its variability in modern German. Yet two and "three" used to be inflected:

Case	Old High German	Middle High German
nom., acc.	*zwene, zwa (zwo), zwei*	*zwene, zwo, zwei*
gen.	*zweio*	*zwei (g)er*
dat.	*zweim*	*zwein*
nom., acc.	*dri, drio, driu*	*dri, driu*
gen.	*drio*	*dri(g)er*
dat.	*drim*	*drin*

The *Nibelungenlied* (Verse 437) says:

Der schilt was under buckeln ... wohl drier spannen dicke.
"Beneath the boss the shield was three spans thick."

Until well into the 17th century, Two still generally had both gender and inflection.

Masculine gender *zween*:

Eber zeugte zween Söhne, "And unto Eber were born two sons" (Genesis 10:25);
Niemand kann zween Herrn dienen, "No man can serve two masters" (Matthew 6:24);

feminine *zwo* and neuter *zwei*:

Und stand auf in der Nacht und nahm seine zwei Weiber und seine zwo Mädge.
"And he rose up that night, and took his two wives, and his two women servants" (Genesis 32:22).

Inflection:

Durch zweier Zeugen Mund wird allerwegs die Wahrheit kund.
"Through the mouths of two witnesses the truth is always made known" (Goethe, Faust I, line 3013).

"*Zweier*" "of two" is still heard occasionally to this day. Again, vernacular dialects which are close to the people are faithful guardians of much that is traditional in speech; in Upper Hesse, for example, people still say:

zwien Osse, zwoo Käu, zwaa Kinner, "two oxen, two cows, two children."

The word *zwo* commonly used in German today serves merely to avoid misunderstanding (*zwei, drei*).

Four as an Old Limit of Counting. Let us pause here for a moment. What significance does the use of number words as adjectives have in the development of the number sequence? It represents the first detachment of the number from the object counted. A number word can now be used to count anything; it already stands free and independent in the realm of language and can associate itself with every object, although not yet purely counting, but still "describing an attitude."

But here is a striking question: Why do only the first four numbers, "one," "two," "three" and "four," appear as inflected adjectives? Why not "five" or "seven" or "twenty"?

We can very well answer this ourselves: Because they were the earliest number words (not counting the very first step forward, from One to Two, which had still taken place completely within the realm of the mind). A word that agrees with its subject in gender and case is more intimately bound up with it than one that does not; and the word does this because the concept for which it stands does, and hence because the number is, as yet, more intimately connected with the thing counted. But this is, as we have seen, a sure sign of prehistoric times.

One may then ask: Of course this is natural enough, but why should the break come just after Four? Why not after Seven? Two reasons may be given. The first is that the hand has four fingers, not counting the thumb. What happened to the thumb here was like what happened to One — it was not regarded as being equal, it was not a "finger" like the others. The handsbreadth, measured without the thumb across the knuckles, was used as a basic measure by almost all ancient civilizations.* The Greek and the Egyptian *ell*, for instance, had 6 handsbreadths, or $6 \times 4 = 24$ fingers; likewise the Roman foot (*pes*) was made up of 4 *palmae* and of $4 \times 4 = 16$ *digiti*. A second reason might be that a quantity larger than four, or even three, can no longer be directly apprehended. If we ask, "How many people were there?", the answer is "three or four," not "nine or ten," for that would already be "a large number." And in those early primitive times only clearly perceptible numbers were apprehended as words, as we have seen in the case of the dual, which expresses a distinct Two, the "other" that goes with the One.

Four, then, was certainly another limit of counting.

A series of unexpected pieces of evidence reveal this very ancient break at Four, even in complete, "mature" number sequences. In

---

* *Translator's note:* In English a horse's height at the withers is still measured in "hands" of 4 inches each.

the original Indo-European language, *octō(u)*, "eight," is grammatically a dual, as can be clearly recognized in the Greek *októ* and the Latin *octo* (which have the *–o* ending of the dual; see p. 14). It must therefore mean a doubling of four (2 × 4), although the number word Four cannot be recognized in it linguistically. The hypothesis of this ancient counting by fours is confirmed by the startling similarity between the Indo-European terms for "nine" and "new":

"nine": Sanskrit *nava*   Latin *novem*   Gothic *niun*   Tocharian *nu*
"new": Sanskrit *navas*   Latin *novus*   Gothic *niujis*   Tocharian *nu*.

The explanation is that after Eight, when the breadths of both hands have been used up, there follows a "new" number, which is Nine. On the other hand, the distinction given to the number 13 and its superstitious burden cannot be explained by the supposition that is a "new" number after the third Four.

Of the still living languages, the Slavic tongues have preserved the age-old break at Four, in many cases quite sharply:

The Czech says (p. 98):

"one and one *are* two" *jedno a jedno jsou dvě*
"two and two *are* four" *dvě a dvě jsou čtyři*
but "three and two *is* five" *tři a dvě jest pět.*

The word *jest*, "is," is always used when the sum is greater than four — that is, from five on. We can readily see the connection, for we need only add the things counted:

1 cow and 1 cow *are* 2 cows; 2 cows and 2 cows *are* 4 cows; 3 cows *and* 2 cows *are* 5 cows, and so forth.

When the objects counted are named, the plural of the verb is invariably used. Thus up to the point where the number was always coupled with the objects counted, that is, up to 4, this peculiarity has persisted even today with abstract numbers. This is another remnant of prehistoric times and can be understood only with reference to them.

The break after Four is still more characteristic in Russian (p. 98). Whereas the number word One agrees with its noun, the things counted after 2, 3, and 4 in Russian are in the genitive singular; from 5 on, however, the number in Russian governs the genitive plural. Thus 2, 3, 4, 5, and 100 houses are:

odin dom   *dva (tri, četýrě) doma*   *pjatj (sto) domov*
1 house   2 (3, 4) of house   5 (100) of houses.

How did this strange rule come about? This form is by no means the meaningless genitive singular, "two of house," but an old dual that has now disappeared; the form now used resembles the genitive singular only superficially. One can easily understand how in speech the noun *house* was put in the dual, not only after Two, but also after Three and Four, but *only* through Four! In the case of a compound number, by analogy, the noun follows the same rule: "24 of the house," but "25 of the houses." What happened here? In Russian, as in Old Slavic generally, the number words up to and including Four were inflected adjectives; beginning with Five they became rigid, nondeclinable nouns followed meaningfully by the genitive plural ("five of houses").

This old limit of counting at four, however, is not restricted to the units; under their influence it extends to all the other ranks, through the *Děsjatj*, *sót*, and *týsjač* according to the rule are genitive plurals; see also the Czech number sequence (p. 98).

Tens		Hundreds		Thousands	
10	*děsatj*	100	*sto*	1000	*týsjača*
20	*-dzatj*	200	*-sti* (Dual)	2000	*-týsjači*
30	*-dzatj*	300	*-sta*	3000	*-týsjači*
40	(*sorok*, p. 185)	400	*-sta*	4000	*-týsjači*
50	*-děsjatj*	500	*-sót*	5000	*-týsjač*

Except for a single construction, inexplicable to the uninitiated, the ancient break after Four has persisted. To the question, "How old are you?" the Russian child will answer:

mne dva (*tri, četýrě*) *goda* — "to me [are] 2 (3, 4) of the *year*";

but thereafter:

mne pjatj (*děsatj*) *let* — "to me [are] 5 (10) of the *summers.*"

Thus up to the age of four a person is "years" old, and after that "summers."

Now comes a surprise: The Romans had the very same custom. They counted the years of a person's age, "two, three, four, ... years old," as follows:

bimus, trimus, quadrimus, but then *quinquennis, sexennis, ...-ennis.*

Here again we have a distinct break after four. At the same time the languages have here preserved the ancient Indo-European counting of the years by winters; *bimus* < Indo-European *bi-himus,* "two-wintered," is akin to the Sanskrit *himas,* Greek *cheimṓn* and Latin *hiems,* "winter." In the Germanic languages we have the corresponding Anglo-Saxon *anwintre,* "one-wintered," just as even today in the Lower Rhine region yearling cattle are called *Einwinter* "one-winter." The Greeks, moreover, had the same term, *chímaira,* a "one-winter" yearling, a male goat (with the body of a fish) from which the mythical chimaera arose. *Twalib-wintrus,* "twelve-wintered," is the term applied to the twelve-year-old Jesus in the Gothic Bible (Luke 2:42; see also p. 27).

The habit of counting by "years," as in the Latin *annus* < *-ennis,* is of more recent origin and begins only after the number Four. The old break after Four was also preserved in two further Roman customs: in the naming of children and in counting the months. The Roman father gave his children numerical names only from the fifth child on: *Quintus, Sextus, Septimus.* No *Quartus* or *Tertius* has been recorded in Roman history.

The Roman Calendar. The Roman year originally began on March 1. The first four months had the proper names *Martius* (31), *Aprilis*

(30/29), *Maius* (31) and *Junius* (30/29); thereafter, however, following the fourth month, there were *Quintilis*, (31), *Sextilis* (30/29), *September* (30/29), *October* (31), *November* (30/29) and *December* (30/29).

On the subject of the Roman calendar, which — somewhat modified, to be sure — is also our own, it is interesting to note that in the very earliest epoch the Romans used an old lunar year of ten months, from March through December, with 4 × 31 + 6 × 30 = 304 days in all (the first numbers cited after the names of the months, cf. p. 37). The legendary king Numa Pompilius is said to have replaced this, following the Greek example, with a year of 12 months and 354 days, which was later changed to 355 because the odd number was considered sacred. The new calendar had four months of 31 and seven of 29 days each (the second number after certain of the months' names above), and one month of 28 days. Of the new months, Numa Pompilius placed *Januarius* (29) at the beginning and *Februarius* (28) at the end of the year. Then in 450 B.C. the Decemvirs for the first time placed January in the position of the 11th month, just before February, where it still is today. To make this revised lunar year consistent with the solar cycle, intercalary days were inserted at the year's end, in February (cf. *bissextile*, p. 174).

The Roman year began, significantly enough, in the spring, on March 1, on which day the two new Consuls were inaugurated. For a completely different reason, in the year 154 B.C. January 1 was adopted as the official beginning of the year, and has remained so until the present day. A rebellion broke out in Spain at the end of the year 154. To avoid a change of command on March 1, 153 B.C., the year 154 was allowed to last only ten months and the new year 153 was begun on January 1. With this shift the previous numerical names of the months became literally false and lost their meaning, so that October (the 8th month) should now have been December (the 10th) and others to correspond.

Caesar then undertook to rearrange the calendar, which had become distorted with the passage of time; in his honor *Quintilis*, the month in which he was born, was renamed *Julius* in 44 B.C. And *Sextilis* became *Augustus*, because that emperor, in the year 8 B.C., also had corrections made in the calendar, which had again fallen into disorder.

We may also point to some special kinds of number words that in several languages have preserved a distinct boundary after Four (Latin) or Three (Greek, English); for example, in the adverbial forms:

Latin:	*semel*	*bis*	*ter*	*quater*	but then *quinqu-ies*, etc.
Greek:	*hápax*	*dís*	*trís*		but then *tetrá-kis*, etc.
English:	*once*	*twice*	*thrice*		but then *four times*, etc.
German:	*ein-*	*zwei-*	*drei-*	*vier-*	*fünfmal*, etc.

Thus in German the break has disappeared from view.

Linguistically: Greek *hápax*, "once," in which *ha* < Indo-European *sm̥-*, "one" (see p. 146), because Indo-European *m̥* > Greek *a* and Indo-European *s-* before a vowel > Greek *h-*; cf. Indo-European *septm̥* > Greek *heptá*, "seven." Latin *–ie(n)s*, "–times" > *–iens*, afterwards forms *quot-iens, tot-iens,* "how many times," "so many times."

Let us conclude with two more examples from different cultures. In addition to their own number words for the numbers 1 ... 10, the Japanese also have words for the same numbers borrowed from the Chinese (see Far Eastern Systems, p. 449). Thus there are certain compound words, for example those compounded with "evening," which the Japanese counts "1, 2, 3, 4 evenings" only with his native Japanese number words, but from "5 evenings" on he uses the borrowed words. This break after Four, which occurs so often in the number sequences of primitive peoples, may also help us to understand the special status enjoyed by Four in the Turkish language. If a Turk wishes to intensify the force of an assertion, for example to say that he has worked *very, very* hard, he says that he has done it "with four hands" (*dört elle*). He can also "wait with four eyes," and when he rides at a gallop he rides "with four horseshoes" (*dört nal*). Perhaps this intensifying sense of the word Four is rooted in early astronomical ideas, but this together with the "numerical" reason may have led to the unique position of Four in Turkish.

Before we carry our discussion further, we must make the following important observation: All these pieces of evidence show that Four was a very ancient limit in counting. Will Five be the next such boundary? Five, of course, is practically offered by the human hand with its fingers. No, surprisingly enough, in our Indo-European culture this was not the case. Five was not a limit of counting but rather one of the essential members of the number sequence. This very knowledge of how to *arrange* the numbers following after Five is, indeed, the reason why beyond Five the number sequence no longer pauses but runs on continuously. The first four number words are not members of the sequence; they are the first steps forward, made gropingly and without any sense of a general plan, which — although it too contains numerical breaks — was finally attained with the number Five.

The Units as Adjectives. After this excursion on the subject of Four as a limit of counting, let us now go back to our main subject, numbers as attributes, and examine number words as adjectives.

Celtic has a remarkable manner of differentiating the units from the higher numbers in a very characteristic fashion. In the case of a compound number like 23, for instance, it pulls out the unit 3, places it as an adjective directly in front of the thing counted, and then puts the higher number after the object:

12 hours Celtic *di huair deec,* "2 hours 10"
23 sons Celtic *tri meib ar hugeint,* "3 sons and 20"
185,000 Celtic *coic mili ochtmugat ar chet,* "5 thousand 80 and 100."

This pattern expresses the ancient, intimate relationship to the units, the very earliest of the numbers; they must stand next to the noun and are governed by it, to the extent that they are still declined like an adjective. The new, "larger" numbers, which are held far more vaguely and remotely by the imagination, are merely hooked on afterward. The same phenomenon appears in Old Norse, in which language the age of 48 is expressed as follows:

hafdi atta vetre en fimfta tigar,
"he had 8 winters in the fifth decade."

We find the peculiar position of the unit again in Old High German, not quite so strikingly, but still distinctly enough (Luke 15:7):

niuni inti niunzug recte, "nine and ninety just persons,"

in which the unit 9 in the compound number 99 is governed grammatically by the noun and appears in the plural; by contrast, in Gothic we find

niuntehundis jah niune garaihtaize, "90 and 9."

Adjectives and Nouns in the Number Sequence. We have already encountered more than one change in the classification of number words, the last in the case of Four as a limit of counting, where the words "one" through "four" are adjectives but the following number words are not. This is the most convincing proof that the verbal number sequence was not built up all at once, in a single move. As an example, let us consider Latin (see p. 92).

Using the symbol "(adj.)" to indicate that the number word is an adjective, we cite the following Latin numbers in the dative case (so as to show their inflection as well):

(1) to the 3 men	*tribus viris*	Literally: "to the three (adj.) men"
(2) to the 4 men	*quattuor viris*	Literally: "to the four men"
to the 10 (100) men	*decem (centum) viris*	Literally: "to the ten (hundred) men"
(3) to the 200 men	*ducentis viris*	Literally: "to the two hundred (adj.) men"
(4) to the 1000 men	*mille viris*	Literally: "to the thousand men"
(5) to the 4000 men	*quattuor milibus virorum*	Literally: "to the four thousand *of* men."

Thus we see in this number sequence the succession of adjective (1); indeclinable number word (2); adjective (3); indeclinable number word (4); inflected noun (5) that governs the partitive genitive case in the object counted.

Let us examine these examples more closely:

1. *Uno, duo, tres,* and originally also *quattuor* were inflected (see p. 18); the last of these later became indeclinable through the influence (acting in reverse) of the subsequent number words 10, 9, ..., 5.

2. *Quinque,* 5, ... *decem,* 10, and the rest through *kentum,* 100, are all indeclinable number words. They stand "dumb," so to speak, next to their noun ("men" in our case), which alone assumes a relationship within the sentence through its inflection in the appropriate case (here in the dative). Since *decem* and *centum* are derived from the same Indo-European root *dekṃ* (see p. 126), both these number

words may have been formed simultaneously. Thereafter the language appears to have driven the number sequence beyond 100 into a new direction:

3. The hundreds, 200, . . ., 900 are again adjectives. Why? Because of the remote effect of "two" and "three." *Centum* was felt as a new One. As later formations, however, the hundreds do not go through the change in word form after Three, but rather remain adjectives all the way through 900. And yet the old break after Three can still be seen in the sound shift from *–centi* to *–genti*: *du-* and *tre-centi*, but then *quadrin-genti*, *quin-genti*, . . ., *non-genti*.

4. *Mille*, 1000, is an old obstacle.

5. In its multiples, 2000, 3000, . . ., however, *mille* is not reduced to an adjective as are the hundreds; they remain nouns and are themselves counted, like any other nouns: *duo milia*, "two thousands." As a noun *mille* is itself declined, takes on a relationship to other words in the sentence, and itself takes the genitive case in the object counted: *quattuor milia virorum*, in contrast to the uninflected number words in paragraph 2. This procedure is followed through *centum milia*, 100,000

6. The higher thousands are all formed from this "last" number rank, "a hundred thousands." In the course of their formation *centum* is pluralized to *centena*, "every hundred," and the count now makes use of the "times" form:

decies centena milia, "ten times every 100 thousand" = 1 million.
vicies centena milia, "twenty times every 100 thousand" = 2 million.

The Roman, accustomed to dealing with very large amounts in public affairs, for example, pronounced the adverbial word "times" only once. In his History of the Caesars, the writer Suetonius relates that the Emperor Vespasian, upon succeeding to office, ordered the amount of money (in sesterces) in the state treasury to be counted and reported to him. The amount was colossal, but it could be expressed vaguely in only two words:

quadringenties milies, "400 times 1000 times,"

but this referred to *centena milia*, meaning $400 \times 100 \times 100,000 = 4 \times 10^{10}$ sesterces.

The Emperor Tiberius once deliberately overlooked the implied *centena milia*, thus making a "mistake" in his own favor. In her will, his mother Livia had bequeathed to his relative Galba

sestertium quingenties,

literally "of sesterces 500 times" (to be read: "times 100,000") = 5×10^8 sesterces. Tiberius, however, who was residual heir and moreover had not been on the best of terms with his mother, used the customary omission of the 100,000 to his own advantage and paid out to Galba only 500 sesterces — *quia notata non praescripta erat summa*, "because the intended sum had not been written out in full," according to the Roman narrator. In terms of Roman numerals, Tiberius read $\overline{\text{D}}$ instead of $\boxed{\text{D}}$, suppressing the brackets (for more on this subject, see p. 44).

In this thenceforth undisturbed and consistent use of the adverbial "times" formation to express the largest number we distinctly sense the use of a conscious, synthetically created device, whereas the beginnings of the number sequence, through 1000, grew up laboriously and at the same time unconsciously, without direction, from still obscure conceptions of number.

Note: in medieval manuscripts the expression *mille milia*, "1000 thousands" = 1 million, often took the place of the Roman expression "ten times a hundred thousand"; thus *septies mille milia*, "7 million."

It is of interest to recall that Adam Riese knew only Thousand as the highest numerical rank and thus would have expressed 40 million as "40 thousand times a thousand" and reported the wealth of Vespasian's treasury as "40 thousand thousand times a thousand" (see p. 142).

With the Roman word for "sesterce" we encounter the same case of a number slipping into the guise of a noun in its grammatical form. *Sestertius* < *sem(is)tertius*, "half of the third," meaning two and a half, was abbreviated *II S*, for *duo et semis* (for Roman coins and their inscriptions, see pp. 28, 78, 355). "Hundred sesterces" was properly written *centum sestertii*, and thus "2000 sesterces" was *duo milia sestertium* (not -*orum*). Now, however, this third-declension genitive plural ending came to be felt as a neuter singular — *sestertium*, genitive singular *sestertii* — and then dropping the *milia*, "2000 sesterces" was expressed as *duo sestercia*, literally "two sesterces." Moreover "500,000 sesterces" was correctly said as *quinquies centena milia sestertium*, but after omitting the hundred thousand it becomes *quinquies sestertium*, in which the numismatic inscription was thought of as neuter nominative singular, "five times a sesterce." Thus "he bought it for 2 million sesterces" was expressed as *emit sestertio vicies*, literally "for twenty times a sesterce." As a result, the one word, depending on its form, could have three different meanings:

sestertius, -ii, — *sestertia, -orum,* — *sestertium, -ii*
one sesterce   1000 sesterces   100,000 sesterces.

To sum up: there is no doubt that the inflected adjective (such as *duo, tres*) is more closely tied to the things it counts than is the indeclinable number word in noun form (e.g., *decem, centum*), which does not depend on the objects counted. The noun (*milia*) made itself completely independent, and it, not the object counted, expressed the grammatical relationship within the sentence (the case). The object counted followed it as an appendage (in the genitive case). Thereafter, with the "times"-word forms (*decies centena milia*), such a high degree of confidence in dealing with numbers and also such a degree of consciousness had been attained that henceforth the number sequence ran on without disturbance or change in form. Thus we have, in the changes in the form of the number words, a remarkable means for tracing the way in which the concept of numbers was gradually refined and the number words became increasingly detached and abstracted from the objects counted.

Special Number Sequences and Classes of Numbers

Having concluded our discussion of numbers as attributes, let us now speak of two more characteristic ways in which numbers have been used.

There are some primitive peoples who at times employ a special number sequence for various things — for example, for animate objects, round things, days, and so on. Thus, for instance, a tribe of North American Indians (in British Columbia) counts as follows:

	1	2	3
Animate things	menok	maalok	yatuk
Round things	menkam	masem	yutsqsem
Long things	ments'a	mats'ak	yututs'ak
Days	op'enequls	matlp'enequls	yutqp'enequls, etc.

In the language of this tribe, "two men" are called "*maalok* men," but "two days" are "*matlp'enequls* days." What a difference from our own single, abstract number sequence, that is used for counting everything!

This becomes understandable if we recall that early man experienced things in his environment far more intimately and immediately than we do, that to him a rabbit is a single, particular, real, living rabbit and that he never thought of abstracting from his concrete reality any general concept of "animal." Similarly the object also places its imprint on the number word by which it is counted.

Incidentally, vestiges of this are still extant in our own speech. For example, we still have different expressions for two of various objects. We speak of a *yoke* of oxen, a *pair* of shoes, of a *brace* of partridge, of *twins*, or of a *duet*, but we do not say "a duet of shoes," or "a yoke of children," or "an ox twin," even though each of these expressions means two. Not only the number 2, but also something of the actual object has gone into the numerical expression. Similarly, pen nibs are bought by the gross and buttons by the dozen; in Germany eggs were formerly sold by the *Mandel* (15) and the *Stiege*, "score" (20). One does not go to the store to buy a *Mandel* of buttons.

The same is true of words for groups of unspecified number, like a *herd* of deer, a *pride* of lions, a *bevy* of girls, a *flock* of sheep, a *pack* of wolves; we do not say "a bevy of sheep" or "a pride of deer." The form of the number is closely bound to the concept. Ancient measures are likewise strongly tied to the things they measured: cloth was measured in *ells*, height in *feet*, depths are stated in German folk tales in *Klaftern* and by seamen in *fathoms*. All these are measures of length, of about the same order of magnitude, but there exists no system of conversion (such as 1 ell = 2 feet). The fact that a "last" in English (*last, load*) may have greater or less value according to the thing specified (1 last of meal or herring amounted to 12, of salt to 18 tons, of gunpowder to 24 kegs, of bricks to 500, of tiles to 1000, of hides to 144 pieces, of wool to 12 sacks) is fairly easy to understand, but that the number word used as a measure should have different values — that a "hundred" herring meant actually 120, codfish 124, or salt 126 tons — is surprising. Other ancient traditional measures (such as the quarter of grain, the carat, the bushel, or the scruple) are similarly tied to the substance measured. The only universal measures are our "abstract" decimal measures, which were artificially devised at the time of the French Revolution.

Money is, in a sense, a kind of "thing-free" universal measure. Yet even here some primitive peoples have exchange media that are tied to the objects whose value they express, such as the shells that can be used to buy coconuts but not a chicken, which must be paid for with porpoise teeth, there being no system of conversion between the two forms of "currency."

An intermediate stage between the abstract number sequence and the concrete sequence tied to the things it counts is represented by the number classes, developed primarily by the Chinese (some 100 such classes) and the Japanese (about 50), but are also known in Persian (20) and Turkish (2). A number-class word of this kind is placed between the number word and the named object just like in our own expression 8 "head" of cattle. Thus it happens that a certain number word cannot be joined directly to the thing, whatever it may be, but only to a few classes of objects. Perhaps tables, for instance, cannot be counted directly, but must first be arranged in classes, in Chinese in the class *chang*, "spread," and in Japanese in the class *kyaku*, "foot":

4 *chang* tables,   tables 4-*kyaku*.

Since all cylindrical objects are reckoned in Japanese by the class *hon*, "root," and all poems by the *shu*, "head," 4 trees and 4 poems are counted as follows:

trees, 4-*hon*   poems, 4-*shu*.

We can understand this classification, in the case of some classes of objects, perhaps because of their physical form. Sometimes the part stands for the whole, as in Chinese "mouth" stands for "man"; in other instances the relationship is not at once discernible, as when the Japanese count automobiles by the "tower." Since they can also be represented by the seat or chair in which one sits and is carried, an automobile can be thought of as a "moving chair." But if, in contrast, the Chinese *t'iao*, "small twig," is the classification word for all elongated things (trousers, belts, strings), we can no longer understand why oxen and even murderers should be counted thus. Curious and cynical speculation is also aroused by the fact that both medicines and the boards used to build coffins are counted in Chinese by the class word *fu*, "gift."

The more classes of numbers there are, the more primitive and object-bound is the number system. It still isolates the world of physical objects and strongly emphasizes the uniqueness of each separate thing. When, by way of contrast, Turkish differentiates only between the two classes "man" (*nefer*) and "other than man" (*tane*, "piece"), as in

yüz nefer asker,   *yüz tane at,*
100 men soldiers,   100 pieces horses.

the only relevant question is whether the object has the attribute of being human or not human. We can see that if this last difference were also omitted, the number word would stand directly in front of the word for the object, and the number sequence could be used to count everything.

The number classes once more reveal clearly how closely the number was involved with the object in early man's conception, how strongly things dominated numbers. They also illustrate the mental obstacles that primitive man had to overcome not only in liberating the number sequence from these preliminary stages but also in building upon the earliest beginnings. In doing so, he obtained help from a different source — and that is what we shall look into next.

Expansion of the Number Sequence by Means of Supplementary Quantities

ARRANGEMENT BY ELEMENTS

On the island of Ceylon live the Wedda, a primitive tribe with a very low level of culture. If a Wedda wishes to count nuts, for example, he collects a heap of sticks. To each coconut he assigns not a number word but a stick: one nut — one stick; and each time he does so, he says, "that is one." So many coconuts, so many sticks; for he has no number-words. Does this mean he is unable to count? Not at all! He translates the pile of coconuts he has laid out into the auxiliary quantity of sticks. Can he then tell whether anyone has stolen one of his coconuts? Yes: he again arranges the nuts and sticks in order, one to one, and if there is one stick left over, one coconut must be missing. Can he say how many nuts he has? No, for he has no words for that. He can only point to his pile of sticks and say, "That many!"

Thus, we are here dealing with a translation of the counted things into a nameless, supplementary quantity. By arranging similar pebbles or sticks element by element, early man thus succeeded in doing away with the multifarious nature of things; he achieved a clearer view of what he wanted to know about them: namely, their number.

There is a very ancient law, inscribed in the original letters and words, according to which the senior Praetor on the Ides of September [the beginning of the Etruscan year] is to drive a nail into the right side of the Temple of Jupiter, where the sanctuary of Minerva is. It is said that this nail was to be a sign of the year's number, for letters and numerals were then very rare. Because numbers were invented by Minerva, this law was ordained in her temple. Cincius, a reliable authority, asserts that in Bolsena, too, a similar nail was driven into the temple of the Etruscan goddess Nortia as a means of reckoning the year.

With these words the Roman historian Livy describes a supplementary quantity in the form of nails used to number the years.

Among primitive peoples, the victor in battle of the hunter carries about with him the number of his vanquished enemies or slain animals in the form of scalps or boar's tusks. Young unmarried girls of the Masai, a warlike tribe of herdsmen living on the slopes of Mt. Kilimanjaro, each year add one heavy brass ring around their necks, so that their precise age can be known from this extraordinary necklace of annual rings (see Fig. 4).

We, ourselves, still continue to count with the aid of supplementary quantities. The innkeeper or bartender "tallied up" the bill, in earlier days, on a slate and today on the beer plate; in Spain at one time he would toss a small pebble (*chinas*) into the hood of the guest's cloak for each drink, whence the expression *echar chinas* "to throw pebbles," still means "to chalk something up to someone's account." The Greek sculptor Lysippos (4th century B.C.), according to Pliny,

Fig. 4 Masai girl wearing annual rings, which show her to be 23 years old.
Institute of Ethnology, Göttingen.

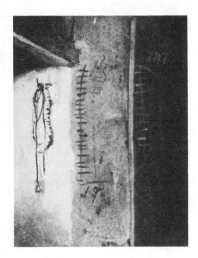

Fig. 5 The Poor Poet in Spitzweg's painting has "tallied up" something on the walls of his garret — perhaps the number of days' rent he owed. By kind permission of the Staatsgalerie, Munich.

produced a total of 1500 statues during his career; Pliny deduces this from the fact that he would always set aside one gold denarius from the money he was paid for each work of art, and his heir inherited 1500 denarii. What was the poor poet in Spitzweg's painting tallying on the wall of his garret (see Fig. 5)? Most likely not gold coins, but perhaps the number of days' rent he owed. Similarly, the soldier marks off the days to his next furlough on his calendar. Adalbert Stifter, the German poet, counted the days until he was to see his betrothed again by eating apples: "There are still 14 apples lying there, and yet they are only 14. When I wrote my last letter to you there were 21; and that's a comfort — tomorrow there will be only 13! Finally only one apple will be left, and when I have eaten that, I shall shout for joy" (excerpted from a letter). Each of us is familiar with tallies of this kind, and has often made them himself.

When supplementary quantities are illustrated graphically in this manner, the number need not be expressed at all, or can be apprehended merely conceptually. A chieftain on the island of Celebes was sentenced by the colonial authorities to pay a fine of 20 buffaloes. Someone expressed surprise at the severity of the punishment. Quite astonished, the chieftain asked: "Do you consider the fine that high?" and began to count out nuts from a pouch, one for each buffalo. Only when he had "grasped" the number in the truest sense of the word did he become incensed at his punishment.

By this means of aligning objects and translating them into a supplementary quantity, primitive people who possess very few number words are able to count far beyond the highest number they can express verbally. It is a mistake to think that a tribe that has only 3 number words can count only to 3. By physically representing the number by a quantity of pebbles arranged element by element, they can count much further than that.

The Damara in South Africa, for instance, have only 3 number words. An ethnologist tells how he bought a number of sheep from them at the price of 2 twists of tobacco per sheep. In carrying out the transaction he had to hand over 2 twists at a time, sheep by sheep; when he tried to hand over 4 twists of tobacco at once, for 2 sheep, the Damara tribesman became confused and lost track. He put down 2 twists side by side, turned to one sheep, then took the next two and turned to the second sheep. Rigorous ordering, trading, the exchange of a quantity far exceeding the verbal number sequence, even computation, if you please — and all without specifying the number!

But this is a path leading to written numerals sooner than to spoken number words. The words needed to specify numbers are not arrived at in this manner. They do, however, arise from the practice known as body-counting.

Body-counting. The term "body-counting" designates the number sequence arrived at by some primitive peoples, in which the parts of the human body — the head, the eyes, the arms, and so forth — are arranged in a certain order. A tribe on the island of Papua counts as follows:

1	*anusi*	right little finger	12	*medo*	nose
2 ⎫		right ring finger	13	*bee*	mouth
3 ⎬ *doro*		right middle finger	14	*denoro*	left ear
4 ⎭		right index finger	15	*visa*	left shoulder
5	*ubei*	right thumb	16	*unubo*	left elbow
6	*tama*	right wrist	17	*tama*	left wrist
7	*unubo*	right elbow	18	*ubei*	left thumb
8	*visa*	right shoulder	19 ⎫		left index finger
9	*denoro*	right ear	20 ⎬ *doro*		left middle finger
10	*diti*	right eye	21 ⎭		left ring finger
11	*diti*	left eye	22	*anusi*	left little finger

One might suppose that this was a sequence of number words extending through 22. There is, however, a more profound difference. What does "*doro* days" stand for — 2, 3, 4, or 19, 20, 21 days? This sameness and the mirrorlike repetition in reverse make it possible to understand a count only if the whole sequence is run through from the beginning or if the meaning is specified. This number system thus merely assigns names but does not organize and does not build on itself: it has no governing principle. Nevertheless, we can see in it the first step toward the kind of numbering that does lead to a true number sequence.

NUMBERING BY FINGERS AND TOES

Finger Signs and Number Words. Apart from teeth, fingers and toes are the only parts of the body which man possesses in any abundance. They are supplementary quantities that he always has conveniently "on hand." Thus it is not surprising that virtually all primitive peoples count on their fingers, in that they assign number values to their fingers, which they use as a supplementary quantity. Words are not absolutely necessary in this process; and with many primitive tribes which "point" the number, each number is actually a gap in the flow of speech that is filled by a finger gesture.

For the most part this assigning of numbers takes the form of extending or bending the fingers of one hand, just as we might reckon to ourselves how many months there are from May to September. Now we can understand how the Dene-Dinje, a tribe of American Indians, built up the following sequence of number words:

one   "the end is bent" (little finger)
two   "it is bent once more" (ring finger)
three   "the middle is bent" (middle finger)
four   "only one remains" (thumb)
five   "my hand is finished."

The verbal designations of their finger signs have become their number words: Four days are "only-one-remains days." Such is the case with the majority of primitive peoples; their number words, to the extent that they are expressed at all, refer to finger signs. More dramatic than "my hand is finished" are such frequently encountered expressions as "my hand dies" for 5, "my hands are dead" for 10, "my hands are dead and one foot is dead" for 15, and "a man dies" for 20. We can visualize the hand held out: the fingers stand upright and then sink down, one after another: they "die."

Some statements of numbers thus sound very strange. In translating the Bible for one tribe of Papuans, the passage (John 5:5): "And a certain man was there, which had an infirmity thirty and eight years" had to be expressed as "a man lay ill one man (20), both sides (10), 5 and 3." Even more picturesque is the expression for the number 99 in British New Guinea: "four men die (80), two hands come to an end (10), one foot ends (5), and 4." Very rarely, on the other hand, are the actual names of the fingers used as number words. An isolated example of this occurs among the South American Kamayura tribe: Here 3 is called "peak finger" (middle finger), and "three days" is thus "peak-finger days."

In comparison, number words containing the roots "clap" (10) or "jump across" (6) refer to number *gestures*: To indicate 10, the hand is clapped against the palm, and for 6 one hand is passed rapidly over the other. The number of possible variations here is beyond belief. Certain African tribes can even be identified and classified ethnically according to whether they begin to count with the right or the left hand, whether they extend or bend the fingers, or whether they turn the palm toward or away from the body. For our purposes, however, the important thing is that *man learned to count on his fingers*.

A charming instance from the recent past is provided by an anecdote related by the Englishman R. Mason about the last world war (". . . *and the Wind Cannot Read*"):

Sabby was a Japanese girl in India, which was then at war with Japan. Her friend therefore introduced her as Chinese to an Englishman who had been living for a long time in India: "Miss Wei." "Really?" He stretched his face forward and examined Sabby from close up, as if he were nearsighted. "Nonsense," said he. "Count with your fingers! Count to five!" Sabby looked shocked; she wasn't quite sure whether this extraordinary man was joking or mad. Hesitantly she raised her hand: "One, two, three, four, five," she said uncertainly. Mr. Headley burst out delightedly: "There you are! Did you see that? Did you see how she did it? Began with her hand open and bent her fingers in one by one. Did you ever see a Chinese do such a thing? Never! The Chinese count like the English. Begin with the fist closed. She's Japanese!" he cried triumphantly.

The Romans, who were very poor at computations and therefore faithfully preserved the early methods of reckoning, used such turns of speech as *numerare per digitos*, "to count with the fingers," and *novi digitos tuos*, "I know your fingers," which means "I recognize your skill in reckoning," thus furnishing additional proofs of this (see also the quotation from Ovid, p. 37). Throughout the middle ages the units were called *digiti*, and in English they are still called *digits*.

Organization of the Number Sequence. The fundamental significance of fingers and toes in the development of the number sequence is revealed by the following consideration:

Whereas the custom of assigning number values to pebbles or to parts of the body yields only a continuous, undifferentiated supplementary quantity, the quantity derived from the fingers and toes is already classified and grouped by nature: 5 fingers make a hand, 10 is represented by two hands, 20 by hands and feet. It is quite natural that languages should also show these breaks: five "1 hand," ten "2 hands," twenty "one man"; this happens not only with primitive peoples but even in the number sequences of civilized societies. In addition to the familiar 10-group, we can also encounter remnants of an ancient 20-grouping, and we have already demonstrated the peculiar break after Four as a representation of the "finger-hand," held with the thumb concealed (see p. 25 and Fig. 12, p. 50).

What is the unexpected significance of this arrangement? It lies in the fact that now the path is clear to permit the number sequence to extend further, beyond the first few words: Once we have counted through "the man," the second such series can begin exactly like the first, followed by the third, the fourth, and so on. In this way one level is built up on the next; this classification ensures the ordered progress of the number sequence. In a very real sense, this is nature's gift to mankind.

I know of no more delightful expression of this idea than Ovid's verse (*Fasti* III) that alludes to the intimate union of the number Ten with nature, and in the final line even mentions the organization of the number sequence, although it may not be correct in every detail:

Annus erat, decimum cum luna receperat orbem,
Hic numerus magno tunc in honore fuit.
Seu quia tot digiti, per quos numerare solemus:
Seu quia bis quino femina mense parit:
Seu quod adusque decem numero crescente venitur,
Principium spatiis sumitur inde novis.

"A year was past when the number of full moons was ten,
This number was held in great honor by men in those days:
Because that is the number of fingers on which we reckon,
Or a woman gives birth to her child in two times five months,
Or the numerals increase to ten, and from there
We again being a new round."

While our attention was focused on other matters, we completely failed to notice that something else has happened, something that early man did not achieve: the number sequence has now finally become completely— we might almost say painlessly — detached from the objects counted. It is no longer things that are counted, but the now familiar supplementary quantity, to which all objects, of whatever kind, can be transferred.

At first the supplementary quantity insinuated itself between the amount of objects to which numbers were applied and the quantity of number words; once these have been formed, the supplementary quantity has done its job and disappears. The number sequence is born! To be sure, it has not yet acquired all the characteristics that we expect of it today, such as infinite progression. But it does at last possess the outlines from which it can develop to its logical conclusion.

Thus far we have traveled a very important path. We began by speaking about our abstract number sequence, which can be used to count anything; we then descended into the dawn of man's intellectual development, when mankind took its first timid steps in counting. From there we climbed back up, step by step, until we again arrived at the abstract number sequence with which we began.

This reconstruction is not to be taken to mean that every individual number sequence, such as our own, must have followed exactly this path; if that were the case, these intellectual steps would form the rungs of a historical ladder, the only one, in fact, which every number sequence would have to climb in order to achieve its full growth. No, these steps are merely stages of development which we can discern now in one and now in another number sequence. They also illustrate the How, and especially above all the What, of an evolution through which every number sequence has had to pass.

Our next task is to examine the manner in which our own number sequence is organized.

Principles of the Number Sequence

SUCCESSION AND GROUP

Ordering and Grouping

Let us return to our Wedda tribesman counting his coconuts (p. 33): he assigns sticks to them, one by one; then he gathers all his sticks into a pile and knows that he has so many sticks and hence so many nuts.

Members of the South American Bakairi tribe do it a little differently. They put down two coconuts together, three times; thus there are 3 pairs. Then they form these pairs on their own fingers, putting together the left little finger with the left ring finger, then pairing the middle and index fingers together, and then the left thumb with the right little finger. Thus through the supplementary quantity represented by his own fingers, he has grasped the number 6. Anything beyond that they no longer group in pairs but merely line up as does the Wedda tribesman.

This is a good example of the organization of counting which we shall call *ordering* and *grouping*. Up to 6 objects the Bakairi *groups* by pairs; from there on he merely *orders* them. Which then was the earlier stage? Ordering, of course: without first ordering, there can be no grouping. The Bakairi stands, as it were, with one foot on the lower step of ordering and diffidently places the other on the next higher step of grouping by twos: another picturesque instance of how numbers first became clear to primitive man through the use of a supplementary quantity.

But why grouping? Let us ask ourselves: why do we always divide a bundle of postal cards by twos? Why do we break up a pile of coins into little heaps of five? To count them better! "Better" means more clearly, more confidently, and therefore more quickly.

Just the same with primitive man. The Fiji Islander carved notches into his club to count the animals or the enemies he brought down with it. When the number of notches became too unwieldy for him to visualize clearly, he grouped them: After every 9 notches he carved one large notch. The large notch stands for 9 small ones, and thus constitutes a 10-group. Now he can envisage the number 54, even though he may not be able to express it verbally because he has no number word for it. After all, number words are not absolutely necessary for counting (see Fig. 6).

Counting by groups of three is exemplified on a sword from the Philippines, on which the native warrior has recorded the number of his victims by riveted silver nails (see Fig. 7).

The Germanic equivalent of this is *Hildebrand's Dirge* with its characteristic groups of eight:

"At my head stands   the broken shield,
eru thar taldir   tiger ens atta
on it are tallied   ten times eight
manna theira   er ek at mordi vard
proud men   whose slayer I became."

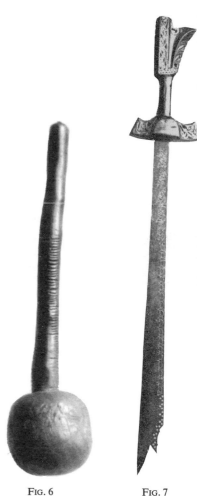

FIG. 6   FIG. 7

Fig. 6 Fiji Island club, with number notches. 40 cm long. Museum of Ethnology, Frankfurt/Main.

Fig. 7 Philippine sword, on whose blade the owner has recorded the number of his victims with silver nails inlaid in groups of three. 94 cm long. Linden Museum, Stuttgart.

(This song is an Old-Norse version of the 8th-century Old High German *Hildebrand's Song*, which was probably brought to the north by German traders in the 12th century.)

There are innumerable examples of number grouping in our own daily lives. The bartender chalks up groups of 5 ticks each, the supplementary quantity standing for the number of glasses of beer consumed by a customer, by drawing the life for the fifth slantwise through four vertical strokes, forming a "gate." The truckman unloading sacks of merchandise, the bank teller counting money, the card or billiard player marking down the points he wins, all use the method of grouping. We can even group things with our eyes, without touching the things themselves. The set of dots : : : : can be counted as two horizontal rows of four each, as squares of four each, or as four pairs lined up in a row.

Commercial and Natural Groupings

Shakespeare's Shepherd in *The Winter's Tale* (Act IV, Scene 4) very astutely sets forth his complicated reckoning:

Clown. Let me see: every 'leven weather tods; every tod yields pound and odd shilling: fifteen hundred shorn, what comes the wool to? ... I cannot do't without compters.

He makes groups of 11 sheep each and assigns a weight of wool to each group — not 11 weights as the Bakairi tribesman would have done, but only one, a form of ordering that we shall speak of presently. But first let us take note of what the passage from Shakespeare also testifies: *For Primitive Man, computing was identical with counting!*

This is a very important insight. When the Damara tribesman (see p. 34) calculated his sheep transaction with the buyer, he was counting off, a simultaneous ordering of sheep and of twists of tobacco. Thus one sheep can, as here, be lined up with every 2 twists, and hence with a 2 group; or, as above, 11 sheep with one "tod" of wool and this in turn with one pound and some shillings. This unit exchange value need be established only once; the continuous ordering will then complete the transaction without recourse to any further calculations.

With this we come to the types of grouping that we wish to designate as either natural or commercial. Shakespeare's eleven sheep are a *commercial* grouping that translates directly to a unit of money, one pound. Eskimos tied their pelts together in bundles of four to prepare them for sale, since 4 pelts will make one fur coat. In many parts of Africa cowrie shells serve as currency; the natives string together 40 shells, so that 1 string = 40 cowries and 50 strings = 1 head.

Here the size of the groupings is determined by commercial expediency or by mutual agreement. But there is no such consideration in counting and grouping postal cards or coins. It would never occur to anyone to arrange some 97 dimes in groups of 11 or piles of 40. Here we revert to the *natural* arrangements by groups of 5, 10, or 20, and perhaps also 4 (see Fig. 12, p. 50). The 2 group, the pair, also belongs to this category.

Examples can be dispensed with, even though this new insight will later yield surprising conclusions; some of the instances cited previously can also serve for the groups of 5, 10, and 20 we are discussing. But two examples are pertinent just the same: one because it has preserved for us the former state of a now highly civilized, "mature" people. A native member of this people was asked by an ethnologist how he would count some given quantity of objects. "Dear Sir, we do it this way: When we have counted up to five, we set them aside in a small heap and start again from the beginning." Thus he "fives," just as Homer's Old Man of the Sea "divided into groups and counted" his herd of seals (*Odyssey* 4, 412):

αὐτὰρ ἐπὴν πάσας πεμπάσσεται ἠδὲ ἴδηται
λέξεται ἐν μέσσησιν, νομεὺς ὣς πώεσι μήλων.

"When he has seen them and counted them up by fives,
He will lie down among them, like a shepherd among his sheep."

What does the word *pempázein* mean? Since the Aeolic Greek *pémpe* is "five," it means nothing other than "to five," to divide up into groups of five. No word could more faithfully reveal the early stages of a people's culture, its counting by groups of 5.

The second example shows that even in this day and age the ancient custom of grouping can be used very prudently, for instance when one has to count an unwieldy quantity of something. Karl Bädecker, publisher of the world-famous red travel guides, wanted to know, for his volume on Italy, the exact number of steps leading to the Cathedral of Milan. He counted them by removing one pea from his vest pocket every 20 steps and placing it in his trousers pocket.

GROUPINGS OF HIGHER RANK

The Elements of a Sequence are Groupings

In the old days the German army was organized as follows: 120 men = 1 company, 4 companies = 1 battalion, 3 battalions = 1 regiment, 2 regiments = 1 brigade, 2 brigades = 1 division, 2 divisions = 1 army corps. What sort of counting is this? Nothing more than a *grouping* of soldiers arranged in order. Not only are men grouped in companies, but companies into battalions, battalions into regiments, and so forth. Thus they are "grouped groupings."

A grouping whose elements are again groupings is called a *grouping of higher order* or rank (second, third, . . ., etc.). All systems of measurement are actually groupings in ascending rank:

60 seconds = 1 minute   1000 mg = 1 g
    60 minutes = 1 hour   1000 g = 1 kg
        24 hours = 1 day   1000 kg = 1 metric ton
            30 days = 1 month
            12 months = 1 year

Measures are groupings of elements of known magnitude in a sequence. Are there also similar groupings of abstract numbers? Of course there are; there is no need to bring up examples from our daily life or from the lives of primitive peoples. A native counting sea shells puts down one small stick for every 10 shells; for every 10 small sticks he has one large stick, equivalent to 10 times 10 shells. Our currency system is nothing but a grouping or a division into groups.

Groupings and Numerals

Grouping is the important advance over the unorganized counting done by merely placing the objects to be counted in series; once this step has been taken, the unlimited progression of the number sequence is initiated: first ordering → first grouping → second ordering → second grouping → third ordering → third grouping, and so on. The significance of this progression is apparent from the important fact that the symbolic representation of these progressive groupings leads to the earliest number symbols used by various peoples.

As an example of pure 10-groupings, I have chosen the pictographic numerals of the Egyptians on temple walls and monuments; I have placed them here next to the discontinuous Roman method of counting by 5-groups (Fig. 8).

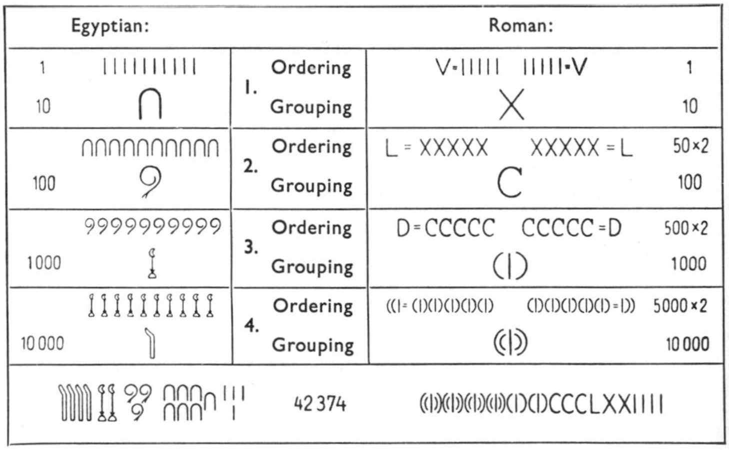

Egyptian:			Roman:		
1 10	IIIIIIIIII ∩	I. Ordering Grouping	V·IIIII IIIII·V X	1 10
100	∩∩∩∩∩∩∩∩∩∩ 9	2. Ordering Grouping	L = XXXXX XXXXX = L C	50×2 100
1000	9999999999 ↓	3. Ordering Grouping	D = CCCCC CCCCC = D (I)	500×2 1000
10 000	↿↿↿↿↿↿↿↿↿↿ ↾	4. Ordering Grouping	((I= (I)(I)(I)(I) (I)(I)(I)(I)(I) =I)) ((I))	5000×2 10 000
)))) ⌇⌇ 99 ∩∩∩ III ∩∩∩ I		42 374	(I)(I)(I)(I)(I)(I)(I)CCCLXXIIII		

Fig. 8 Egyptian and Roman numerals compared.

Let us cite a historical example (see Fig. 9). The Egyptian king Sahure commanded the herds of cows, oxen, asses, goats, and sheep which he had taken as booty on a Libyan campaign to be represented on a large inscription. Our illustration shows a detail of the herd of asses and a fragment of the herd of goats. Even in those days large numbers made an impression; he therefore had the number of head inscribed above the last animal in each herd: 232.413 goats and 2234.. asses (the last two digits have been broken off). The tadpole is the symbol for the number 100,000 (see p. 122).

There is evidence that many other peoples formed their early number symbols in a similar fashion. Their *early* number symbols, for in many cases in the course of their development they went far beyond this simplest rule of ordering and grouping. Thus the clarity of

Fig. 9 Egyptian numbers in pictographs, 2500 B.C. We can read 223,4.. and 232,413. Each donkey is about 50 cm high.
German Institute for Antiquity, Cairo.

earlier Egyptian numerals was later obscured in the so-called "hieratic" script of the papyri, which developed from the old hieroglyphic writing. The Romans themselves later developed different rules for representing higher numbers, although for the lower numbers they continued to use their original systems.

An arresting example of the earlier Roman numerals, and one we shall refer to often hereafter, is the *Columna rostrata*, the commemorative column which C. Duilius had erected at Mylae in the year 260 B.C. to celebrate the naval victory over the Carthaginians. It was termed *rostrata* (< *rostrum*, "beak," by transfer from the ram in the bow of a warship) because it was ornamented with iron ship's rams of this type. On the fragment of the inscription illustrated here (see Fig. 10) our attention is caught by the bottom few lines with the number symbols:

(*Auro*)*m captom: numeri 1000 1000 1000 500*
(*Argen*)*tom captom: praeda: numei 100,000*
(*Omne*) *captom aes:* (more than 20 times) *100,000*

"— gold captured: 3500 coins
— silver captured: booty: 100,000 coins
— total loot in all: (more than) 2 million *aes*."

(The *aes*, "bronze," was a bar of bronze weighing one pound, the basic monetary unit in ancient Rome.)

One surprising thing is the array of 22 symbols representing a hundred thousand in a row; actually, as we know, there were even 32 of them. Thus a higher grouping for 10 hundred thousands, for 1 million, had not yet been devised. This is a very choice proof of the *limitation* of the Roman number sequence and number symbols. Pliny also confirms this:

Non est apud antiquos numerus ultra centum milia; itaque et hodie multiplicantur haec, ut decies centena milia aut saepius dicuntur.

Fig. 10 Inscription from the *Columna rostrata*, containing, among other things, more than 20 symbols for a hundred thousand. The column was set up in Rome to commemorate a naval victory over Carthage in 260 B.C. This fragment measures about 80 × 80 cm, with letters 2.5 cm high.
Palazzo dei Conservatori Rome.

"The ancient Romans had no number higher than a hundred thousand; hence, even today [in the 1st century A.D.] these numbers are multiplied by saying ten times a hundred thousand, or the like."

We know this already from our discussion of the Roman number words (p. 28); now we have the documentary proof, as it were, on the memorial column: at one time 100,000 was the "last" number known to the Romans. Later, as the Million became more current — we recall the state indebtedness under Vespasian (p. 28) — the Romans wrote it as 10 × 100,000 or $\boxed{\text{X}}$, exactly according to their number word for this sum. The frame around X thus stands for 100,000. Even if up to now we were not sure just how this came about, there is no longer any doubt: the frame is the terminating sign for the symbol for 100,000 which we see written out in full on the *Columna rostrata*. Used as an abbreviation, it now stands for the whole symbol; *decies*, "10 times," appears under it as X.

The Romans thereby created a numerical symbol that quite graphically reflects the number word, but with it they abandoned the rule governing their earlier writing of numbers: They no longer arranged the hundred thousands in a row, but rather counted them (Fig. 11).

Fig. 11 Large Roman numerals.

The amount that Livia left to Galba was $\boxed{\text{D}}$ sesterces; Tiberius, however, read the number without the frame around it (see p. 28). The 40 billion sesterces in Vespasian's indebtedness were readily expressed verbally and easily written down in words (p. 28) but not in numerals: A double frame would have had to be placed over the CCCC, and this the Romans did not do. All this is a compelling proof that number words are more powerful than number symbols and advance more readily to higher numbers.

Grouping and Language Speech

Now that we have recognized grouping as an important source of written numerals, the question arises: What significance does it have for spoken number words? Is the number sequence also built up by alternate ordering and grouping? Are number words also groupings, like the X among the Roman numerals?

Let us run through our own number sequence once more. "One, two, three, ... nine, ten, eleven, ... nineteen, twenty, twenty-one, ... thirty, ... ninety-nine, a hundred, ... a thousand ...." Is this ordering? Grouping? It is certainly not mere ordering "one, one, one ..," but "ten" and "–ty," "four hundred," "five hundred," "six" and "seven thousand" — are these not groupings? At any rate, the number sequence is not built up from one word alone.

GRADATION BY STEPS

The Number Steps

"One, two, three, ..., nine, ten" is a sequence whose individual elements have names. If a quantity of coins is divided into groups of 5 each, the coins are heaped together indiscriminately into the small piles without being first named individually. The coins are combined "on even ground," like bundles of grain into shocks whereas in counting we have the sensation of ascending the steps of a stairway: 1, 2, 3, ..., 9, 10, after which we come to a landing, where we can pause for breath. Then the steps continue on up: 11, 12, ..., 19, 20 — here again there is landing. Thus we climb higher and higher up the tower; as we look way down we see how the early numerals are always ordered and grouped on the floor; the stalks are tied together into sheaves, the sheaves are piled up into shocks and the shocks into wagonloads, but always on the mowed field of stubble (compare Fig. 14, p. 58 with Fig. 8, p. 42).

The reader should be guided by these illustrations. He will then recognize the essentially different arrangement of our verbal number sequence, which we have called its *step gradation*. The number sequence is graduated. This is achieved by giving names to the elements of the sequence; these elements no longer remain anonymous and undifferentiated but form the steps of a staircase: Seven is "higher" than three. In this staircase "ten" plays the part of a threshold, or a landing, as we have called it. So do "twenty," "hundred," "two hundred," and "thousand." In the Middle Ages the 10-step and every subsequent multiple of 10, such as 30, 80, or 1960, was called *articulus*.

But why is just 10 a threshold — why not 7 or 11? The answer is that we have ten fingers; after that, the hands are "finished." But the hands themselves are groupings: "hands" = 10 fingers. As we learned earlier, in New Guinea 99 is called 4 men and 2 hands and 1 foot and 4. Was this a gradation or a grouping? A grouping, of course.

But the early concepts of grouping *lead* to gradation. We can even say that the "landings" on our number staircase *are* the old groupings. They have merely lost the marks of their origin in the graduated number sequence; they faded out as groupings when the spaces between them were filled up with number steps.

Yet if the stair landings, in particular the steps 10 — 100 — 1000, are groupings, then this signifies that, in the composition of the number sequence, at first all the ranks are found by grouping — 10 tens make a hundred, 10 hundreds make a thousand — and not until later are the numbers between them, like 78 or 543, put into place. The ranks are set up ahead of time as a framework, 10 ... 100 ... 1000, and the steps are then built in from landing to landing. To put it in a nutshell: Early man possessed the higher number 1000 before he grasped the lower number 543, for 1000 was found through grouping, whereas 543 was found by counting in ascending steps. Our "theoretical" formation of the number sequence, proceeding from unity by steps of one ("plus one, plus one, ...") in building up a whole structure of numbers, does not correspond to historical reality. The true situation is far better understood from the analogy of the staircase of numbers whose chief pauses (the landings) were first erected and then followed by the intermediate steps, which were interpolated one by one.

Even today, if we contemplate the number sequence in our minds, 1000 seems clearer, more "available" to us than 543, a number of which we can really say only that we must certainly arrive at it if we keep counting long enough. We can visualize it directly only with difficulty, if at all. We are fascinated in recognizing the prominence of the landings and the secondary nature of the intermediate steps in the changes in word forms, such as we have seen, for example, in the Latin number words (p. 27): *kentum — mille* are indeclinable, solid "blocks" between which the more malleable, inflected step words for the multiples of 100, *ducenti*, etc., have been inserted. For a Germanic example of the same thing, see p. 47.

The old groupings, which we have come to recognize as the rule governing the early writing of numerals, also play a part in the formation of spoken number words. Let us examine this surprising fact through a few more examples.

GROUPINGS IN THE GRADUATED NUMBER SEQUENCE

The Ten-steps
If those whose native language is German or English run through the French names for the multiples of ten (see p. 99), they feel a sudden jolt at 80: the even progress of the number sequence is suddenly broken by the completely alien word formation *quatre-vingt(s)*. What has happened here? In our view the graduation of the

number sequence has been interrupted by an old 20-grouping ("four twenties," not "four times twenty" — the *-s* plural ending allows no other interpretation). This is an excellent example of the grouping principle breaking into the graduated number sequence.

In Latin, "twenty" is *viginti, vi-ginti,* "thirty" is *tri-ginta*; in Greek, "thirty" is *triá-konta* (see p. 92) — words formed with the endings *-ginta* (*-ginti*) and *-konta.* Seemingly this is connected with "ten," but, as we shall see later, it is actually means "a tenned," a grouping of 10 units. *Vi-ginti* is then "two tenned"; the *-i* ending is the grammatical dual. *Triginta* and *triakonta* are "three tenned"; the *-a* ending is the inanimate plural.

These graduations of ten thus betray their origin in groupings. Once we have opened our eyes to them, we can discern them in many languages. The reader may look for them himself in the tables of number-words (pp. 92–99). Let me point them out in Gothic, for example (p. 92): *taíhun* is 10, yet 50 is not *fimf-taíhun* but *fimf-tigjus,* "five tens," and so on; similarly in Old Norse 10 is *tío,* but 50 is *fimm-tiger* rather than *fimm-tio.* We shall show later that the meaning "tenned from tens" is concealed in the word *Hundert* (see p. 128).

In modern German the multiples of ten are distinctly marked with the suffix *–zig.* But in the case of the higher ranks German still uses word forms that clearly emphasize the grouping: *Viele Hunderte* (*Tausende*) *von Menschen* ..., "Many hundreds (thousands) of people ...." If we had said instead, *Viel hundert* (*tausend*) *Menschen,* "Much hundred (thousand) people," the semantic meaning would have disappeared along with the grammatical group form. We use the first form only for large but unspecified quantities; Latin, however, says *duo milia militum,* "two thousands *of* soldiers" (genitive case), and many other languages do the same. Until World War II, for example, the numbers from 200 to 2000 in Danish were capitalized: *hundrede* (100), but *to Hundrede, to Tusind* (200, 2000). Wherever the number-word (*milia*) is a noun and the quantity specified is in the genitive case after it, the meaning is always that of a grouping.

When after one rank of numbers (e.g., 100) the next higher one is indicated as "large," "big," or "strong," we clearly recognize the first as a grouping, for a number (100) does not actually grow by being termed "large." A significant example of this is provided by the Germanic word for "thousand": Gothic *pushundi* < *pus-hundi,* "the strong hundred" (see p. 132); the best-known example (although it is an artificial one) is *milli-one,* "the large 1000" (p. 134). From the word *šar* (= 60^2) the Sumerians formed their highest rank of numbers, *šar-gal,* "the great *šar*" (see p. 165). Sanskrit also uses the "great" formation for numbers of higher ranks, as in *padma,* 10^{10} — *maha-padma,* 10^{11} (*maha* = "large"). Similar word formations occur among primitive peoples: the Hottentots say *disi* for 10 and *gII-disI* for 100, "big ten"; the Welsh gypsies say *baro deš,* "great ten" for 100 (see p. 191). A nice example of the way large size is indicated is provided by the Sioux Indians: *crabrah,* "10," *crabrah-hugh-tougah,* "10-herds-masculine," 100.

Here is one more Germanic example of the precedence of the (grouped) graduation and the subsequent insertion of the intermediate steps. In this connection the reader should look at the Gothic number sequence from 10 to 100 (p. 92). Gothic 10 is *taíhun* and 100 is *taíhunt-e-hund*, "tenned from tens" — clearly a grouping! The words for multiples of ten up to 60 are formed with *-tigjus*, and from 70 to 90 with *-tehund*. There can be no doubt that this break in the sequence of tens did not arise from a step-by-step building of the numbers into a "tower of numbers," but was due to the reverse influence of 100 (*taíhunte-hund*) working back to 60, which must have existed as a gradation (grouping) before the successively ascending steps of ten were inserted.

Collective Number Words

Some languages have the capacity of forming nouns from number words; this is conceptually equivalent to grouping. Thus Latin from the "landings" *decem*, 10, and *centum*, 100, derives the nouns *decuria*, "ten-ness" and *centuria*, "hundred-ness" or "century," the military detachments of 10 and 100, equivalent to squad and company, respectively; *quinque centuriae militum* does not stand for "500 soldiers" but "5 groupings of 100 soldiers each." But because the plural *milia* of *mille*, 1000, is already a noun that takes the genitive case in the quantity counted, *milia* is felt to be a grouping itself and thus does not form such a collective number word.

In the same way, the Greek language formed the collective nouns *dekás* ("decade"), *hekatontás*, *chiliás* and *myriás* ("myriad"), from the numerical ranks *déka* 10, *hekatón* 100, *chílioi* 1000 and *mýrioi* 10,000. The Bible makes frequent use of these words: "... and the number of them was ten thousand times ten thousand, and thousands of thousands " (Rev. 5:11):

ναί εἶν χο ἀριθμός αὐιόν μηριάδες μηριάδων ναί χιλιάδες χιλιάδον
literally: "... myriads of myriads and thousands of thousands."

In the Vulgate this passage has only the last expression, *milia milium*, since Latin lacks a word for a higher rank.

Based upon these coinages derived from the number ranks, Greek has created collective words corresponding to every number: *monás* ("monad, one-ness"), *dyás*, *triás*, and so forth. This generalization of the groupings applied to every number is palpably artificial and self-conscious, whereas the words for the numbers of successive ranks evolved out of number concepts lost in antiquity.

The French similarly use the suffix *-ain(e)* to form a collective noun for each separate number, and then readily employ the collective word as a numerical measure governing the genitive case in the object counted: *six-ain* or *six-aine*, "6 pieces, a six"; *douzaine*, "dozen"; *quinzaine*, "15 pieces" (German *Mandel*); *vingtaine*, "20 pieces"; *quarantaine*, "40 pieces" (from the latter is derived the term for the 40 days of detention ("quarantine") to which travelers coming from a diseased area must submit); *centaine*, "100 units"; etc. One man belonging to a group of thirty is called *un trentain*, "one of the thirty." We can quite clearly trace the grouping in these collective number words if we compare the expressions *une douzaine de livres*, "a dozen books," with *douze livres*, "12 books"; in the latter case the books are felt to be counted one by one, 1, 2, 3, ... 12, whereas the former groups them together so that they are felt in their totality rather than counted individually — so much so that the words *douzaine* or *dozen* often need not denote exactly 12 items.

20-GROUP "MAN"

We already touched briefly on the 20-group, *quatre-vingt*, in the French number sequence. This is not the only example. Before we go any further into this subject, let us discuss the abstract *20-Group "Man"* as it has grown out of the number sequence to form special words in a variety of folk languages. Only then can we understand why this peculiar 20-grouping has become an inherent part of the various number sequences.

It is difficult for an illiterate peasant to tell how many years he has lived; the Piedmontese, for instance, would answer that he carries *quat borla*, "4 large loads," on his bent back, meaning 4 × 20 = 80 years. This is far easier for him to visualize; he does not climb the steps of his years age, 77, 78, 79, 80, but remains standing on the ground and, like a mule, takes up the "loads" of his years grouped in bundles of 20. The Sicilian dialect has no special word for the 20-group, so that in Sicily the peasant says *tri vintini et deci*, "3 twenties and 10," for 70 years; thus he uses the number-word *vinti*. In some regions of Spain the monetary units (1 *duro* = 20 *reales*) are used in counting, so that a peasant there would reply that he is "3 *duros* and 10 *reales*" old.

This may remind us of Charlemagne's old coinage: He had 12 *denarii* of 20 *solidi* each struck from one pound of silver. This monetary system lasted throughout the Middle Ages into modern times; France did not give it up until the French Revolution, and it is still in use in England, where 1 pound = 20 shillings, and 1 shilling = 12 pence. The abbreviations £1 = 20s = 20 × 12d are merely the initial letters of the Latin words *libra, solidus, and denarius*. To indicate shillings the long s, or ∫ was also used; this has been shortened to a slanting stroke in expressions such as 5/6 (5s 6d).

On the whole, the English language has most faithfully and widely preserved the vernacular 20-group in the word *score*. This was originally a "notch" that the farmer would carve for every 20 animals as he counted his cattle or sheep (we shall have more to say about this in connection with exchequer tallies, p. 236), but today it has taken on the strict meaning of "twenty." The English Bible, just like the Piedmontese peasant, counts a person's age by groups of 20 years (Psalm 90:10):

"The days of our years are threescore years and ten; and if by reason of strength they be fourscore years, yet is their strength labour and sorrow."

Shakespeare also counts by scores. A passage in *The Merry Wives of Windsor* (Act 3, Scene 2) reads:

"as easy as a cannon will shoot print-black twelve scores."

Prince Hal says of his fat friend Falstaff (1 Henry IV, Act 2, Scene 4):

"I'll procure this fat rogue a charge of foot; and I know, his death will be a march of twelve scores."

Our oldest documentary evidence dates from around 1050, when the inventory of a monastery's possessions translated the vernacular expressions into Latin as in a dictionary:

V *scora scoep*   *quinquies viginti oves*   "5 × 20 sheep,"
VIII *score oecer*   *octies viginti agri*   "8 × 20 acres."

Modern English no longer has a word for the German *Schock* (a mass of sixty units), derived from the same root, and thus uses the far more expressive *threescore*. In some trades *score* serves as a unit of measure, e.g., in mining for a basketful of coal. In Ireland it also serves as a weight of 20 pounds.

In expressions like *scores of times* ("very often"), *scores of years ago* ("many, many years ago") or *scores of persons* ("dozens of people") the numerical meaning of 20 has lost its force, but the 20-grouping still distinctly comes to the surface. How strong the concept of 20 continues to be can also be seen quite picturesquely in the way numerals were written. A 14th-century inventory of the Scottish treasury writes the sum £6896 as

m   c   xx
vj viij iiij xvj — "6 thousand 8 hundred 4 *scores* 16."

As a rare example of a 20-grouping in an early system of numerals, I shall cite the Indian "Kharosti" numerals, which exhibit three groupings lower than 100: a 4-group, a 10-group, and a 20-group. These numerals could represent the French *quatre-vingt-dix* exactly, as could the Celtic series of multiples of ten (see Fig. 12 and p. 97).

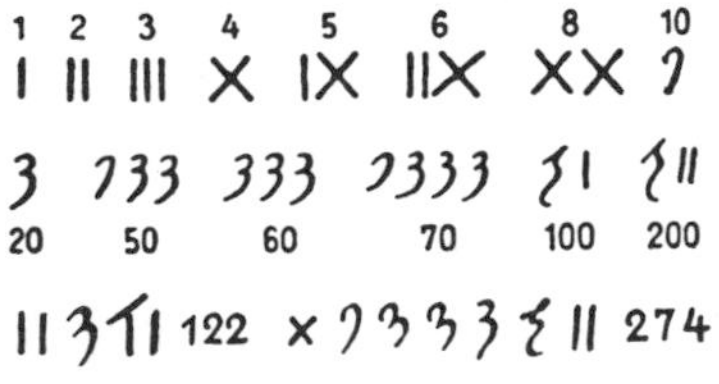

The Low-German word *Stiege* (High-German *Steige*, "ladder, series of steps"), which is likewise a 20-group like *score*, has not acquired the same force. Previously, however, it was very widely used, as the equivalent word *stega* in Crimean Gothic testifies. *Stiege* were used in counting sheep and sheaves of grain in a field, but also eggs and linen cloth. An old book contains the words:

Lasst von dem Linnen mir für Geld nur eine halbe Stiege,
"Let me have just half a *Stiege* of the linen for money,"

and a 16th-century poet extolls the virtues of a marriageable girl in the words:

Zwei stieg Schilling sie auch hat,
davon sie rent kan kriegen.

"She also has two *stieg* of shillings,
from which an income she can get."

The way in which this word acquired its numerical meaning is lost in obscurity. The Middle High German *stiga* is a corral for small animals like sheep or goats; perhaps the amount of space designed to hold 20 sheep came to be thought of as a measure of quantity. The expression *Steige Salat* in the local German dialect of Hesse refers to a lightly built crate of laths with air spaces between (about 50 × 40 × 20 cm in size), which grocers use for shipping vegetables.

Fig. 12 Indian Kharosti numerals, ca. 200 B.C., with three groupings of 4, 10, and 20. The symbol for 20 which resembles "3" consists of two 10-symbols superimposed. 20-groups occur very rarely in numeral systems. These numerals are read from right to left. The hundreds are no longer grouped but merely counted.

In contrast, the common Germanic word *Schneise*, "glade, a cutting of trees" (Anglo-Saxon *snas*, Old Norse *sneis*, Danish *snes*, Dutch *snees*), has the original meaning of a "rod" or "twig" used to line things up in a row. Fishermen in the Baltic use it to indicate the sticks on which they impale "stockfish" (such as cod, salmon, haddock, etc.) to dry. An old monastic register from the year 1186 includes in its annual income

24 *snesas anguillarum* — 24 "*Schneise*" (480) eels.

The expression for an indefinite quantity, "as many fishes as will go on a stick," very early became associated with a 20-group, "20 fishes," and then detached itself from this to become an expression for the abstract number "twenty." A *schnasse Zwiefel* (*Schneise Zwiebel*) "a string of onions" slang for a twist of straw with 20 onions attached. In addition to fish, eggs are still counted in Holland by *snees*, and the Swedish farmer often says *fyra sneser*, "four *snes*," when he means "eighty." All the examples here enumerated are connected with *Schneise*. Middle High German *sneise* is a string on which something (such as berries or pearls) is strung, and Bavarians today still speak of certain things in a long row as a *Schnaisen* without necessarily implying that there are 20 of them.

A remarkable evidence of the force of the 20-group in Scandinavia is the absorption of the Germanic *tiu*, "ten," as "twenty" (!) by Finnish, which after all does not belong to the Indo-European family of languages (see Fig. 21, p. 90); today in Finland it is used exclusively for eggs: *tiu munia* = "20 eggs." A similar case is the Russian word *sorok*, 40, which constitutes an exception within the Russian number sequence, for *sorok* goes back to the Old Norse *serk(r)*, "hide" or "pelt," since furs were tied together in bundles of $40 = 2 \times 20$ (see p. 185).

In the *Odyssey* (I, 431) Laertes buys a slave girl for 20 head of cattle.

Although all the Romance languages derived from the Latin a well-ordered number sequence based on gradations of 10 (see p. 99) — Latin does not place any special stress on the number 20 — traces of the 20-group have nevertheless survived here and there in popular speech, not as special words like *score*, *Stiege*, or *Schneise*, but in peculiar uses of the word "twenty." Thus, for example, money in Portuguese is counted in *vintens*, "twenties." The elided expression *dois cinco* is short for *dois vintens e cinco*, "2 twenties and 5" (= 45); *seis menos cinco* does not mean "six minus five" but "six twenties less five" (= 115). As we have said this phenomenon has been preserved in the vernacular speech of Romance countries, in some places as isolated cases and in others more commonly. Later, we shall have more to say about this and about the persistent survival of the 20-group in Celtic and French (see p. 64).

It is quite striking that the 20-grouping is most strongly rooted in the north of Europe, in Iceland, Denmark, and England. When we think of gestures symbolizing numbers, we might well suppose that these would be most commonly used and firmly established in areas where people go barefoot and thus can actually count on their toes as well as on their fingers. Yet the Eskimo, who very readily counts in terms of "men" (1 "man" = 20 fingers and toes) can scarcely do without shoes. The habit of counting by actually touching objects is, to be sure, one path leading to the vigesimal grouping, but not the only one. The mere knowledge of possessing an illustrative supplementary quantity consisting of 20 elements, to which things can be assigned in successive order, is also quite sufficient. It also "invisibly" dominated early man's conception of numbers. It may also be due to the favorable magnitude of the number 20 and its convenience in counting larger quantities; it is neither too small nor too large. This, rather than any "recollection" of the 20-group "man," is undoubtedly what prompted Herr Baedecker to count the steps in front of the Cathedral in Milan in groups of twenty (see p. 41). The latter, however, looms darkly in the background.

Let us conclude with two more instances of groupings by 20. When Pliny the Elder says in his *Natural History* (13, 77) that more than 20 sheets were never used for one scroll, this statement contradicts the fact that scrolls of more than 20 leaves existed. But perhaps he only wanted to imply the papermakers counted "bales" of papyrus in units of 20 leaves each. This ancient fashion of counting paper has been preserved even in our own day, for bales of paper are now reckoned at 10 reams of 20 quires each, so that the bale is made up of 10 × 20 quires ("folios" of 24 sheets); similarly in French 1 *rame* = 20 *mains* of paper ("hands" consisting of 25 sheets) or = 20 rolls of wallpaper. Both the words *ream* and *rame* are derived, through the Medieval Latin *risma*, from the Arabic *rizma*, "bale, bundle."

Of the many examples to be found among primitive peoples (see p. 35), I shall cite the number sequence of the Indonesian Enggano, a language spoken on the island of the same name (just south of Sumatra). This number sequence is clearly built up of groupings of 20, of which the following formations expressions are of interest:

kahaii ekaka   ariba ekaka   kahaii edudodoka
one man (20)   hand man (100)   one our body (400).

There is thus no special word for 100, and 100 is not a stage or gradation in counting — so strongly is the 20-step felt; *ariba* is derived from *lima*, "hand," which is the number word for "five" throughout the whole Austronesian linguistic region. The last number word for 400 may be explained as "so many times man as our body yields by its fingers and toes," and hence "man times man" (20 × 20 = 20^2).

Let us now sum up our ideas about the rules governing the structure of number sequence. We have seen how, in counting, the principle of grouping arises from an ordering of the counted elements. The groups thus derived are themselves ordered in turn, and are again collected together in groupings of higher rank. In pure counting the natural divisions of 5, 10, and 20 — "hand, hands, man" — are predominant. The physical representation of this process leads to the earliest numerals used by primitive peoples, which we thus designate as *row-characters*.

The natural groupings also enter into spoken number words, but here they become gradations because the number sequence no longer recognizes a succession without individual names. The vigesimal grouping above all has been especially persistent among various peoples, in that the very ancient organization of "man" coincides with a component quantity convenient for use in actual counting.

Before we use this new insight to investigate various historical number sequences, we must make sure not to overlook a fruit that is now ready to fall ripe into our lap, but which we are for the most part inclined to forget without noticing it: the recognition that spoken number words and written numerals, curiously enough do not coincide. Written number symbols are definitely not mere representations of number words.

What should we expect? The sequence of number words in a culture is built up long before any mature system of numerals becomes established. Thus we may think that numerals are simply "copies" of number words.

But our researches thus far have shown us that the laws that govern early numerals, ordering and grouping, do not in fact correspond to the rule of the number sequence, which is stepwise gradation. Hence, the writing of numerals is *not* merely the representation of the number-word sequence.

And thus one of those paradoxes of cultural history comes about, in that the Romans, who possessed such a subtly and precisely ordered, flexible verbal number sequence, which differs from our own only in minor details, used a system of numerals so crude and cumbersome that it is hard to see how it could be a product of the same culture:

four	thousand	eight hundred	seventy-	nine
quattuor	*milia*	*octingenti*	*septuaginta*	*novem*
(I) (I) (I) (I)		DCCC	LXX	VIIII

as contrasted with 4879.

NUMBER LANGUAGE AND NUMBER SYMBOLS

It is a dramatic documentation of cultural history that various cultures throughout the world have always felt this discrepancy between speech and writing, and have thus strived to invent or borrow a good system of written numerals. Our modern numerals, so clearly and readily representing the large number 4879, are not the offspring of our own intellect. And although our forefathers had just as good a verbal number sequence as the Romans, they still did not at first have their own universally valid numerals.

On this whole planet, only two nations have succeeded in devising numerals that represent their verbal numbers, after first overcoming earlier levels of writing: the Chinese and the Indians; to be really precise, only the Chinese. This is because Chinese writing is a conceptual or pictorial system that provides a separate pictographic character for every object or idea (see the summarizing discussion of this subject pp. 450 ff.). If for example, we symbolically wrote numerals for the units, and T, H, and Th for tens, hundreds, and thousands, our number would be written:

4 Th   8 H   7 T   9,

and it would exactly represent the manner in which we speak this number ("ten" of course is shortened to "-ty"). This is precisely what Chinese does; it writes the number words with an ideogram that at the same time is its numeral (Fig. 13).

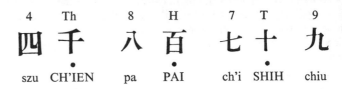

Fig. 13 Chinese representation of the number 4879.

We have written the number horizontally, instead of vertically in the proper Chinese fashion; the gradations or orders of magnitude are capitalized and are represented by the dots beneath the Chinese characters.

Whereas for us the written word "four" and the numeral 4 are two distinct and separate ways of representing a single concept, Chinese has only the one: Number word and numeral are identical. This accordance of spoken number words and written numerals is unique, the only known instance in world history.

The Indian numerals do not indicate the rank of a number by a special symbol but merely by its position: 4879. Thus the number ranks are not named, they are a nameless "positional" system of writing, so to speak; the Chinese numerals, on the other hand, make up a "named" system, as if we substituted symbols of numerical rank for position: 4 Th 8 H 7 T 9. Unlike the Chinese, the Indian numerals are not, at least not directly, a representation of the verbal number words; we shall see later that they were actually derived from the counting board. The essential difference between the two systems of numerals lies in their possession, respectively lack, of a symbol for the rank, or order, of a digit. To express the quantity 4079, the Indian number system must expressly designate the missing hundreds (by the symbol for nothing, zero) — something that the spoken number words do not do. The Chinese, on the other hand, can omit it altogether by writing 4 Th 7 T 9.

In regard to their internal structure, however, the two systems of numerals are alike. They follow the law of gradation, no longer that of ordering and grouping; thus they have risen to the highest state of perfection that a system of numerals can achieve.

In conclusion, let us touch upon another idea that is often overlooked: Our numerals do not have the same origin as the phonetic alphabet we use in writing and printing. Our letters are not akin to our numerals. One would naturally be inclined to suppose, in all innocence, that the human mind, when it took the trouble to record its ideas and concepts, would have devised similar systems of writing words and numbers, "seven" and "7." But this did not happen, neither in our western culture nor anywhere else in the world. The early system of writing numerals is everywhere the older of the two sisters. We shall learn this in the course of our voyage through cultural history, but even now the reader may recognize the extraordinary paths along which the human mind has traveled from one culture to another, as are epitomized in the following fact: Our language is Germanic, our writing is Roman, our numerals are Indian!

Historical Gradations

REPRESENTING THE GRADATION

Levels of Rank

By now we have learned to understand the essence of gradations in the number sequence; that was the burden of the preceding section of this book. But before we look into some of the numerical gradations that have appeared in history, let us say a little more about the pictorial and written representation of such gradations, so as to visualize these concepts as clearly as possible.

Our number sequence "one, two, ..., ten" is a gradation of 10's; in other words the number 10 governs its general structure. Let us, as an example, write out $2463 = 2 \times 1000 + 4 \times 100 + 6 \times 10 + 3 = 2 \times 10^3 + 4 \times 10^2 + 6 \times 10^1 + 3 \times 10^0$. In this expression 10 has appeared as 10^3, 10^2, 10^1, and $10^0 = 1$; 2 thousand 4 hundred 6-ty 3. The exponents of 10 we call its *rank*; the "pure" levels of 10, 10^3, 10^2, 10^1, and 10^0, are thus its *levels of rank*. $1000 = 10^3$ is therefore the third level of rank. Then what are the ranks of the gradations of 20? In ascending order: 20^0, 20^1, $20^2 = 400$, $20^3 = 8000$, and so on. The pauses, the orders of magnitude in ascending this particular numerical ladder, are 20, 400, 8000, etc. As we climb upward, our steps become larger and larger at each successive rung.

The number sequence based on gradations of 5 has the following ranks: 5, $5^2 = 25$, $5^3 = 125$, etc.

In the various gradations named, the corresponding levels of rank are:

gradations of 5	5	$5^2 = 25$	$5^3 = 125$
gradations of 10	10	$10^2 = 100$	$10^3 = 1000$
gradations of 20	20	$20^2 = 400$	$20^3 = 8000.$

If we write the same number "eighty-nine" in terms of these various gradations, it would appear

in the 5-base system as	$3 \times 5^2 + 2 \times 5 + 4 = 324$
in the 10-base system as	$8 \times 10 + 9 = 89$
in the 20-base system as	$4 \times 20 + 9 = 49.$

Expressed in terms of pure gradations,

324 would be the 4th level on the 2nd level of rank 1 on the 3rd level of rank 2

89 would be the 9th level on the 8th level of rank 1

49 would be the 9th level on the 4th level of rank 1.

The reader may be able to visualize these strange forms of expression with the aid of the figure (see p. 58). If we were to write them in numerals, the vigesimal system, based on gradations of 20, would create some difficulties, for it would require nineteen different symbols and words in order to count the steps of each rank; our decimal system, based on gradations of 10, uses nine, the nine numerals 1, 2, 3, ..., 9; and the system based on gradations of 5 would have only four numerals. In all cases, a missing rank is indicated by the zero symbol 0.

In algebraic, or generalized, notation, a number expressed in a system with x levels would look like

$$(1) \quad z = a_n x^n + a_{n-1} x^{n-1} + a_{n-2} x^{n-2} + \cdots + a_1 x + a_0,$$

or

$$(2) \quad z = a_n R_n + a_{n-1} R_{n-1} + \cdots + a_1 R_1 + a_0.$$

Thus x is the base number of the gradation, the number on which it is built (in our case, 10); its powers or exponents, x^1, x^2, x^3, ... (10^1, 10^2, 10^3, ...) are the ranks R_1, R_2, R_3, ... (10, 100, 1000, ...). For example, in our decimal number sequence the third rank level $R_3 = 10^3 = 1000$. (For the sake of uniformity, the zero ranks $x^0 = R_0$ in expressions (1) and (2) have been combined in the last term, $a_0 R_0$.)

The coefficients a are the units (the digits) that give the number of each rank. There are always $(x - 1)$ units a_k, where a_k is one of the numerals 1, 2, 3, ..., $(x - 1)$; again, in our own case, where $x = 10$, there are nine digits. When the ranks are expressed by the positional values of the digits, the zero symbol 0 must appear.

In reckoning by multiples of the units, the rank levels, $a \times R$, the individual rank groups $(a \times R) + (a' \times R')$ are counted together. In precisely the same manner our language uses one number word less for the units that count the ranks than the base number of the number system; hence in the system based on ranks of

10 there are 9 such number-words: one, two, three, ..., nine
 5 there are 4 such number-words: one, two, three, four
20 there are 19 such number-words: one, two, three, ..., nine, and so forth.

Beyond 9 we would have to find additional number words, which could not, of course, be formed as compounds with "ten" (like "thir-teen").

In addition, the language has words for the ranks of ten, hundred, thousand and so on, and correspondingly for the other systems of rank.

A few additional examples should serve to make this general method of writing numbers more familiar. In the system based on gradations of 5, where $x = 5$ and $a_k = 1, 2, 3, 4, 0$, the number $z = 178$ would be written as follows:

$z = 1 \times 5^3 + 2 \times 5^2 + 0 \times 5^1 + 3$, or
$z = 1 \times 125 + 2 \times 25 + 0 \times 5 + 3$; written positionally, $z = 1203$.

In the 20-base system the number $z = 2657$ is

$z = 6 \times 20^2 + (12) \times 20 + (17)$
$ = 6 \times 400 + (12) \times 20 + (17) = 6\ (12)\ (17)$.

Here the digits above 9, which we do not possess in our decimal system, are enclosed in parentheses. The Mayan language, however, had individual words for all the numbers in the vigesimal system (see p. 59), so that the number 2657 in Mayan would be read as

6-bak (12)-*kal* (17).

If there were no (12)-*kal*, the number 2417 in the vigesimal positional number system would look like this: 6 0 (17).

The Babylonians had a sexagesimal or 60-base system, yet they wrote their units 1 ... 59 by gradations of 10, as we do (see p. 167). Thus the number 1955 would appear in Babylonian as

$$z = 32 \times 60 + 35, \text{ i.e., } (32) \text{ sossos } (35).$$

The number 1920 is (32) 0; but the Babylonians had no zero and would merely have written (32), thus leaving the order or rank of (32) completely unspecified (on this subject, see p. 167).

Representations of Levels
Now let us graphically represent a number written to the base 5 and to the base 10 (see Fig. 14).

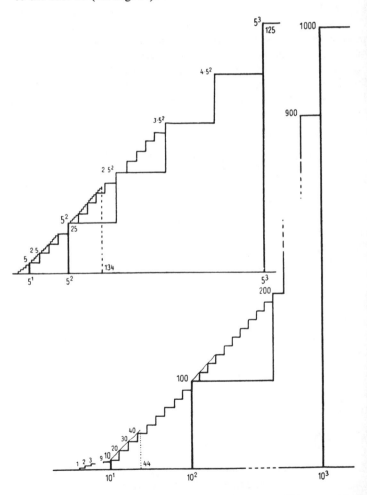

Fig. 14 Number systems based on levels of 5 and 10. The ranks 5^1, 5^2, 5^3 and 10^1, 10^2, 10^3 rest on the "floor." At each level the steps of the preceding rank also continue upward. The number 44 is represented in both systems: $1 \times 5^2 + 3 \times 5 + 4 = 134$; $4 \times 10 + 4 = 44$.

In the 5-base system (upper diagram) we see the five unit steps ascending from the bottom up to the first rank level (5^1). From there the 5-levels of the first rank move upward by larger steps to the second rank level, 5^2; in the meantime, however, we should not forget the units still climbing by small steps from one level to the next. The diagram shows them only for the first few ranks. From the second "landing" up, the five larger steps of rank 2 ascend to the level of the third rank, 5^3, but meanwhile on each of these the small unit steps of rank 1 also continue to ascend, and so forth. Thus the number system based on gradations of 5 continues to build upward. The 10-base system ascends by much more powerful strides, as the reader can see for himself from the diagram.

The rank levels are shown from the ground up. In addition, we also see the number 44 represented graphically:

$$44 = 4 \times 10 + 4 = 1 \times 5^2 + 3 \times 5 + 4 \,(= 134 \text{ in the 5-base}$$
system).

Thus we can now understand visually what is meant by "the 4th step on the 3rd step of rank 1 on the 2nd step of rank 2." The diagrams serve to impress upon us the manner in which such a system of gradation is built up conceptually.

If we climb up once more to the threshold of the third rank on the decimal staircase and from there survey the grouping of numbers (see Fig. 8, p. 42), we can look down directly on the "expanding" essential difference between the verbal spoken number sequence and the earlier systems of numerals which keep their groupings "on level ground."

NUMBER SYSTEMS BASED ON GRADATIONS OF 20

In the following pages we shall discuss a few number sequences that have been to a greater or lesser extent based on ranks of 20. We shall encounter two somewhat disrupted but very sharply etched 20-base number systems, which constitute a remarkable convergence of the cultural histories of two pre-Columbian American peoples, the Mayans and the Aztecs of Mexico. From these we shall go on to certain peculiar number sequences of Europe (Celtic, French, and Danish) in which a 20-base structure is still distinctly traceable but could not prevail in decimal terms, although it persisted in a few noteworthy formations as vestiges of the most ancient times. Then we shall examine the number sequence of a simple and primitive people, the Ainu of northern Japan, to see how early man built up a unique and highly original tower of numbers, combining the very convenient 20-base system with a very simple 10-base system.

The Mayans
This name is applied to the Indian tribes of Central America (Yucatan and Guatemala) with their related languages, who at one time possessed a very high level of culture, including their own indigenous systems of writing and of numerals. The ruins of their enormous temples show that the Mayan priests were the real bearers and guardians of this culture. By the time the Spaniards invaded in

1524, the Mayan empire had already gone into decline, after flourishing for two periods, from the 5th to the 7th and from the 10th to the 12th centuries A.D.

The Mayan number sequence was based on the number 20. Its successive ranks

$$R_1 = 20, \quad R_2 = 20^2 = 400, \quad R_3 = 20^3 = 8000, \quad R_4 = 20^4 = 160,000,$$

were called

    hun,   *bak,*   *pik,*   *calab,*

respectively, and thus had their own particular names. Quite a surprising gradation! But instead of 19 different unit designations it had names only for 1 ... 10 and used these to form the remaining necessary units 11 ... 19. Thus the 20-gradation was interrupted by a decimal gradation. Their full number sequence was as follows:

R_0 units 1 ... 19			$R_1 = 20$ to 399		
1	*hun*		20	*hun-kal*	$1'20 = 1'R$
2	*ca*			30 *lahu-cakal*	10 (of) 2'20 see below
3	*ox*		40	*ca-ikal*	2'20
4	*can*			50 *lahu-y-oxkal*	10 (of) 3'20
5	*ho*		60	*oxkal*	3'20
6	*uac*			70 *lahu-cankal*	10 (of) 4'20
7	*uuc*		80	*can-kal*	4'20
8	*uaxac*			90 *lahu-y-hokal*	10 (of) 5'20
9	*bolon*		100	*ho-kal*	5'20
10	*lahun*		120	*uac-kal*	6'20
11	*buluc*		....		
12	*lah-ca*	10'2	200	*lahun-kzl*	10'20
13	*ox-lahun*	3'10	....		
14	*can-lahun*	4'10	300	*ho-lhu-kal*	5'10'20
15	*ho-lahun*	5'10			
....					
19	*bolon-lahun*	9'10			

$R_2 = 400$ and up	
400 *hun-bak*	1'400
500 *ho-tu-bak*	5 (20) + 400 (see below)
600 *lahu-tu-bak*	10 (20) + 400
700 *holhu-tu-bak*	15 (20) + 400
800 *ca-bak*	2'400
1000 *lahu-y-oxbak*	10 (20) (from) 3'400
....	
1200 *ox-bak*	3'400
1600 *can-bak*	4'400
....	

thereafter $R_3 = 20^3 = 8000$ *pic* and $R_4 = 20^4 = 160,000$ *calab*

In this table the number words are followed by their literal translations into numbers; for example, 13 *ox-lahun* 3'10 (read "thir-teen"), 60 *ox-kal* 3'20. The reader will easily find the combinations 3 + 10 and 3 × 10. In the cases of 300 and 700 the word *lhu* is a contraction of *lahun*, 10; with 500, which should properly be *ho-kal-tu-bac* and all higher numbers, the *-kal*, 20, is omitted. For the meaning of the *-tu-* in 500, see below.

This is an unusual number sequence, one which without any doubt did not arise from the needs and experience of the people but was an artificial and conscious creation of the priests, perhaps designed for calendar computations. For what other purpose would such high numbers have been required? Certainly not in ordinary life, nor in trade and commercial transactions. From $20^2 = 400$ on up, the ranks of 20 become so vast and extensive that even we can visualize and represent them only with difficulty.

It is highly instructive to observe the stages in the growth of this number sequence. Instead of nineteen different names for the units, as we have said, we see only ten; the number words for 11 ... 19 are formed like our own. There can be no doubt that this is an ancient feature, deeply rooted in the people.

Now we may be inclined to think that these 19 units are then used to progress upward in steps of 20: 21, 22, ..., 20 + 10, 20 + 11, ... 20 + 19. But this is not what happens. A different and very remarkable mode of formation appears:

20	*hun-kal*	1′20	
21	*hun-tu-kal*		1 (on the) 20 (-level)
22	*ca-tu-kal*		2 (on the) 20 (-level)
	. . . . . .		
29	*bolon-tu-kal*		9 (on the) 20 (-level)
30	*lahu-cakal*	10′(2′20)	10 in the 40-interval!
31	*buluc-tu-kal*		11 (on the) 20 (-level)
	. . . . . .		
35	*holhu-cakal*	15′(2′20)	15 in the 40-interval!
36	*uaclahu-tu-kal*		16 (on the) 20 (-level)
	. . . . . .		
39	*bolon-lahu-tu-kal*		19 (on the) 20 (-level)
40	*cakal*	2′20	

Thus the method of forming number words in the interval from the first to the second twenty is not uniform. The units are added in their cardinal form with *-tu-* to the next lower twenty-step; thus 21 is the first number on the 20-step. But at 30 the next *higher* twenty-level, *cakal*, or 40, is referred to, and the same is true of 35. This is the rare "counting from above," which now no longer exists anywhere except in the Teutonic north of Europe. The Mayan concept of these numbers is thus the following: 1 ... 20 is the first and 21 ... 40 the second level; 30 lies within the second level. The level (or interval) is thus referred to in terms of its final number 40 — whence "thirty is 10 in the 40-interval" (see p. 76).

It is characteristic, moreover, that from 40 on up the Mayan number sequence uses exclusively this mode of counting from above to form its number words, and thus proceeds from 40 up without change or disturbance. As an example: 60 is *oxkal*; thus

41 is not *hun-tu-cakal*   but   *hun-tu-y-oxkal*
1 (on) 2′20   1 (in the) 60 (-interval).

The syllable -*y*- in the above is merely a connective. Accordingly,

185 is *ho-tu-lahun-kal*   and   386 is *uac-tu-hunbak*
5 (in the) 10'20-   6 (in the) 1'400(-interval).

We shall not go astray if we interpret this peculiar number sequence as follows: The units 1 ... 9 each have their own names and then combine with 10 to form the 10-level interval from 11 to 19. The ancient decimal grouping was here in use among the common people and made itself felt: The number sequence proceeds undisturbed from 20 to 40; 30 retained no name of its own. This we can readily understand; we recognize the same thing in the 60 ... 80 interval of the French number sequence, where 78, for example, is *soixante-dix-huit*, 60'10'8. But then comes a sudden break. The sequence begins to count from above, and from this point this principle governs the structure of the whole number sequence without exception; moreover, the principle of counting backward has also a reverse effect on the forms of 30 and 35. These are no longer natural outgrowths of the language but artificial and conscious verbal formations. The same thing is evidenced by the great extent of the number sequence, up to $20^4 = 160,000$. Although calendar computations may well require the use of very large numbers, they do not seem to have been the only reason for the enormously high numbers. The latter also bespeak reverence for the gods and a yearning to climb ever higher and higher, closer and closer to them, step by step. It was both a holy thing and a mystery of the initiated to ascend such a tower of numbers, to count up to such heights that numbers were no longer felt as quantities. The striving to mount ever upward was realized in the law of numbers. We shall encounter similar "holy" towers of numbers in India (see p. 137).

The Aztecs
The ancient Mexicans possessed a number sequence clearly based on ranks of 20 along with — and this was its amazing feature — a sequence of units ordered on the number 5; thus 5, 10, and 15 had their own separate names, like the levels marking the successive ranks of 20. The reader can see this for himself from the following table:

1 *ce*			11 *matlactli-on-ce*	10 + 1	
2 *ome*			12 *matlactli-on-ome*	10 + 2	
3 *yey*			13 *matlactli-on-yey*	+ 3	
4 *naui*			14 *matlactli-on-naui*	+ 4	
5 *macuilli*			15 *caxtulli*		
	6 *chica-ce*	5'1	16 *caxtulli-on-ce*	15 + 1	
	7 *chic-ome*	5'2	17 *caxtulli-on-ome*	+ 2	
	8 *chicu-ey*	5'3	18 *caxtulli-on-ey*	+ 3	
	9 *chic-naui*	5'4	19 *caxtulli-on-naui*	+ 4	
10 *matlactli*			20 *cem-poualli*	1'20	
30 *cen-poualli-om-matlactli*			1'20 + 10		
40 *ome-poualli*			2'20		

· · · · · ·

100 *macuil-poualli*   5'20, thus not a separate word

further $20^2 = 1'400$ *cen-tzuntli* $20^3 = 1'8000$ *cen-xiquipilli*.

Example: 39 = *cen-poualli-on-caxtulli-on-naui* 1'20 + 15 + 4.

The various ranks of 20 have the following names and meanings:

1, *ce*;   20, *poualli*, "counted group";   20² *tzuntli*, "hair,"

the latter a word that once, according to an ancient counting limit, had designated an unspecified multitude or "many," but later became attached to the specific meaning of 400 as a number word; and finally,

20³, *xiquipilli*, a "pouch" containing 8000 cocoa beans.

The word *ma-itl*, "hand," gave rise to the number words for "five" and "ten," with the use of *-on-* or *-om-* "on."

There is no doubt but that this number sequence was built up more naturally and with less mental strain than the Mayan one. It contains a 5-count that appears very ancient and does not count backward from higher levels. A significant document in this connection would seem to be an old Spanish report, according to which the Indians' fighting forces were organized on the principle of this vigesimal grouping: In our terms, their army consisted of 20 battalions, each having 20 companies, each of which had 20 men — hence 20³ men in all. From this document it seems likely that this old and convenient ordering and grouping by 20's — for this military organization just mentioned is nothing but that — gave rise to the organization by ranks of 20, in which the 20-groups became the levels of rank. Thus the Aztec number sequence provides a unique example of this very ancient procedure, already mentioned (see p. 47), by which the groupings were first set up as independent supports, between which the actual number steps themselves were inserted.

We can see from our diagram (Fig. 15) that here the 5-count is not a ranking based on 5 — it goes no further than the second rank level 5² — but rather that a short ladder containing three steps of 5 numbers each has been "superimposed" on the large gradations of 20.

Thus we find that no matter how many number sequences we examine, there are absolutely no pure vigesimal rankings, either in verbal number sequences or in written numerals.

Yet are there any gradations based on 5? If the size of the 20-intervals is too large for convenience in counting, the intervals of five are too small. If the human mind wishes to comprehend a larger number, it stumbles over the short intermediate steps and, once it has achieved a sequence with gradations of 10, simply climbs from one 10-level to the next (see Fig. 15). Just as there is no purely vigesimal gradation, there have also been no examples of number sequences based purely on gradations of 5.

Ever since the earliest attempts at counting, which succeeded in forming only the first four or five number words, systems based on stacks of fives have risen here and there; but it never happened that a people or culture arrived only at the second rank, 5² and then stopped. In most cases they continued by way of 2 × 5 = 10 to the larger steps of 10 (see the number sequence of the ancient Sumerians

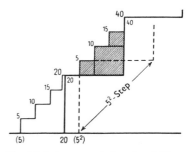

Fig. 15 Aztec numerical ranks of 20, with steps of 5 up to 20; thereafter the sequence climbs simply by large steps of 20. The fact that this is not a gradation of 5's is shown by the cross-hatched steps, which are to be seen as the second level of rank. The vigesimal ranking is clearly dominant.

p. 165). Therefore there is no justification for speaking of a "quinary" system or, in our terms, a gradation of 5; at most we can speak of traces or vestiges thereof. Likewise there are no number sequences based on gradations of 2. Some have attempted to see such a system in the fact, for example, that the Bushmen of South Africa have only the two number words *a*, "one," and *oa*, "two," and that they say "four" as *oa oa* and "five" as *oa oa a*. But this is an instance of mistaking one numerical principle for another: for this is not a gradation but merely an ordering, a much earlier phase, especially if it occurs in speech.

Now let us look at a few peculiarities and traces of 5-gradations in mature number sequences. Irish, a Celtic language, regularly says *deich* for "ten" (< Latin *decem*; see p. 97, first column); yet it forms the interval between 10 and 20 not on the principle of 1 + 10, 2 + 10, etc., but says *oin deec*, *da deec* and so forth, where *deec* (< Celtic *da-coic*, 2 × 5) means not 10 but "two fives."

Moreover Cymric, the Celtic language spoken in Wales, contains an example, occurring in no other European language, of piling up units on 15 *bymthec* 5'10 (see p. 97, column two);

16 *un ar bymthec* (1 + 15)
17 *dau ar bymthec* (2 + 15)
19 *pedwar ar bymthec* (4 + 15)
20 *ugain*,

a method of counting that would be easier to understand if the units 6, 7, and 9 were formed as 5 + 1, 5 + 2 and 5 + 4, as they are in the Aztec number sequence discussed earlier.

VIGESIMAL GRADATIONS IN EUROPEAN NUMBER SEQUENCES

If we now examine number sequences in our own culture for signs of gradation or steps of twenty, we shall find some extremely revealing examples. The Celtic languages have maintained and disseminated the vigesimal gradations with a persistence equaled by no other group of tongues; in Danish and French the vigesimal system has infiltrated in a few places into the basically decimal number sequence and has overcome the latter for considerable intervals. But once we have found that the original Indo-European language possessed not the slightest traces of any form of counting by intervals of 20, we naturally begin to wonder: Where did it come from, and how did it enter these particular languages?

The Celtic Vigesimal Numbers

The individual tongues that belong to the Celtic group are treated together, so far as the Celtic number sequence is concerned; the reader is asked to look these over, especially in the interval from 10 to 100 (p. 97). The adjoining map will make the over-all picture clear (Fig. 16). (No number sequence has come down to us from the now extinct language of ancient Gaul.)

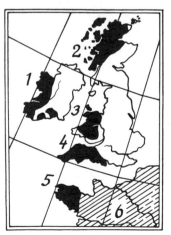

Fig. 16 Where Celtic was and is spoken: 1, Irish; 2, Gaelic in Scotland; 3, Welsh; 4, Cornish (extinct); 5, Breton; 6, Gallic (extinct).

In examining these tongues, we are suprised at the regularity with which the vigesimal principle recurs. In present-day Irish (as well as in Scottish Gaelic, which differs but little from it), the sequence 20, 40, etc., runs on by intervals of 20 up through 180, and the intermediate tens, 30, 50, 70, etc., do the same up to 190:

20	*fiche*		30 *deich ar fiche*	10 on 20
40	*da fiche*	2'20	50 *da fiche's a deich*	2'20 on 10
60	*tri fiche*	3'20	70 *tri fiche's a deich*	3'20 on 10
80	*ceithre fiche*	4'20	90	4'20 on 10
......			......	
160	*ocht fiche*	8'20	......	
180	*naoi fiche*	9'20	190 *naoi fiche's deich*	9'20 on 10

Examples (for the units, see the tabulation on p. 26):

deich mbliadma ocus ceithri fichit,
10 years   and   4'20 = 90 years;
deich mnaa secht fichit,
10 women   7'20   = 150 women.

Surprisingly, Old Irish did possess a distinct gradation based on 10, which was completely displaced by vigesimal gradations in modern Irish! The Old-Irish tens were formed from the words for the units:

10	*deich*	60	*sesca*
20	*fiche*	70	*sechtmoga*
30	*tricha*	80	*ochtmoga*
40	*cethorcha*	90	*nocha*
50	*coica*	100	*cet.*

A remnant of this old 10-count is the number word for 100, *ceud* (*cant, kans,* cf., Latin *kentum*), which shows its influence in the Cornish and Breton 50, *hanter-cans,* "half-100"; another inconsistency is the Breton 30, *tregont* (p. 97, fourth column).

The Danish number sequence

This is a further example from northern Europe (p. 95). Here 10 is *ti,* 20 *tyve,* 30 *tredive* and 40 *fyrretyve.* The word *tyve,* "tens," was originally the plural of "ten" and comes from an old word *tiugh,* corresponding to the Gothic *tigum,* from which the German *-zig* developed: *tre-tyve* > *tredive,* "thirty"; cf. *drei-ssig.* This plural is regularly used to form the first four tens; then, however, *tyve* takes on the meaning of 20:

60 *tre-sinds-tyve,* "3 × 20"   80 *fir-sinds-tyve,* "4 × 20."

Thus the same number sequence contains both these meanings of *tyve*:

tyve 20 — *fyrre-tyve* is not 4'20, but 4'10; *fir sinds-tyve,* again 4 × 20!

The meaning behind this word formation is, for example in the case of 70, "half of the fourth 20-group upon three 20-groups." In common speech the number words are shortened (see p. 95).

   The Danish sequence of tens is thus a rich deposit of primitive forms: formations based on 20, break after 4, 20-group, 20-levels

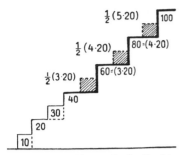

Fig. 17 The half-steps in the Danish number sequence, based on gradations of 20.

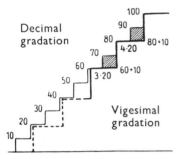

Fig. 18 Invasion of the vigesimal gradation into the French decimal number sequence.

that show up through the dominant 10-levels and, finally, the extraordinary "half-levels" (Fig. 17). These make use of a rare form of connection, that of dividing the last gradation. They correspond exactly to the German or Russian expression, "It is $\frac{1}{2}$ 4 o'clock," meaning not 2 o'clock but half of the time interval between 3:00 and 4:00. Thus in Danish "half eighty" is not "forty" but the half-step between the vigesimal levels of 60 and 80 (Fig. 18). Later we shall have more to say about this form of "overcount" (p. 76).

The Danish number sequence also provides a beautiful example of the dispossession of the vigesimal grouping *fem-sinds-tyve* (5×20), which is still retained in the expression for $90 = \frac{1}{2}(5 \times 20)$, by the decimal gradation *hundrede*.

The French Number Sequence

Here we find evidence of one of the most fascinating cases of the infiltration of the primitive vigesimal gradation (see p. 49). The French number sequence progresses smoothly upward by levels of 10; and not until 70, *soixant-dix* ($60'10$) is there any intrusion of the idea of counting by twenties, which then at 80, *quatre-vingt(s)*, "four twenties" (*not* 4×20) clearly thrusts itself into the decimal gradation. But it is at once forcefully tied down, and as soon as it forms the next half-step, like a low stool set in front of the excessively high 20-level, it immediately becomes fully established and is then literally suppressed.

But it was not always thus. Fortunately, we can reconstruct the history of the French number sequence. Nothing is known of any Gallic numbers, but if they had possessed a strong vigesimal gradation, some traces of it ought to have been preserved in French personal or place names, or in Old French poetry. The Romans, who for centuries imposed their language and cultures on the region, had a decimal system of counting and reckoning which showed not the slightest hint of any grouping or gradation by twenties.

The 20-gradation suddenly appeared in the 11th century in northern France, spread from there to the south and until well into the 17th century formed a number sequence that ran from 60, *trois-vingt* (old form *vint*), through 120, *six-vingt*, and 140 *sept-vingt*, all the way to 360, *dix-huit-vingt* ($18'20$). *Les Onze-vingts*, "the eleven-twenties," was the name for a constabulary force consisting of 220 men. And the old hospital built by Louis IX in the 13th century in Paris is still called the *Quinze-vingts* from its 300 blind inmates, whence the colloquial expressions *un quinze-vingt*, a blind person, and *entrer au quinze-vingts*, "to sleep." In a few isolated local dialects (as in Savoy), the old steps of 20 are still used in common speech up to 160, *huit-vingt* ($8'20$). It is also remarkable that whereas Old Provençal regularly counted upward by tens, the modern dialect has adopted the vigesimal count, from *tres- vint* ($60 = 3'20$) up to 380, *dès-e-nou-vint*, ($10 + 9)'20$.

It is significant that the number sequence breaks off just at $19'20$ just before the second rank level, $20'20 = 20^2$, a vigesimal gradation. This shows that this vigesimal count was thought of only in terms of the popular 20-groups, without leading up to any consciously thought-out gradation of twenties, like the Mayan number sequence (see p. 59). This is confirmed by the fact that the words for 100 and 200 in French are formed with *cent*, for the most part, the expression *cinq-vingt* occurring only in vernacular speech. Above all, it is shown by the fact that from the 17th century on the widespread 20-groupings completely disappeared from all written number words. Of the original, powerful, vigesimal invasion, there now remains only the interval from 60 to 99, with 80, *quatre-vingt*, the only number word incorporating levels of twenty. The expression for 120 = 6′20, as well as the peculiar method of writing numbers by 20s, continued in use into the 18th century. In Molière's *Le Bourgeois Gentilhomme* (Act 3, Scene 4) the creditor pulls out his account book and reads aloud to his debtor:

"... *donné a vous une fois deux cents louis*" — "*Cela est vrai*" — "*Une autre fois six-vingts*" — "*Oui*" — "*Et une autre fois cent quarante.*"
"... lent you on one occasion 200 louis d'or (2 hundred)" — "That's right" — "On another occasion 120 (6 twenties)" — "Yes" — "And another time 140 (hundred 40)."

Almost certainly the second amount was written down in the account book as *six-vingt*, VI^{XX}; and IIII^{XX}, *quatre-vingt*, continued to be so written until some time in the 18th century. Did *six-vingt* survive so long because it was a French echo of the Germanic "great hundred" (120)?

Now, however, only *quatre-vingt* remains, like a large protruding boulder in the midst of a smoothly progressing number sequence; in its immediate vicinity it has displaced the forms *septante* 70, *huitante* or *octante* 80, and *nonante* 90, which still live on here and there in a few spoken dialects. The *Septuagint*, the Greek text of the Old Testament, is called *La Bible des Septante* in French (see p. 186). These all show the force of the ancient, popular 20-group, "man," in counting. Today the ground has been leveled out, and the decimal ladder proceeds unperturbed past this alien intrusion.

Now arises the question that has been on the tip of our tongues for so long: What was the origin of this rare manifestation of the vigesimal count, which so oddly broke into an alien number sequence, dominated it for a while and then disappeared almost completely? Why was the Roman-French decimal gradation suddenly put aside between the 11th and the 17th century?

The Normans

Although much still remains obscure, it seems to have been the Normans who brought the vigesimal habit of counting into the French number sequence. We have already said that the original Indo-European language contains no traces of a vigesimal grouping or gradation. But the Celtic tongues are Indo-European, and we have seen that Celtic incorporated and held on to the vigesimal count with a tenacity shown by no other language of our general culture. Were the pre-Indo-European inhabitants of the subcontinent the original bearers and transmitters of the 20-count to the Celts? We do not know.

But we do know that the vigesimal system of counting was in constant and common use in the North. We recall the groupings such as the *Stiege*, the *Schneise*, and the *score* (see p. 49), all of them words for groups of twenty elements which the Germanic languages invented in addition to the regular number sequence. These special 20-forms, curiously enough, did not follow the usual Indo-European formation with the syllable *vi-* (e.g.: Old High German *zwein-zug* as against Latin *vi-ginti*, p. 96).

From the North, primarily from the great realm of the ancient Danish kings, the Normans spread as raiders, conquerors, and traders to Iceland, to the lands of the Angles and Saxons, to the French coast and thence deep into France itself, along its rivers. In A.D. 911, at the mouth of the Seine, they founded a kingdom of their own, Normandy, where they adopted the French language and French customs. From Normandy they invaded and conquered England (1066); and they sailed along the Atlantic coast and into the Mediterranean, where they established the Norman Kingdom of Sicily that was to play such an important role in western history.

The Normans were not merely conquerors but also traders, whose area of commercial activity stretched from Greenland all the way into the Mediterranean. It is understandable that Norman customs were introduced into areas with whose inhabitants they came into contact. Thus the northern habit of counting by twenties came into France. How could this custom, which had not been used for many centuries, if at all, suddenly come into common currency, except through some stimulus from outside? This is suggested not only by the period in which it was introduced, the 11th century, but also by the fact that it was thereafter more commonly used in the north of France than in the south. But how was it able to gain a foothold at all? Because the 20-group is so deeply rooted in the numerical concepts of primitive man that once it was given its initial impetus, it took root. Moreover it did not have to be derived linguistically from the Norman word *skor*, "score"; the indigenous number word *vingt* could immediately serve as the designation for a 20-group.

The Portuguese System

The Portuguese vigesimal count, which we mentioned earlier (p. 51), is more difficult to recognize as being Norman in origin; on the other hand, there is no doubt of the Norman influence in Sicily, where eggs, fruits, and people's ages are still counted by twenties in ordinary speech:

40 *du vintini* (2 twenties) — 50 *du vintini e deci*, etc., up to 100 *cincu vintini* (5 twenties).

Moreover, the decimal number sequence was always available. For the illiterate peasant, who has occasion to count more often than to calculate and whose computations for the most part take the form of counting, the 20-count, as we know, is quite convenient.

The Albanians

Against this "native" background the Albanians, whose language is also considered to belong to the Indo-European family, have taken the only indigenous twenty-step. An Albanian counts as follows:

10 *dhiete*
20 *nji-zet*
30 *tri-dhiete*, 3'10
40 *dy-zet*, 2'20, but from then on again by 10-steps:
50 *pese-dhiete*, 5'10
60 *gjashte-dhiete*, and so on

Albanian *-zet* < Sanskrit 20 (*vim*)-*šatih* (see p. 93).

Another endemic feature is the vigesimal counting up to 80 = 4'20 of the non-Indo-European Basque language, which we shall say more about presently (see p. 110).

THE AINU NUMBER SEQUENCE

We include the Ainu number sequence in our discussion not so much because it is a rare instance of an almost pure vigesimal count but because it has been used to build up a unique system of number words.

Today the Ainu are a dispossessed and dying people of the Far East (Sakhalin Island), who seem to belong to the white rather than the yellow race. In creating their number sequence they piled up the ancient 20-group "men" one atop another, higher and higher, not with any idea of arriving at a series of gradations of 20 with levels of 20^2, 20^3 and so on, like the Mayans and the Aztecs, but with the thought of having the numbers proceed by levels of ten each. Thus the Ainu number sequence has no special designation for the ranks 100, 1000, or 400.

1 *shi-ne*	6 *i-wan*	4'10 ('4 from 10')
2 *tu*	7 *ar-wan*	3'10 ('3 from 10')
3 *re*	8 *tu-pesan*	"2 steps down"
4 *ine*	9 *shine-pesan*	"1 step down"
5 *aschik-ne*	10 *wan*	

20 *hot-ne*		30 *wan e tu*	*hotne*	10 von 2'20	(overcount)
40 *tu hotne*	2'20	50 *wan e re*	*hotne*	3'20	(overcount)
60 *re hotne*	3'20	70 *wan e ine*	*hotne*	4'20	(overcount)
80 *ine hotne*	4'20	90 *wan e ashikne*	*hotne*	5'20	(overcount)
100 *ashikne hotne*		110 *wan e iwan*	*hotne*	6'20	(overcount)

200 *shine wan hotne* 1'10'20
300 *ashikne hot-ikashama* ("*on*')- *shine wan hotne* 5'20 *on* (1'10'20)
400 *tu-shine wan hotne* 2'(1'10'20)
500 *ashikne hot-ikashama-tu-shine wan hotne* 5'20 *on* 2'(1'10'20)
. . . . . .
1000 *ashikne-shine wan hotne* 5'(1'10'20').

A truly Cyclopean tower of numbers! The 20-group "man" (*hot*, "the whole") is the building stone that forms all the levels of rank. And the tens do not even disrupt the progression at 100 or 1000! A surprising feature is the overcount from the next step above at the odd-numbered steps, such as saying 70 as "10 from 80" in the form of "10 in the fourth twenty" (see p. 76).

Another feature peculiar to this sequence is the series of units: *shi*, "beginning"; *-ne*, "to be," helps to forms the words for 1, 5 and

20; 4 *ine*, "much," 5 *ashikne*, "hand," 10 *wan*, "two-sided" standing for "both hands." And the words for 6, 7, 8, and 9 are governed by 10. The One and the 20-group "man" are the only building blocks, and this language puts them together according to a very simple rule. To the Ainu anything more than 1000 is "many"; from here on the number words become too large and cumbersome for him.

When we compare this number sequence with our own, we can readily see the flexibility and simplicity of the latter, and the way it proceeds quickly and easily up slender number steps to the very highest numbers. Compared to ours, the Ainu number sequence stands as a rigid unbending tower, heavy and massive, but impressive because of its unvarying use of the ancient number-group as its building stones.

This insight prepares us for our next problem: How have peoples of various cultures formed the individual words of their number sequences?

The number 18, for example, occupies the same position in all number sequences, but we shall see that it is not, contrary to what the uninformed may think, generally expressed in the same manner in all languages.

First let us sum up our last discussion:

Early formations of number words lead occasionally to breaks after the number 5.

In some languages gradations based on levels of 20 still persist from the earliest grouping of numbers.

But within the framework of European culture, and as long as counting was tied to the practical life of ordinary people (unlike the Mayan), systems based on steps of 5 or 20 never took hold. The emphasis is upon gradation; that is, a clear structure erected with certain levels of rank on which the steps of the lower ranks mount upward according to some definite rule.

The gradation based on the number 10 was the only one to achieve a pure and full development. Ten lies within the natural area between intervals that are too narrow (the 5-gradation) and too wide (20-gradation); the former hinders the process of counting, the latter makes it conceptually and linguistically too hard to grasp (the Ainu and Mayan sequences). Yet one finds dramatic documentations of very early intellectual history in the way the ancient, primitive number concept of the 20-group "man," has succeeded, to a greater or a lesser extent, in breaking into the otherwise dominant decimal number sequence.

Laws Governing the Formation of Number Words

SUCCESSION OF MAGNITUDES

Thus far we have been speaking about the way number sequences were built up; now let us take a look at the formation of the number words themselves. Here, too, we shall encounter forms that seem very strange to us, but that we can understand in the light of early man's conception of numbers.

Eighteen

In the "conceptual" number sequence and hence in the succession of abstract numbers . . . 17–18–19 . . ., the number 18 stands in the same position in the minds of all peoples. It would be a mistake, however, to suppose that the number word for 18 is therefore formed in the same manner in all languages. If we glance at various verbal number sequences, we find the following:

German	*acht-zehn*	8'10
French (Classical)	*dix-huit*	10'8
Greek (Modern)	*oktō-kai-deka*	8 and 10
Greek	*deka-okto*	8'10
Latin	*decem et octo*	10 and 8
Latin	*duo-de-viginti*	2 from 20
Lithuanian	*aštuno-lika*	8 over (10)
Irish	*ocht-deec*	8'(2'5)
Breton	*tri-ouch*	3'6
Welsh	*deu-naw*	2'9
Mexican	*caxtulli-om-ey*	15 and 3
Finnish	*kah-deksan-toista*	2 (from) 10 (in the) second (tens)
Ainu	*tu-pesan-ishama-wan*	2 climb down added to 10

Truly a bewildering assortment! We may be surprised at the variety of ways in which languages can form this one number word. Yet all these formations (except the Breton and Welsh) proceed in a natural succession from 5, 10, or 20: from one of these numbers they move either forward or back (18 = 20 − 2). The units and the rank levels are thus the building blocks; the four simple arithmetic operations of addition, subtraction, multiplication, and division are the mortar that holds them together.

The ways in which 72 is formed from levels of 20 is also fascinating:

Danish	"half (4 × 20)'2"	see p. 65
Modern Irish	"(3'20 + 10)'2"	see p. 65
French	"60'12"	
Ainu	"12 from 4'20"	see p. 69
Mayan	"12th on the 4'20(-level)"	see p. 61

We shall be able to understand these and other formations, however strange they may seem to us, if we remember that early man, who devised the first number words, was more intimately bound to his environment than modern man. Even today the peasant is more closely and tightly dependent on wind and weather, man and beast, tree and soil, than is the city dweller, to whom the world is wider, to be sure, but also more remote. And numbers are a part of the world. Early man wants to *see* numbers; they must remain visible to him, and he must be able to touch them if he is to grasp them with his mind.

For this reason he breaks down larger numbers into smaller ones, if he can. The two Celtic wordforms for the number 18, "3 × 6" and "2 × 9," are to be understood thus, as is the answer given by an aged Sicilian peasant woman when asked how old she was:

tre vvote vinti cincu anni, "3 times 20'5 years" (= 75).

In Otfried's Old High German Book of the Gospels there are the following decompositions of fairly large numbers:

zwiro sehs (jaro), $12 = 2 \times 6$
einlif stunton sibini, $77 = 11 \times 7$
thria stunton fintzug ouh thri, $153 = 3 \times 50 + 3$
thrizzug stunton zehinu, $300 = 30 \times 10.$

Units and Groupings

In addition to these, early man had yet another way of breaking down numbers. How was he to conceive a large number like 47? It is above all the units 1, 2, 3, . . . , 9 — in this case 7, which are understandable; whatever is over and above the units (the 40) becomes clear as clusters or groupings. *Quat borla,* "four loads," said the Piedmontese farmer when asked his age (see p. 49); and the Old Spaniards used to say, *tres vent medidas de jarina,* "3'20 measures of grain." The visualization of numbers leads to their grouping, as we have seen especially with the 20 group "man" (p. 49).

There is a striking documentation of this: Old Norse, like Celtic, puts the counted object right after the units, and behind these places the "block" (= grouping) of the higher numbers. Thus in Old Norse, 364 days are called *fiora dagar ens fiorþa hundraþs,* "4 days in the 4th (great) hundred (120)"; in Celtic, 11 horses are called *un march ar dec,* "1 horse and 10."

The reader may also refer to the examples on pages 27 and 65.

Why did these curious turns of speech arise? Because they are easier to visualize. 47 sheep = 7 sheep and 40; 40 is a round number, a grouping whose magnitude is already familiar in the form 2'20; but here the 7 is that which must be emphasized. The number thus falls naturally into two easily visualized parts: units and groups. In contrast, the undivided number 47 is much more alien. It is merely one of so many numbers, . . . 46 — 47 — 48 . . . , which are met with far less often than the unit 7 and the group 40.

This interpretation also explains the reversed positions of the units and the ten-group in the German number words. If we have seen in this nothing more than a kind of inexplicable verbal caprice, now we understand the reason for it; it is just the ancient precedence of the units. As long as the units are small, this seems to make sense. With big numbers like 4385, for example, with which early man had no occasion to cope, the sudden jolt at the end, ". . . five and eighty," seems quite obvious to us. It is doubtless confusing, but now that we know its history we can tolerate it. This is not to say, however, that we ought not to eliminate it in this day and age.

In Swedish, for example, today one says 364, *tre-hundra sextiofyra*, in the proper order of magnitudes, having given up the Old-Norse metathesis of the units and tens. This did not happen for reasons of logic or convenience, but only because the native number words were later modeled after those of the Roman church, which were introduced into Scandinavia at the same time as Christianity (see p. 79).

The law of the succession of magnitudes explains the descending order of the levels of rank in 4385: symbolically 4 Th   3 H   8 T   5 U. Numbers are built up from the rank levels (R) of thousands Th, hundreds H, tens T and units U, together with the "pre-units" a that count them (in this case 4, 3, 8, and 5), as we may call them for the moment in order to differentiate between them and the rank of units U. Thus we shall call aR, here 4Th, a rank-group. The individual rank-groups are generally connected either without words or by the word "and":

symbolically   4—Th   3—H   8—T   5,
English   four-thousand three-hundred eighty-five,

and in Modern German with the order of the last two reversed, "... *fünf-und-achtzig*, ... five-and-eighty." Old High German, by contrast, often retained the logical and consistent order:

drizug inti ahto jar,   38 years,
zehnzug inti finfzug inti thriu,   100 and 50 and 3.

Many languages, such as Russian, Arabic, etc., reverse the order of the units and the tens, like modern German; others observe the order of magnitudes without exception, like Chinese.

The reverse succession of magnitudes or ranks, from lowest to highest, can also be documented in some languages, as in the Classical Greek expression for 56,404 talents:

téssera tetrakósia hexakis-chília pentakis-mýria tálenta,
   4   400   6 times 1000   5 times 10,000.

Number Symbols and the Succession of Magnitudes
Let us now turn our attention to the discrepancy between written numerals and the succession of magnitudes, which occurs quite commonly. This happens especially in German: we pronounce 4385 as "... 5 und 8-zig," so that a child first learning to write is likely to put down 4358. The spoken number words are older than the numerals, which have come into their domain from outside, as a stranger. In Old Norse, when the Roman numerals were introduced along with Christianity, the spoken phrase prevailed even in writing:

III *vetre oc XX*, "3 winters and 20" (= 23 winters),

a form of notation that follows the spoken word, exactly like the French IIII^{xx} for *quatre-vingt* (4'20).

I shall cite a little-known example from Arabic: Here, too, the units precede the tens in speech, while the normal succession is otherwise observed:

641 = 600′1′40 *sittu mi'atin wa-ahadun wa-arba'una*
6   100   and 1   and 40.

(Besides the usual metathesis, Arabic also uses the ascending "reverse succession," 1′40′600.)

Persian, in addition to the common Indian-East Arabic numerals, also has the so-called *Siyaq* script, which runs from right to left and which is frequently used by tradesmen and merchants in bazaars, see Figs. 109 and 110. Here one writes down the tens and units in reverse order, in our notation as 600′1′40, and thus exactly reflects the way the number is spoken (Fig. 19).

How did this unique situation in the history of writing come about? A system of writing numbers that consists of pure numerical symbols does not reverse the order of the rank-groups. The answer is surprising: The symbols in the Siyaq script are not pure numerals at all but abbreviated number words, and as such they reflect the way the numbers are spoken. Had this not been suspected for other reasons, it might have been deduced from the reverse order of the units and tens alone.

There is another curious complication in the history of these numerals: The Turks first adopted the Arabic Siyaq script in their financial administration, and along with it the reversed order of the units and tens, which the Turkish language itself does not possess. Thus the Turks would read the number 641 as *alti-yüz-kirk-bir*, 6′100′40′1, but they would write it as 600′1′40. This, of course, is the exact opposite of what happens in German: The numerals observe the proper order of the ranks and the spoken words do not, while with the Turks the language follows the correct order but the Siyaq numerals do not (only the Siyaq, not the East Arabic numerals, which is commonly used in writing).

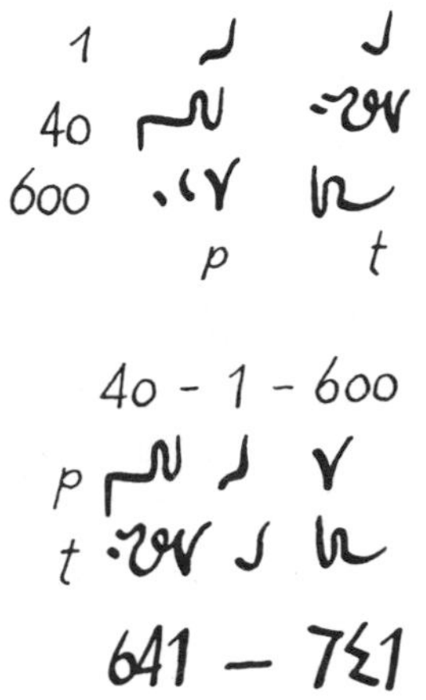

Fig. 19 The number 641 as written in Persian (p) and Turkish (t) Siyaq script. At the top are the symbols for 1, 40, and 600; below the number is written from right to left with the units and tens reversed; and at the bottom in Indian–West Arabic and Indian–East Arabic numerals.

BACK-COUNTING

As a rule, number words are formed by addition and multiplication. But the forms ("difference numbers") which arise by subtraction, or more precisely backward counting, are more interesting. Our examples show how widespread these are:

Latin 19 *un-de-viginti*, "1 from 20"
Latin 58 *duo-de-sexaginta*, "2 from 60";
Ancient Greek 58 *dyoîn déontes hexḗkonta*, "2 lacking from 60."

The Biblical passage, "Of the Jews five times received I forty stripes save one" (II Cor:11, 24) is, in the original Greek text,

(*pentákis*) *tesserákonta parà mían*,
5 times   40   less 1,

and in Bishop Wulfila's Gothic Bible

(*fimf sinþam*) *fidwor tiguns ainamma wanans*,
5 times   4   tens lacking 1.

The passage (John 5:5): "And a certain man was there, which had an infirmity thirty and eight years" reads, in Old High German:

wankta zuein, thero jaro fiorzug ni was,
lacking   2,   which years   40   not was.

Moreover Anglo-Saxon 19 is *anes wona twentig,* "1 less than 20"; similarly, Old-Norse "29 nights" is

nott midr enn þritögr,
1   night less than   30,

and in a very odd fashion of counting, "54 men" is

man midr en halfr setti tögr,
1 man less than half   six   tens.

Among the more remote languages, I shall cite only the examples of

Sanskrit 19 (*ek-*) *una-vimsati,* "(1) lacking (from) 20,"

and so on through all the tens, so that

Sanskrit 39 is *una-čatvarimsat,* "lacking 40."

Just as here one omits the number (1) that is subtracted, so in some Roman words for fractions one omits the number from which the subtraction is made:

$\frac{11}{12}$ *de-unx* "(as = $\frac{12}{12}$) from this 1 ounce,"
$\frac{3}{4}$ *dodrans* < *de-quadrans* "(*as*) from it $\frac{1}{4}$."

Similar formations are cited on p. 83.

Most important, however, are the examples of counting back from ten. These are to be found among many primitive peoples, sometimes even as far down as 6, as in the Ainu *i-wan* (4'10) (see p. 69), but also occur in the Finnish words for 8 *kah-deksan* (2'10) and for 9 *yh-deksan* (1'10). Perhaps the original Indo-European language may also have formed 9, 8, and 7 by counting backward from 10 (see p. 170). If we recall the Dene Dinje expression for 4, "only one is left," we have an example of back-counting with reference to the number 5 (see p. 35).

Another peculiar word formation is the Russian *děvja-no-sto,* "9 before 100" for 90, which derives from the early number concept, to distinguish it from "9 to (or before) 10". This is a beautiful example of how the ten series is basically seen as a sequence of units, but in this case "*before* 100."

How are we to interpret all these word formations? The next higher rank exerts its influence backward, dominating the numbers just preceding it, just as the full hour does the few minutes before it: "10 minutes to 6:00" rather than "5 hours 50." The rank levels, the old number-groupings, are numbers that early man could visualize. We have seen this already: 38 is just one of many numbers that are hard to grasp, but 40 is clear and palpable and so is 2; hence "2 from 40" can be understood, whereas 38 cannot. This conception of numbers as bundles is distinctly discernible in the following examples (see p. 51):

Portuguese 115 *seis (vintens) meno cinc*, "6(20) − 5,"
Sicilian   85 *cente me quinci*, "100 − 15."

On the other hand, the back-counting that occurs in the writing of numbers by some peoples using special symbols (like the Babylonians) or changes in the normal succession of magnitudes (like the Romans), is no more than a kind of shorthand:

Babylonian 19 ⟨ 𒌋 10′9 as compared to ⟨⟨𒁹𒕌 20 − 1; "less" is 𒁹𒕌 (p. 164)

Roman IIII (4) as compared to IV (1′5), or XVIIII (10′5′4) as compared to XIX (10′1′10), or 400 CCCC as compared to CD (100′500), or 89½ SXC (½′10)′100, where S stands for *semis*, "half." These abbreviations are late Roman; in medieval Latin the full form of the numeral was written out as often as the abbreviation.

A "classic" passage in which the Roman IV occurs in Goethe's *Roman Elegies* (15):

"Endlich zog sie behende das Zeichen der römischen Fünfe / und ein Strichlein davor . . . (und verwischte alles wieder) — / aber die römische Vier blieb mir ins Auge geprägt."

"In the end she deftly wrote down the sign of the Roman five / And a small line before it . . . (but erased ev'rything quickly) — / yet the Roman sign 'four' remains engraved in my mind."

OVERCOUNTING

Now we come to a very remarkable manner of counting, which once prevailed in two areas of the world, the Germanic north of Europe and ancient Mexico; today only traces of it remain here and there (as in the Ainu number sequence; see p. 69). This method of counting expresses 24, for example, not in the usual manner as either "4 and 20" or "4 and 2 tens," but as "4 from the 3rd ten"; for previously cited examples in Old Norse, see pp. 27 and 72, and Mayan, see p. 61.

The Meaning of Overcounting
The common method of counting places the units (in this case 4) upon the next lower rank level (in this case on the 2nd ten): this may be called counting from the lower level, or undercounting. But the other method places the units in the interval of 20 . . . 30, between the two rank levels, and thus within the third interval of tens: this is counting from the upper rank level, or overcounting, because the rank-level number of the interval (the third) is determined by the next higher level, which in our case is 30.

Thus in the Old-Norse script we read:

102 men   II *menn hins ellifta tigar*
            "2 men in the 11th ten,"

in the year 969 *a niunda are hins sjaunda tigar ens tiunda hundrads*
            "In the 9th year of the 7th ten of the 10th hundred
            (*fra holdagan vars herra*)
            (after the birth of the Lord)"

If we think of the numbers shown along the line of Fig. 20 as proceeding in order from 0 running left to right, we can readily see how the method of undercounting really does build up the number from below in descending order: 9 H   6 T   9 U, whereas overcounting begins with the units 9 and then suspends these in ascending order from the next level above: 9 U in the 7th T of the 10th H. In so doing it reverses the succession of ranks. Undercounting sees the number 969 by itself, resting upon the number line; overcounting, on the other hand, regards 969 as a point within an expanse of numbers arranged one inside the other, the tens within the hundred interval. It indicates these intervals verbally by the order in which the number word is arranged.

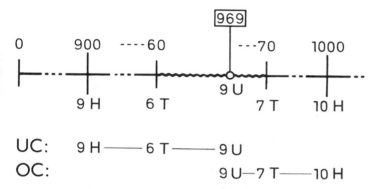

Fig. 20 The number 969 as expressed by the methods of undercounting (UC) and overcounting (OC).

Why this peculiar way of counting? For no other reason than to make numbers (such as 969) as easily visualized as possible. It begins with the unit 9, which lies in the 7th ten-interval, and the tens in turn in the 10th hundred-interval. Instead of point-like rungs ... 967 — 968 — 969 — 970 ... on a ladder of numbers as in undercounting, the "overcounter" sees stretches of numbers (the intervals) within which the number lies. These are, in our sense, nothing but groupings: 9 units in the 7th ten-group of the 10th hundred-group. The method of undercounting 969 begins with the hundreds, a large number that the mind can scarcely grasp directly, whereas the overcounting takes hold of the number by its "nearest" end, the units, and from there climbs up to the higher 10-and 100-groups.

On the whole, therefore, overcounting has the same purpose — to get a clear view of the number — as the already mentioned fashion of placing the units in front of the object counted (see p. 72). Our example, "two men in the eleventh ten" for "102 men" unites the two, the placing of the units in front and the overcount.

The Half-count

This can be considered a form of overcounting. In German it lives on in the expression *anderthalb*, "one and a half" (literally, "half of the other") — that is, the other (second) half — and more rarely in the expressions for 2½, *drittehalb*, and 3½, *viertehalb*. These word forms for fractions using overcounting occur not only in Norse but in the Germanic languages generally; thus 2½ appears in Old Norse as *halfr þridi*, in Anglo-Saxon as *þriddehalf* and in Dutch as *derdehalf*.

In addition, verbal fractions formed in this manner also appear in other languages that otherwise form their number-words by undercounting:

Latin 2½ = *sestertius* < *semis-tertius*, "half of the third,"

from which was derived the name for the Roman coin known as the "sesterce," which originally had the value of 2½ *as* (see p. 29). Similarly *epi-deúteros* "on the second," was the Classical Greek expression for two and a half. But the Slavic language as well as Finnish and Hungarian also make use of the word formation (perhaps because they are the eastern neighbors of German) for example:

Russian 1½ *pol-tora* < *pol-vtora*, "half of the second,"
Finnish *puoli-toista*, *puoli-kolmatta*, "half of the second," resp. "third,"
Hungarian *masad-fel* (< *amasodik*, "the second" and *fel*, "half"),
Hungarian *harmad-fel* and *negyed-fel*, "half of the third," resp. "half of the fourth."

This manner of forming fractions and its general application to other fractions suggests to me that they again arise from the desire to make a difficult numerical expression (as the fraction is) easy to visualize: "one, and added to that half of the second (group)."

The "half" in this overcounting is, however, not the same as the half that is left over as a remainder in reckoning, as the latter is expressed in the Celtic *hanter-kant*; this is really "half a hundred" and thus 50; likewise, Sanskrit 50 *śat-ardha*, "100-half" = 50, or Sanskrit 25 *ardha-pañčasat*, "half 50."

On the other hand, Old-Norse *halfuhr tiunde tughr*, "half of the 10th ten," is not 50, but 95! One must pay close attention to the arrangement of the number. Similarly,

Danish *half-tred-sinds-tyve*, "half 3 × 20" or "half 60"

is not 30 but 50, in the sense of the overcount "half of the third twenty" (see p. 65). The same thing is familiar in the German way of stating the time of day: "*halb* 4 *Uhr*" is not "two o'clock" but 3:30. In contrast, all the Romance languages (and English) "undercount" the hours, as in the French *trois heures et demie*, English "half past three."

A very large number, such as 55,950,000, for example, which would be impossibly cumbersome if expressed by overcounting, is divided up into

½ of the 6th T × (ThTh) and ½ of the 10th HTh:
  halftum setta tigi sinna þusund þusunda ok   halft tiunda
half of the 6th 10 times   1000 ′ 1000   and half of the 10th
hundrat þusunda,
   100 ′ 1000.

The Nordic Names for Fractions

These names, of which the expressions for half are a particular case, are formed by the same procedure:

Icelandic (ca. 1300) $5\frac{2}{3}$ hundred *tueyr hlutir   ins VI. hundrad,*
                       2   parts of the 6th hundred;
Icelandic $26\frac{2}{3}$ ells *tveir hlutir siaunda elnar oc XX alnar,*
            2   parts of the 7th ell and 20 ells.

(The different spellings of the same word, like *tueyr* and *tveir* in the above examples comes about because they were taken from different manuscripts.)

In Old Norse it was the custom, in the case of fractions whose denominator is 1 more than the numerator, to omit the rest of the fraction: Thus "2 parts" is understood to mean $\frac{2}{3}$, "2 parts (of three)." Quite surprisingly a number of other languages do the same thing; this of course applies only to fractions commonly encountered in daily life, not to artificial fractions that occur in computations, such as 28/29. Thus the only ancient Egyptian non-unit fraction $\frac{2}{3}$ was called "both parts"; the same fraction in Akkadian-Babylonian was termed "both hands." The curious Roman name *bes* for $\frac{2}{3}$ *aes* is also explained in this manner:

Latin *bes* < *duessis* < *duo* (*partes*) *assis*, "2 (parts) of the *aes*."

Modern Greek says for $\frac{2}{3}$, $\frac{3}{4}$, and $\frac{4}{5}$

tà dýo (*tría, téssera*) *mére*, "the two (three, four) parts."

Old Norse then has an odd manner of expressing a fraction whose denominator is larger than the numerator by more than 1 part: "$\frac{5}{8}$ (of the fish caught in the River Grima belong to the Church)" is

fim hluter   en þri huerta   undan,
5   parts and   3   go   underneath.

In this connection let us quote an interesting explanation of the value of the coefficient $\pi \approx 3\frac{1}{7}$ from an Icelandic manuscript dated about 1300:

ummaeling hrings hvers þrimr lutum lengri en breidd hans
the circumference of the circle is some 3 times longer than its breadth
ok sjaundungr of enni fiordo breidd,
and a seventh of the fourth breadth (= diameter).

The History of the Norse Custom of Overcounting

The development provides fascinating evidence of the way in which number-words changed and migrated. Overcounting is not common to all the Germanic languages, however. For example, Bishop Wulfila's Gothic Bible translates the "99 just persons" (Luke 15:7) by undercounting, as *niunte-hundis jah niune garaihtaize*, "90 and 9."

This Nordic manner of dealing with numbers gradually disappeared under the influence of Christianity. Along with the Latinate ceremony and worship of the Church, the Roman method of undercounting infiltrated into the native manner of expressing numbers.

Nordic and Roman ways of writing numbers can be reliably distinguished in old manuscripts. In place of the 364 days in the Norse overcounting,

fiora dagar ens fiorþa   hundraþs,
  4   days in the 4th (great) hundred,

we find CCC *fagar oc* IIII, where only the non-Roman great hundred C (= 120) betrays the scribe's Nordic background. He has already abandoned the Norse reverse word order, "4 days and 360." The Norse habit of overcounting gradually died out between the 12th and the 14th centuries.

Thus it is remarkable that the Finns, the Esthonians, and the Lapps, neighbors of the Germanic Swedes, follow the custom of overcounting to this day. They borrowed this from their neighbors, as they did many other Germanic habits (see p. 112). In Finnish it especially dominates the second interval of tens, from 11 to 19; *yksi-toista,* "1 (in the) 2nd (tenned)" is 11, and so forth (see p. 113). Moreover Finnish has, besides the *under*counted 21, *kaksi-kymmentä-yksi,* 2'10'1, also the *over*counted 21, *yksi-kolmatta,* "1 (in the) 3rd (ten)." The overcounting formation is used especially in stating the days of the month between 20 and 30; for instance, the 27th is not the 2'10'7, but "the 7th in the 3rd."

Because of the strong influence of the Swedes on the Finns until around A.D. 700, the origin of Finnish overcounting is suspected to be due to their Swedish neighbors, among whom, however, it can no longer be found. It was the Swedes who settled Norway, where they also brought their overcounted number words, which migrated thence to Iceland (see p. 109). Here it resisted with difficulty the alien Christian fashion of undercounting and did not disappear from Iceland until around 1500.

Overcounting is now a rare phenomenon in the world. It was clearly developed by the Mayans as well (p. 61), but the Mayan overcount, of course, had no connection whatever with that of the Norse. Moreover, despite their similarity, there are very strong differences between the two (one of them being an artificial vigesimal gradation not tied to the objects counted), so that the Mayan overcounting cannot justifiably be interpreted as being founded on an attempt to visualize numbers more readily. On the other hand, this was surely the basis for the overcounting among the Ainus (p. 69).

ADDITIONAL WAYS OF FORMING NUMBER WORDS

Now that we have seen the great variety of ways in which number words are formed by overcounting and by using the arithmetic operations of adding, subtracting, multiplying, and dividing, let us look at three more instances that may show further paths that languages can follow, for example in forming the tens from 20 through 90.

Semitic

The Semitic tens, for example, are the grammatical plurals of the units. In the Arabic and Hebrew number sequences (see the complete number sequences, p. 115) we find:

Arabic 4, *arbaʿun* — 40, *arbaʿun-a*, Hebrew 4, *ʾarbaʿ* — 40, *arbaʿim*,
Arabic 6, *sittun* — 60, *sittun-a*, Hebrew 6, *šeš* — 60, *šišš-im*.

The Arabic *-a** and Hebrew *-im* are the respective plural endings of these languages, as in Hebrew *sepher*, "book," *sepharim*, "books." In Semitic, 40 is thus literally "the forties." The word for 20 is the only exception; it is formed not from 2 but from 10 (Arabic *ʿasrun*, Hebrew *ʿeśer*); 20 *išrun-a* and *ʿeśr-im*, respectively, are literally "the tens." We have already indicated (see p. 14) that this is a reflection of the ancient primitive counting limit after ten. Originally the words for 20 had not a plural but a grammatical dual ending, which however was later driven out by the analogy with the other tens; this kind of linguistic "remote control" is quite common in number words arranged in order.

Although the Semitic number sequence is already remarkable for this manner of forming the tens, it is even more so in the curious gender and case inflections that characterize the number words. Each number word has both a masculine and a feminine form, e.g.:

Arabic 7 (m) *sabʿun* — (f) *sabʿatun*; Hebrew 7 (m) *šebaʿ* — (f) *šibʿah*.

But if the things counted are masculine, such as "7 men," the feminine form of the number word is used, and vice versa; thus we have 7 (f) men and 7 (m) women! Why this happens, no one knows. Have some conceptions of mythology influenced these number words? Only the Semitic words "one" and "two" are treated as adjectives agreeing in gender with the nouns they modify (as are all numbers, such as 102, that end in 1 or 2).

The changes in form of the number words in Arabic are also extremely peculiar: "one" and "two" are adjectives agreeing in gender with their nouns: "two-ish (m) men";

3, 4, ..., 10 take the genitive plural and are thus nouns: "5 (f) of the men";

11, ..., 99 take the accusative singular: "30 (f) the man";

beyond 100 the number words are again nouns and thus regularly govern the genitive case, but now in the singular: "250 of the man";

1001 nights is *alf laila wa laila*, "a thousand night and night." Certainly an odd number sequence! How could anyone wishing to create a new number sequence artificially arrive at such a one! But the very fact that it does not proceed consistently and without disruptions shows how closely it was bound to the thought processes of its people.

Sanskrit

The Sanskrit tens and those of the related Avestan language are, from 60 up, units raised to the status of nouns (see the full sequences on p. 93):

šaš, "six"; *šaš-ti-h*, "the six-ed, six-ness, six-ity."

The number word becomes a noun by acquiring the ending *-t* (see

---

* *Editor's note:* The Arabic plural is *ūna* in the nominative case. The essential feature is the lengthening of the vowel.

p. 126). Linguistically this is the expression of a grouping, a "six-ness of tens." Sanskrit even says *daśa-ti-h*, "a ten-ness (of tens)" and hence "a hundred."

Strange things occur in the life of a language: The same *t* formation that exists in Sanskrit also occurs in Slavic from the number 5 on, as for instance the Czech *pĕ-t, šes-t, devĕ-t, dese-t*, but here it is not the tens but the units 5, 6, 9, and 10 that are treated as nouns: "five-ness, six-ness, nine-ness, ten-ness." The first four Slavic number words, on the other hand, are ancient adjectives, whereas the higher numbers are felt to be groupings; we have already spoken of this earlier (see p. 23).

In Gothic the last three tens have the form of 70, *sibunt-e-hund*, "the seven-ned ten" in the sense of "10 seven-nesses" (see p. 92).

Verbal Numeration

This is the name we shall use for the peculiar custom of an African language spoken in the Sudan, in which the words for the tens are not related in any way to those for the units and separate names are given both to the units and to the tens:

1 *telu*			6 *yeres-ko*	60 *kiyi*
2 *in-di*	20 *colo-ro*		7 *telo-ro*	70 *timba-ga*
3 *yas-ko*	30 *dere-go*		8 *ton-go*	80 *kandeln*
4 *de-go*	40 *gama-ro*		9 *inde-go*	90 *tschimbar*
5 *o-go*	50 *inde-ro*		10 *me-go*	100 *andai.*

The Turkish words for 30, *otuz*, 40 *kirk*, and 50, *elli*, likewise have no connection whatever with the names of the units (see p. 113).

Among the various known systems of numerals, such numeration was also used by the Egyptians, whose so-called hieratic script had separate symbols for each of the tens, and thus wrote the numbers up to 100 with 19 different symbols, not with 10 as we do.

Ordering is characteristically lacking in systems of number words and numerals of this kind. In their structure they have reverted to the level of counting by the parts of the body: they do not simply assign numbers to the parts of the body, to be sure, but their words and numerals nevertheless lack any internal organization (see p. 34 above). This should again caution us that our clear insight into the number sequence as we know it today — after a long succession of detours — cannot be assumed to have existed at earlier stages of cultural history.

The freezing and borrowing of number words could likewise be classified under the heading of "additional ways of forming number words." Freezing: A word with the indeterminate numerical meaning of "much" or "many" (like the Turkish *on*) is *frozen* into a definite number word with a fixed meaning ("ten"). Borrowing: if the number sequence of a people stops at 100, for example, but needs a word for 1000 and then *borrows* this from a neighboring language, as the Finns took over the word *tuhat* from the Germans, this could also be considered an alternative method of forming number words. We shall encounter such freezings and borrowings time and again and where necessary shall refer to them where they occur (see p. 179).

Why do the words for "ten" deviate so much from the rule? Because "ten" is the first landing that rises above the foundation walls of the earliest number words, the units. The latter are expanded to become the tens, as it were, not just in some form of computational manner (e.g., 20 = 2 × 10 as in Old High German *zwein-zug*) but also through grammatical transformations of the words themselves (plural as in the Hebrew *eśrim*, dual as in the Latin *vi-ginti*, and 20-group "man," as in the Ainu *hot* and Japanese *hata-chi hito*, "man"). This enormous variety of ways of forming number words, which we find so intriguing, is an important proof of the slow and natural growth of the number sequence.

THE FIRST STEPS BEYOND TEN

We count ten, eleven, twelve and then thir-teen, four-teen and so on: why the odd names for just the two numbers following "ten"? Let us look into the history of these words.

Both these words are of common Germanic formation:

Gothic	*ain-lif*	*twa-lif*
Old Norse	*ellifu*	*tolf*
Anglo-Saxon	*anleofan*	*twelf*
Old Saxon	*elleban*	*twe-lif*
English	*eleven*	*twelve*
German	*elf*	*zwölf*
Dutch	*elf*	*twaalf*
Old High German	*einlif*	*zwelif*
Middle High German	*eilf* (Goethe)	*zwolf* (Luther)

Examples: A passage in the Gothic Bible, I Cor. 15:5: "And that he was seen of Cephas, then of the *eleven* (King James Version: 'twelve'; Luther: '*Zwölfen*')" reads ... *þata þaim ainlibim* (dative case!) Another passage (Mark 4:10): "... they that were about him with the twelve asked of him the parable" reads ... *miþ þaim twalibim*. From Otfried's Book of Gospels we have: "... which was one of the eleven" — Old High German *er ein thero einlifo uuas*; and Notker of St. Gall describes the wild boar as a charging beast with "teeth twelve ells long" — Old High German *unde zene sine zuuelif elnige*.

The number words we have just enumerated are combinations of "one" or "two" with the syllable *-lif*. What does this syllable mean? If we look into some related languages, we find Latin *re-linquere* (cf. "re-liqu-ary") and Greek *leípō*, "to leave remaining," Gothic *leiban* and English *leave*. The Greek word is familiar as the root of the word "ecliptic" (from Greek *ékleipsis*, "omission, nonoccurrence"), the path traveled by the sun on the celestial sphere. It is so called because the moon, when it approaches, darkens the sun so that the sun is "missing." The word "el-lip-se" also belongs to the same group (according to the Greek explanation, an ellipse is a curve that can be used to draw a desired right angle, some definite portion of which is afterwards cut off or "left out"). The Indo-European root from which all these words are derived, after appropriate phonetic shifts, is *leiqu*. Hence, eleven and twelve mean nothing more than "one-left" and "two-left." Left from what? Left over from "one-ten" and "two-ten," of course, after the ten has been subtracted! The omission of the cardinal number 10 is analogous to the omission of the denominator in the names of Roman fractions (see p. 159).

This is striking evidence that the Germanic number sequence at one time ran only as far as ten. Anything above that was "more." One and two more than ten were still counted, but anything beyond them was perhaps, as so often happens among primitive people, merely considered "many." Then along with the later clear conception of numbers, the subsequent computational number-word formations arose: 3'10 thir-teen, 4'10 four-teen, 5'10 fif-teen, and so on.

I shall cite two interesting Chinese expressions that seem to belong to this category: the phrase that says literally "not reaching 10" is used with the meaning of "incomplete," and "12 parts" has the meaning of "to be in excess." These both make sense if we consider 10 as a limit of counting: Whatever does not reach 10 is "incomplete," whatever goes beyond 10 is "excess."

Lithuanian also has fully adopted these Germanic *lif*-formations for the number words from 11 to 19; in Lithuanian these are (*lika* = -*lif*):

11 *wienó-lika*	16 *šešió-lika*
12 *dvy-lika*	17 *septynió-lika*
13 *try-lika*	18 *aštunió-lika*
14 *keturió-lika*	19 *devynió-lika*
15 *penkió-lika*	20 *dvi-dešimtis.*

There is no doubt but that in ancient times these number words migrated from the Germanic people to the Lithuanians. Whether it replaced some older Lithuanian one or whether, as may also have happened, constructed the number words of this interval for the first time in Lithuanian cannot now be ascertained. At any rate, this word formation now occurs only in Lithuanian and in the Germanic languages.

An interesting analogue of the Lithuanian number sequence is to be found in a language spoken by a Negro tribe of the Sudan, whose numbers, however, are built upon the number 5:

6   *wal-ta*, "1 and"
7   *lena-ta*, "2 and"
8   *segua-ta*, "3 and"
9   *ses-sa*, "4 and."

The Interval from 11 to 19
There are some languages that form the number words of this interval entirely without disturbances, thus without "eleven" and "twelve." Among the Indo-European languages this applies especially to those of the Slavic group, which set the units like building stones "upon" (*na*) ten:

Czech: 10 *deset*, 2 *dva*; whence 12 *dva-na-deset*, "2 on 10" contracted to *dvanást*, followed by 13 *trinást*, etc.
Russian: 10 *děsatj*, 2 *dvě*; 12 *dvě-ná-dzatj*, 13 *tri-ná-dzatj*, etc.

The "teens" are counted similarly in Albanian (*nja-mbe-dhiete*, "1 upon 10") and in Rumanian (*un-spre-zece*, "1 on 10"; see below). When the number sequences in the languages of less "mature" peoples run through this interval without disruption, one can suspect — so long as there has been no borrowing as in the case of Finnish and Lithuanian — that the number words have been formed after a pattern (as in Hungarian *tizen-egy* 10'1 and Turkish *onbir* 10'1; see p. 113). On the other hand, the Semitic sequence (Hebrew *ahaḅ'-esreh* 1'10, etc.) and the Chinese (*shih i* 10'1, etc.), to name only two, are quite independent.

We have already encountered the unorthodox Celtic examples for $18 = 2 \times 9 = 3 \times 6 = 8 + (2 \times 5)$ and the Welsh manner of counting 15, 15 + 1, 15 + 2, 2 × 9, 15 + 4 (see pp. 64 and 71).

This is also the place to point out the unexplained break in the second interval of tens in the Romance languages, which up to or through 16 set the units in front of the "ten" but thereafter place them behind it. Classical Latin does not have this variation, although the inverted terms *decem et septem*, "10 and 7," *decem et octo*, "10 and 8," etc., have been documented in place of the more usual *septendecim* 7'10 and *duodeviginti*, "2 from 20." But the Latin spoken by the soldiers and settlers in the outlying provinces of the Roman Empire used the former words, which came from them into the Romance languages. Yet although the origin of this metathesis is thus clear enough for the case of the Romance languages, it is not explained for the case of Latin itself: why did the sequence in popular spoken Latin run to 16 according to rule and then suddenly reverse the order? Moreover, in Spanish and Portuguese the inversion takes place after 15, while Romanian lacks it altogether. In the following table the number words from 11 through 20 are given side by side in Classical Latin, Italian, French, Spanish, and Romanian, so that the reader can see this peculiar anomaly for himself:

The Interval from 11 to 19 in the Romance Languages

	Latin	Italian	French	Spanish	Romanian
10	*decem*	*dieci*	*dix*	*dies*	*zece*
11	*un-decim*	*un-dici*	*on-ze*	*on-ce*	*un spre zece*
12	*duo-decim*	*do-dici*	*dou-ze*	*do-ce*	*doi spre zece*
13	*tre-decim*	*tre-dici*	*trei-ze*	*tre-ce*	*trei spre zece*
14	*quattuor-decim*	*quattor-dici*	*quator-ze*	*cator-ce*	*patru spre zece*
15	*quin-decim*	*quin-dici*	*quin-ze*	*quin-ce*	*cinci spre zece*
16	*se-decim*	*se-dici*	*sei-ze*	*diez-y-seiz*	*šase spre zece*
17	*septen-decim*	*diciasette*	*dix-sept*	*diez-y-siete*	*šapte spre zece*
18	*(octo-decim)*	*diciotto*	*dix-huit*	*diez-y-ocho*	*opt spre zece*
19	*(undeviginti)*	*dicianove*	*dix-neuf*	*diez-y-nueve*	*noua spre zece*
20	*viginti*	*venti*	*vingt*	*veinte*	*douazeci,* shortened to:
				y = and	11 *unsprece*, etc.

Rom. *spre* < Lat. *super*, "above."

Underlying the break at 17 we may suspect an old 4-count, such as $2 \times 4 = 8$ (a grammatical dual; see p. 14), $3 \times 4 = 12$, and $4 \times 4 = 16$, which was then followed by the new manner of counting with the word order reversed. But it seems unlikely that the 4-count could have been pushed so high without leaving any direct traces of itself in the spoken numbers. If, on the other hand, the backward-count 10 − 1, 10 − 2, 10 − 3 for 9, 8, 7 really did exist in the Indo-European language, it could well have rubbed off on the higher interval and marked out the numbers 17, 18, 19 before the others (see p. 170). There are plenty of convincing examples of the reflection of one verbal peculiarity in a different interval of numbers; some of these have been described earlier (see p. 14).

In Classical Greek, to be sure, the difference in formation of the number words 11 and 12, on the one hand, and the remaining numbers up to 20, on the other, is not so very great, but it is still discernible:

11 *héndeka* 1′10,   12 *dṓdeka* 2′10,

but

13 *treîs kaì déka* 3 & 10   14 *téttares kaì déka*, "4 and 10," etc.

Thus "eleven" and "twelve" were once felt by the Greeks to be "other" or "closer" numbers than the ones immediately following them; the latter were tacked on with "and," just as two discrete concepts can always be joined together with this conjunction. Modern Greek, however, now counts 13, *dekatreîs* (10′3), 14, *dekatésseres* (10′4), and so on.

Retrospect

With this last remark, we come to the end of our investigation of the nature and origin of the number sequence. Now let us take one last look back over the road we have traveled.

At the end we directed our attention to the laws characterizing the formation of the number words in various languages. Number words had to be combined, of course, unless an impossibly large quantity of them were to be invented. What did we learn? That the combinations of number words have not followed any thought-out, clearly recognized plan, but quite often have had to be put together awkwardly from words already at hand. In addition, we found word formations that reflect the force of the very oldest number groupings. We recall how these groupings systematically invaded graduated number sequences, and we have seen how strongly they still live on in everyday popular speech.

Let us therefore return to the ideas we encountered in the course of our wanderings through linguistic history and which emerged from the ordering and grouping of the number sequence, its gradation, supplementary quantities, spoken numbers, and written numerals, and the attached number sequence, which became abstract by gradual detachment from the objects it counted — then we shall recognize what is the most significant conclusion of our study:

A people's number sequence is not a system created fresh out of the pure workings of the mind; it is rooted in the same soil as the people. Like culture itself, it grows up slowly over the millennia, and even in its mature form it reveals the history of its people through the successive deposits of the passing years.

Our Number Words

The Indo-European Family of Languages

"Number words are among the most change-resistant words of any language; they lose their original meanings very early, and they change scarcely at all with the passage of time and the rising level of culture. If our number sequence stood isolated in the sphere of language, we should never be able to learn anything about its origin."
(See pp. 100 and 125)

Up to this point we have been speaking about the general development of the number sequence; we shall now turn to our own number words. What can they tell us about their history and the history of our own forefathers?

Our number words are linguistically related to most of the other European and to the Sanskrit-Iranian number words. Thus we can shed light on them from the widest variety of different languages and gain insights that would otherwise have been lost to us forever.

Now I must ask the reader to travel with me through the regions of linguistic history that lie beyond the sphere of numbers themselves, in order to learn something about the nature of *linguistic affinity*. Our efforts will be rewarded by allowing us to take part in one of the most magnificent dramas in the history of peoples and their cultures — one which language alone has preserved for us.

To keep the reader who is unfamiliar with linguistics from losing his way in a sea of minute details and wandering aimlessly in a dark wasteland, let us take a quick look at the route we are about to travel. We shall begin with a rapid survey of the Indo-European languages and their number words, with which the reader should gradually become familiar, so that he can perceive, test, and work out for himself the conclusions that we draw from them. Next we shall learn, from the example of the Romance languages, what is meant by linguistic affinity. Last, we shall obtain a general idea of the basic Indo-European languages and of the individual languages that grew out of it.

Thus prepared, we shall turn to the *meanings* of our number words. We shall speak of their flexibility and their fading out. We shall begin our discussion of these words with the ranks of ten, hundred, thousand, and higher numbers (as used in India, by Archimedes, and by ourselves), and continue with the units and the tens. Through them, the remarkable break in our sequence after 60 will lead us to a discussion of the great hundred (120), of the number 12, and of the Babylonian sexagesimal system. There our study of number words will end, and after a last look at *hidden number words* we shall leave not only these, but the entire subject of spoken numbers.

SURVEY OF LANGUAGES AND NUMBER WORDS

Before we arrive at the number sequences, we come first to the languages themselves, in the order in which they are treated in the study of Indo-European linguistics. They can be readily visualized from the accompanying linguistic map (Fig. 21). Here is the succession in which we shall meet them:

Kentum and *Satem* languages (p. 91) and their number words (pp. 92–93);

Germanic languages (p. 94) and their number words: old Germanic (p. 94), new Germanic (p. 95) and Modern German (p. 96);

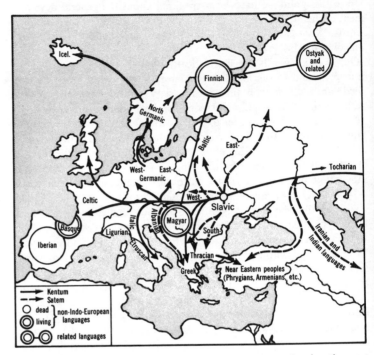

Fig. 21 Map of the Indo-European family of languages, showing the geographic distribution of the *kentum* and *satem* branches and of a few non-Indo-European languages.

Celtic languages and number words (pp. 96–97);
Balto-Slavic languages and number words (pp. 97–98);
Romance languages and number words (p. 99).

The reader will profit greatly from a comparison of these Indo-European with non-Indo-European languages. Of the latter, we shall present:

the Basque number sequence (p. 110);
the Etruscan number sequences (p. 110);
the Finno-Ugric family of languages (p. 111) with the Finnish, Hungarian, and Turkish number words (p. 113);
the Semitic family (p. 114) with the Arabic and Hebrew number words (p. 115);
the Sumerian number sequence (p. 165);
the Chinese and Japanese number words (pp. 447 ff.).

Linguistic affinity involves not only a similarity of words, but above all identical linguistic structures. In the case of the last three of these language groups, the differences in their structures are shown briefly by means of examples, so that we can thereby recognize the peculiar characteristics of our own language.

"Affinity" of Languages

Let us begin our discussion with the number words in the Romance languages: Italian, French, Spanish, and Romanian (see table on p. 99). Even a superficial glance will reveal the similarity of these words, for example those for 5 or 40 or 100. Is this resemblance just accidental? No; even if we did not know the histories of these words, this would be impossible to believe. It is contrary to the agreement of almost all the words in these number sequences, even those of peoples who live at great distances from one another, such as the Spaniards and the Romanians.

All the languages of this group are descendants of the mother-tongue, Latin, the language of the Romans. The Roman Empire and its civilization embraced the provinces of Spain and Gaul in the west; its northern boundary ran along the Danube River, north of whose lower reaches was the Roman province of Dacia. To these frontier regions the Romans sent soldiers, traders, and settlers, who along with Roman culture also introduced their Latin folk speech. And the forms of vernacular Latin continued to live on in the far corners of the Empire, long after Rome had ceased to be the Empire's capital and its only source of culture and civilization. Left to its own devices, popular spoken Latin developed into separate languages, in both pronunciation and form: French *homme* and Italian *uomo*, both of which however, like two sisters, betray the common heritage of their Roman mother: Latin *homo*, "man" or "person." Hence, they have been called "Romance" languages.

Here we have a case of related languages or linguistic affinity, the explanation for which is provided by recorded history.

Once again, let us test the relationship of the languages listed below by comparing the corresponding number words in the tables on pages 92 and 93: in the German *drei* ("three"), for example, from the Germanic *þreis*, and in the same word in Greek, Latin, Celtic, Tocharian, Baltic, Slavic and even the Sanskrit *trayah* — in all these we see the same combination of sounds, *tr-*, which German preserves even to this day. A striking similarity!

The words for "two" and "six" also maintain their resemblance through all these various languages. Other numbers, like "eight" and "nine," are similar only between the languages listed on p. 92, while the words for "eight" in Baltic and Sanskrit seem more closely

Kentum Languages		*Satem Languages*
	In Europe:	
Greek*		Slavic*
Italic (Latin)*		Baltic*
Celtic*		Albanian (see p. 68)
Germanic*		
	In Asia:	
Tocharian*		Sanskrit*⎫ Aryan
Hittite		Iranian ⎭
		Armenian and other Near-Eastern languages

* Asterisks indicate languages whose number words are given in the tables following.

Kentum Languages

	Greek	Italic	Celtic	Germanic	Tocharian
		Latin	Irish	Gothic	
1	*heîs, mía, hén*	*unus, -a, -um*	*oin*	*ains*	*sas, sam*
2	*dýo*	*duo, -ae, -a*	*da*	*twai, twos, twa,*	*wu, we*
3	*treîs, tria*	*tres, tria*	*tri*	*þreis, þrija*	*tre, tri*
4	*téttares, -a*	*quattuor*	*cethir*	*fidwor*	*śtwar*
5	*pénte*	*quinque*	*coic*	*fimf*	*päñ*
6	*héx*	*sex*	*se*	*saihs*	*sak*
7	*heptá*	*septem*	*secht*	*sibun*	*spät*
8	*oktố*	*octo*	*ocht*	*ahtaú*	*okät*
9	*en-néa*	*novem*	*noi*	*niun*	*ñu*
10	*déka*	*decem*	*deich*	*taíhun*	*śak*
20	*eíkosi*	*vi-ginti*	*fiche*	*twai-tigjus*	*wiki*
30	*triá-konta*	*tri-ginta*	*tricha*	*þreis-*	*tary-ak*
40	*tettará-*	*quadra-*	*cethorcha*	*fidwor-*	*śtwar-*
50	*penté-*	*quinqua*	*coica*	*fimf-*	*pñ-*
60	*hexế-*	*sexa-*	*sesca*	*saihs-*	*säksäk*
70	*hebdomé-*	*septua-*	*sechtmogo*	*sibunt-ehund*	*säpt-uk*
80	*ogdoé-*	*octo-*	*ochtmoga*	*ahtaút-*	*okt-*
90	*emené-*	*nona-*	*nocha*	*niunt-*	*ñm-*
100	*hekatón*	*centum*	*cet*	*taihunt-ehund*[1]	*känt-*
200	*diakósioi, -ai,*	*ducenti, -ae,*	*da-cet*	*twa-hunda*	*we-känt*
300	*tria-* [*-a*	*tre-* [*-a*	*tri-*	*þrija-*	*tri-*
500	*penta-*	*quingenti*	*coic-*	*fimf-*	*päñ-*
800	*okta-*	*octingenti*	*ocht-*	*ahtaú-*	*okät-*
1000	*chílioi, -ai, -a*	*mille*	*mile*	*pusundi*	*wälts*
2000	*dis-chílioi*	*duo-milia*			*we-wälts*
3000	*tris-chílioi*	*tri-milia*			*tre-*
10,000	*mýrioi, -ai, -a*	*decem-milia*			*tmam*
	For 11 to 19, see p. 86 For 30–90, see p. 150	For 11 to 19, see p. 85 *c*: hard k; thus 100 is "kentum," not "sentum."	For 11 to 19, see p. 64 For 20–90 in Old Irish, see p. 65 *c*: hard k *ch*: ch as in German "*ach*"	[1]also *taíhuntaíhund* *-h*: ch as in German "*Buch*" (6,8,10,70) *aí*: short *e*, open; 10 is "teh-khun" *aú*: short *o*, open; 8 is "ah-khto" *ei*: long *ī*, pronounced like "ee" in English "bee"	*ñ*: ny as in "canyon"

Satem Languages

	Indian	Slavic	Baltic	
	Sanskrit	Old Church Slavonic	Lithuanian	Basic Indo-European
1	ekab, eka	jedinu, -a, -o	vienas	oi-nos; oi-qos sems, smiǝ, sem
2	dvi, dve	dva, dve	du, dvi	duụo; duo
3	trayah, tisrah,	trije	trys	trejes, trie
4	čatvarah, čatasrah	četyre	keturi	quetụor (es)
5	panča	pētj	penki	penque
6	šaš	šestj	šeši	sụeks; seks
7	sapta	sedmj	septyni	septm̥
8	ašta, aštau	osmj	aštuoni	oktou
9	nava	devētj*	devyni†	neụn, eneụen
10	daśa	desētj	dešimt	dekm̥; dekm̥t
20	vim-śatih	dva-desēti	dvi-deszimt	
30	trim-śat	tri-	trys-	
40	čatvarim-	četyri-	ketures-	
50	panča-	pētj-desētj	penkes-	compounds formed with -(d) kom̥-t-
60	šaš-tih	šestj-	šešes-	
70	sapta-	sedmj-	septynes-	
80	aši-	osmj-	aštuones-	
90	nava-	devētj-	devynes-	
100	śatam	sụto	šimtas	km̥tóm
200	dvi-śatam[1]	dve-ste	du-šimtu	
300	tri-	tri sta	trys-šimtai	
500	panča-	pētj sụtj	penki-	
800	aštau-	osmj-	aštuoni-	
1000	sahasram	tysēšta	tukstantis	sm̥-gheslo-m

ś: light sh as in German *Stern* (10)	š and č same as in Sanskrit	for 11–19, see p. 84	m̥: syllabic sound (am, em, om, or um)
š: full, heavy sh as in German *Schwamm* (6)	ē: nasal	†see p. 147	ụ: short *u*
č: ch as in "child" (4)	j: final palatal "y" for Slavic ь and ъ		qụ: as in English "quite"
v: as in "vow" (4)	ụ: short u for ъ		; indicates that both variations are possible (2, 6)
-im-: nasal i (20)			
[1]also *dve sate*	*see p. 147		

Germanic Group (p. 106)

West	North	East
Anglo-Saxon (from 12th century) English* Frisian (from 14th century) Low German Dutch* Low Saxon ("Heliand," 9th century) Plattdeutsch — 2nd Phonetic Shift (see p. 127) High German (table on p. 96) Old High German (8th cent. to 1100)* Middle High German (1100 to 1500)* New High German (after 1500)*	Old Nordic* Norwegian* Icelandic Danish* Swedish	Gothic* Wulfila's Bible (4th century)

Old Germanic Languages

	West Germanic		North Germanic	East Germanic
	Anglo-Saxon	Old Saxon	Old Nordic	Gothic
1	an	en	einn	ains
2	twegen, twa	twene, two twe	tveir, tvan	twa
3	þri, þreo	thria, thriu	þrir, þriu	þreis, þria
4	feower	fiuwar, fior	fjorer	fidwor
5	fif	fif	fimm	fimf
6	six	sehs	sex	saihs
7	seofon	sibun	siau	sibun
8	eahta	ahto	atta	ahtaú
9	nigon	nigun	nio	niun
10	tyn	tehan	tio	taihun
11	endleofan	ellevan	ellefo	ain-lif
12	twelf	twelif	tolf	twa-
13	þri-tene	thri-tehan	þrettan	(þreis-taihun)
19	nigon-tyne	nigen-	nitian	(niun-)
20	twen-tig	twen-tig	tuttugu	twai-tigjus
30	þri	thri-	þri-tiger	þreo-
40	feower-	fiwar-	fjorer-	fidwor-
50	fif-	fif-	fim-	fimf-
60	six-	sehs-	sex-	saihs-
70	hund-seofontig	ant-sibunda	siau-	sibunt-ehund
80	-eahtatig	-ahtoda	atta-	ahtaút-
90	-nigontig	-nigonda	nio-	niunt-
100	-teontig[1]	hunderod	tio-	taihunt-[1]
110	-endleofantig		ellefo-	
120	-twelftig		hundraþ	
200	tu hund	twe hund	—	twa hunda
1000	þusund	thusundig	þusund(raþ) = 1200	þusundi
	[1]also hun(dred)			[1]also taihun- taihund ai, aú (see p. 92)

Modern Germanic Languages

		West Germanic	North Germanic		
			Northwest	Northeast	
	English	Dutch	Icelandic	Danish	Swedish
1	one	een	einn	en	en, ett
2	two	twee	tveir	to	tva
3	three	drie	þrir	tre	tre
4	four	vier	fjorir	fire	fyra
5	five	vijf	fimm	fem	fem
6	six	zes	sex	seks	sex
7	seven	zeven	sjö	syv	sju[1]
8	eight	acht	atta	otte	åtta
9	nine	negon	niu	ni	nio
10	ten	tien	tiu	ti	tio
11	eleven	elf	ellefu	elleve	elva
12	twelve	twaalf	tolf	tolve	tolv
13	thir-teen	der-tien	prettan	tret-tan	tret-ton
19	nine-	negen-	nitjan	ni-	nit-
20	twen-ty	twin-tig	tuttugu	tyve	tjugo[2]
30	thir-	der-	þra-tiu	tredive	tret-tio
40	for-	veer-	fjörn-	fyrretyve	fyr-
50	fif-	vijf-	fim-	halvtres[1]	fem-
60	six-	zes-	sex-	tres	sex-
70	seven-	zeven-	sjö-	halvfjers	sjut-
80	eigh-	tach-[1]	atta-	firs	at-
90	nine-	negen-	niu-	halvfems	nit-
100	hundred	honderd	hundrad	hundrede	hundra
1000	thousand	duizend	þusund	tusind	tusen
		ij: ey as in "hey!" (5) *z*: s (6) *ui*: öi (1000) [1](see p. 152)		[1]For the full forms of 40–90, see p. 65	[1]"shoo" [2]"chügu" *å*: oh (8)

German (and Gothic)

	High German			Gothic (for comparison)
	New	Middle	Old	
1	*eins*	*eins*	*ein*	*aina; ain, ainata*
2	*zwei*	*zwene, zwo, zwei*	*zwene, zwa, zwei*	*twai, twos, twa*
3	*drei*	*dri, driu*	*dri, drio, driu*	*þreis, þria*
4	*vier*	*vier*	*fior*	*fidwõr*
5	*fünf*	*fünf*	*fimf; finf*	*fimf*
6	*sechs*	*sehs*	*sehs*	*saihs*
7	*sieben*	*siben*	*sibun*	*sibun*
8	*acht*	*ahte*	*ahto*	*ahtaú*
9	*neun*	*niun*	*niun*	*niun*
10	*zehn*	*zeben*	*zehan (zehen)*	*taihun*
11	*elf*	*eilf; einlif*	*einlif*	*ainlif*
12	*zwölf* (p. 83)	*zwelf*	*zwelif*	*twalif*
20	*zwan-zig*	*zweinzic (-zec)*	*zwein-zug*	*twai-tigjus*
30	*drei-ßig*	*dri-*	*driz-*	*þreis-*
40	*vier-zig*	. . . .	*fior-*	*fidwor-*
50	*fünf-*	. . . .	*finf-*	*fimf-*
60	*sech-*	. . . .	*sehs-*	*saihs*
70	*sieb-*	. . . .	*sibun-zo (-zug)*	*sibunt-ehund*
80	*acht-*	. . . .	*ahto-*	*ahtau-*
90	*neun-*	. . . .	*niun-*	*niun-*
100	*hundert*	*hundert*	*zehan-zo*	*taihun-taihund*
200	*zwei-hundert*		*zwei hunt*	*twa hunda*
1000	*tausend*	*tusent*	*dusunt; thusunt*[1]	*þusundi*
	ß: ss (30)		[1]also *tusent* older forms also 200 *zwiro* *zehanzug* 500 *finfstunt* *zehanzug* 1000 *zenstunt* *zenzech,* "10 times 10-ty"	For pronuncia- tion, see p. 92

Celtic Group (see p. 104)

MAINLAND CELTIC:	*Gallic* (now extinct)
ISLAND CELTIC:	*British* (from the 8th century)
	Walisian (Welsh)*
	Cornish*
	Brythonic (Breton)
	Gaelic
	Irish*
(see Fig. 16)	Gaelic (in Scotland)

Celtic Languages

	Irish	Welsh	Cornish	Breton
1	oin	un	un	eun
2	da	dau	dow	diou
3	tri	tri	tri	tri
4	cethir	þetwar	þeswar	þevar
5	coic	þimþ	þymþ	þemþ
6	se	chwe	whe	chouech
7	secht	seith	seyth	seiz
8	ocht	wyht	eath	eiz
9	noi	naw	naw	nao
10	deich	dec; deg	dek	dek
11	oin deec	un ar dec	ednack	unnek
12	da-	dour ar dec (deudec)	dewthek	daou-zek
13	tri-	tri ar dec	trethek	tri-
14	cethir-	þetwar ar dec	þuzwarthak	þevar-
15	coic-	hymthec	þymthek	þem-
16	se-	un ar bymthec	whettak	choue-
17	secht-	dou ar-	seitag	seit-
18	ocht-	deu naw	eatag	tri- (ch)ouech
19	noi-	þedwar ar bym-thec	nawnzack	naou-zek
20	fiche	ugeint	ugans	ugent
30	deich ar fiche*	dec ar ugeint	dek warn ugans	tregont
40	da fiche	de-ugeint	deu ugens	daou ugent
50	deich ar du fiche	dec ar de-ugeint	hanter-cans	hanter-kant
60	tri fiche	tri ugeint	try ugens	tri ugent
70	deich ar tri fiche	dec ar tri-ug.	dek warn try ugens	dek ha tri ugent
80	ceithri fiche	þedwar-ugeint	þeswar ugens	þevar ugent
90	deich ar ceithri [fiche	dec ar pedwar-u.	dekwarn þesw. ug.	dek ha pevar ugent
100	cet	cant	cans	kant
1000	mile	mil	myl	mil

The differences in spelling are due to the different ages and origins of the documents. For the numbers 20 to 90, see p. 65.

ch: ch as in German "ach"
c: hard k* (pp. 65 & 150)

ou: oo
ch: ch as in German "ach"

Balto-Slavic Group (see p. 105)

PRUSSIAN (died out in 16th century)
LATVIAN
LITHUANIAN (see table on p. 93)
	South Slavic
	   Old Church Slavonic (extinct; = Old Bulgarian, 9th century; see table on p. 93)
	   Serbo-Croatian
	East Slavic
	   Russian*
	West Slavic
	   Polish
	   Czech*

Slavic Languages

	West Slavic	East Slavic
	Czech	Russian
1	jeden, -na, -no	odín, -ná, nó
2	dva, dvě	dva, dvě
3	tři	tri
4	čtyři	četýrě
5	pět	pjatj
6	šest	šestj
7	sedm	semj
8	osm	vósmj
9	devět	dévjatj
10	deset	děsatj
11	jede-náct	odin-na-dzatj
12	dva-	dvě-
13	tři-	tri-
19	devate-	děvjat
20	dva-cet	dvá-dzatj
30	třia-	tri-
40	čtyři-	sorok[1]
50	pa-desát	pjatj děsjatj
60	še-	šestj-
70	sedm-	semj-
80	osm-	vósemj-
90	deva-	devjanósto[2]
100	sto	sto
200	dvě stě	dvě sti
300	tři-sta	tri-sta
400	čtyři-	četýrě-
500	pět set	pjatj-sót
600	šest-	šestj-
1000	tisíc	týsjača
2000	dva tisíce	dva týsjači
5000	pět tisíc	pjatj týsjač

ě: "ye" (2) ř: rsh (3) š: sh (6) c: ts (11, 20)	č: ch (4) v: v as in "vow" (2) ě: ye (2) -j-: Russian ь, "soft sign," indicating that final consonant is palatalized. [1](see p. 185) [2](see p. 75)

Romance Group (see p. 91)

Italian* and Sardinian (spoken on Sardinia)
French* and Provençal
Spanish*, Portuguese, and Catalan
Rhaeto-romanic ("Churwelsh" in the Upper Rhine region)
Romanian*

Romance Languages

	Italian	French	Spanish	Romanian
1	*uno, una*	*un, une*	*uno, una*	*uno*
2	*due*	*deux*	*dos*	*doi, doua*
3	*tre*	*trois*	*tres*	*trei*
4	*quattro*	*quatre*	*cuatro*	*patru*
5	*cinque*	*cinq*	*cinco*	*cinci*
6	*sei*	*six*	*seis*	*šase*
7	*sette*	*sept*	*siete*	*šapte*
8	*otto*	*huit**	*ocho*	*opt*
9	*nove*	*neuf*	*nueve*	*noua*
10	*dieci*	*dix*	*diez*	*zece* (*< diece*)

11–19		(see p. 85)		

	Italian	French	Spanish	Romanian
20	*venti*	*vingt*	*veinte*	*doua-zeci*
30	*trenta*	*trente*	*treinta*	*trei-zeci*
40	*quaranta*	*quarante*	*cuar-enta*	*patru-*
50	*cinquanta*	*cinquante*	*cincu-*	*cinci-*
60	*sessanta*	*soixante*	*ses-*	*šase-*
70	*settanta*	*— -dix*	*set-*	*šapte-*
80	*ottanta*	*quatre-vingt*	*och-*	*opt-*
90	*novanta*	*— — -dix*	*nov-*	*noua-*
100	*cento*	*cent*	*ciento*	*o suta*

	Italian	French	Spanish	Romanian
1000	*mille*	*mil*	*mil*	*o mie*
2000	*due mila*	*deux mils*	*dos mil*	*doua mii*

	Italian	French	Spanish	Romanian
	c before *i* and *e* pronounced ch (5, 10,100)	**huit* < Old Fr. *oit, wit*	*c* and *z* before i and *e* pronounced like Engl. *th* (100) *c* before a, o, u is hard: k (5 = "thinko") *ch*: as in "child" (8)	*c* as in Italian *z*: voiced s *š*: sh; Rom. ș

related; the words for "nine" in Baltic and Slavic, however, do not fit the pattern at all. Likewise in the case of "ten" and "hundred" we can find some individual languages that resemble each other more closely (Greek *déka*, Latin *decem*; Sanskrit *daśa*, Slavic *desētj*), and others in which these words do not seem to be even remotely related: Germanic *taíhun* and Sanskrit *daśa*; Latin *centum* and Slavic *sto*.

But these similarities overlap inconsistently: one number word seems to be almost the same in two given languages, while some other is quite different. The question becomes increasingly complicated: Are these languages related to each other, or are they not?

They are indeed! When their similarities were first noticed, of course, other words besides those of the number sequences were compared as well. If a certain word is the same or nearly the same in several languages, we speak of a "word-equation": Sanskrit *pitár* = Greek *patếr* = Latin *pater* = Gothic *fadar* = English *father* = German *Vater*. This particular "word-equation," however, does not extend into Baltic and Slavic. Only in a very rare cases does such an equation hold true for all the languages, but these include most of the number words. A few exceptions, like the Lithuanian "nine," can be readily explained (see p. 93).

Number words are among the words of a language that most strongly resist change; they lose their original meanings very early, and they change scarcely at all with the passage of time and the rising level of culture. Consider the Gothic word for 5, *fimf*, on p. 96, a word dating from the 4th century A.D.: The Modern German word *fünf* has remained virtually unchanged for more than a millennium and a half. Thus it is the number words that directly reflect the relationships of the Indo-European languages in their purest form. Now let us turn to these relationships.

Kentum and Satem Languages

From our survey (tables on pp. 92 and 93) we saw that the relationship between these languages is not quite so obvious as the affinity of the Romance languages to each other (p. 99). A much graver difficulty, however, is that recorded history has nothing whatever to tell us about the relationships between the Indo-European languages. Which was the ancestor of this great variety of languages?

Since history remains silent, linguistics must take the whole burden up its shoulders. Students of comparative philology have succeeded in discovering the phonetic laws behind such differences as between the Latin *centum* and the Sanskrit *śatam*, "hundred." One of the most important criteria for classifying the Indo-European languages has turned out to be the fact that some languages have softened the hard consonant *k* to a sibilant *s* (or *sh*).

This division is manifested in the words for "hundred" just cited above: Latin *centum* (*k!*) — Sanskrit *śatam* (*sh!*). This ties together a whole cluster of differences we noted earlier: Latin "four," *quattuor* (*kw!*) versus Sanskrit *čatvarah* (*ch!*); Greek, Latin, and Celtic "eight," *ok-* (*k!*) versus Sanskrit and Baltic *aš-* (*sh!*); Greek, Latin, and Celtic "ten," *dek-* (*k!*) versus Sanskrit, Slavic, and Baltic *des-* or *deš-* (*s, sh!*). Using this consistent phonetic feature, linguistics has differentiated the *Kentum* from the *Satem* languages (*satem* is the Old Iranian word for 100), which we find, arranged accordingly, on the tables on pages 92 and 93.

Germanic is numbered among the *Kentum* languages, although its word for "hundred" begins with a *t* and "four" with an *f*. We shall find the explanation for this later.

The languages belonging to the two main branches are:

Kentum group: Greek, Latin, Celtic, Germanic, Tocharian, Hittite;

Satem group: the "Aryan" languages of the Indo-Iranian group, the languages of the Baltic and Slavic groups, and Albanian and Armenian (see map on p. 90).

Now that he can readily cross the *Kentum-Satem* bridge from one branch to the other, the reader should be able to recognize the over-all linguistic similarity of the Indo-European languages; and I ask him to do just that.

THE ORIGINAL INDO-EUROPEAN LANGUAGE

If all these languages are sisters, they must have a common ancestor, an original language from which they have developed. But we know of no people that spoke or wrote such a mother language nor have we any direct evidence or written documents concerning it. Thus comparative philology has had to build up the basic Indo-European language, going back into prehistory, just as if Latin had completely disappeared and had had to be reconstructed from French, Italian, and Spanish. This was done by applying all the laws of phonetic and linguistic change that have been discovered and to which some-times one, sometimes another language has contributed. Languages that have become extinct, such as Sanskrit, the language of the Indian scholars and priests, as well as languages that were either left or pushed aside and have escaped attrition by being far away from the cultural mainstream, such as Lithuanian, preserve the old Indo-European features and forms more faithfully than those that are still living and changing. Which of us, for instance, could understand Old High German without specially studying it?

The basic mother language was named after its two geographic end members: Indian (Sanskrit) in the east and Celtic in the west. But since the existence of the Celtic group had not been fully established at the time this great discovery was made, and Germanic was then thought to be the westernmost group of this family of languages, the Indo-European languages are still known in German as *Indo-germanisch*, "Indo-German." Today, moreover, it is known that Tocharian was the easternmost member of the family, so that we should properly speak of "Tochario-Celtic" when referring to the primeval mother language by the names of its outermost descendants.

A language is used and developed by people. Who spoke the original language? The "Indo-Europeans," of course. But this is purely a name for the people who spoke the language. Students of Indo-European prehistory have quite readily inferred the occupations,

customs, and way of life of the Indo-Europeans from their reconstructed vocabulary, but so far it has not been possible to locate
their place of origin with any degree of certainty. According to one
older view, they (initially) inhabited a long belt extending from the
Central Asian highlands across the steppes north of the Caspian
Sea into Europe itself; another view confines their initial home to the
regions of Central Asia; and according to a third hypothesis there
are solid reasons for supposing them to have originated in the region
between the Baltic and the Black Seas.

Let us take another look at our linguistic map (Fig. 21), which
shows the present geographic distribution of the *Kentum* and *Satem*
languages, both of which derive their common origin from the Indo-
European language. It does *not* show — and I ask the reader to take
careful note of this — the migrations of the Indo-Europeans, about
which we know nothing whatever. All we do know is a little about the
wanderings of some individual peoples during historical times. Our
map is thus a purely linguistic one, and at the same time a kind of
family tree of the Indo-European languages; when so used it will,
taken together with the table of Indo-European languages on p. 91,
clarify their geographical distribution.

We do know, to be sure, that the Aryans, who were the first Indo-
European people to appear on the stage of history, migrated around
the middle of the third millennium B.C. from the northeastern regions
through Persia (Iran) toward India, where they partly displaced and
partly intermingled with the aboriginal inhabitants of the subcontinent. That the Greeks and the ancient Italian tribes migrated south-
ward to their present homes from the north has also been ascertained.
To that extent and with respect to these peoples, the arrows on the
linguistic map can also be regarded as general routes of migration.

THE INDIVIDUAL INDO-EUROPEAN LANGUAGES
Now that we have gained at least a general picture of the *Kentum*
and *Satem* languages, we can turn to the individual languages of the
Indo-European family and try to learn who once spoke them or still
speaks them the periods from which their first linguistic documents
originate.

Indic Languages
The language of the oldest collection of hymns in existence, the Rig
Veda (second millennium B.C.) is Vedic. In a broader sense this forms
a part of Sanskrit, the written language in which the great popular
epics (the Mahabharata and the Ramayana) were composed. The
Sakuntala, by the Indian dramatist Kalidasa (5th century B.C.) was
also written in Sanskrit. Sanskrit then gradually became a purely
learned language, like Latin in western Europe during the Middle
Ages, and thereafter no longer took part in the evolution of the
vernacular (see table on p. 93). In addition to the languages already mentioned, this group included many common vernaculars,
from which Prakrit and Pali, the written language of Southern

Indian Buddhism, were derived. The language spoken by the gypsies can also be included among the innumerable dialects of modern Indic (see p. 190).

Iranian Languages

These are so closely related to the Indic that they have been classified together with them in an "Aryan" group. The language of the writings concerning Zarathustra's religion is Avestic. Somewhat younger is Old Persian, which during the period from Cyrua (as well as Xerxes, Darius and others) to Alexander the Great was the official language of the ancient Persian Empire and has been preserved in inscriptions written in cuneiform characters. Middle Persian, the so-called Pahlavi, was the language of the second great Persian empire (of the Sassanidae), which lasted until it was overthrown by the Arabs in A.D. 641. Afghan (see p. 191) is also one of the Persian dialects. Since the Iranian tongues are so very similar to Indic, only the Sanskrit number sequence has been cited here (see p. 93).

Near Eastern Group

We have likewise omitted the number sequences of this group, which among others has included Armenian (spoken south of the Caucasus), Phrygian (interior of Asia Minor), and Hittite. The latter is one of the *Kentum* languages, and was written in cuneiform on clay tablets. Excavations by German archaeologists (carried on since 1906) have uncovered the archives of the Hittite kings of the middle of the second millennium B.C.; thus, Hittite is today considered, along with Sanskrit, to be the oldest preserved Indo-European language. The Hittite Empire was overthrown by Phrygian invaders in the 12th century B.C.

Tocharian

We did, however, present the number sequence of the easternmost Indo-European language, discovered around the turn of the century by German scholars. The written documents of this extinct people come from the areas of Kashgar and Turfan in eastern and northern Turkestan. It has been possible to distinguish two related yet sharply differentiated Tocharian dialects A and B, which are thought to have flourished in the first millennium A.D. The encroachments of the Turkic peoples put an end to Tocharian, which belonged to the *Kentum* and not to the *Satem* branch, as did the Aryan languages. Thus was destroyed the hypothesis that all Indo-European languages of the interior of Asia belonged to the *Satem* branch (see p. 92).

Greek

Among the Indo-European languages of Europe, this is the oldest that has been documented (dating from the 8th century B.C.). It consisted not of just a single, common language but of a series of dialects, of which the most important are *Ionian-Attic* (Ionian on the west coast of Asia Minor was the language of Homer, Hesiod, and

Herodotus, while the Attic spoken in Athens and elsewhere was used by the classical poets and tragedians and by Thucydides, Plato, and Aristotle; "four" is *téttares* in Attic, *tésseres* in Ionian); *Aeolian* northern Asia Minor (written by Sappho of Lesbos; here "four" is *péssyres*); and *Doric*, spoken in the Peloponnese, Crete, Sicily, and southern Italy ("four" is *tétores*).

Only after Alexander the Great and the political overthrow of Greece did the Attic and Ionian dialects give rise to the Hellenistic common language known as the *Koiné*, which continued to be the literary, political, and legal language of the Eastern Roman Empire until its final conquest by the Ottoman Turks in 1453. *Koiné* is also the language of the New Testament (see p. 92).

Latin

Latin has been known since about 300 B.C.; as the language of the Romans, it won out very early over the other dialects of the ancient Italic branch (Oscan, Umbrian, Sabellian and others). Classical Latin is still in use today as the language of the Roman Catholic Church and was until quite recently the universal language of scholarship; it was not this, however, but the Vulgar Latin spoken in the early Middle Ages which gave birth to the modern *Romance languages* (see p. 91).

Celtic

This is both linguistically and geographically connected with Italic. Around the middle of the first millennium B.C., when Greece and Italy had "settled down" to begin their political and cultural dominance of the then civilized world, the mounted Celtic nomads in the north restlessly wandered and raided here and there. Their Late Iron Age culture (Latène) was distinguished for the richness of its ornamentation. Whether Britain and Gaul were their original homes cannot now be determined; at any rate, from there they migrated to Spain, where they mingled with the aboriginal Iberians to form the Celtiberian culture. In 390 B.C. they sacked and destroyed the city of Rome; a century later they plundered the Greek sanctuary of Delphi. One group of Celts migrated to Asia Minor, where they still live, surrounded by alien peoples; the region of Galatia (St. Paul's Epistle to the Galatians in the New Testament) takes its name from them.

In prehistoric times the Germanic peoples invaded western and southern Europe, generally along the Danube valley. Somewhat later the Celts abandoned these sites, and the Germanic peoples settled there and took over the old Celtic place names: Rhine, Main, Danube, Worms, Melibokus, Mainz, Walchensee, and Chiemsee are all names of Celtic origin. Toward the end of the first millennium B.C. these Germanic movements broke the power of the Celts. The Germanic tribes penetrated to the west, where their Gaul was conquered by Julius Caesar. In the course of the next several centuries the Roman culture and language became so deeply and thoroughly rooted that today, apart from a few place names, virtually nothing is left of Gallic, the mainland Celtic language. On the other hand,

Celtic is still extant in the British Isles (Island Celtic), in the forms of Irish in Ireland, the somewhat different Gaelic of Scotland, and Walisian (the language of Wales, also called Welsh, and not to be confused with the Romance dialect known as Churwelsh, spoken in the ancient province of Rhaetia in the Upper Rhine region); Celtic also remains in use in the form of Breton in Brittany and until recently as Cornish in Cornwall. The former region was, to be sure, a part of ancient Gaul, but its inhabitants came from Britain in the 5th century A.D., as the present names of the language and the peninsula testify. Except for Cornish, which died out some 100 years ago, all three Celtic languages are still spoken today. The oldest documents (the Irish of the so-called *Ogham script*) are all from the Christian era and go back to the 8th century A.D. (see p. 97 and Fig. 16).

Slavic and Baltic

These are the eastern neighbors of the Germanic languages. The individual languages of this group and their geographic distribution may be seen from Fig. 21 and the table on p. 98.

The oldest Slavic document is the translation of the Bible by Cyril and Methodius in the 9th century; its language is known as Old Church Slavonic or Old Bulgarian (see table on p. 93). In the Orthodox Slavic countries this has become the language of the Church, as did Latin in the Roman Catholic regions.

Together with Baltic, whose principal language is Lithuanian, Slavic forms a closely interrelated major group similar to the Aryan of Persian and India; the Balto-Slavic languages (along with Albanian) are the *Satem* languages of Europe. But whereas the Aryan

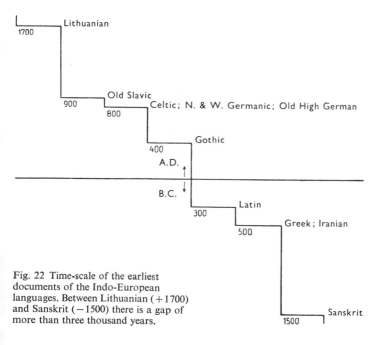

Fig. 22 Time-scale of the earliest documents of the Indo-European languages. Between Lithuanian (+1700) and Sanskrit (−1500) there is a gap of more than three thousand years.

group, through the vicissitudes of history, developed further and further away from the original Indo-European language, present-day Slavic, in contrast, and especially Lithuanian, surpasses all the other Indo-European languages in the profusion of ancient Indo-European forms it has preserved. Yet the oldest Lithuanian documents date only from the 17th century A.D.! The reader will see from the "time ladder" (Fig. 22) that the word equations previously discussed (see p. 100) do not all lie on the same chronological level, but from Lithuanian back to Sanskrit run through more than three millennia.

The Germanic Languages

Figure 21 shows the Germanic group split up into the West Germanic languages spoken between the Rhine and the Elbe Rivers, the East Germanic along the northeastern shore of the Elbe, and the closely related North Germanic tongues. They are all wedged tightly between Celtic to the west and south (see p. 104) and the Balto-Slavic group to the north and east on the continent. Nevertheless they are sharply differentiated from both these neighboring language groups. And if we refer to their number words and compare them (see p. 92) — "ten": Celtic *deich*, Lithuanian *dešimt*, Slavic *desětj*, but Greek *taíhun* — we find that we cannot at first glance assign the Germanic languages to either the *Kentum* or the *Satem* branch. According to the Roman historian Tacitus, the Germanic tribes intermingled less than any others with alien peoples who had occupied the areas they inhabited since the earliest times. Since there is also further evidence to this effect, it is at least not impossible that the Germanic peoples initially acquired an Indo-European language as a primitive people and that they were later "Indo-Europeanized," perhaps as a result of the dominance of the Celts, who at one time penetrated deep into what is now Germany. This hypothesis may at least explain the break, or gap, in linguistic relationship whose sharpness is unique among all the Indo-European languages: The Germanic tongue has adapted the new language to its own idiosyncrasies, since originally it was not "suitable," just as Germans today have trouble pronouncing Swiss dialects or Russian, and Englishmen notoriously speak poor French and colorless Italian.

Philology has been able to establish the rules governing this Germanic modification of the Indo-European sounds and thus determine which sounds have been "displaced"; for they have not all been changed, as the Germanic words for the numbers 2, 3, 7, 8, and 9 testify (see tables on pp. 92 and 93). This "Germanic phonetic shift" was the furnace in which the Indo-European language was "melted down" into "Germanic." If we examine the accompanying *dekm* table (Fig. 23), we see how the original Indo-European word *dekm* appears in the various *Kentum* and *Satem* languages to the right and left.

While all the other branches derive directly from the main trunk, the Germanic group undergoes two phonetic shifts, the "Germanic" and the "German," indicated by the circles on the diagram. As the word *dekm* runs through the Germanic branch, it changes phonetically to such an extent that it appears in Gothic as *taíhun*.

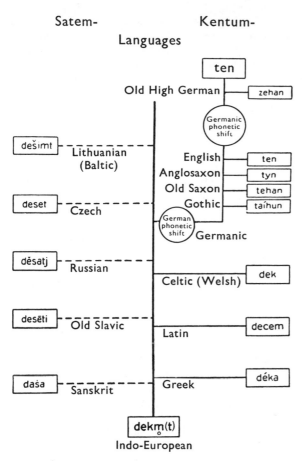

Fig. 23 Changes of the Indo-European word *dekm*, "ten," in the *Kentum* and *Satem* languages; observe the Germanic and the German phonetic shifts (indicated by the circles).

If we pursue this path still farther, we shall come, beyond the West Germanic languages (Old Saxon, Anglo-Saxon, and English), to a second circle with yet another phonetic shift: Once again the Germanic languages show a phonetic shift (in this case from *t* to *z*, pronounced "ts"), giving rise to the Modern German *zehn*, "ten." Thus the original Indo-European *dekm* has had to undergo two phonetic shifts, stronger in Germanic and milder in German, before it finally adjusted to the German ear in the form of the word *zehn*. Later we shall discuss the specific rules that have governed these phonetic shifts (p. 127).

Thus the individual Germanic languages differ from each other only as dialects. The oldest Germanic language that has been preserved is *Gothic*. Bishop Wulfila (A.D. 310–383) translated the Bible into the language of his own people and thereby created the oldest and most significant Germanic document. Fragments of this translation, written in the characteristic Gothic script in silver on purple-colored parchment and bound in a silver cover (the *Codex argenteus*,

"Silver Book"), have come down to us and are now preserved at Uppsala in Sweden (see Figs. 27 and 28; also *Gothic Bible*, Figs. 92 and 93). Just as Boniface was the "Apostle to the Germans," so was Wulfila the Apostle to the Visigoths. Originally the Visigoths were united with the Ostrogoths in the Gothic nation, which together with the Vandals and other peoples formed the cultural community of the East Germans (see p. 94).

Land hunger and pressure from the Slavs set the Germanic tribes into motion in the 1st century A.D.: The previously quiescent sea of northern peoples now set out on the migrations that surged forcibly against the bulwarks of the Roman Empire. The strongest and boldest of these barbarian floods were the East Germans; they were the true migrants, adventurers driven not only by hunger but also by a desire to make conquests in sunny and shining faraway lands where they could settle.

In Dacia in the Lower Danube region they were stopped by the resistant wall of the Roman military. A part of them came to an agreement with the Romans, stayed on, and adopted Christianity (in the Orthodox Greek form, whose missionary was Bishop Wulfila). Other remnants of these tribes persisted for a long time on the Crimean Peninsula, where their "Crimean Gothic" language became known in the 16th century (see p. 50); they and their tongue, however, died out in the 18th century. But the rest of the East Germans, now driven on by the pressure of the Huns, broke forcibly into the world of Roman culture. For a time the Romans succeeded in deflecting the streams of Goths into Spain, where it was dammed up briefly into a Visigothic kingdom, but this ultimately collapsed, having reached the end of its restlessly squandered strength a few centuries later, under the blows of the Arab invaders.

After that, however, a second flood of barbarians broke into the very heart of the Roman Empire, halting briefly when the leader of the Ostrogoths, Theodoric the Great, consolidated his young nation on Italian soil (with his capital at Ravenna). Not until after Theodoric's death in A.D. 526 did dissensions among the Germans and their damned-up lust for adventure again overflow the ordered channels of the Gothic Kingdom; now, wave upon wave, the Gothic floods swirled onward into the enticing south. They burst forth once more at the slopes of Mt. Vesuvius, then flowed back to the broadly curving, seductive Bay of Naples, which had already absorbed so many other peoples and a scant millennium later was to be enriched by the blood of the last Hohenstaufen Emperor.

The Gothic tongue came to an end with the Gothic people. Their example shows how difficult it is to shed light on the dim past of the Indo-Europeans and to trace their way of life and the course of their wanderings; for the peoples who themselves were the bearers and transmitters of the original Indo-European language vanished without a trace.

The West Germanic peoples were more peaceful. In the West the walls of the Roman Empire held fast and quite early forced the migrating tribes to settle down and support themselves by a more highly developed agriculture. Thus at the time when, far from their northern home, their Gothic and Vandalic cousins were doomed to extinction, they gradually consolidated the nations and kingdoms which, after the Roman rule had been overthrown, were destined to become the bearers of Western culture. To be sure, there were advances and retreats among the West Germanic peoples too, but these movements were less violent and destructive. The Angles and the Saxons, who left the Elbe River area to invade England in the 6th century, ultimately combined with the later invaders, the Normans, to become the true "Englishmen," next to whom the native inhabitants (the Celts) from then on played only a subordinate role in British history. But the vacuum which the departed East Germans had created along the Elbe River could not be filled adequately by those who remained, and so a Slavic people (the Prussians) pushed their way to the Elbe itself. Deliberate and forceful policies ultimately succeeded in stemming their invasion, and in the centuries to come the territory was regained step by step.

The oldest documents of the languages of the West Germanic group (English, Frisian, Low and High German; see p. 94) date from the 8th century; they include the Low-Saxon Biblical hymns of "Heliand," of the 9th century. High German is commonly broken down into Old (to 1100), Middle (to 1500) and New High German (from 1500 on; see table on p. 96). Charlemagne spoke Old High German; the language of the *Nibelungenlied* and of the Minnesingers was Middle High German.

The North Germanic subgroup lay to one side of the more rapidly flowing mainstream. With the passage of time it deviated from the remaining Germanic groups and by A.D. 700 had formed the original Nordic language, which has been preserved only in isolated runic inscriptions. Later it became divided into West Nordic (in Iceland and Norway) and East Nordic (in Denmark and Sweden); the language prior to about 1500 was known as Old Norse. The Old Germanic tongue lived on in its truest form on the remote island of Iceland (settled by Norwegians around 900 and Christianized around 1000). The Edda, a collection of Old Norse songs about gods and heroes, was first written down in the 13th century.

In the tables on pp. 94–96 the reader will find the old and modern West and North Germanic languages compared, with Gothic included in the first of these groups. He can not only compare these languages directly at a given period of time, but also learn which changes, English, for example, went through as it developed from the Anglo-Saxon.

Thus we have presented a general survey of the Indo-European languages, represented visually in the linguistic map (Fig. 21), the time-ladder (Fig. 22), the *dekm* table (Fig. 23), and the number sequences of the various individual languages, past and present (pp. 91–99).

The Non-Indo-European Languages

We have yet to discuss the non-Indo-European languages, which appear within the circles on our linguistic map (Fig. 21): the now defunct Iberian, Ligurian, and Etruscan tongues and the still spoken Basque language and Hungarian (Magyar), the latter of which, along with Finnish and Ostyak, belongs to the so-called Finno-Ugric branch, which grew out of the stem of the Uralic-Altaic family of languages from which Turkish and Mongolian have also sprouted.

The number sequences of these non-Indo-European languages spoken in Europe are given in the next few pages. To these, as examples of a broader and stronger linguistic relationship, are added the Semitic number sequences, Arabic and Hebrew, which we shall encounter repeatedly in the course of our study. (The reader will find the Chinese and Japanese number words and numerals on pp. 447 ff.)

Basque

Basque today is spoken by about a million persons and is thought to be a possible successor of the now vanished Iberian language:

1 *bat*		
2 *bi*	20 *oguey*	20
3 *hiru*	30 *oguey-t-amar*	20 + 10
4 *lau*	40 *ber-oguey*	2'20
5 *bost*	50 *ber-oguey-t-amar*	2'20 + 10
6 *sei*	60 *hirur-oguey*	3'20
7 *cazpi*	70 *hirur-oguey-t-amar*	3'20 + 10
8 *zortzi*	80 *laur-oguey*	4'20
9 *bederatzi*	90 *laur-oguey-t-amar*	4'20 + 10
10 *amar*	100 *eun*	
	1000 *milla.*	

The Basque number sequence, whose number words all the way to 100 have not the slightest resemblance to the Indo-European (except for the word for "six," which by coincidence sounds familiar), is clearly built up by levels of 20. The old Basque limit of counting was evidently 100, for the word *milla* for 1000 has been taken from the Romans.

Etruscan

The Etruscan and Ligurian peoples, who have now disappeared, were the Romans' preceptors in many things pertaining to government and religion, and maintained their highly respected special status well into the period when Rome had outwardly overtaken them. Originally the Etruscans occupied large areas of Italy; later their home was confined to Tuscany, which takes its name from them (*Tusci = Etrusci*). They were probably one of the peoples who came over from Asia Minor, appearing at the borders of Egypt with other maritime nations around the middle of the second millennium B.C.; driven away from Egypt, they crossed the sea and set foot in Italy. Their language is preserved only in the form of a few not yet reliably deciphered remnants. The Etruscan number words up to 90 were:

1 *thu*	10 *sar*
2 *zal*	20 *zathrum*
3 *ci*	30 *ci-al-ch*
4 *huth*	40 *huth-al-ch*
5 *mach*	50 *muv-al-ch*
6 *sa*	60 *se-al-ch*
7 *cezp*	70 *cezp-al-ch*
8 *semph*	80 *semph-al-ch*
9 *nurph*	90 *nurph-al-ch*

As in the case of Basque, the words are completely foreign to us and show no similarity at all to those in any of our familiar languages. But we can see that the tens, apart from 20, were formed from the units with the suffixes *-al-ch*.

Philologists see in the *-al-* a genitive ending that serves as the equivalent of an adjective; the *-ch* then turns this into a noun: thus 30, *huth-al-ch*, is "three-like, the" and hence "the three-like." The numbers just before the gradations, such as 17 or 29, are formed by back-counting: "3 from 20," *ci-em-zathrum*, or "1 from 30," *thun-em-cialch*.

Finno-Ugric

The Finno-Ugric group of languages, which together with Samoyedic belongs to the so-called Uralic branch, is represented in Europe by its two largest offshoots, Finnish and Hungarian. Along with some smaller groups of languages it stretches from the East Slavic region as far east as the Urals. Philology has divided the Uralic group into the following branches:

1. Lapp;
2. Baltic Finnish (Finnish, Estonian, Livonian, Carelian);
3. Volga Finnish (Mordvinian and Cheremyshian, both belonging to small groups spoken along the bends of the Volga River south of Moscow);
4. Permian (Zyryanian and Votyakian) spoken in the region of Perm, now the city of Molotov, in the Central Urals;
5. Ugric (Vogulian and Ostyakian on the Ob River; Magyar or Hungarian west of the Carpathians).

All of these languages trace their descent back to an original language whose native home was somewhere in what is now eastern Russia. From the brief survey in the table on p. 113 we see the wide gap between Finnish and Hungarian, the latter of which is far more closely related to languages spoken east of the Urals. Their affinity consists in having the same or similar grammatical structure and vocabulary, and in the case of the basic words also in regular phonetic changes; there is less of the obvious similarity of words that the laymen can recognize, which we are accustomed to find among various Indo-European languages.

To mention but a few of the features peculiar to both these languages: They have no grammatical gender, inflect all nouns in the same manner, form the plural with the same symbol (Hungarian -*k*-, Finnish -*i*-, in the nominative and accusative cases -*t*-), and have the same case endings for the singular and the plural. Hungarian has 24 cases and Finnish 15, expressions that we would treat as prepositional phrases as "in, into, to, on, from the house" all being considered as "cases." A few examples will illustrate the difference between these and our Indo-European verbal forms:

Hungarian:	*a haz,* "the house"	*a haza-k* "the houses"
	a haz-ban, "in the house"	*a haza-k-ban,* "in the houses"
Finnish:	*talo,* "the court"	*talo-t,* the courts"
	talo-ssa, "in the court"	*talo-i-ssa,* "in the courts."

Both Finnish and Hungarian are agglutinating or amalgamating languages, not inflected like modern German, because they leave the root (e.g., *talo*) undisturbed and represent their grammatical indications by suffixes "tacked on" (agglutinated) to the ends of the roots (for example, *talo-i-ssa* contains the suffixes -*i*- for the plural and -*ssa* for the locative).

Now let us consider the number sequences on p. 113. Surprisingly, both the Finnish and the Hungarian number sequences are built up uninterruptedly on the basis of the number 10. All the tens are formed by adding suffixes to the units, $30 = 3'10$, etc.; even 20, which in so many other languages takes a special form, is here $2'10$. The numbers 11, 12, ..., 19 have no special words for "eleven" and "twelve" but run consistently straight through, in Hungarian taking the form $10'1$, $10'2$, etc. and in Finnish using the overcount borrowed from the Germanic languages (see p. 80). Does this testify to a rare case of a regular, undisturbed progression on the part of a highly mature number sequence possessed by the Finno-Ugric peoples? Let us examine the root meanings of the number words; we find that the 2nd and 3rd rank levels have been borrowed from Indo-European:

Finnish	Mordvinian	Vogulian	Zyryanian	Hungarian	
sata	*sada*	*sat*		*szaz*	< Indo-Iranian *sata*
tuhat		*sotr*	*surs*	*ezer*	< Indo-European

and, to be sure, Finnish *tuhat* < Baltic *tukstantis*, Hungarian *ezer* < Indo-Iranian *hazar*.

Thus the Finno-Ugric number sequence was built up independently only up to ten. In fact all these languages have in common only the number words from 1 through 6; the Hungarian word for 7 comes from Indo-European (*setpm*, Vogulian *sät*, Hungarian *het*, whose initial *h* was created by analogy with 6, *hat*).

At the very first rank threshold, 10, the Finno-Ugric languages begin to deviate from each other, whereas the Indo-European languages have all the numbers through the 100s in common and do not go their own separate ways until 1000. The basic Finno-Ugric language has a word meaning "number," which in the Lapp *lokk* is the number word for 10; Finnish has it in *luka*, "to count, to reckon." Finnish 10 is *kymmenen* (Mordvinian *kemein*), whereas Hungarian borrowed the Indo-Iranian *tiz* (Syryenian *das*). Lapp also uses the word *tseke*, which means "notch, groove," for 10 — an exact analogue of the English word *score* for a group of 20.

The Permian tongues form the tens with *-mis*, Hungarian *-mints*, which we recognize as a degenerate remnant in the Hungarian 20, *hu-sz*, and 30, *harmin-c*, as well as in 8, *nyol-c* and 9, *kilen-c*, the latter

	Finno-Ugric Languages		Turkish
	Finnish	Hungarian	
1	*yksi*	*egy*	*bir*
2	*kaksi*	*kettö*	*iki*
3	*kolme*	*három*	*üč*
4	*neljä*	*négy*	*dört*
5	*viisi*	*öt*	*beš*
6	*kuusi*	*hat*	*alti*
7	*seitsemän*	*hét*	*yedi*
8	*kahdeksan*	*nyolc*	*sekiz*
9	*yhdeksan*	*kilenc*	*dokuz*
10	*kymmenen*	*tiz*	*on*
11	*yksi-toista**	*tizen-egy*	*on bir*
12	*kaksi-*	*-kettö*	*on iki*
13	*kolme-*	*-harom*	*on üč*
20	*kaksi-kymmentä*	*búsz*	*yirmi*
30	*kolme-*	*harminc*	*otuz*
40	*neljä-*	*négy-ven*	*kirk*
50	*viisi-*	*öt-ven*	*elli*
60	*kuusi-*	*hat-van*	*altmis*
70	*seitsemän-*	*hét-ven*	*yetmis*
80	*kahdeksan-*	*nyolc-van*	*seksen*
90	*yhdeksan-*	*kilenc-ven*	*doksan*
100	*sata*	*szaz*	*yüz*
200	*kaksi-sataa*	*két-szaz*	*iki yüz*
300	*kolme-*	*harom-*	*üč-*
1000	*tuhat*	*ezer*	*bin*
2000	*kaksi-tuhatta*	*ket-ezer*	*iki bin*
3000	*kolme-*	*harom-*	*üč-*
10,000	*kymmenen-*	*tiz-*	*on-*
	For 11–19, see p. 80 *(see p. 78)	*sz*: ss (20) *c*: ts (8) *g*: dj (1, 4) *z*: soft s (10)	

two having the meanings "2 from 10" and "1 from 10," respectively. The same syllable also appears in the Turkish 60 and 70.

Finnish forms these same back-counted 8 and 9 with *-deksan*. The Hungarian tens from 40 to 90 have the "10" syllable *-van* or, after phonetic shift attraction, *-ven* in 70 and 90 (Syryenian *-min*, Vogul *-men*).

The surprising thing is that with this multiplicity of words for 10, the Hungarian word form *hu-sz* < *ke(ttö)-(min)ts*, 2'10, runs through most of the Finno-Ugric languages: Ostyak *ko-s*, Vogul *ku-s*, Mordvinian *ko-ms*. This surely is due to the influence of the 20-group "man" concept, here expressed verbally not as a grouping but computationally in the form 2'10 (like the French *vingt* for the Norman *skor*; see p. 68).

What have we learned from our study of these number words? It has demonstrated clearly that an undisturbed progression of the number sequence is not due to any higher degree of maturity, but to a less developed number sequence hidden in the background. This then modeled itself on the pattern of the already developed neighboring languages (Slavic and Germanic) and thereby was able to avoid all the incomprehensible anomalies and disturbances in the progression, which it would have undergone had it developed independently.

Turkish

Similar circumstances have impressed themselves on the *Turkish number sequence*. Turkish is part of the Altaic subfamily of languages, a sister family of the Uralic, while the Finno-Ugric group belongs to the Uralic division. Thus Turkish is a "cousin many times removed." They share, however, the same clear grammatical structure. The Turkish word *on*, "many," for 10 is an old limit of counting. In the Turkish 60 and 70 we recognize the Finno-Ugric 10 ending *-mis*; *-san* (or *-sen*) is also a form of "ten." In contrast, the words for 20, 30, 40, and 50 are irregular, and no one has yet succeeded in explaining them satisfactorily; they are not formed from the units but are "numerational" and thus preserve a very early state of the number sequence that extended only to 50 (see p. 82).

Semitic

The Semitic languages, like the Indo-European, make up a large family of related tongues, of which we shall cite only the Arabic and the Hebrew number sequences.

Philologists have broken down the Semitic family into the following classification:

1. Eastern Semitic (Babylonian from the third millennium B.C.; Assyrian from 1100 B.C. on; see p. 165);
2. Western Semitic
   a. Northern Group (Aramaic and Canaanitic, including Hebrew and Phoenician)
   b. Southern Group (Arabic and Ethiopic including Abyssinian).

Semitic Languages

	Arabic		Hebrew	
	Masculine*	Feminine*	Masculine*	Feminine*
1	*ahadun*	*ihda*	*ehad*	*ahaþ*
2	*iþnani*	*iþnatani*	*šənayim*	*šittayim*
3	*þalaþun*	*þalaþatun*	*šaloš*	*šəloša*
4	*arba'un*	*'arba'atun*	*arba'*	*arba'a*
5	*hamsun*	*hamsatun*	*hameš*	*hamišša*
6	*sittun*	*sittatun*	*šeš*	*šišša*
7	*sab'un*	*sab'atun*	*šeba'*	*šiv'a*
8	*þamanin*	*þamaniyatun*	*šəmonē*	*šəmona*
9	*tis'un*	*tis'atun*	*teša'*	*tiš'a*
10	*'asrun*	*'asaratun*	*'ešer*	*'asara*
20	*'išruna*		*'eśrim*	
30	*þalaþuna*		*šəlošim*	
40	*'arba'una*		*arba'im*	
50	*hamsuna*		*hamiššim*	
60	*sittuna*		*šiššim*	
70	*sab'una*		*šiv'im*	
80	*þamanuna*		*šəmonim*	
90	*tis'una*		*tiš'im*	
100	*mi'atun*		*mē'a*	
200	*mi'atani*		*ma'þáyim*	
300	*þalaþu mi'atin*		*šaloš me'oþ*	
1000	*'alfun*		*elef*	
2000	*'alfani*		*alþayim*	
3000	*þalaþatu-alafin*		*šəlošeþ'alafim*	
10,000	*'ašarat-'alfan*		1) *'ašereþ'alafim*	
			2) *alafim 'aśara*	
			3) *rəbaba* or *ribbo*	

The case endings -*un* are no longer always pronounced. *h*: ch as in German *Buch*	The murmured vowel is almost silent: "30" is thus almost "shloshim."

' is a guttural; ' is a glottal stop: German "Post-amt," not "Pos-tamt."

* *Editor's note:* In traditional grammar, as well as in usage (both spoken and written), columns 1 and 3 are considered "masculine," and columns 2 and 4 "feminine."

Hebrew, developed since the migration of the Jews into Canaan (14th century B.C.), is the language of the Old Testament. As a result of the destruction of Jerusalem (by the Assyrians) and the Babylonian Captivity (in the 6th century B.C.), Hebrew from that time on became increasingly displaced by Aramaic, and was virtually extinguished after the great westward migration of the Jews during the Hellenistic epoch, beginning in the 2nd century B.C. Jesus and His Apostles spoke not Hebrew but Aramaic. As a ritual language, however, Hebrew has been maintained down to the present. Next to it, around A.D. 200, arose the New Hebrew of the Talmud, the compilation of Judaic laws. Finally, as a result of Zionism and the return of the Jews to modern Israel, Hebrew has again become an official and commercial language in that part of the world.

Arabic

Arabic and its many dialects were spread by the great conquests of the 7th and 8th centuries A.D., of which we shall have more to say on pp. 406 ff.; through the Koran alone, which Moslem believers may read only in the original Arabic text, it was introduced into all the lands where Islam predominates. Thus, for example, fully half the vocabulary of Persian, which as Avestan was once a purely Indo-European language, now consists of words of Arabic derivation.

The essential feature of the Semitic languages is that they attach the root meaning of the word to a framework generally of three consonants (like the English *drnk*). Words of related meanings are then created by inserting vowels (as in *drink, drank, drunk*). This vowel gradation, or *ablaut*, which as our examples show can also occur in German or English (apart from words formed by adding prefixes and suffixes), is much more important in Semitic than in the Indo-European languages.

An example from Hebrew is the word for 10,000, *rəbaba*; its root is *rbb*, from which *ribbō* is immediately derived, as the number sequence shows. But the stem *rb* is also part of the following family of words: *rab*, "much, strong: Lord" — *rob*, "multitude, size, abundance" — *raba'b*, "to be many, numberless" — *rabba'* (> Rabbi), "leader." Thus we see, as these words have documented, that the Hebrew word *rəbaba* has been condensed from the indeterminate meaning of "many" to the strict concept of 10,000. Compare the example on p. 133. Another peculiarity of the Semitic number words, their habit of changing gender and word class throughout the sequence, has already been mentioned (p. 81).

Herewith the Arabic and Hebrew number sequences, whose affinity, in contrast to the Finno-Ugric, can be readily demonstrated by the great similarity of the number words.

The Meaning of Our Number Words

IMAGES BEHIND THE CONCEPTS
OF SIZE AND NUMBER

If we address ourselves directly to our number words, "one, two, three ... nine, ten, hundred, thousand," they remain strangely silent about their inner meanings; in this respect they behave like but few other words of our language. It is remarkable that in our daily lives we encounter them constantly and use them as the most reliable bearers of concepts in our possession. Yet we allow ourselves to be content with their usefulness while knowing them only "by name." They pass by us mute, like alien slaves valued only for their services, and we do not dignify them by inquiring into their "person" or their homeland. And yet they do have "personalities," chosen by early man out of his colorful, chaotic environment to be the bearers of his concepts of numbers. We are aware, of course, that the number sequence did not spring from a single brain, as a perfected system, but that it developed slowly, like a tree, keeping pace with man's gradual development. Thus we cannot view the first few number words as being artificially contrived formations invented, like Esperanto, by one person alone and then adopted by others.

But however much we may rattle and shake the word "hundred," tap it or squint at it through a magnifying glass, it remains silent. And no matter how diligently we may search among all words for its relatives, none will admit belonging to the same tribe. Anyone will reject the word "hound" as a cousin of "hun-dred." Thus we have no alternative but to seek evidence for our hypothesis that number words originally had graphic or figurative meanings. Once again we must turn to common speech and look for words that designate magnitudes, which is, of course, what number words are.

Pictorial Quantities

A German farmer does not cultivate so many square meters of acreage, but so many "mornings" or "days" — that is, as much land as he can plow in one or more days. Elsewhere he speaks of a day's work, of a *Juchert* (the amount of land a "yoke" of oxen can plow each day) and of a *Mannshauet* (as much as one man can "hew" in one day). In brief, out of the things that surround man day after day, from what he sees and knows and feels, he creates the units of measure that convey ideas of size or quantity to him. From his own body he took the *ell*, which has been used as a measure of length by virtually all civilized peoples: Latin *ulna* > French *aune*, Greek *ōlénē*, Old High German *elina*, modern German *Elle*, Italian *braccio*, "arm"; and also the foot, the handbreadth (Latin *palmus*, Greek *dóron*), the fingerbreadth (English *digit*, "$\frac{3}{4}$ of an inch"; French *pouce*, "inch" < Latin *pollex*, "thumb" and "inch" or $\frac{1}{12}$ of a Roman foot); and in addition, all the measures of extent: the span, the pace, and the fathom (the distance a man can "fathom" with his two arms extended); the Swiss have a *Chlupfel*, a "*Ge-lupf*," an armful of hay, just as in Swabian a *Hämpfli* is a "handful." The sheaf also belongs to this class (< Indo-European *ghreb*, "to grasp, grip"), since it is a quantity of stalks "grasped" together into a bundle. The German *Klafter* is the English *fathom*, from Anglo-Saxon *fathmos*, "both arms stretched out"; the fathom later became specifically a measure of depth, but in northern Germany it once also measured a volume of wood, like a "cord." *Faden*, "thread"* is thus as much yarn or thread as can be measured with both arms, in sewing. From the same root comes the word *Fuder*, a large wine measure called the "cartload," which was orginally also a fathom's length, as was the related Latin *passus*, "pace," a thousand of which (*mille*) came to be called a *mile* (see p. 186).

The Greek coin known as the *drachmé* likewise was derived from a fathom or grasp measure; the word originally meant "handful" and was worth 6 pieces of iron (Greek *óbolos*), which had a monetary value in early Greece (see the story of the slave girl Rhodopis in Herodotus' *History*, II, 35). The English apothecaries' weight *dram* is a descendant of the drachma. On the other hand, it is purely fortuitous that the Russian *kopek* also derived its name, like the Greek obol, from the word for a spear or lance (Russian *kopje*), since the image of a rider carrying a lance was stamped on the early kopeks.

Since the first money used by early man did not consist of coins with commonly accepted arbitrary values, but of objects of barter — cattle, sheep, iron and so on — the names of many coins of antiquity and even of the Middle Ages were originally those of objects. Who would have thought that the French equivalent of the "dollar," the *écu*, has an animal hide as its ancestor? Yet if we trace the development of this word back to antiquity, the connection becomes quite clear: *écu*, Spanish *escudo*, Italian *scudo*, derived through the Latin *scutum*, the leather-covered long shield carried by Roman soldiers (compare Latin *ob-scu-rus*, "hidden") from the Greek *skýtos*, "leather, hide"; this animal skin was originally traded and bartered, and thus was used as money. Later on *écu* was also inscribed on the stamped *escutcheon* of the master of the mint. These designations of quantity and monetary value grew directly out of the ordinary life of early man.

Whenever a measure did not immediately confront him pictorially, he created one that did: He took a *rod* as a measure of length; the English still use the *yard*, a word cognate with the German *Gerte*, "rod." Another example is an obsolete German measure of capacity, the *Simmer* (Old High German *sümber*), literally a basket. Similarly, *Schäfflein* (a small trough), from Latin *scapha*, "boat, skiff" > Old High German *scaf*, English *skiff*, became a bushel; and an *Eimer*, "bucket," (Greek *ámē* > Latin *ama* > Middle High German *ame*) became an *ohm*, a liquid measure of about 35 gallons. The modern German word *nachahmen*, "to imitate, counterfeit," which Luther still spelled *nachohmen*, literally means "to measure again, to verify," and was later changed to "become equal to a measure." The *Malter*, "quarter" of grain, roughly equivalent to a "bushel" is a *Mahl-ung*, a "grinding" or "milling," and the *Metze*, "peck" (Old High Ger-

---

* *Translator's note: Faden* in German means both "fathom" and "thread," thus revealing the relationship between its meaning.

man *mezzan*) of grain was originally the "measured quantity" or "measure." These dry measures gradually evolved into weights. The best example of this is the "ton" (cf. archaic English *tun*, "barrel"), which like the others was originally a measure of grain; "a ton of land" was as much land as could be planted with a ton of seed grain. This later became a measure of salt, coal, and the cargo capacity of a vessel. Today it is our largest unit of weight.

But even for weights, there were things which people took over directly from nature to use as measures: the carat is the small red kernel in St.-John's-bread, which looks like a small horn: Greek *kerátion*, from *kéras*, "horn" (as in rhino-ceros, "nose-horn"). This was worth four barley- "corns" (Latin *granum*), from which our apothecaries' weight, the *grain*, is derived. The Babylonian weight measure *ma-nu*, which entered the Greek scale of weights as *mnâ*, "mina," is probably a date pit. The old German weight, the *Lot*, is related to English *lead* (löten = solder) which could be poured out in measured amounts as weights.

Even for something as hard to measure as time, primitive man drew on his observations of his surroundings: mid-day, sun-down, sun-up. The ancient Sanskrit "forenoon" was *samgava*, "the time when the cows are driven together," just as the Greek quitting time in the evening, the *bu-lytós*, was the time for "unharnessing the oxen" (< *bous*, "ox"; *lyein*, "to release"; *bulýtonde*, "in the afternoon"). An early Germanic example is the Anglo-Saxon *undorn* (going back to the Latin *inter*, "between"), the "between-meals time." The Russian peasant divides up his whole day after this fashion: first, second, and third *upovod* "between," just as he also breaks up the night into the first and second "cockcrow."

Wherever we look we see that the early measures of size and quantity were not deliberately synthesized and artificially given new names especially invented for them. Instead, early man simply took some events or objects which most conveniently designated magnitudes and introduced them as measures. Today we still make the same use of objects of fairly definite and commonly known magnitude: We speak of a "ream" and a "quire" of paper, a "skein" of wool, a "pack" of cigarettes, a "bale" of cotton; as informal measures of time, we have "before you can say 'Jack Robinson'," and a human generation.

Descriptive Number Words

But images have been substituted even for abstract numbers. We need only recall our 20-group, the *score* "notch," the *snas* "rod," and the *borla*, "load" (see p. 49), not to mention the Lapp *tseke*, "notch" (see p. 113); other examples are provided by words for the rank levels (as on pp. 63 and 141).

Primitive cultures also offer some excellent illustrations. Things that commonly come in twos are often called "wings," "eyes," or similar paired objects. In the Carib language the word for 2 has the root "break, split," indicating two created out of one; an expression still used in modern German is *ent-zwei*, "into two". In the language of one Negro tribe "four" is called "two forks"; the Abipone Indians say "rhea toes" for "four" because the kind of rhea (ostrich) with which they are familiar has four claws.

The hands and fingers have become measures and number words all over the world. We recall how the finger gestures became the Dene-Dinje number words, how the "hand" is almost universally used to indicate five, and how 10 was called "both hands," 15 "both hands and one foot," and 20 "one man" (p. 35).

Symbolic Numbers of India

In India the custom arose of expressing numbers by symbols: 1 is "moon"; 2 is "eyes" or "arms"; 4 is "throw of the dice," or "brothers" (because Rama in Indian poetry had three brothers), or "world's age"; 7 is "head," because the head has 7 apertures; 9 is "cattle," also in reference to Hindu myths, and so forth. Other cultures also provide numerous examples, such as the "Muses" representing 9 or the "Apostles," 12. For the numbers 1 and 2 there are more than three hundred different symbolic words, including even "rhinoceros."

Although the Hindus possessed a fully formed number sequence and thus had no need for such substitutes, such symbols were still readily and without difficulty understood as constituting numbers. They were used not only in poetical works, to express a "naked" number couched in symbolic terms, but predominantly in scientific texts on mathematics and astronomy. Most of these were written in verse, and in order to fit numbers into the rhythm of the lines and perhaps also to enable the reader to grasp them better pictorially, the verbally fixed number words were expressed by appropriate verbal symbols. This practice extended throughout India, Tibet, and Southeast Asia all the way to Java where in fact it is still used when a number must be expressed without the slightest ambiguity, e.g., in administrative affairs or in historical inscriptions on public monuments.

Since the 7th century the custom of expressing numbers symbolically has seen some rather remarkable uses. For example, the number 1021 is put together from symbols used in the same manner as digits, according to the place-value principle which after all had its origin in India. The only difference here is that the number is written and read in ascending order from left to right:

	one	two	zero	one
Hindi:	*śaśi*	*paksa*	*kha*	*eka*
	"moon"	"wing"	"hole"	"one."

The *kha*, "hole" or "empty space," stands for zero, likewise a device of Indian invention in its written form "0." In this place-value notation the numerals 1, 2, ..., 9 were represented by the meaning of a word rather than by a fixed written symbol. A number consisting of two digits can also be given by a single word, such as "sun" for 12 (in the zodiac) or "teeth" for 32. To make up an example in our own language:

$$3125 = \text{"hand"} - \text{"Apostles"} - \text{"the Fates"} = 5 - (12) - 3.$$

Here we note for the first time that peculiarly Indian delight in numbers which is manifested in their playful habit of substituting symbolic words for numbers. It springs from a uniquely Indian gift for numbers which has often been an object of wonder and admiration.

Images as Number Symbols

We have mentioned pictorial measures and number words; now for the view that early man drew on images from his environment to express his ideas we shall call on a third witness: writing, language's younger sister. Here we shall make a distinction which has, quite surprisingly, seldom been made before: the distinction between the writing of words and that of numbers.

How many people are aware that our letter A is an oxhead and the letter G a camel's neck? The letters of our alphabet have a colorful history: they have come down to us from Rome, whither they had migrated from Phoenicia by way of Greece. Yet even Phoenicia was not their origin, but Egypt. The image painted and chiseled on the walls of tombs and temples by the scribes of the Pharaohs were the true ancestors of our alphabet. If we turn the *A* upside down and write the *G* in the form of the Greek gamma, *Γ*, we may recognize in them the horned cow's head and the neck and head of the camel. Their names make this absolutely clear, however: Hebrew *álef* > Greek *alpha*, "cattle"; Hebrew *gimel* > Arabic *djamal* and *gamal*, our word "camel."

We may also turn to China, where still today pictures of objects are considered "written" symbols, or ideograms; the character for "tree," for example, is in fact a tree with trunk, root, and crown (see Fig. 1). Thus, no further proof is needed for the assertion that the first written symbols were pictures of the objects they represented.

But numerals are much closer to us in time. They can be of various different kinds, as we shall see in the course of our discussion; those of one kind are definitely of pictorial origin. A glance at Fig. 8 will show that the ancient Egyptian symbol sign for 10 was a horseshoe-shaped stanchion for fettering animals; that for 100 a measuring cord of 100 ells, shown rolled up as depicted in illustrations of Egyptian surveyors; that for 1000 a lotus blossom; and that for 10,000 a finger. We are not now concerned with the significance of these symbols apart from their pictorial quality, but this leaves nothing to be desired.

An excellent example of the way measures, coinage, and pictorial representation are united is provided by the so-called marriage tablets from the Kai Islands, south of New Guinea (Fig. 24). On these carved tablets the groom and the father of the bride inscribed the price that had been paid for the bride years before, when the bride was purchased as a child; thereafter she remained in her parents' house until the purchase price was fully paid. On the tablets a disc stands for a bronze gong, a ring for a bracelet, a pair of "tongs" for an earring, a supine animal for a goat worth 10 Dutch guilders, and a toothed bow for a golden comb. The second tablet (on the right) shows very clearly how the bride price was paid off in

Fig. 24 Marriage tablets from the Kai Islands. The groom and the father of the bride mark the bride's price on the tablet. An example of the interconnection between measure, coinage, and pictures. Size 41 × 16 cm and 86 × 14 cm.
Linden Museum, Stuttgart.

Fig. 25 Symbols for common fractions. Babylonian half and half, Egyptian half and quarter.

Fig. 26 Number beasts. *Top*: tadpole, the Egyptian symbol for 100,000; *bottom*: scorpion, the Chinese sign for 10,000 on the obverse side of a 6th-century coin.
Schlösser Collection, Hannover.

installments: the objects already paid have been cut out, leaving blank spaces.

Early signs for fractions also derived from visual symbols and developed from measures. Thus the Babylonian sign for one half was a measuring vessel whose contents are shown as cut in half (Fig. 25). The ancient Egyptian half is clearly the "half," i.e. one side of "the whole," just as our own word "half" also originally indicated "one of two side(s)" and in this manner first acquired its meaning of half; the Low German expression *up der halwe lîn* means "to lie on one's side" and hence "to be ill." The Egyptian sign for one quarter is a recumbent cross, which as a hieroglyph means both "to break" and also stands for a "doubled half," or one quarter, represented by the four parts, the four arms of the cross.

This category also includes the high "number beasts" whose pictures and names are used by some peoples as numerals or number words. Thus, for instance, the Egyptian symbol for 100,000 is a tadpole, such as wiggle by countless numbers in the mud of the Nile when the water retreats within its normal banks after the yearly flood (see Fig. 26 for tadpole symbol in the hieratic script and Fig. 9 for the same sign as a hieroglyph). Its name, *hfn*, also means "innumerable." Similarly the scorpion, because of the large numbers in which it occurs and perhaps also because of its many feet and segments ("thousand-foot," milliped), in China became the numeral, and its name *wan* the number word, for 10,000. Our illustration of its appearance on an ancient coin of the 6th century shows the creature, with its two antennae and its barbed tail, in a form that is easier to recognize than its counterpart in modern Chinese (see pp. 450 ff.).

Another fascinating instance is the reverse of the process in the Indian name for the lac louse (*Tachardia lacca*) which, like an aphid, always occurs in "countless" multitudes and whose secretions form a hard surface (shellac) on certain plants — its name derives from the Sanskrit number word *lakša*, lakh, 100,000.

To these examples we can also add the Greek word for 10,000, which from *myríos*, "countless," was condensed down to the specific number word *mýrioi* and was probably derived ultimately from the Greek *mýrmex*, "ant." The shift of the accent was made by grammarians who wanted to differentiate between the specific number word and the indefinite word "countless." We may also wonder whether the Avestan *baevar*, "beaver," which also means 10,000, acquired its numerical meaning because of the abundance of these animals.

Two number concepts grew out of the Hebrew *alef*, "ox," by coalescence. The letter *alef* (Greek *alpha*), originally an oxhead, had the value of 1 as the first letter of the alphabet (we shall have more to say about this later: see pp. 262–267, especially the table of Fig. 95). The Hebrew word for 1000 is *elef*, obviously the same root as *alef*. The connection between the two is probably this: The letter's original form was the picture of an oxhead, and so it was called *alef*, "ox." Once the indefinite "many" was visualized as a herd of oxen (a quantity), people also began to say "cattle," Hebrew *alef*, for "many," and the word after a time came to be the specific number word "thousand."

As a complement to our numerical bestiary, we have a "number plant" in the Egyptian lotus blossom, which was used as a sign for 1000 (Figs. 8 and 9).

In summary, we can say that early man, in speech as in writing, drew on his environment for symbols to indicate sizes and quantities; he did not fashion artificial words or symbols to apply only to specific numbers and measures but had no intrinsic connection with them.

Now we can hope for insight into the meaning underlying our own number words "one," "two," and "three."

SEMANTIC FADING

Eene deene Bohnenblatt,
unsere Küh sind alle satt . . .
"*Eene deene* bean and leaf
All our cows are satisfied."

Here and there German children still count things off with this jingle. But what are children's nonsense rhymes doing in a scholarly study of written numbers? We are faced with a conundrum: We know what beans, leaves, and cows are, but what does "eene deene" mean?" Some might say smilingly that this is simply childish gibberish and nothing more.

But then we find a cousin of this German rhyme in America:

Eeny meeny miny mo,
Catch a nigger by the toe.
If he hollers let him go;
Eeny meeny miny mo,

and we are forced to admit that this American children's verse is very much like the German one. And if we now look at still another example, which English children recite in India and South Africa as well as at home:

Une dune des
Catlo wuna wahna wes
Each, peach, muskydom
Tillatah, twenty-one,

then, in spite of some differences, we must again acknowledge the similarity, and we can no longer dismiss *eene deene* as mere nonsense and beneath the notice of serious students.

Perhaps we can get at the meanings of these obscure words if we look on them as number words. The English *one, two, three* do not get us very far; but let us try the French:

Une, deux(-ne), trois
Quatre-cinq, six-sept, huit-neuf, dix;

then if we rewrite the last two lines of the English rhyme as

Each speech must be dumb
Till I tell twenty-one,

we have one of the best and clearest examples of a counting rhyme.

Our English rhyme makes sense only as far as *catlo*. The following explanation may be found plausible. Because of the rhythm, *deux* has also taken on the *une* ending *-ne*; "one-ah, two-ah, three." *Catlo* has the same meter as *quatre-cinq*, but the subsequent syllables do not fit the pattern at all. Yet *wahna* has the same ring as *huit-neuf*. Then *wuna* may be understood as an altered *six-sept* changed by remote influence of the words next to it, as often happens in children's verses:

Eins zwei drei,
hicke (?) hacke hei!

Similarly, *wes* and *des* are explained as syllables exchanged for number words: *des < dix* and *wes < (tr)ois*. Now the two first lines of nonsense words reveal themselves as number words whose meanings have faded out and been lost.

What is the history of this rhyme? We know that in the year 1066 the Normans invaded England and settled there as lords over the earlier Anglo-Saxon and Celtic inhabitants, and centuries passed before they mingled completely with the people they had conquered. Thus the *un, deux, trois* came to England, where its sound remained as alien as its sense; for the English count *one, two, three*. In this way the number sequence in the children's verse became eroded to the point of unintelligibility; if this were not a counting rhyme, no linguist would read the French *deux* into the syllables *duna*.

Two more examples show the connections between general history, children's verses, and the number sequence. In the German province of Hesse, in Mainz, this verse is still in circulation:

Eene deene dus manee
Riwwele rawwele sondernee.
Ecke Brot, sonder Not,
Dusee.

The first and last lines sound French: *dus < douce?* ("sweet"), *manee < monnaie?* ("money"), and *dusee < douze* or *douzain*, "twelve" or "dozen" — are there twelve number words? The second line is metrical gibberish, but the third line can be understood: if a man has an *Ecke*, "corner" (piece), of bread, he will not be in want (*Not* = want). Mainz was occupied by the French many times in its history; the children heard a strange language and composed a jingle whose beginning and end were the foreign number sequence but whose middle came from their own language. Here too we see the surprising longevity of these unintelligible foreign phrases.

There is still another impressive example. In the county of Lincolnshire, which today is a purely English-speaking region in the eastern part of the Midlands, sheep in the fields and stitches in knitting are counted in a very peculiar fashion:

1	*yan tan tethera pethera pimp* 5
6	*sethera lethera hovera covera dik* 10
11	*yan a dik tan a dik — — bumpit* 15
16	*yan a bumpit — — — figitt* 20.

What is this? The numbers at the ends of the lines, 5, 10, 15, and 20, show quite clearly that this is Celtic; we need only compare these with the Welsh number sequence (p. 97). But how does Celtic suddenly pop up right in the middle of an English-speaking area? There is no doubt that Celts previously lived here who were later driven westward, into Wales. Thus the traces of a historical event, for which there is perhaps no other documentary evidence, have been tenaciously preserved in this number sequence. Here, where the number sequence was not officially preserved in its pure form but was left to the common people, we have a beautiful example of the way adjacent number words and others in the sequence work on each other: 4, *pethera* < *pedwar*, has expanded 3, *tri*, into *tethera*; the dactylic rhythm, however, has also implanted itself in the numbers 6 through 9; the vowel change from *e* to *o* in 8 then extends to 9 as well, and so on. This vernacular number ditty suggests a whole series of linguistic observations that can equally well explain the "accepted" number sequences of a language.

The old ways of counting of course live on in children's verses too, sometimes altered to the point of unintelligibility, as in

Eetern feetern penny pump,
All the ladies in a lump.

The words of the second line, which is governed primarily by the requirements of rhyme and rhythm, do not make much sense; but in the first line, *eetern* < *een—teen*, 1, 2; *feetern* < *tether-em*, 3, *penny pump* < *pethera pimp*, 4, 5.

Modern Greek provides an example of the erosion of number words that is taking place right now: In the popular speech of Greece today, the tens in the number sequence are *triánta, saránta, penénta, hexénta, hebdoménda, ogdónta*, and *ennenénta*, rather than *triákonta*, etc. as was shown in the table on p. 92.

Against the background of our larger subject, these small children's ditties are outstanding examples of how in the course of time words can have their forms and meanings eroded to unintelligibility and yet for this very reason continue to live on and move about without change, because fear of the unknown keeps them from being meddled with.

OUR INDO-EUROPEAN NUMBER WORDS

In the course of our search for the meanings of our own number words, we have found both encouragement: that images enter into concepts of number and magnitude — and disappointment: that words may gradually slough off their meanings until they become mere gibberish. The resistance shown by our number words to changes in form unfortunately tends to show an early fading out of their pictorial quality.

There is one hope for enlightenment, however, and only one: the relationships among the various Indo-European tongues. If our number sequence were a solitary orphan in the world of languages, with claim to neither ancestors nor relatives, we should be able to learn nothing at all about its origin and its past. But when we assemble all languages into a family gathering and compare them, we see a common cast to the features of each number word that we had scarcely noticed. This may now help us to discover their previously unrecognized similarities.

The First Two Ranks: the Tens and the Hundreds

Dekm-kmtóm. We must turn once more to the table on p. 107, to recall the similarity of the number word "ten" in the *Kentum* and *Satem* languages, derived from the original word *dekm*. Since two phonetic shifts intervene between *dekm* and the Modern German *zehn*, we shall eliminate the Germanic element in the table that follows (see pp. 92–93). In this comparison the reader may be surprised

	ten	hundred
Kentum (c = k)		
Greek	*de-ka*	*(he)* **ka-to-n**
Latin	*de-cem*	**cen-tu-m**
Celtic	*de-c*	*can-t*
Satem		
Sanskrit	*da-śam*	*śa-ta-m*
Old Slavic	*de-sē-ti*	*s-to-*
Lithuanian	*de-šim-t*	*šim-ta-s*
Original Indo-European forms:		
	dé-km	*km-tó-m.*

to see the curious recurrence of the terminal syllable of "ten" as the initial syllable of "hundred."

The original Indo-European language has the root *-km-*: this consists of the laryngeal consonant *k* which is softened in the *Satem* languages to *s* or *š* (see p. 100), and the nasal *m*. The dot beneath the latter indicates that a short vowel *a, e, i, o,* or *u* is sounded with it: for example, in the German root *nmmt* a suppressed *i* turns the consonant *m* into a syllable and the root into the word *nimmt,* "takes." The changes which the voiced *m* of the original Indo-European language has undergone in the individual languages are known precisely: Indo-European *m* becomes Sanskrit *a,* Greek *a,* Latin *em,* Lithuanian *im,* Slavic *ē,* and, let us note for future reference, Germanic *um.* Having confirmed this from our table, the reader will understand the formulation *dekm — kmtom.*

Now, without going to the trouble of listing all the individual cases, we can simply say: The Indo-European word for 100 is derived from the word for 10: *dékm* → *kmtóm.*

Just as the second rank is intrinsically bound up with the first in the arrangement of the number sequence, the two are linguistically connected as well. But in which sense? All the "hundred" words contain the syllable *-to-,* which also occurs in the Slavic and Lithuanian for "ten." We have already mentioned this phenomenon (p. 81):

it turns the number word "ten" into the noun "ten-ness." The initial syllable *dé-* has dropped out of the (*d*)-*kṃtóm*; the stress accent has shifted to the last syllable. The Balto-Slavic "ten" form *desēti*, *dešimt* and the Sanskrit *daśatih* (p. 81) likewise mean the same as "hundred": a "ten-ness" of tens; the corresponding respective "hundred" forms have merely dropped the *de-* (see the table on p. 93).

Thus we arrive at the surprising conclusion that in the Indo-European languages a "hundred" is a "ten-ness" (of tens). This means that the second rank is not just somehow connected to the first, but that this relationship is the original ancient form, the grouping. This is the strongest evidence for the view stated previously (p. 46) that the first major gradations were built into the number sequence as groupings.

Now, what has happened with the German words *zehn* and *hundert*? These fail to fit the general pattern that is condensed in the table on p. 126, to a degree that no connection between them, such as can be seen in the other languages, would ever be suspected. This is the result of *phonetic shifts*.

The Two Germanic Phonetic Shifts. The *first* of these shifts, which took place sometime between 400 B.C. and A.D. 100, set the Germanic group of languages apart from the original Indo-European (and hence is called the *Germanic* phonetic shift); the *second*, which came to an end around A.D. 600, differentiated High German from the Low German, which preserves the old Gothic sounds. The line of separation between *Hochdeutsch* and *Plattdeutsch* runs through the center of Germany, from Aachen through Düsseldorf, Kassel, and Magdeburg to Frankfurt on the Oder (and has been called the *dat-das* or the *Benrath Line*, because it crosses the Rhine at the town of Benrath).

A phonetic shift, generally speaking, registers a regular change in a phoneme, for example, in the case with which we are already familiar, the alteration of the Indo-European laryngeal *k* into the sibilant *s* in the *Satem* languages. The most important of these phonetic laws apply to the so-called plosives, namely the hard consonants *k*, *p*, and *t* and the soft *g*, *b*, and *d*. The following tables present these phonetic shifts visually. Because the phonetic structure of Latin is quite

Table of Phonetic Shifts

	Indo-European	I Germanic	II High German initial	medial
1	k	*ch, h*		
2	p	*f*		
3	t	*þ*	*d*	
4	g	*k*		
5	b	*p*	*pf*	*f, ff*
6	d	*t*	*z*	*ss*
7	(Greek th)	*d*	*t*	

similar to that of Indo-European, I have chosen to start with the more familiar Latin words instead of their Indo-European equivalents. The material has been made as palatable as possible, so that the reader will not shy away from the effort of acquainting himself with these phonetic shifts. His reward will be a clear insight into the ways in which both German and English are thoroughly interconnected with the whole Indo-European linguistic family. These few phonetic equivalents are the most important threads by which German, for example, is drawn out from its isolation and woven into the great tapestry of Indo-European languages.

It will be seen that:

1. The Germanic phonetic shifts I — 1, 2, 3 in the table changed the three hard Indo-European plosives *k*, *p*, and *t* into the aspirated consonants *ch* (as in German "*ach*"), *f* and *þ* (*þorn*, the *th* in the English thorn), and the soft plosives *g*, *b*, and *d* into their hard equivalents (I — 4, 5, 6).

2. The second, or German, phonetic shift, labeled II, changed only some of the sounds, including the Germanic *d* into the German *t*, and in many cases differentiated between the initial and medial plosives (phonetic shifts II — 5, 6).

The next table gives examples in which the Germanic group is represented by Gothic and English, whose phonetic structures are closer to the original Germanic than those of other Germanic languages (the phonetic shifts referred to are those in the preceding table):

Indo-European >	Germanic	>	High German		
Latin	Gothic	English	Old	New	Phonetic Shift
cord-	*hairto*	*heart*	*herza*	*Herz*	1; 6
pater	*fadar*	*father*	*fater*	*Vater*	2; 7
turba	*þaurp*	*thorp*	*dorf*	*Dorf*	3; 5
ager	*akrs*	*acre*	*ackar*	*Acker*	4
papa	—	*pape*	*pfaffo*	*Pfaffe*	5
edere	*itan*	*eat*	*ezzan*	*essen*	6
Greek *thýra*	*daur*	*door*	*tor*	*Tor*	7
tincta	—	—	*tincta*	*Tinte*	No phonetic shift.
praedicare	—	—	*predigon*	*predigen*	

The last two examples shown are words that first entered into the German language after the second (German) phonetic shift and thus preserved their original sounds (*p-d* in the second) without change (to *f-t*). The German number words from *eins*, "one," to *neun*, "nine," which obey these phonetic laws, can be seen readily from the Table on p. 146.

The Germanic 10 *and* 100. Now let us change the Indo-European *dekṃ* and *kṃtom* into their Germanic forms according to the phonetic laws we have observed:

Indo-European *dekm̦* > Gothic *taíhun* (phonetic shift I — 6; 1),
Indo-European *km̦tom* > Gothic *(c)hund* (phonetic shift I — 1).

Taíhun is pronounced "tä-chun," the *ch* being voiced as in the German *acht*, "eight." Instead of the *-d* in *hund*, one would expect to see *-þ* from *t* (according to phonetic shift I — 3). This change holds true when the *t* follows an accented syllable (similarly with *k* and *p*). If however it precedes the stressed syllable as it does here in *km̦tom*, it is changed to the soft voiced *d* (as are *k* to *g* and *p* to *b*). Thus in front of a stressed syllable the hard plosives *k*, *p*, and *t* become the soft voiced *g*, *b*, and *d*, respectively (according to the rule known as Verner's Law). Still another example: Indo-European *fráter* > Gothic *broþar* (phonetic shift I — 3), but Indo-European *pət̆er* > Gothic *fadar* and Greek *dekás* > Gothic *tigus* ("a ten, ten-ness").

Thus phonetic shifts have given us the Gothic rank levels *taíhun* and *hund*, the latter used only from 200, *twa hunda*, on up. These are the words used in the Gothic Bible (John 6:7; see Fig. 27):

twaim hundam skatte hlaibos ni gano (hai),
"Two hundred pennyworth of bread is not sufficient."

Fig. 27 "Two hundred," *twain hundam* (·) in the Gothic Bible. Lines are 14 cm long.

The Gothic word for "hundred," however, is not *hund* but is represented by two full word forms: *taíhun-taíhund*, "a ten of tens," in which the grouping "ten" is counted. Hence, it testifies to the primitive idea which clearly formed the "hundred" from the ten. This is a remarkable find of historical evidence, preserved for us by the Gothic language alone.

Besides this, however, there was also the (probably original) Gothic form *taíhunt-e-hund*, meaning something like a "ten-ness" of tens (not of hundreds!). This form affects the tens in the Gothic sequence all the way back to 70 (see table on p. 92). In the Biblical parable of the unfaithful steward (Luke 16:6 and 7), we see all three formations peculiar to the Gothic number sequence:

jah gasitands sprauto gamelai fim tiguns,
"... sit down quickly, and write fifty;"
taíhuntaíhund mitade kaurns,
"... an hundred measures of wheat;"
nim þus bokos jah melei ahtaútehund,
"Take thy bill, and write fourscore."

Later we shall have more to say about the curious break in the Gothic formation of the tens between 50 and 80 (see p. I/163). Here we are concerned only with the formation of the word *ahtaút-e-hund*. The first part is the grouping "eight-ness"; the *-hund* is likewise quite clear. The connecting *-e-* is also found in the Greek tens, such as *hex-é-konta*, 60, for example, Because of this formation the Indo-European "ten" is also given with a final *-t*, as in *dekm̦(t)*, so that the

Gothic ten-form *-hund,* and above all the Greek *-konta* and the Latin *-ginta,* can be derived regularly from it.

This interpretation sheds a great deal of light on early ideas of numbers, for it shows not only the close connection between "ten" and "hundred," but far more — that the second rank level, 100, was formed verbally by a repetition, "ten-ten," of the already existing rank "ten." And the Indo-European *kṃtom,* 100, is thus also derived from "ten-ten":

$$dkṃt\text{-}dkṃt > dkṃt\text{-}(dk)om(t) > dkṃt\text{-}om > kṃtóm.$$

The German "zehn" and "hundert." Now we can arrive at these number words in German by applying the second, or German, phonetic shift, to the Gothic *taíhun:*

taíhun > Old High German *zehan* = New High German *zehn* (phon. shift II — 6)
hund > unchanged New High German *hund.*

Luther once wrote: *Die andern straffen und penen | solt man tzehenn ell tief begraben in die erden,* "... buried *ten* ells deep in the ground" — putting the Low German *t* and the High German *z* together in one word, as it were. Many German dialects preserve the palatal consonant in the word "ten," as in the Carinthian *zögn.*

But where does the German form *hund-ert* come from? The Gothic word "to count" is *raþjan; raþjo* means "reckoning, number" (see Latin *ratio*). In telling the story of the loaves and the fishes, the Gothic Bible (John 6:10) says (Fig. 28):

wairos raþjon swase fimf þusundjos
men number about five thousand
"So the men sat down, in number about five thousand."

The Gothic word was borrowed from the Latin *ratio,* "calculation," with the expanded meaning of "consideration, deliberation, reason," from *reor, rei, ratus sum,* "to reckon, consider, estimate, appreciate, believe." The numerical sense of the word is present in "rate" (< *rata pars*), the "calculated portion," and mathematicians use it in their rational — (that is, proportional) numbers. From Gothic, Old High German derived the word *redja* (Modern German *Rede,* "speech, discourse"); the expression *jemand Rede stehen,* "to answer to someone, give someone an account of ..." still has the connotation of setting down an account or reckoning.

Fig. 28 The words "number," *raþjon* (·) and "five thousand," *fimf þusundjos* (:) in the Gothic Bible. Lines are 14 cm long.
Figs. 27 and 28 from *Codex argenteus* Folio 48.

The Modern German *gerade,* in the sense of an "even" number, comes from the Gothic *ga-raþjan,* "to count."

Now if we follow the word "hundred" through other Germanic, Northern European, languages, we go back via the English *hund-red,*

Old Saxon *hunde-rod,* Old Norse *hund-rad* (*d* for voiced *þ*) to a Gothic word *hundra-raþ,* which means "hund(red) number" or "the number hundred."

The German word *hundert* first came into use in the 12th century, before which people said *zehan-zo* or *zehan-zug,* following the Gothic pattern of repeating "ten of tens" (*-zo, -zug,* see p. 96). Thus *zehanto-herosto* in Old High German means "leader of 100" or "captain," one who acquired his title after the model of the Roman *centurio,* commander of a "century," a military unit of 100 soldiers. The writer Notker of St. Gall (ca. A.D. 1000) once explained the name of the Greek goddess Hekate, who was among other things the goddess of growth or increase, by the number 100 (Greek *he-katón*):

Heizet si ouk Echate daz chit centum, uuanda der erduɰuochr ofto chumit zen-ze-faltiger
"She is also called Hekate, that is *kentum* ["Hundred"], because through her the fruit of the earth often grows ten-ten-fold."

The simple Old High German *hunt,* which as in Gothic is used for the numbers from 200, *zwei hunt,* on up, was relatively late in replacing the word *zehanzo.*

The Image Latent in "Ten." Now that we have explained the origin and development of the first two rank levels, there is one more question to be answered: If "hundred" is actually a "ten of tens," what then does "ten" itself mean? What sort of image or idea does this number concept embody?

Unfortunately here we are not on such solid ground as before. The meaning of the Indo-European *de-km̥(t)* probably is "two hands." There are some quite substantial supports for this view. One is the role played by the hands in forming number words, as we have seen in various primitive cultures (see p. 35). In addition, the attached syllable *de-,* which in all Indo-European languages contributes to the formation of the number word "ten," could well be the shortened word *dųo,* "two," whose *d-* is likewise very resistant to cnange in all languages. The dropping out of the semivowel *ų* (= *w,* Gothic *t(w)ai-*) has analogues in the Greek and Latin two-forms *dis-,* respectively, *dia-* (see p. 175) in the Greek *dô-deka,* 12, and also in the pronunciation of the English word *two* as "too."

Further support is provided by the phonetic shift in Gothic:

de-km̥(t) > *taí-hun(d).*

In Gothic, *handus* means "hand" and has virtually the same sound as the same word in modern German or English.

Thus it may not be entirely wrong to recognize *zwei-hand,* "two-hand," in the German word *zehn,* Old High German *ze-han.* And since the Gothic word *handus* is derived from *hinþan,* "to touch, grasp" (in many languages the word for "hand" means the "grasper," as in Old Norse *greiþ*), the word *de-km̥(t)* can be understood in its root meaning of "two grips" or "two handfuls." If this is so, moreover, then the English words *hunt* and *hound* and the German *Hund* "dog" — although at first this would have seemed ridiculous — may well be related, if very distantly, to the numbers "ten" and "hundred."

The Third Rank: Thousand
The Germanic "Thousand." For the number 1000, our tables on pages 92 and 93 show Greek *chílioi*, Latin *mille*, Germanic *þusundi*, Tocharian *wälts*, Sanskrit *sahasram*, Old Slavic *tysēšta*, Baltic *tukstantis* — a variety that makes it hard to believe that there could be any relationship between these words like that which was so strikingly revealed between the words for "hundred." And, in fact, with the "thousand" such a relationship no longer exists.

The linguistic bridge constituted by the first phonetic shift led us from the Latin *centum* without strain to the Germanic *hund*; yet in crossing the stream from the Latin *mille* we find no foothold on the Germanic shore: *þusundi*. Thus the Germanic tongues must have turned away from Indo-European sometime before the number sequence reached the 1000 level.

As our number sequences indicate, *tausend* or *thousand* is a Germanic word (see tables pp. 94–95). Although the word itself at first does not suggest any connection with *hundred*, "thousand" definitely resembles the Old Norse *þushundrad*. Hence "thousand" is a *þus-hundrad*. But what is a *þus*? If we trace the sound *þ* back from the Germanic languages, we arrive at an Indo-European root *tu*, which means "swelling" (phonetic shift I — 1, p. 127). From this root are derived, among others, the Sanskrit words *tavas*, "strong" and *tuvi*, "much"; Avestan *tuma*, "fat, plump"; Latin *tumere*, "to swell" (cf. "tumor"), as well as the Anglo-Saxon *þuma*, Old High German *thumo* and English *thumb*, which is literally the "stronger finger" and is thus distantly related to "thousand." Hence our "thousand," which can be traced through the intermediate steps of Modern German *tausend* to Middle High German *tusend* < Old High German *dusunt* and *thusunt*, Gothic *þusundi* > English *thousand*, is nothing more than "many hundred," a "strong" or a "great" hundred.

If we go back to the root meaning of "hundred," the first three rank levels in the Germanic languages can be arranged in ascending order semantically as well as grammatically:

German: *zehn > Gezehnt   > starkes Gezehnt*
English: ten > ten of tens > strong ten of tens.

Even if we disregard the perfectly plausible meaning of "ten" as "two-hands," we can still observe here a beautiful example of how the perfected number sequence grows slowly and gradually out of the simplest and most basic conceptions of ordinary people: no conscious planning, no insight into the laws of gradation, no knowledge of infinite progression — in other words not a deliberate intellectual act but a gradual advance of the number sequence, an unconscious accretion from one level to the next, induced by nothing more than expanding necessity. Languages provide remarkable testimony to this slow and tedious growth, for beyond "ten" they do not devise an essentially new number word but follow along in the same tracks and derive the new "hundred" from the old "ten," which has merely been grouped into a "ten of tens."

"Many hundred" at first is "many" or "countless many," because it proceeds through that stage to the numerical hundred. Then the limit of counting is gradually pushed farther outward so that it increasingly displaces the wavering shadow of the indefinite "many." The next step brings us to a knowledge of the — but now there is no word for it! As the fluctuating, indeterminate "many hundred" was absorbed into the established number sequence, it was condensed to the *specific* number word "thousand."

Semantic Specification. We must now deviate from our main course for a moment and say something about this important piece of evidence for the step-by-step growth of the number sequence. The phenomenon of semantic specification occurs in many languages. A word for "many" or "numberless," such as the Greek *mýrios,* whose semantic content is unspecified or variable, becomes a number word with a clear and fixed meaning like the Greek *mýrioi,* 10,000. The intangible and tenuous idea is *condensed* or *specified* into a fixed, solid numerical concept.

Homer was familiar with *chérados myríon,* "pebbles without number" (*Iliad* XXI, 319), but not with the specific meaning of this word, for he wrote *déka chílioi,* 10'1000. The word *mýrioi* first appears in Hesiod and Herodotus. We mentioned the probable derivation of this word from *myrmex,* "ant," when we spoke of the "number beasts" whose names or pictorial symbols have been condensed into number words or numerals; the reader may refer to these again on p. 122. Here we shall cite a few more illustrative examples, beginning with the following one from Hebrew. In the Old Testament (I Samuel 18:7,8) it is said that

"Saul hath slain his thousands, and David his ten thousands. And Saul was very wroth ... and he said 'They have ascribed unto David ten thousands, and to me they have ascribed but thousands'"—

nathnu   l-Dawid   rebaboth   w-li
"they have ascribed unto David ten thousands and to me

nathnu   ha-alafim,
they have ascribed thousands."

The sense of this passage is an augmentation: If Saul killed a thousand, David slew many, many more, "thousands without number." In fact the word *rebaboth* was condensed from this indefinite original meaning to the fixed 10,000, as we have already mentioned (see p. 116). Luther translated the number in Psalm 68:17 into the old indefinite meaning of "many": "The chariots of God are many thousand times a thousand":

recheb   elohim   ribbothayim   alfe   šin'an,
literally: the chariots of God two-ten-thousand thousands repeat.
(King James version: "The chariots of God are twenty thousand.")

Here the grammatical dual very clearly expresses the plural "many," and semantic specification thereby takes a further step away from antiquity (see p. 14).

The original meaning of the Tocharian *tmam,* 10,000, is also "innumerable"; this is related to the Old Slavic word *tuma* (Russian *tma*), which means "dark." The Tocharian word *wälts*, 1000, is likewise contained in the Old Slavic *velije*, "great, mighty."

That the indefinite plural of a number word can be specified to a fixed meaning, something like the manner in which "many tens" became "twenty," is shown by excellent examples from both Hebrew and Danish, which we have already seen in a different connection (p. 14). Moreover, the Akkadians took their number word *meu*, 100, from the Sumerians, in whose language it served, in the form of *me*, to indicate the plural of any word. Thus the number sequence of the Akkadian invaders had not yet advanced to 100, whereas the Sumerians were already counting up to 60^3 — a striking example of the way cultural events in the history of two peoples can be recognized in their number sequences.

An uncommonly fascinating instance of such specification is presented by the Japanese word for 10,000, *yorozu*. In the Japanese number sequence *yo-tsu* stands for 4, the *-tsu* being merely a suffix that identifies it as a number word (see table on p. 450 and pp. 452 ff.). *Yo* originally meant "many" or "numberless" and was condensed from this to the specific "four." But then the original meaning was revived, so to speak, as the number sequence advanced, and higher number words had to be found: *Yo* was joined to the syllable *ro*, which indicates that the objects exist in an unspecified number. Thus *yo-ro* literally has the tautological meaning of "indefinitely many" and then, specified by the number suffix *-tsu*, became the fixed number word for 10,000.

There is an instructive analogue to our "thousand" in the language of the gypsies. Here *deš* is 10 and *šel* is 100 (see p. 191). For 100 the gypsies in Wales say *baro deš*, "great ten," and for 1000 those in Russia say *baro šel*, "great hundred" — exact parallels of the expression "strong hundred."

Although this extension of an existing number word and its subsequent specification took place in prehistoric times in the case of our "thousand," the same process in the case of a million is much closer to our own time: what is a "million?" nothing but a blown-up Latin *mille*!

Compare the Italian *padre–padrone*, *sala–salone*; the ending *-one* has an augmentative effect. *Padrone* is a patron or protector, and thus a heightening of "father"; *salone* is a "great" hall. Similarly *milli-one* is a "many-thousand," the exact counterpart of the "many hundred" in our "thousand." Only one thing is different: "million" is a synthetic word. It may be natural to an Italian, but not to an Englishman. The further steps "billion," "trillion," and so on, are just as artificial in Italian.

Examples of such specification can be found not only in the history of language, but also in that of writing. To the "number beasts," which should be recalled here (see p. 122), we must add the Babylonian numeral for *šar*, $3600 = 60^2$, and thus for the second rank in the sexagesimal system (see Fig. 35). This is a large circle, formed by four cuneiform wedges, and originally represented "universe" or "cosmos" and thus stood for an indefinite, limitless abundance. Another analogue is the Egyptiah *hh*, "infinitely many" (see Fig.

29), represented by the god of space *Hh*, whose pictorial symbol acquired the fixed meaning of million, but later, when the culture of the ancient Egyptian empire declined as a result of foreign invasions and its range of numbers shrunk, it was again relegated to the meaning of "infinity."

Fig. 29 The Egyptian god of air space — the symbol for million.

Other Forms for 1000. Now let us return once more to the linguistic history of the third numerical rank. We shall look for explanations of "thousand" in the other Indo-European languages.

The only other forms belonging to the same class as the Gothic *þusundi* are the Balto-Slavic forms: the Lithuanian *tukstantis* and Old Slavic *tysēšta*; these may have been borrowed from the Germanic languages. Number sequences have advanced by taking over foreign number words just as often as by semantic specification of native terms. Besides specification, the *borrowing* of a rank is another important piece of evidence for the gradual advancement of the limit of counting; we shall have more to say about this in its proper time (p. 190).

But what is the situation with the other 1000 forms (Sanskrit, Greek, Latin, Tocharian, and Celtic), which have nothing whatever in common with the Germanic "thousand?" Those of the first three languages listed, *sa-hasram, chílioi,* and *mille,* are all interrelated and are derived from a single Indo-European ancestor, *sm-gheslo-m,* in which the element *sm* means "one" (see p. 93). From this derives the ancient Sanskrit *sa-hasra-m,* in which the *l* has changed to *r* (and also the Avestan *ha-zanra* > Persian *ha-zar*). The prefix "one" occurs again in the Greek *he-katón,* "one hundred" (Indo-European *s-* > Greek *h-*; cf. p. 146). Then from the Indo-European root *-gheslo-* came the Greek *chésloi,* with the dialectic variants: Lesbian *chéllioi,* Doric *chélioi,* and Attic *chílioi.* The Indo-European feminine form *(s)mi-ghsl-i* gave birth to the Latin *mixli,* which later was smoothed out to *mille.* We know of similar verbal polishing in such words as Latin *seni < sex-ni,* "every six," and *ala,* "wing" < *axilla,* "axle" < Indo-European *aksla.*

Thus the Sanskrit, Greek and Latin, "thousand" actually mean "one *gheslom.*" But what is a *gheslom*? That remains obscure. Attempts have been made to connect the Old-Sanskrit *sahasra* with the Sanskrit *saha-,* "power, strength, victory" (Gothic *sigis,* "Sieg"-fried, "Vic"-tor?, from an Indo-European root *segh,* meaning "victory"), which would render the "thousand" a "force"-hundred and thus to draw it into the same semantic area as the Germanic *þus-hundi.*

The Tocharian *wälts,* 1000, seems to point in the same direction if it is connected with the Tocharian word *wäl,* "prince."

If we add that the Celtic *mil* is borrowed from the Roman *mille,* we then have traced all the Indo-European words for the rank level of 1000 back to their root meanings.

The Higher Ranks

The Number Towers of India. If the words for the third numerical rank show little relationship among the Indo-European languages, there is none at all in the case of the higher ranks. Some individual languages still form words for the fourth rank, 10,000: e.g., Greek *mýrioi*, Tocharian *tmam*; in the Germanic group, however, "thousand" remained the highest indigenous number-word, for "million" and the subsequent ranks are alien and artificial creations that did not grow out of the vernacular.

Only a few peoples in the world have devised words for the higher ranks of numbers out of their own vocabularies. We have seen the Egyptian example (p. 134), but we also saw how this likewise lost the impulse to climb higher up the ladder of numbers. The ancient Egyptians went no higher than a million, and they later retreated from this last rung. But one Indo-European people, in contrast to all others, uniquely erected the boldest tower of numbers in existence: the Hindus. It is curious that one people is inclined in a particular direction while another follows a quite different tendency. The Indians have shown a unique gift for abstract numbers. They knew of no better way to honor sacred divinity than counting, which they esteemed far above all merely human endeavors. Here is the "manifestation" of Gautama as the Buddha (the Enlightened One):

The Buddha will be recognized by 32 principal and 80 secondary symbols, his mother by 32, and the house in which he will be born by 8. His mother, the Queen Maya-Devi, will be attended by 10 million ladies-in-waiting. Hundreds of thousands of holy men and hundreds of thousands of millions of the enlightened will pay homage to the Buddha. His throne is constituted of good works performed during hundreds of thousands of millions of *kalpas* [i.e., mythical periods of 4320 million years]. But the great lotus which blooms during the night of the Buddha's conception opens its blossom to an extent of 68 million miles."

These sacred numbers, except perhaps for the lowest ones at the beginning of the passage just quoted, are not mythical or mystical numbers like 7 or 13, but they are sanctified by their very magnitude, which leaves all mortal things far behind. There is no more striking example of the Indians' intrinsic affinity for numbers than Buddha's Temptation. This legend is contained in the book of the Lalitavistara, which tells the story of the Buddha's life. It is the strongest proof of the Indians' knowledge of the infinite progression of the number sequence, a knowledge which they did not obtain from experience and which was not forced on them, like the first three ranks, by economic demands.

When Buddha reached the age of manhood he courted Gopa, the daughter of the Prince Dandapani. But he would be rejected unless he could show public proof of his abilities. And so, together with five other suitors, he was put through trials in writing, wrestling, archery, running, swimming, and number skills. In all these contests he brilliantly defeated his rivals. After the competition, Gopa's father commanded him to pit himself against the great mathematician Arjuna, who was to be the measure of Buddha's knowledge. Arjuna instructed him to list all the numbers (that is, the numerical ranks) above 100 *kotis*. *Koti* was the name of the seventh rank, meaning 10^7 or 10 millions. Beyond *sahasra* (10^3), India at an early date had *ayuta* (10^4), *niyuta*, also called *lakša* (10^5), and *prayuta* (10^6 = a million).

Buddha answered:

koti (10⁷, abbreviated 7)	*hetuhila* (31)
ayuta = 100 *kotis* (9)	*karahu* (33)
niyuta (11)	*hetvindriya* (35)
kangkara (13)	*samaptalambha* (37)
vivara (15)	*gananagati* (39)
akšobhya (17)	*niravadya* (41)
vivaha (19)	*mudrabala* (43)
utsanga (21)	*sarvabala* (45)
bahula (23)	*visandjnagati* (47)
nagabala (25)	*sarvasandjna* (49)
titilambha (27)	*vibhutangama* (51)
vyavasthanapradjnapti (29)	*tallakšana* (53).

In Buddha's numerical ladder *ayuta* and *niyuta* refer to 10^9 and 10^{11}, respectively, not 10^4 and 10^5 as elsewhere. Indian religious poetry contains various number ladders of this kind, modeled after Buddha's, but the number words and their meanings show little consistency from one scale to another (see p. 47 for the formations with *maha-*). Their importance lies not in any clear conception of numbers (for who can visualize a number like 10^{53}?) but in their underlying recognition that a tower can be built up, level upon level, and that the tower of numbers rises high above the world of mortals to the realm of the gods and the "Enlightened." For us, however, their greatest significance is the fact that their gradations clearly build up to invisible, superhuman heights, like the successive levels of an Indian temple (Fig. 30); and — most important of all — the ranks are named! Up to *koti*, 10^7, they are true decimal gradations; from there the numbers rise by levels of 100. The ranks up to *koti* must have been more useful and practical than the subsequent, artificially invented ranks. In other numerical ladders the decimal gradation is used throughout. (In this connection we may recall the Mayan "sacred" tower of numbers; see p. 62).

But Buddha is far from finished: this number tower is only the first, the *tallakšana* count; superimposed on this is the *dvadjadravati* count, above that there is the *dvajagranisamani* and above the latter there are still six more number sequences! What is the ultimate number of the last of these sequences?

Fig. 30 Indian temple: clearly superimposed, the structure rises up level by level.

First count:

$$10^7 \cdot \underbrace{10^2 \cdot 10^2 \cdots 10^2}_{23 \text{ times}} = 10^7 \times (10^2)^{23} = 10^{7+2 \times 23} = 10^{7+46} = 10^{53}.$$

Second count:

$$10^{53} \times 10^2 \times 10^2 \cdots \times 10^2 = 10^{53} \times 10^{46} = 10^{7+2 \times 46} \text{ etc.,}$$

to the ninth count which in the end leads to $10^{7+9 \times 46} = 10^{421}$, and thus to the monstrous number 1 followed by 421 zeros.

Yet Arjuna was still not satisfied, and he gave Buddha a further test: to name all the divisions or atoms of a *yoyana* (a mile). Buddha began:

7 atoms make a very minute particle,	
7 very minute particles make 1 minute particle, and hence	(7^2) atoms
7 minute particles make one that the wind will still carry	(7^3)
7 such particles make 1 rabbit track	(7^4)
7 rabbit tracks make 1 ram's track	(7^5)
7 rabbit tracks make 1 ram's track	(7^5)
7 rams' tracks make 1 ox track	(7^6)
7 ox tracks make 1 poppy seed	(7^7)
7 poppy seeds make 1 mustard seed	(7^8)
7 mustard seeds make 1 barleycorn,	(7^9)
7 barleycorns make 1 knuckle	(7^{10})
12 knuckles make 1 handbreadth	$(12 \cdot 7^{10})$
2 handbreadths make 1 ell	$(2 \cdot 12 \cdot 7^{10})$
4 ells make 1 bow	$(4 \cdot 2 \cdot 12 \cdot 7^{10})$
1000 bows make 1 *krosa* (a measure: 1 "cry")	$(10^3 \cdot 4 \cdot 2 \cdot 12 \cdot 7^{10})$
4 *krosa* make 1 *yoyana* (mile)	

One mile therefore contains $4 \times 10^3 \times 4 \times 2 \times 12 \times 7^{10} = 384{,}000 \times 7^{10}$ atoms, a number I shall gladly let the reader work out for himself. To this Buddha added that just as he had numbered the atoms in a mile, one could also number all the atoms in all the real and mythical lands in this world and even in the 3000 thousand worlds contained in the universe.

This enumeration of particles is very likely the oldest example of calculation with vast numbers; it was known long before the Lalita-vistara was written (300 B.C.) and was perhaps included in one of the sacred books of India precisely because of its popularity. The Indians' adeptness and familiarity with such vast magnitudes is evidenced by the large numbers of such numerical towers in southern and northern India. We shall content ourselves with Buddha's tower of numbers.

After stripping away all the legendary elements, we are left with the Indians' successful attempt to devise a tool with which they could grasp such huge numbers verbally and thus also mentally. But conceiving such numbers verbally does not mean visualizing them; it means mastering the undefined, the innumerable, the obscure "many" by using the clear, verbally defined concept of the counted, of the number which falls within the scope of the short number sequence created by the requirements of ordinary life. The significance of such an intellectual attitude for the entire culture of a people, and the difference between the Hindu approaches to numbers and those of other peoples can be seen here very distinctly. The contrasting attitude of the Romans, for example, is revealed by Cicero, who with surprising perceptiveness compared his fellow countrymen with the Greeks:

In summo honore apud illos geometria fuit, itaque nihil mathematicis illustrius. At nos metiendi ratiocinandique utilitate huius artis terminavimus modum.

"By them [the Greeks] geometry was held in the very highest honor, and none were more illustrious than mathematicians. But we [the Romans] have limited the practice of this art to its usefulness in measurement and calculation."

We have already touched on the importance of having a name for each and every numerical rank in developing the perfected number sequence, but it may be stated here again:

Sanskrit 5 *koti* 7 *prayuta* 3 *lakša* 2 *ayuta* 6 *sahasra* 4 *šata* 3 *daša* 2

expresses the identification of the numerical ranks in exactly the same manner as the Chinese (see p. 54). If we eliminate the names of the ranks, the units alone will suffice to express this number according to the positional principle: 57,326,432.

Archimedes' Counting of the Sands. The European peoples did not generally have the Indian propensity for the poetry of numbers. The Greeks were more attracted by things that can be visualized and sensed than by abstract numbers. Yet it was a Greek who once "counted the sands," almost certainly after the Indian model: Archimedes, who lived in Syracuse in the 3rd century B.C. He was the most original and creative of all the Greek mathematicians. Foremost among his achievements was the "infinitesimal" analysis, related to the concept of infinity, by which he arrived at a large number of new insights into the measurement of the circle, the volumes of bodies bounded by curved surfaces, the areas of plane figures with curvilinear edges, and many others. Some two thousand years later the ideas with which he broke new ground became a logical part of the integral calculus developed in western Europe. The historian Plutarch relates fanciful stories of Archimedes' skill in engineering and of his mechanical genius (it was he who, among other things, discovered the principle of the lever and the law of fluid displacement). During the siege of Syracuse by the Romans he lifted ships into the air and set them afire with huge magnifying lenses. The story of Archimedes' death is famous. After the fall of Syracuse, as he was struck down by one of the Roman invaders while meditating on the diagrams he had drawn in the sand, he cried out: "Do not disturb my circles!"

In his "Sand-reckoner" (*Psammítes* < *psámmos,* "sand") Archimedes proposed to count the seemingly uncountable by creating a number sequence similar to that of the Indians, which could count all the grains of sand contained in a sphere the size of the universe:

Many people believe, King Gelon, that the grains of sand are without number. Others think that although their number is not without limit, no number can ever be named which will be greater than the number of grains of sand. But I shall try to prove to you that among the numbers which I have named there are those which exceed the number of grains in a heap of sand the size not only of the earth but even of the universe.

With these words, in which he clearly stated his intention, Archimedes began his treatise. Despite its resemblance to the Indian example, this was essentially a different undertaking: He did not wish to demonstrate any superhuman skill but merely to show by counting the sands that the ordinary Greek number sequence, which went only as far as *mýrioi,* could in principle be extended without limit, and that any imaginable number could thereby be expressed verbally. Archimedes did not create poetic fantasies out of his feeling for numbers but cogitated soberly about a number sequence that was to do nothing but count. This was the essential difference. It was also the difference between the solitary thinker contributing to the advancement of science and the religious tradition of the Indians with its powerful ties to the people.

In presenting the arrangement and structure of the Archimedean ladder of numbers, we leave it to the reader to find the similarities and differences between it and the "Testing of Buddha." The numbers from 1 to $10^8 = 10^4 \times 10^4 = a$, the product of *mýrioi* × *mýrioi*, was called by Archimedes the *first eight-power series*; *mýrioi* is the highest numerical rank in Greek. The final $a = 10^8$ becomes the unit of the second eight-power series, which in turn results in still another eight-power series:

$$a, a + 1, a + 2, \ldots, a + a = 2a, 3a, \ldots, 10a, \ldots, 10^8 a$$
$$= a \times a = a^2,$$

and so forth. Thus eight-power series succeed each other up to $10^8 = a$ powers:

$$a^{a-1}, a^{a-1} \times 2, \ldots, a^{a-1} \cdot a = a^a = 10^{8 \times 10^8} = p,$$

in which the largest number p is 1 followed by 800 million zeros.

According to Archimedes, all the numbers from 1 to p belong to the first period. This is followed by the second period, in which p is the lowest number:

1st eight-power series $p \times 1, \ldots, p \times 10^8 = p \cdot a$,
2nd eight-power series $p \times a, \ldots, p \times a^2$, and so forth, up to the ath eight-power series $p \times a^{a-1}, \ldots, p \times a^a = p^2$.

These in turn are followed in the same manner by the periods p^3, p^4 and so on, to the $10^8 =$ the ath period:

1st eight-power series $p^{a-1}, \ldots, p^{a-1} \cdot a$, etc. up to the ath eight-power series $p^{a-1} \times a^{a-1}, \ldots, p^{a-1} \times a^a = p^{a-1} \times p = p^a$.

This last number, the pinnacle of this vast tower, is the ath-number of the ath eight-power series of the ath period or, as Archimedes himself calls it,

hai myriakis-myriostâs periódu myriakis-myristôn arithmôn [eight-power series] *myríai myriádes,*

"the myriad of myriads of the myriad-myriadth eight-power series of the myriad-myriadth period."

Whereas the final number p of the first period is 1 followed by 800 million zeros, the largest number of all, p^a, has 10^8 times more, 8×10^{16} zeros.

Now Archimedes can begin his actual counting of the sands: how many grains of sand can be contained in the celestial sphere, at the center of which stands the earth and over which the sun and stars trace their paths? The problem is to compute how many times a particle of sand S with a diameter s can be divided into the space H of the heavenly sphere whose diameter is h, or $H \div S$. Since the volume of a sphere is a function of the cube of its diameter, $H \div S = h^3 \div s^3$.

Then Archimedes expressed the diameter of the celestial sphere in terms of the diameters of the grains of sand s:

$h = 10^4$ diameters of the earth of 10^6 stadia, each of 10^4 finger-breadths, each of 40 poppyseeds, each of 10^4 sand-grain diameters s, or

$$h = 10^4 \times 10^6 \times 10^4 \times 40 \times 10^4 s = 4 \times 10^{19} s.$$

Hence the number of grains of sand contained in the sphere of the heavens

$$H:S = (4 \cdot 10^{19})^3 = 64 \times 10^{57} < 10^{59} = 1000 \times 10^{8 \times 7};$$

that is, it is smaller than the 1000th number in the 7th eight-power series of the first period, or 1 followed by 59 zeros, and thus much smaller than even a single period!

But even this is not enough. Archimedes continues: The astronomer Aristarchos of Samos taught that it is not the earth but the sun which stands at the center of the universe, that the earth rotates around the sun, and that everything is contained in a hollow sphere upon which the stars move and which is much larger than the earth-centered sphere. Yet even for this sphere the grains of sand can be counted; their number is $10^{8 \times 7 + 7} = 10^{63}$, that is, 1 followed by 63 zeros, and thus smaller than the 8th eight-power series 10^{64}.

"I suppose, King Gelon," Archimedes concludes, "that all this seems unbelievable to one who is not skilled in mathematics, but to the mathematician it is quite clear. Therefore I have thought it worth your while to learn of it."

Great as the last number is, it is still far smaller than the last number of Archimedes' sequence. The whole point of Archimedes' discussion is that even so inconceivably large a number as that of the grains of sand contained in the universe can not only be clearly understood and verbally expressed, it can even be easily exceeded: the limitless progression of the number sequence has finally been recognized!

Scientifically recognized, we should add, looking back on the Indian tower of numbers, for the latter merely gives colorful names to the individual stories or ranks: *koti*, "peak"; *vardha*, "sea"; *padma*, "lotus-blossom." Archimedes, on the other hand, expresses only the gradation, without the colors and textures that are inherent in the Indian names; here, finally, we have the concept of abstract numbers put into words.

This number sequence, which was thought out in isolation by one individual, bears no traces of a slow, unsystematic growth, as it would if it had evolved gradually and anonymously out of the culture of a whole people; it is an artificial creation with an astonishingly logical and coherent structure.

The great mathematician Gauss had so much admiration for Archimedes that he said regretfully: "How could he have missed [the discovery of our present positional system of writing numerals?] To what heights science would have risen by now, if only he had made that discovery!" For Archimedes made his computations with the peculiar Greek letter numerals, about which we shall speak later.

But this, too, shows how even Archimedes, who in one respect was fortunate enough to rise to a level of thought far above that of his century, in other respects remained enmeshed in the intellectual fabric of his own time.

The Higher Ranks of Numbers in our Own Culture. Equally artificial is the continuation of the number sequences in western Europe through the higher ranks, the million, billion, trillion and so forth. Its genesis is one more striking documentation of the essential difference between the individually created and the evolved number sequence.

By the time the Germanic peoples first appeared on the scene of recorded history, they had already formulated the first three ranks, the tens, hundreds, and thousands, of their number sequence. The broadening of their commercial activity was naturally followed by an expansion of the domain of numbers; numbers above a thousand now cropped up everywhere. But popular language no longer kept up: the thousand was the last and highest named rank, and so it remained throughout the whole Middle Ages. "Ten thousand, hundred thousand, and thousand thousand" wrote the calculators when employing their vernacular scholars, who spoke Latin, followed the Roman fashion: *decem milia, centum* or *centena milia, mille milia* or *decies centena milia* (see pp. 28–29).

Adam Riese's book on arithmetic contains the number

86789325178

"This is six and eighty thousand thousand times a thousand / seven hundred thousand times a thousand / nine and eighty thousand times a thousand / three hundred thousand / five and twenty thousand / one hundred eight and seventy."

Symbolically: $86T(T \times T)$; $7H(T \times T)$; $89(T \times T)$; $3\,H\,T$; $25\,T$; 178. The most curious aspect in all this is the isolation of the hundreds, perhaps to set them off from the units and the tens in their inverted word order.

We see very clearly that 1000 is the highest level of the number tower in both our own and the Latin languages; if a higher number is required, another, identical, tower is imposed all over again upon the first. The Indians, however, kept on building the same tower higher and higher, level after level:

8 *mahapadama* 6 *padma* 7 *vyarbuda* 8 *koti* 9 *prayuta* 3 *lakša* 2 *ayuta* 5 *sahasram* 1 *śatam* 7 *dašan* 8.

Imagine what the number of particles contained in the Indian mile would have looked like in medieval German (see p. 138)!

"Numeration" — the reading and writing of numbers — was therefore a separate branch of the art of computation in the Middle Ages. The number was broken up into "triads" by the use of periods, strokes, or commas. Adam Riese recommends:

Seind aber mehr dann vier ziffer vorhanden | so setze auff die vierdte ein pünctlein | als auffs tausendt | Vnd heb gleich allda widerumb an zu zehlen | eins | zehen | etc. biss zum ende. Als dann sprich auss | souiel punct vorhanden | so manchs tausendt nenne. Das hundert ist die dritte figur nimb allein in benennung | Als dann die erste vnd ander mit einander,

"But if there are more than four digits / then above the fourth place a small dot / as being over the thousand. / And there take up counting again / one / ten / and so forth, all the way to the end. Then say / how many dots there are / so many thousands are to be named. The hundred is the third figure; name this separately. / Then take the first and the others together,"

as in the example we have just seen; compare this with the corresponding instructions for the medieval counting board (see pp. 319 ff. and 332 ff.). Others, unlike Riese, placed not one dot but, going from right to left, first one, then two, and then three dots, and so on, over the respective 1000 digits. This division of the number into groups of three when reading the number, a custom we still follow, originated at a time when the language had formed only three numerical ranks and is thus a feature rooted deep in the history of our number sequence. The Greeks, whose language has a fourth rank (M) in the word *mýrioi*, break up long numbers into "tetrads" or groups of four each:

867 MM 8932 M 5178.

Quite understandably, men who concern themselves scientifically with the study of numbers are apt to need and find names for the higher ranks; for people in general the "thousand" is more than enough.

The word "million" first made its appearance in Italy, probably in the 14th century, as the word itself testifies (see p. 134). Its invention has been ascribed to the Venetian merchant Marco Polo, who had been in China and described his travels among its people, saying that he saw "many thousand" (literally: *milli-one*) persons there. As a specific number word, *millione* was first used by Italian merchants. By the end of the 15th century it appeared in printed texts on arithmetic. In his great treatise on mathematics, published in 1494, Luca Pacioli says:

mille migliara che fa secondo el volgo el millione,
"a thousand thousands, which in popular speech is called a million."

The word first appeared in German textbooks on arithmetic a generation later (around 1530), and was occasionally used by Adam Riese, but took a long time to become established. Riese himself, as we have seen, used the old 1000 × 1000 formation, which lasted well into the 18th century. When the new word was mentioned, it was distinctly pointed out: "And when a million is written in numerals as 1000000, this is ten times a hundred thousand."

The "million" entered commercial parlance from the Dutch language as a large unit of money: 10 tons of gold = 1 million guilders. The "ton of gold" was worth 100,000 units of the national coin in use at the time. Hans von Schweinichen, the Duke of Liegnitz's steward, tells of being invited to visit the wealthy Fugger family at Augsburg in 1575: "There was a sideboard extending along the entire hall, loaded with drinking vessels and beautiful Venetian glass; this was said to be worth far more than a ton of gold." Thus a million was not the fourth rank, the one following a thousand, but the sixth.

In a Spanish text on computations of 1505, whose author was Philip II's teacher in arithmetic, the word *million* is used even for the twelfth numerical rank, while a special word, *cuento*, was adopted for the sixth (literally: "counting" < *contar*, "to count, reckon"); in Portuguese *un conto de reis* is "one million reis" (a unit of coinage); today both words have changed their meanings in Spanish.

These examples show that this term arose less from the scientific need to extend the number sequence by decimal gradations than from the desire for verbal abbreviation and the need to visualize higher numbers. In contrast to the Indian method, such a verbal projection of the number sequence, in which whole gradations are omitted, hardly resembles the positional system of writing numerals.

However convenient the shortened expression "million" may be, Italy, where the word was invented, was very slow and very late in adopting the word "billion" and always regarded it as a curiosity. Verbally this is a contraction of *bis-million*, "two times a million," but with the meaning of "a million times a million." In America it was used into the 18th century for the ninth, not the twelfth rank, and in France it still has this significance: 10^9 is called either *billion* or *milliard*. What better proof can there be of the synthetic character of this formation?

As far as we know from documents still preserved today, Nicolas Chuquet near the end of the 15th century was the first to use a whole series of such words to extend the number sequence:

745324·804300·700023·654321 . . . *le premier point peult signifier million |
le second point byllion, . . . tryllion, quadrillion, quyllion, sixlion . . . non-
yllion et ainsi des aultres se plus voultre on vouloit proceder.*

"The first dot indicates a million, the second a billion, . . . trillion . . . nonillion and so on, as far as one may wish to go."

Later the series was continued by others in the same or similar fashion, either independently or following in Chuquet's footsteps; some writers used the forms *bi-million, tri-million*, etc., instead of the contraction "billion." In general, all these sequences proceed upward by steps at every sixth rank: 10^6, 10^{12}, 10^{18}, and so forth

Milliard is a French word which appeared for the first time in the 16th century. Originally it designated the 12th and later came to stand for the 9th rank; it did not come into common use in France until the 19th century, and in Germany not until after 1871, when it was made popular by the "5 milliard" of war reparations imposed on France after the Franco-Prussian War.

The history of written numerals has an exact parallel of this artificial expression, which in one stroke removed the then existing limits but provided only a colorless extension of the sequence of numerals. For their first three numerical ranks, the Romans initially used the symbols X, C, and CIↃ; these were all different and had different origins. But for the subsequent 10^4, 10^5, . . ., etc., the last symbol was artificially expanded, as in ((|)) and (((|))), by adding more of the curves which initially, in the 10^3 symbol CIↃ, did not have any numerical meaning but were merely strokes of the pen (see Fig. 8). This corresponds exactly to the artificial extension of the number words *bi-*, *tri-llion*, etc. from *mille*. Because of the rarity of the higher Roman number symbols, I shall illustrate three Roman denarii, which bear numbers indicating expenditures and not monetary values (see Fig. 31; note the "half" symbols on the first and third).

The artificiality of this extension of our old, naturally evolved number sequence is perfectly obvious. It is shown by its origin in the scholar's study, by its omission and skipping over some of the ranks of numbers, by its ambiguity, by its word formations and, finally, by the verbal "statelessness" which reveals itself at the end of all such synthetic devices: the word "thousand" is Germanic, but what is "billion"?

Along with their native homes, words and things also lose their character. They become pale and barren and featureless, they are shrunk down to gray abstract concepts which now have unique, specific meanings but no longer the resonance of history and culture. *Katzengold*, "fool's gold," has been boiled down to FeS_2. Whereas such concepts, because of their colorlessness and lack of ambiguity, enter immediately into the language of science, "billion," "trillion," and their fellow orphans were not so fortunate. We have become accustomed to billions in financial statements, and the milliard still plays a role here and there in France and Germany, but the trillions and their successors have completely disappeared from scientific communication. "One cubic centimeter of gas at $0°$ C and 760 mm pressure contains 28×10^{18} atoms," says the scientist; not twenty-eight trillion atoms. He needs no number words other than the commonest and most ancient ones; the higher ranks of numbers he counts just as Archimedes once did.

With this we have once and for all finished with the history of the higher numerical ranks. Now let us go back to our starting point once more.

Fig. 31 Three Roman denarii of C. Calpurnius Piso (64 B.C.) with the symbols for 5000, 10,000, and 50,000 shown behind the head of Apollo. The symbol for 100,000 is inscribed on the *Columna rostrata* (see Fig. 10). British Museum, London.

Meanings of the Units and the Tens

The Units. Although we shall find some very intriguing nuggets of information in this area, the total yield will be quite meager. Many hypotheses about the original meanings of "one," "two," or "three" have been advanced: it has been suggested that the personal pronouns "I," "thou," and "he" were incorporated into the first few number words; but no matter how much philological acumen has been lavished on this subject, it has thus far been fruitless and will probably always remain so. The few suppositions that have any merit at all are stated below. We shall begin by reviewing the connection between our Germanic words for the units and the Indo-European words; the latter, along with the Latin equivalents, reveal their relationship to our number words through the first and second phonetic shifts. We shall draw on our tabulation of these shifts (p. 127), referring especially to the last column. An asterisk (*) in the following table refers to the discussion of the units. There we shall also see what can be plausibly said at present about the meanings of the number words themselves in the other Indo-European languages.

Relationship of the Germanic to the Indo-European Units

	Indo-European and Latin	>	Gothic and English	>	German	Phonetic Shift (p. 127)
1	*oi-nos, oi-qos* *sems, smiə, sem*	*unus*	*ains*	one	*eins*	—
2	*duuo, duo*	*duo*	*twai*	two	*zwei*	6
3	*trejes, trie*	*tres*	*þreis*	three	*drei*	3
4	*quetuōr*	*quattuor*	*fidwōr*	four	*vier*	*
5	*penque*	*quinque* (Greek *pénte, pémpe*)	*fimf*	five	*fünf*	2
6	*sueks, seks*	*sex*	*saihs*	six	*sechs*	1
7	*septm̥*	*septem*	*sibun*	seven	*sieben*	*
8	*oktou*	*octo*	*ahtaú*	eight	*acht*	1
9	*neun*	*novem*	*niun*	nine	*neun*	—

One. For "one" there are two primeval Indo-European forms:

1. *oi-nos*, which leads readily to the Latin, Celtic, and Germanic "one." In the word *oiné* the Greek language recognizes "the One on the die," which also exists in the dialect form of *oinós*, "wine"; the Lithuanian *w-ienas* belongs to the same category. The Slavic *jed-in* > Russian *od-in* may perhaps go back to an expanded Indo-European form *e-dhi-no*, the linguistic explanation of which would lead us too far away from our proper course. The Indo-European *oi-qos* > Greek *oîos*, "solitary, alone" is also related to the Sanskrit *e-ka*, "one." Other cognate words probably mean "the only, the very."

2. In contrast to this specific meaning, there is also the "collective" meaning of "one": Indo-European *sem*, from which the Sanskrit *sam* and the German *zu-sam-men*, "together," *sam-meln*, "to collect," and *säm-tlich*, "jointly, collectively," are derived. From the same root have come the Tocharian word *sa* and the Greek *heîs*, *(s)mía*, *hén*, "one," which is already familiar to us in *he-katón*, "100" and *há-pax*, "once" (see p. 25). If we recall that the Indo-European *s-* before a vowel becomes Greek *h-* (as in *septm̥* > *heptá*, "seven"; Indo-European *su*, Latin *sus* > Greek *hys*, "pig, sow"), we shall arrive at Greek *háma*, "at once, at the same time," *hamâsthai*, "to collect," *hómalla*, "sheaf, pile," and also *homos*, "the same, common, mutual," and *homoîos*, "like, similar" — that which is "one" with something else. In Latin we find *sim-ilis*, "similar," *sim-plex*, "simple" (cf. German *ein-fach*), *sin-guli*, "unique," *sem-el*, "once, one time," and *sem-per*, "once and for all, always"; examples in the Germanic group are Gothic *simle*, "one," Gothic *sama*, and Old High German *samo*, "the same," (cf. Russian *sam*, reflexive pronoun with the same meaning) Gothic, *samana*, Old High German *saman* and *zi-samane*, Modern German *zu-sam-men*, "together," as well as Gothic *sums*, "any," which reappeared in the English word *some*.

We have already discussed the connection between *sem* and the Latin *mille* (see p. 135); the Latin word *semester* has no relation to this (see p. 181).

Two. This word is clearly discernible through all Indo-European languages. We have already discussed at length the early, primitive ideas reflected in this word (see pp. 12 ff.).

Three. The possible connection of "three" with the notion of "beyond, trans-, on the other side" has also been mentioned (see p. 17). One curious feature worth noting here is that only Sanskrit and the Celtic languages have preserved an ancient feminine form for three and four: Indo-European *t(r)i-sores* and *quete-sores* > Sanskrit *tis-rah* and *čatasrah*, Celtic *teoir* and *cetheoira*, which are perhaps related to *sor*, "wife, woman." This 3 form may have taken its name from the "third wife," who followed the two principal wives; the feminine 4 form may have resulted from the remote influence of the feminine "three," if it did not itself originate in the same way. Analogously the "husband, man" appears to have entered into the masculine 4 form *čat-verah* (Sanskrit *virah*, Latin *vir*, Gothic *wair*, "man"; cf. Germanic *wer* as in *Wergold*, the fine imposed for killing a man).

Four. Derivation: the Gothic *fidwōr* calls for an Indo-European form *petwor* in place of *quetuor*, beginning with *p-* instead of *q-* (see Phonetic Shift I—2, table on p. 127). The initial consonant *f-* did not originate in a phonetic shift, however, but resulted from the backward influence of the initial *f-* in *fimf*, "five." Wherever, as in the number sequence, words commonly stand directly adjacent to one another and are eroded by constant repetition, such adaptation or assimilation of the sound or rhythm of the word often takes place. We have seen a beautiful example of this in the Celtic number sequence (p. 125); to mention still another case of phonetic assimilation: Lithuanian *devyni* and Slavic *dēvetj*, "nine," exchanged their initial consonant *n-* for *d-* as a result of the influence of the subsequent number words *dešimt* and *desetj*, "ten" (see p. 93).

In Greek, the Indo-European *qu* may become *k*, *p*, or, as in this case, *t: tettares*. But what is the underlying meaning of the Indo-European *quetuor*? Here we can only offer a conjecture. I shall suggest two new explanations, whose validity the reader must judge for himself.

The first of these bases the number word "four" on the "finger-tip sequence" of one hand (omitting the thumb). This hypothesis derives the Indo-European *quetuor* from a conjectural word *oke-to-uoro*. The latter contains the elements *ok-* (also *ak-*), "tip, peak" (as in the Greek *akris*, "mountain peak"; cf. Acro-polis, literally "city on the heights," and Latin *acer*, "sharp"); *-to-*, the same syllable that occurs in *km-to(m)*, "hundred," with the sense of "sequence, series" as in *oketom*, "row of tips"; and the final syllable, which became the Sanskrit *vara*, "row," intensifies the conception of a sequence. In the case of "four," *quetuor*, the initial syllable *o-* was dropped. In "eight" it was retained, and at the same time the word took the dual form, *oktou*, to express the idea of "two times four finger tips."

It is hard to say how far this linguistic hypothesis can be pushed. Internal evidence indicates that it may well be true, especially since it provides a plausible explanation of the Indo-European dual form *oktou*; the fingers, then the hand without the thumb, which have caused the sharp break after "four," and, finally, the two hands together have, as we have shown, played a decisive role in developing the number sequence. The hypothesis also finds some support in the Indian Kharosthi numerals, of which the 8 has the form of a doubled 4 (see Fig. 12); the Chinese character for the number 8 may also have originated as an ideogram representing two 4-fingered hands (see Fig. 13).

The other interpretation proceeds from the cross with four arms or "tips," probably the most descriptive symbol for the number 4. Latin *tri-quetrus* means "three-cornered," the idea of "peak" or "corner" being embodied in the *quetrum* and derived ultimately from the Indo-European *quetuor*. From *quetrum* (not from *quattuor*) came the Latin word *quadra*, "square, cross"; in this word the old meaning again comes to the surface. "Eight," which is grammatically a dual, seems to have originated in a "doubling" of the cross, in that another arm or beam was inserted perpendicular to each of the four arms, thus forming the hooked cross or swastika that has been used as a primitive symbol by many unrelated cultures.

Now we do sense that for primitive man, when he wanted to count to four, such a symbol as a model for a number word was far too remote, even if it did exist in his environment. Quite aside from this, the swastika is never felt as a "doubling" (dual) of the simple, four-armed cross.

Both these hypotheses connect the number word "four" with "point," or "tip." The German language contains a surprising expression of this connection between "four" and "point": the Old High German word *ort* did not mean "place," as it does in Modern German, but "point" or "corner"; in many regions of Germany the sharp point of a cobbler's awl is still called *Ort*. *Ort wider orte*, "tip against tip," "point against point," says Hildebrand's Song, and Albrecht Dürer once remarked that an artist should draw in "*die allerkleinsten Rünzelein und Ertlein*," the most minute wrinkles and corners." But how is this word related to "four"? Numismatics designates a coin divided by two perpendicular snips as being in *vier Orte*, "four sharp wedges" ⊕, or four "corners"; hence, the quarter-thaler was also called the *Ortstaler*, and the quarter-guilder also an *Örtel* or an *Eckele*. From the domain of coinage, the *Ort* came to be used for a quarter unit in measures and weights; the Dutch word *oord* is a "quarter measure." It is also quite significant that the odd, or "uneven," number was felt to be a "point": *En Stier hab i örti*, says the Bavarian peasant, "I have one 'point' ox," meaning that he has one ox in addition to the even-numbered yoke.

Five. The Germanic word "five" should, since it derives from the Indo-European *penque*, have once had a laryngeal as a final consonant; this is still actually present in the Swabian dialect form *fuch-zk*. Today it has assimilated the initial *f*-, just as, to cite a parallel example, the medial *k*-sound in the Latin *quinque* has transformed the initial *p*- in its Indo-European ancestor to a *k*- (*qu*-) in Latin.

From our previous discussion we know the role which the "5 fingers" or "hand" played in the arrangement of the number sequence. Thus the number word "five" has always been associated with these two images. The Gothic word *fimf* is probably connected with the Gothic *figgrs* (pronounced "fingrs") "finger," just as the Slavic *pētj* is related to the Slavic *pēsti*, "fist." It is also certain that the Egyptian "five" is the same as their word for "hand"; the same thing is true of Austronesian and many other languages of primitive peoples (see p. 52).

Now we come to something quite remarkable. If we compare the numbers 5 and 4 in the *Kentum* and *Satem* languages (see tables on pp. 92–93) — Sanskrit *panča* — *čatvaras*, Greek *pénte* — *téttares*, Latin *quinque* — *quattuor*, Lithuanian *penki* — *keturi*, Irish *coic* — *cethir*, Welsh *pimp* — *petwar* we see that the word for 4 invariably begins with the final syllable of 5. Thus it seems that some root R attached itself (like the Sanskrit *-ča-*) in front of a syllable s in the word for 5 and after a syllable s' in the word for 4; thus, diagrammatically, $5 = s + R$ and $4 = R + s'$. Dare we see more than mere coincidence in this pattern, which extends through virtually all the Indo-European languages? Just what does the root R mean? Perhaps "hand," so that 5 could mean something like "the whole hand" and 4 a "hand less 1"; or else 5 could be "1 more than a (four-fingered) hand" and 4 a "hand less the thumb." Such formations appear elsewhere in the number sequence (cf. eleven, *ein-lif*: pp. 83 and 157). At any rate, one could hardly find a better explanation for this peculiarity, any more than for the interpretation of the Indo-European *penque*, which must likewise have been some actual thing, as "hand and 1," taking the postpositive *-que* to mean "and," as it does in Latin.

Six and Seven. On the derivation of "seven": According to the rules of Indo-European linguistics, the Latin *septem* should have led to *seftem* or, by dropping the *t* between *f* and *n*, to *sifun* instead of *sibun*, And, indeed, we do have the Anglo-Saxon words *seofon*, English *seven*, and Dutch *zeven*.

The resemblance of these two number words to the corresponding words in the Semitic family is striking: original Semitic *šeššu* > Assyrian *šišša*, Hebrew *šeš*; cf. Sanskrit *šaš*, 6; and Semitic *sabu* > Hebrew *šeba'*, Assyrian *sibi*, Arabic *sabun* (cf. Sanskrit *sapta*, 7). Whether this similarity is purely accidental or the result of borrowing (in which case the words must have gone from the Indo-European into the Semitic languages) is hard to tell.

Eight. This number has already been discussed in connection with "four."

Nine. We have mentioned the connection between this word and "new," as well as the interpretation that one "new" number has been added to the two 4 groups of both hands (not counting thumbs) (see p. 23). Ancient Egypt provides some support for this postulated connection between "nine" and "new": the root of the Egyptian word for "nine" is also used to indicate the rising of the sun in the east and the first appearance of the new moon. See p. 170 for a different interpretation of "seven," "eight," and "nine," which may explain the break that exists in the Indo-European sequence of tens.

The Tens

Greek *-konta* and Latin *-ginta*. The Indo-European languages form the tens by coupling the units with the first numerical rank, "ten," usually not in the multiplicatory form "three times ten" which we might expect but rather according to the old grouping, "three tens": Greek *triá-konta*, Latin *tri-ginta*, Slavic *tri desēti* (see p. 126).

The Greek *-konta*, "ten" or "ten-ness." The verbal device used to form the tens (except for twenty), is derived from the Indo-European *-komta*, an expansion of the *dekm̥(t)*, "ten" which in turn comes from *(d)km̥t*. This contraction was still present in the Doric word *Fi-kat-i*, 20 (Indo-European *m̥* > Greek *a*; see p. 126). The final *-i* indicates the grammatical dual; the first syllable prefix *Fi*, Latin *vi-* < Indo-European *dui̯*, *u̯i* is a dual formation of "two," *duo*. The number word for 20, corresponding to the Latin *vi-gint-i*, is thus picturesquely built up from a threefold two form, testifying to the special status and great age of the word for 20. The archaic Greek letter *F*, pronounced "w," is called di-gamma, or double *Γ*, because of its shape. We also find the prefix *u̯i-* in the Attic dialect word for 20, *(w)ei-kosi*, cf. Sanskrit *vim-satih* and Irish *fi-che*, as well as in the Gothic dual *wit*, "we two, both of us" (see p. 13); in the Latin word *di-vi-dere* it appears in the divisive sense of two. For the erosion and elision of the tens in modern Greek, see p. 125.

A curious aspect of linguistic history is the derivation of the Latin *-ginta* from the Latin *konta*: The Etruscans used only the hard final consonants *k*, *p*, and *t*. But after they adopted the Greek alphabet for writing their language, the soft gamma *Γ* became the hard sound *k* in Etruscan. And the Romans, who in turn took their letters from the Etruscans, changed the *Γ* to ⟨ and finally to C, which stood for both the hard *k* and the soft *g* sound. The abbreviations C., for Gaius and Cn., for Gnaeus, and also the interchangeable forms *vi-c-esimus* and *vi-g-esimus* for "twentieth," all document this phonetic ambiguity, which was finally ended in the 3rd century B.C., by the addition of one line, turning the C into G for the representation of the *g*-sound. As a result of this early double meaning of C, however, the 20 form *-ginti* arose from *-konti*. In addition, the *-o-* bracketed by *i* on both sides in *vi-konti* was changed to *i*, so that the form *viginti* finally emerged. All subsequent tens from 20 on have the ending *-ginta*, resulting from the influence of the *-ginti* (the *-a-* sandwiched into the words for 50, 60, and 70 is introduced by analogy with *quadra-ginta*).

German *-zig*. What is the explanation of the ten forms *vier-zig*, *fünf-zig*, etc. in German? The suffix *-zig* has the sense and form of a "ten" ending just like those discussed in the preceding section.

In the tabulation of Germanic and German number words (see p. 96) this ending can be traced back as follows:

German *vier-zig* < Old High German *fior-zug* < Gothic *fidwor-tigjus*. The Gothic *tigjus* is the plural form of *tigus*, "ten, ten-ness," an auxiliary form of *taíhun* (like the Greek *dekás* and *déka*, "decade" and "ten") which is encountered in all the Germanic languages: Old Norse *-tiger*, Old Saxon and Anglo-Saxon *-tig*, English *-ty*, Icelandic

-tiu, Swedish *-tio* and Danish *-tyve*. The German phonetic shift altered *-tig* to *-zug* and later to Middle High German *-zec* — the Nibelungenlied says *drizec tusend* — and finally to *-zeg* and *-zig*.

The force of the noun "ten" can be seen in Old High German from the genitive case that follows it: *feorzug wehhono*, "40 of the weeks" (*wehha*, "week") and *zehenzog scafo*, "ten-ty (a hundred) of the sheep" (*scaf* = sheep).

The Germanic and the German "twenty" and the Balto-Slavic forms of this word are remarkable among the Indo-European languages for their regularity (see tables on pp. 92–96)! The Germanic form is simply *twai tigjus*, "2 tens" and thus not a grammatical dual, to say nothing of a doubled word form like *viginti*; the only exception is the Old Norse *tuttugu*.

The Romance languages derived their words for 20 from Latin, although Rumanian did not, probably influenced by Slavic proximity: "twenty" in Rumanian is *doau zeci* (see table on p. 99).

In looking through the old Germanic forms of the tens, we discover a striking break after 60, as in the Gothic *saihs-tigjus* but 70 *sibunt-ehund* (see p. 94). The reader may be surprised to find that this break after 60 occurs in many of the *Kentum* and *Satem* languages (tables on pp. 92–93). The explanation cannot be found within the Indo-European family alone. We shall learn far more if we extend our research from the narrow sphere of historical linguistics into other domains that appear relevant to this problem — to the "great hundred" and to the special meaning of 12 and 60 in cultural history generally.

First, however, let us pause for a moment, for we have reached the end of the road in our search for enlightenment about number words. What some readers may have hoped for, namely a revelation of the original formations of the number words, has not been realized. But we did uncover, sometimes to our amazement, the pulsating evolution of the number sequence, which sometimes forged ahead vigorously and at other times remained stagnated at a conventional boundary. We have seen a constant shuttling to and fro as the loom of cultural history wove the tapestry of the number sequence. And whenever we carefully pulled one thread or another out of the fabric, in order to trace its convolutions back to the very beginning, we were rewarded with profound glimpses into the early condition of the human mind.

Now, however, let us once again emphasize the two most important conclusions to be drawn from our discussion:

1) The Indo-European number sequence built up around the number 10.
2) The extraordinary interweaving of our own number sequence with those of the other Indo-European languages.

Was There a Babylonian Influence on Our Number Sequence?

THE BREAK AFTER SIXTY

This is too obvious to be overlooked. Indeed, our comparative tables show that it is present in all Germanic number sequences (pp. 94–96):

	Anglo-Saxon	Old Saxon	Gothic	Old High German
50	*fif-tig*	*fif-tig*	*fimf-tigjus*	*finf-zug*
60	*six-*	*sehs-*	*saihs-*	*sehs-*
70	*hund-seofontíg*	*ant-sibunda*	*sibunt-e-hund*	*sibon-zo*
80	*-ahtatíg*	*-ahtoda*	*ahtaút-*	*ahto-zo*

It is also found in other Indo-European number sequences (pp. 92–93):

	Greek	Latin	Celtic	Tocharian	Sanskrit
50	*penté-konta*	*quinqua-ginta*	*coi-ca*	*pñ-ak*	*panča-śat*
60	*hexé-*	*sexa-*	*ses-*	*säks-äk*	*šaš-tih*
70	*hebdomé-*	*septua-*	*secht-mogo*	*sapt-uk*	*sapta-*
80	*ogdoé-*	*octo-*	*ocht-moga*	*okt-*	*aši-*

In Gothic, 60 is called "six tens," but then one finds "the sevenfold, (or 'seven-ness'), the eightfold, the ninefold, and the tenfold of tens" (cf. pp. 150 and 129). The Old High German *-zo* was probably derived from the Gothic *-te-(hund)* after the *-hund* had been dropped. Still more peculiar are the Germanic formations, in which *hund-* or *ant-*, with the original Indo-European meaning of "ten," is placed in front of the number word. Thus *hund-seofon-tig*, for example, means 10-7-10! This is also the explanation for the hitherto incomprehensible Dutch word for 80: *tachtig* < *(an)tachtig*.

The break disappears in the more modern languages, in which the simpler word forms for the tens up to 60 have displaced those for higher numbers (another example of remote influence). Old Norse, to be sure, also lacks this break (or no longer has it? see p. 94) in its number sequence, but the expression "so and so many decades old" is counted in this manner: first from 20 *tvi-tögr*... up to 60 *six-tögr*, but then 70 *sjau-tögr* as well as *sjau-raedr*, 80 *atta-raedr*, ... up to 120 *tolfraedr* (for *raedr*, Gothic *raþjo*, see p. 130). Here the break is immediately apparent.

In Greek, Latin, and Celtic the units in the ten forms from 60 on occur in their ordinal and not their cardinal forms: instead of *hepté--konta*, "7 tens," we have *hebdomé-konta*, "the 7th tens," so that the plural form *-konta* is nonsensical. Its presence can be explained only by analogy with the earlier tens, for the logical form in Greek would be *hebdomé-kas*, "the 7th ten." Even if, as philologists have recently come to believe, the Greek forms from 70 on were not derived from the ordinal forms of the units but from some other phonetic rule, the break after 60 — which is what concerns us here — still remains.

Latin *septua* < Old Latin *septuma* > later *septima,* "seventh," is formed like *decimus* < *decumus,* "tenth"; the *porta decumana,* the main gate of a Roman army encampment, was the point at which the tenth cohort of the legion was always stationed. Similarly *octua* and *nona* < *novena.*

Tocharian also shows a change in the ten forms, *-ak* > *-uk,* in going from 60 to 70.

It is interesting that in Sanskrit the break occurs after 50. Perhaps this arose in going from one hand to the other, if the tens were counted off on the fingers.

In Slavic and Lithuanian, however, we find no change after 60; the break after 40 in these languages has already been mentioned (see p. 23). This, too, is an indication that the break after 60 dates from a very early time, for the Slavic number words are relatively young, as we gathered from borrowings and from the Lithuanian *lika-* formations (see p. 84). Since there can be no doubt at all about the existence of the break at 60 in many or most of the Indo-European languages, is there an explanation for it?

THE ROLE PLAYED BY THE NUMBER SIXTY IN CULTURAL HISTORY

In Greece both 60 and its multiples such as 360 were frequently used as "round numbers" — that is, numbers whose specific meanings are inflated into the indefinite "many," as in the expression, "If I've told you once, I've told you 'a hundred times' that ...." The inflation or semantic fading of a number word is the opposite of the specification of its meaning and usually involves the three ranks 10, 100, and 1000, but it can also happen to any other number which for one reason or another has acquired special significance (as in *seine seiben Sachen packen,* "packing up his seven things").

The Greek historian Herodotus says that Darius, the Persian King, had 60 knots tied in a thong and then ordered the Ionian princes to untie one knot each day (*History* IV, 98); they were to guard the bridge of ships over the Istros River in Darius' absence until the last knot had been untied. Xerxes in fury commanded the Hellespont to be flogged with 300 lashes, because of the destruction the waters had caused in a storm. The oldest known Greek example of the significance of this number is provided by Eumaeus' herd of swine (*Odyssey* 14, 20):

"The swineherd always sent them (Penelope's suitors) the best one of the fatted pigs for them to feast on; and the number of swine remaining was only three hundred and sixty."

Among the Romans, *sexaginta* and especially its tenth multiple *sescenti,* 600, were commonly used as round numbers meaning "many."

The Old High German Song of Hildebrand also once uses 60 as a round number:

Ih wallota sumaro enti wintro sehstic ur lante,

"I spent sixty summers and winters (= many years) outside the land."

In the *Annolied*, which dates from the beginning of the 12th century, wo find the round number 300:

Romere scrivin cisamin	"The Romans wrote down together
in einer guldiner tavelin	upon a golden tablet
driu hunderit altheirin	three hundred old men
die dir plegin zuht und erin.	who render thee homage and honor."

Even to the present day, however, sixty is sometimes used exactly and at other times to mean "a heap" or some indefinite large number. In this latter sense we can make out the original use of *Schock* (a German word meaning a large lot of sixty units) exclusively for "60 sheaves" of grain and hence for a pile of sheaves (cf. the archaic English word *shock* = a pile of sixty sheaves of grain): *tein scok gavarno* reads an Old High German passage dating from the 9th century. In the Swabian dialect *schocken* still means "to pile up grain"; in the dialect of Hesse a building's walls *schucken*, "are made up of," bricks. As a measure of quantity,

1 *Schock* = 60 pices = 3 score = 4 *Mandel* = 5 dozen;

1 *Großschock* ("great shock") = 64 pieces (a shock with an "excess").

In northern Europe the doubled sixty was even more frequently used, and was known as the great hundred.

GREAT HUNDRED

In his novel "The Vagabond," the Norwegian writer Knut Hamson says: "They counted a so-called great hundred for every 120 fishes." This term still measures quantities of fish, even in Germany, where it is used for other things as well. In Lübeck there was a *Hundert Bretter* = 120 items = 10 *Zwölfter* ("twelves") or, as they said in Mecklenburg, 10 *Tult*. English differentiates between the *long hundred* of 120 units and the *short hundred* which has only 100. Today, however, we recognize "hundred" only in its meaning of 100.

The northern Germanic region, primarily Iceland, was the home of the great hundred; there *hundraþ* meant 120 in all monetary calculations and in designating military units, until the introduction of Christianity around the year 1000. Thereafter it came to stand for the small hundred of 100 in ecclesiastical and learned writings (see p. 79), and the two hundreds were sometimes, but not always, distinguished as the 10-(*ti-roed*) or the 12-(*tolf-roed*). A writer around the year 1250 once designated the 360 days of the year in the old manner:

III^c daga tolfroed — "three hundred days by the 12-count,"

and commented: "Nevertheless in the book language (Latin) all the hundreds are reckoned by the ti-count (*oll hundraþ tiroed*), which according to the proper count is *III^c tiroed ok LX daga*."

We may well ask why the writer did not simply use the 100 form corresponding to *tio tiger*, "10 tens" — in this case, "36 tens" — which our number sequence provides (see p. 94)? Because it means ten tens in a 12-count great hundred enumeration, but not 100 as such. We can understand this better if we read from an Old Norse tax roll:

"Whoever has property worth 1 ten shall pay 1 ell of *wadmal*;
"Whoever has property worth 2 tens ... etc.; then
"Whoever has property worth *halft-hundraþ* (6 tens) shall pay 4 ells...
"Whoever has property worth *tio-tiger* (= 10 tens) shall pay 6 ells."

Thus *tio tiger* meant 10 tens in the 12-count hundred system of counting, not an "independent" 100; *tio tiger ok þriu hunderaþ* is 460 = 10'10 + 3'120.

The *wadmal* in the foregoing list is baize or frieze cloth, which was also used as a medium of exchange or "money," in addition to cattle and, later, silver, because the poorer man who had no cattle or silver could weave the cloth himself, and it was easy to divide. The standard of values was originally

1 *hundraþ silfrs* = 120 ounces of minted silver = 2400 ells of frieze-cloth,

whence 1 ounce of minted silver = 20 ells of frieze-cloth,

an equivalence in which we see both the great hundred and the old vigesimal grouping.

The Gothic Bible, however, has only the 10-count *hund* for multiples of 100, perhaps because of the Greek from which it was translated; nevertheless, it must have been in use among the people. Luke 7:41 reads:

There was a certain creditor which had two debtors: the one owed five hundred pence, and the other fifty.

Here the 500 is translated as *fimf hunda*, but the very same number in I Corinthians 15:6 has:

After that, he was seen of above five hundred brethren at once,

expressed in Gothic as *fimf hundam taíhun-tewjan broþre* — that is, "5 hundred by the ten-count"; the Gothic word *tewa* means "series, order." Thus the great hundred was apparently commonly used in popular speech, and was later displaced by the "Christian" 10-count hundred.

Next we come to the great thousand, which contains 1200 units; it has "10 hundreds, but the hundred contains 6 score or 2 shocks," says an arithmetic text of the year 1706. As a "tun," a cask measure amounting to 1200 liters, it is still in use today.

Now we can ask the question that we have wanted to voice ever since we first observed all these bits of evidence: How did the great hundred come into being? Is it made up of 2 shocks ($120 = 2 \times 60$), or does the ancient vigesimal grouping play some part in it ($120 = 6 \times 20$), or was it originally made up of twelves ($120 = 10 \times 12$)? The evidence we have clearly points to the latter.

THE NUMBER TWELVE AS THE BASIC UNIT OF THE GREAT HUNDRED

As we just noted, Old Norse distinguished between hundreds made up of ten tens and those of ten twelves; in Anglo-Saxon 120 was expressed as *hundtwelftig*. The word *tylft*, the "twelft," to which the Mecklenburger term *Tult* is related, is the most widespread small measure of quantity: *tvitylft* = 24, *þrennartylft* = 36 items. "Fifteen years old" in Norse poetry (the 12th-century *Heimskringla*) is:

gammal vetra tolf ok þriggja, "12 and 3 winters old"

Charlemagne's monetary standard of the year 780, which had a lasting influence on medieval European coinage, clearly embodied the basic 12-unit:

(Latin) 1 *libra* or *talentum* = 20 *solidus* = 12 *denarius*,
         1 pound of 20 shillings each of 12 pennies = 240 pennies,
or   1 pound = 8 long (Bavarian) shillings of 30 pennies each
         = 240 pennies,
but also: 1 pound = 12 ounces of 20 pennies each = 240 pennies.

In France the table of equivalents was:

1 *livre* of 20 *sou* (< *solidus*) of 12 *denier* (< *denarius*) each = 240 *denier*, whence the *sou* also came to be called a *douzain*, a "twelver."

This old standard of monetary units is still in force in English currency. The shilling as an old measure of 12 units also derives from it. Luther said in one of his *Tischreden*,

Wenn gott sihet, dass wir undanckbar sind, so lässet er uns durch den teuffel einen guten schilling geben,
"If God sees that we are not thankful, he lets the Devil give us a good shilling,"

a colloquial expression for a drubbing of 12 or even 30 blows. In the dialect of Silesia the word *Schilling* has been shortened to *Schilg*, as in *ein Schilg Eier*, "a dozen eggs," and has been inflated to the indefinite meaning of "many" in the expression *schilgemal* (literally: *Schilg* times means a dozen times).

The various fines and penalties imposed under old Germanic law refer to a basic number: in the Alemannian, Bavarian, Friesian, Saxon, and Burgundian tribes this was 12, in Frankish law it was usually 10, and among the Lombards it was 12 for inflicting wounds or injuries and 10 for other infractions. In the *Lex ripuaria*, the code of the Riparian Franks, a stallion, a coat of mail, and a hunting falcon were valued at 12 *solidi*; a helmet at half this amount, or 6 *solidi*; a sword with its scabbard at one third, or 4 *solidi*; a cow, a mare, and a sword without scabbard at one fourth, 3 *solidi*; an ox, and a shield and lance at one sixth, or 2 *solidi*.

In the *Lex salica*, the law code of the Salian Franks, a fine was once specified thus:

unum tualepti | sunt denari CXX | culpabilis iudicetur,
"the guilty one is sentenced to pay a Twelve, that is 120 pennies"

— a document which again bears witness to the number 12 as the basic unit of the great hundred.

But why precisely 12? The importance of this number in the daily lives of common people, in commercial transactions, and in legal affairs is probably due to its easy divisibility in so many ways (see the schedule of fines of the *Lex ripuaria* on the preceding page). The commonly used fractions of the *tylft* or the shilling could all be expressed in terms of whole numbers of pennies:

1	$\frac{1}{2}$	$\frac{1}{3}$	$\frac{1}{4}$	$\frac{3}{4}$	$\frac{2}{3}$	of a shilling
12	6	4	3	9	8	pennies.

For this reason the north European *tylft* is an original, native measure, and not one that was first brought in by way of the Carolingian coinage system. For, quite apart from its ready divisibility, it was also consistent with the Roman pattern in the table of ounces.

As a measure of quantity we have retained the Twelve in our "dozen" (< French *douzaine* < *douze*, 12) and in our "gross" of 12 dozens (< French *grosse douzaine*, "strong dozen"). The *douzain* was actually another term for the French shilling, the *sou*, and thus came into common usage (see p. 156). In German commercial terminology, to be sure, these words are fairly recent acquisitions. The former made its first appearance in Alsace as *totzen* in the 14th century, while the gross migrated from France through the Low Countries into Germany in the 16th century.

In addition to being readily divisible into so many different fractions, we must not overlook another reason for the universality of the number 12 as a measure of quantity — namely the "excess," which was used by many tribes in the north, primarily in legal writings, in speaking of time limits set by law. It is still used today in commercial dealings. The 1001 "Arabian Nights" give a very clear idea of the "excess," as do such common expressions as the German *acht Tage* for a week, the French *quinze jours* for two weeks, and the English "baker's dozen" for 13.

A term set by law is considered to have elapsed when the period of time following it has been entered upon; thus a portion of the latter was added to fill out the legal term. The Germanic peoples counted time by nights, and therefore defined the legal term of a week as "7 nights and a day," from which comes the week's reckoning as eight days. The expression "after a year and a day" also gives the day as an excess. "For one hundred years and a day" was the old legal formula for eternal banishment.

In complimentary salutations an "excess" is always added on (20 + 1 or 100 + 1), and a "baker's dozen" is actually 13. Examples of this kind are easy to find. But we already have enough to illustrate the customary use of the excess, as a result of which not ten, but twelve as "ten with an excess" may have come into common use. Did this perhaps even give rise to the true Germanic word-forms *ein-lif* for 11 and *zwei-lif* for 12, in the sense of "1 (or 2) left" or "left over"? When we find that the Icelandic legislature, as we know from documents dating from around the year 1000, was made up of

"12 men from each district" — *tylptir manna* —

the number 12 here does not indicate a pointless choice of a readily divisible number, but reflects the customary legal usage of the excess. It is certainly not the "mystical" number 12 (standing for the

Apostles), which could hardly have come into use in Iceland before the advent of Christianity. Now we can also understand the great hundred in terms of 10 such twelves, wherein the excess of 20 over the 100, as the great hundred can also be interpreted, was fully consonant with the sensitivities of the people.

And the *shock* — is it half a great hundred? If we interpret it thus we shall never make sense of it, and it will appear inconsistent with northern European habits of thought; it is more likely to have been a concept of five twelves (or 50 sheaves with an excess of 10).

Is the *shock* borrowed from the Graeco-Babylonian *sossos*, 60? The similarity in both sound and meaning is surprising. To form an opinion on this, we must examine the Babylonian sexagesimal system, one of the most peculiar systems of numbers in history. But before we set out on this path, let us complete our study of the number 12 with the following discussion.

THE ROMAN DUODECIMAL FRACTIONS

A few pages back these offered us some fascinating insights into the primitive state of mind in which men incorporated numbers verbally and mentally into their ordinary lives.

The Roman fractions were originally based on a system of equivalent weights: 1 *as* (or pound) = 12 *unciae* (ounces). Linguistically, *as* is the same word as the Greek *heís*, with the dialect pronunciation *'âs, âs,* "one"; the Latin *uncia* is "unity" < *unus,* "one." Thus a large unit was divided into smaller ones, as in any system of measures. Since money was originally weighed out for transactions, the *as* was an amount of copper in the Roman system of coinage. Its subdivisions were the *unciae* or "ounces." Our illustration (see Fig. 32) shows a quarter *as,* a *quadrans,* valued at 3 ounces. Verbally it has the value of a fraction, but mathematically it is a whole number, as expressed in the three embossed circles at the bottom of the coin. The underlying tendency of all subdivisions of measures is to avoid fractions and to express them instead as whole numbers of smaller units. These fractional measures gradually become separated from their content of weight or quantity and, in terms of the abstract number sequence, become "pure" fractions: a *heres ex quadrante,* for example, is an heir who inherits one quarter of the estate.

In their computations the Romans used no fractions other than these duodecimal fractions (plus a few fractional subdivisions of the *uncia*). If a fraction like $\frac{5}{9}$, for example, turned up in their computations (or in those of the Middle Ages, which adopted the Roman fractions), they either made do with the nearest approximation in terms of a duodecimal fraction:

$$\frac{5}{9} = \frac{20}{36} = \frac{20}{3} \times \frac{1}{12} = \frac{20}{3} \ unciae = 6\frac{2}{3} \ unciae \ \sim \ 7 \ unciae \ (septunx),$$

or else they laboriously translated it into its exact duodecimal fractional equivalent:

$$(6 + \tfrac{2}{3}) \ unciae = \tfrac{1}{2} \ as + 16 \ scruples = 150 \ scruples.$$

Now and then they also expressed a fraction in the Egyptian fashion, as the sum of certain standard fractions. Pliny the Elder, for example,

Fig. 32 A *quadrans* issued by a city in Italy: $\frac{1}{4} = \frac{3}{12}$ *as* = 3 *unciae*. The *quadrans,* or "quarter *as,*" was verbally a fraction but computationally in terms of its monetary value a whole number (3 ounces). Diameter 4.5 cm. Municipal Numismatic Collection, Munich.

in estimating that Europe's area was something more than $\frac{11}{24} = \frac{1}{3} + \frac{1}{8}$ that of the whole world, said (*Natural History* VI, 12):

totius terrae tertiam esse partem et octavam paulo amplius —
"somewhat more than the third and the eighth part of the whole earth."

Thus out of the infinite realm of fractions with all possible denominators, these duodecimal fractions that were tied to measures made up only a very scanty sequence, thus constricting all abstract fractional calculations and causing the computer a great deal of misery whenever he "came up against fractions." From these examples we can readily understand how the introduction of Indian numerals and their method of writing fractions really rescued medieval arithmeticians from a nightmare. But for our present purposes, these peculiar Roman computations with fractions are merely one more instance of the difficulties and complications confronting man throughout cultural history in his search for the key to the proper understanding of numbers.

In the table immediately following, the fractions are shown in their simplest form in column 1, the Roman duodecimal fractions with their denominator in ounces in column 2, their Latin names in column 3, translations of the latter in column 4, the Roman symbols for the various fractions in column 5, and the numerical meanings of these symbols in column 6.

The Roman Duodecimal Fractions

1	2	3	4	5	6
	$\frac{12}{12}$	*as*		I	
	$\frac{11}{12}$	*deunx* (< *de uncia*)	—1 ounce	S $\therefore$	$\frac{1}{2} + 5$
$\frac{5}{6}$	$\frac{10}{12}$	*dextans* (< *de sextans*)	—$\frac{1}{6}$ *as*	S....	$\frac{1}{2} + 4$
		decunx (< *decem unciae*)	10 ounces		
		semis et triens	$\frac{1}{2} + \frac{1}{3}$		
$\frac{3}{4}$	$\frac{9}{12}$	*dodrans* (< *de quadrans*)	—$\frac{1}{4}$ (*as*)	S...	$\frac{1}{2} + 3$
$\frac{2}{3}$	$\frac{8}{12}$	*bes* (< *duo partes assis*)	2 partial *as*	S..	$\frac{1}{2} + 2$
	$\frac{7}{12}$	*septunx*	7 ounces	S.	$\frac{1}{2} + 1$
$\frac{1}{2}$	$\frac{6}{12}$	*semis*	$\frac{1}{2}$ (*as*)	S	$\frac{1}{2}$
	$\frac{5}{12}$	*quincunx*	5 ounces	.....	5
$\frac{1}{3}$	$\frac{4}{12}$	*triens*		....	4
$\frac{1}{4}$	$\frac{3}{12}$	*quadrans* (B 169)		...	3
	$\frac{2}{12}$	*sextans*		..	2
	$\frac{1}{12}$	*uncia*		.	1
			$\frac{1}{3}$ (*as*)		
$\frac{1}{4}$	$\frac{3}{12}$	*quadrans* (see Fig. 32, p. 158)	$\frac{1}{4}$ (*as*)		
			$\frac{1}{6}$ (*as*)		
			1 ounce		

Later on the *uncia* or "ounce" was subdivided in common popular usage into $\frac{1}{2}, \frac{1}{3}, \ldots \frac{1}{24}$ ounces = 1 scruple, and the latter likewise into $\frac{1}{2}$, $\frac{1}{4}$, $\frac{1}{6}$, and $\frac{1}{8}$ scruple $= 1$ *calcus*, resulting finally in the following table of fractions:

As	1	$1\frac{1}{12}$	...		$\frac{1}{12}$										$\frac{1}{288}$			$\frac{1}{2304}$
Ounce	12	11	...		1	$\frac{1}{2}$	$\frac{1}{3}$	$\frac{1}{4}$	$\frac{1}{6}$	$\frac{1}{8}$	$\frac{1}{24}$							
		Scruple		24	12	8	6	4	3	1	$\frac{1}{2}$	$\frac{1}{4}$	$\frac{1}{6}$	$\frac{1}{8}$			
									Calcus	8	4	2	x	1			

The peculiarity that $x = \frac{1}{6}$ scruple can be explained as $\frac{1}{3}$ of $\frac{1}{2}$ of a scruple ($= \frac{1}{3}$ *obolus*). From the table of fractional values above we see, for example, that

1 scruple $= \frac{1}{24}$ ounce $= \frac{1}{288}$ *as* $= 8$ *calcus*, and
1 *calcus* $= \frac{1}{8}$ scruple $= \frac{1}{2304}$ *as*, the last and smallest unit fraction.

In Roman times calculations were generally performed only as far as scruples (the word comes from the Latin *scrip-* or *scrupulum*, "a very small sharp stone" < *scrupus*, "a small sharp stone"; as in English, the Latin word also has the additional connotation of "hesitation, doubt"). The specimens of the Roman counting board or hand abacus which have been preserved have columns only for the half, third, and quarter ounce (see *hand abacus* in the Index on p. 469). It is significant that the later subdivisions of the scruple have names derived from the Greek (*obolus* for $\frac{1}{2}$, *cerates* for $\frac{1}{4}$, whence our word carat, and *chalcus* for $\frac{1}{8}$ scruple).

Such were the Roman fractions, the *minutiae* ("small quantities"). Evidently the uncial fractions or ounces were the chief fractions, the *minutiae usitatae* ("useful or practical small quantities"), which were distinguished in the early Middle Ages from the abstract computational fractions, the *minutiae intellectuales* ("theoretical small quantities").

The system of fractions based on the ounce makes good practical sense. It counts the *unciae* and groups them together, 6 to every *S*, which is the abbreviation for *semis* < *semi-as* (see column 5 above); it is linguistically significant, incidentally, that the root of the word *semis* is related not to "two" but to "break." The names for the fractions are quite original (columns 3 and 4 above). There were three whole numbers of ounces (10, 7, and 5 ounces), four fractional parts of the *as* ($\frac{1}{6}$, $\frac{1}{4}$, $\frac{1}{3}$, and $\frac{1}{2}$ *as*), three subtractions from the whole (-1, $-\frac{1}{4}$, and $-\frac{1}{6}$ *as*), a very ancient sum of standard fractions ($\frac{1}{2} + \frac{1}{3}$) and the equally ancient formation *bes*, which we examined earlier (see p. 79).

In the Middle Ages one more symbol was added to the Roman *S* (6 ounces): an *S* with an additional horizontal leftward stroke at the top for 2 ounces (the *sextula*), which, with a stroke through the middle, became a symbol for a single ounce. The final result was the following sequence from 1 to 11 ounces, in which 11 ounces was written as $6 + 2 + 2 + 1$ ounces:

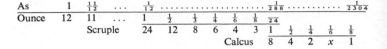

Fig. 33 Medieval Sequence of Ounces.

Since we shall come upon medieval uncial fractions again in discussing the counting board, we cite them here:

Fractions of Ounces

	Scruples	Ounces	Names	Symbols
1	12	$\frac{1}{2}$	*semuncia*, "half-ounce" (*semis-uncia*)	ʃ
2	8	$\frac{1}{3}$	*duella*, "2 sixths" (*duo-sextuale*)	‿‿,2
3	6	$\frac{1}{4}$	*sicilius*, "sicel" (named after the symbol)	)
4	4	$\frac{1}{6}$	*sextula*, "1 sixth"	‿
5	2	$\frac{1}{12}$	*dimidia sextula*, "half a sixth"	Ψ
6	1	$\frac{1}{24}$	*scrupulum*, "very small stone"	℈

The symbols for these fractions may be explained as follows: symbol (1) was the initial letter of the number word, but to differentiate it from *semis*, meaning $\frac{1}{2}$ *as*, it acquired a horizontal stroke, which originally cut the *S* in half; for the same reason symbol (6) was a double *S* with a slanting stroke drawn through both. Symbol (2) was computationally a double symbol (4); the remarkable form, 2, is actually two strokes, one beneath the other, which merged into one (our own numeral 2 arose in the same way; see Fig. 234, p. 413 and Fig. 241, p. 419). Symbols (3), (4), and (5) were derived from the Roman symbol for the *uncia*, which was a small circle (see Fig. 32 on p. 158): the fractions of the ounce were expressed as fractions of the circle, and half the symbol (5) indicated half the value. This list of medieval fractions offers many insights into the origin of early numerical symbols.

At this point we must forgo using these fractions for making computations in the Roman manner. One example,

$\frac{5}{6} - \frac{2}{3} = \frac{1}{6}$ *dextans* − *bes* = *sextans*   "$-\frac{1}{6} - bes = \frac{1}{6}$"

should suffice to show us how the very names of these fractions themselves made abstract computations according to general rules impossible, because they did not produce a conceptual image of the numbers. Thus anyone who came upon fractions in his calculations and wanted to continue using them had to rely on previously prepared computational tables.

The Roman measures of length were similarly divided into twelve subunits:

1 *pes* ("foot") = 12 *pollices* ("thumbs") = 16 *digiti* ("fingers," that is, 4 handbreadths)

so that 1 *digitus* = $\frac{3}{4}$ *pollex*. These Roman measures were adopted in England, where:

1 *foot* = 12 *inches* = 16 *digits* = 12 × 12 *lines*,

and here, too, 1 *digit* = $\frac{3}{4}$ *inch*. The word *inch* < Old English *ynce* < Latin *uncia* thus means $\frac{1}{12}$ and is explained as follows: 1 foot = 12 twelfth-foot = 12 *unciae* (−foot) = 12 inches.

The English system of weights makes use of the Roman ounce:

1 *pound* = 16 *ounces*,
    1 *ounce* = 16 *drams* (drachmas),
        1 *dram* = 3 *scruples*.

In the German system of weights, the ounce lived on into the 19th century as $\frac{1}{12}$ of a pound, with 2 *Lot* (a lead weight) per ounce; in the German system of apothecary weights the ounce has 8 *Drachmen* ("drams") and each dram 3 scruples.

As a tariff or duty, the *uncia* led to the development of the uncial letters: St. Jerome, one of the Church fathers, once inveighed against the *litterae unciales*, the "inch-high letters," in illuminated manuscripts. Today the term uncial is used for a form of medieval lettering.

The duodecimal system of fractions was a purely indigenous Roman growth, for it was unknown to the other three ancient Mediterranean cultures, i.e., the Greek, Egyptian, and Babylonian, which at various times all developed their own fractions. Only the base number 12 can be traced back to Babylonian influence (as in the zodiac and the division of the day and night into 12 hours).* Twelve also occurs very early as a mystical round number. Twelve vultures appeared as an omen to Romulus, the founder of the city of Rome, indicating that it was to last for 1200 years. The number 12 is also significant in Greek history: Homer relates that Ajax and Odysseus each commanded 12 ships; the 12 Ionian cities formed an alliance; 120 Boeotian sailors made up a ship's crew. And when we learn, further, that Menelaus was lord over 60 ships and recall the swineherd Eumeaus' 360 pigs (see p. 153), we realize that the round number 12 belongs to the same category as the other round number 60 and its multiples.

Where did these numbers come from? Here the Greek systems of weights and coinage lead us directly, and with certainty, back to the source. The ancient Babylonian system of weights and coinage was as follows: 1 talent = 60 minae, 1 mina = 60 shekels. The Greeks adopted the first of these equivalents (along with the Babylonian words themselves: Babylonian *manna* > Greek *mnâ*, "mina") and established the following table of monetary values:

1 talent = 60 minae, 1 mina = 100 drachmas, 1 drachma = 6 obols, or 1 talent = 6000 drachmas = 36,000 obols.

The talent and the mina were never issued by the Greeks as coins, but were merely numerical measures, like the pound in Germany during the Middle Ages.

THE BABYLONIAN SEXAGESIMAL SYSTEM

Was the role played by the number 60 limited to the above systems of measures? No, it was an essential peculiarity of the Babylonian number sequence.

---

* *Editor's note:* The division of the day and night into 12 hours is of Egyptian, not Babylonian, origin. Cf. O. Neugebauer, *The Exact Sciences in Antiquity* (2nd ed., Harper Torchbooks: New York, 1962), p. 81.

The Sumerians, who inhabited the southern part of Mesopotamia, soon developed, in addition to a decimal number sequence, an indeterminate positional system of numerals based on gradations of the number 60. Its origin is so unique that it warrants a brief account. This will also afford a contrasting example, from which we shall gain a deeper understanding of the essence of our own positional system of written numerals than we could derive without it.

At the beginning of their history the Sumerians, like all peoples, possessed a primitive system of written numerals — one which had a separate symbol for each and every unit, with several units arranged in order and grouped into a higher unit — a system comparable to the one we noted in ancient Egypt (see p. 42). This is shown on some ancient clay tablets, about the size of the palm of a hand, which served as the Sumerians' "paper" (see Fig. 34).

Fig. 34 Two clay tablets with Sumerian numerical symbols, from Uruk; these are among the world's earliest documents with numerical symbols. They record commercial transactions.

(1) The number at upper left is $600 + 60 + 30 + 8 = 698$;
(2) the large circle indicates 60^2, so that the upper left number is $1 \times 60^2 + 2 \times 60 + 3 \times 10 = 3750$.

Approximately two-thirds natural size.

These tablets show how marks were impressed into the soft clay with a round, tapered stylus which had one thick and one thinner end. The marks on the clay tablets were thus made by pressure, not by drawing or inscribing. When the stylus was held at an angle to the clay surface, the impressions were D-shaped; when it was held perpendicular to the surface the resulting impressions were circular. Our illustration gives their values as 1 and 10; 60 and 10×60 (see Fig. 35, rows A and B).

In the table on the left of Fig. 34, we read $600 + 60 + 30 + 8 = 698$, beneath that $5 \times 60 = 300$, and in the broken corner perhaps again 600, a numeral created by combining the symbols for 60 and 10. The forms of these numerals are extremely simple, since they were not inscribed, but made by pressing the stylus downward into the clay. The form of the large circle, indicating 60^2, can be seen

1		2	3	4	5
Row		Value	Original Form	Cuneiform	Sumerian number word
A	a	1	D	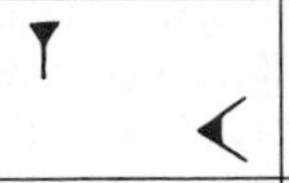 ᵞ	aš
A	b	10	o	⟨	u
B	c	60	D	 ᵞ	geš
B	d	60·10	ⓓ	ᵞ⟨	geš-u
C	e	60^2	O	 ✡	šar
D	d	60^2·10	◎	 ✡⟨	šar-u
D	f	60^3		✡	šar-gal

Fig. 35 Early Sumerian numerals and number words dating from the third millennium B.C.

in the tablet on the right in the same illustration, on which we read 36,000 + 2 × 60 + 3 × 10 = 3750, and also 270, 30, and 70.

All tablets of this kind have been discovered in the vicinity of temples. Hence they are probably records of livestock and grain brought as offerings, for they consist only of numbers placed next to the names of the persons making the offerings and the nature of the offerings themselves (sheep, chickens, fish, etc.). The reverse side often shows the totals in each category itemized on the obverse. These clay tablets from the third millennium B.C. are the oldest known documents that show numerals in actual use.

These numerals have the usual properties: 10 units, represented by a small D, make a small circle O (a ten-group), just as 10 large Ds make the next group symbol, which stands for 10 × 60; this is made very clear by the number words themselves (see Fig. 35). But the grouping changes surprisingly from Aa to Bc, for it is not 10, but 60!

Now we come to a critical question: Why this peculiar change from a decimal to a sexagesimal grouping? At first glance, it would seem that the symbols in rows A and B originally formed two independent groupings of ten: one set of units represented by a small numerical symbol and the other by a similar but larger symbol. The former group referred to a small and the latter to a large measure. The two groups were widely disparate in scale, like the mile and the ell in Germany and the mina and the shekel in Babylonia.

The situation has been correctly explained on the basis of the significant number 60 by A. Neugebauer in 1934:* this is the connecting link between the two groups of measures. The requirements of ordinary life in Babylonia, as everywhere else, demanded the use of the common fractions of a measure, in this case $\frac{1}{2}$, $\frac{1}{3}$, and $\frac{2}{3}$. As time passed it became necessary to express the fractional parts of the larger measure as whole numbers of the smaller, for example $\frac{1}{3}$ mina

* O. Neugebauer, Vorgriechische Mathematik, Berlin 1934.

in shekels, and thus to "bond" the two separate groups of measures. The smallest common number is 6; yet this would have made the measures too close to each other (about $\frac{1}{4}$ mile = 2 ells!), so that the distinction between the large and the small measures would have lost its purpose. Hence the number 60 was chosen because it is a multiple of ten; then 1 mina = 60 shekels and $\frac{1}{3}$ mina = 20 shekels.

Now let us analyze the Sumerian number sequence:

1	*aš* (also *geš*)	10	*u*
2	*min*	20	*niš* (< *ni-aš*)
3	*eš*	30	*ušu* (< *eš-u* 3'10)
4	*limmu*	40	*nin* (< *ni-min* 20'2)
5	*ia*	50	*ninu* (< *ni-min-u* 20'2'10)
6	*aš* (< *ia-aš* 5'1)	60	*geš*
7	*imin* (< *ia-min* 5'2)	120	*geš-min* 60'2
8	*ussu*	180	*geš-eš* 60'3
9	*ilimmu* (< *ia-limmu* 5'4)	600	*geš-u* 60'10 (Akkadian *ner*)

This number sequence does not differ essentially from any other "normal" sequence, except perhaps that it shows an unusually large number of very "early" characteristics. Thus there is an old counting limit following the Three, for the first three number words for 1, 2, and 3 literally mean "man," "woman," and "many" respectively: *eš* is the plural ending. There are additional counting limits after 5, after 10, and after 20. The succeeding tens are made up of vigesimal groupings. Then *geš*, "the great unit," also becomes a numerical rank. But the Sumerians had another old number sequence that went up to 100, and which was tied to a specific measure for cereal grains.

Thus nothing in this rare early number sequence indicates any "deeper insight" by the Sumerians for favoring the number 60. There was some external reason, perhaps the system of weights and measures, for giving this number its special status. It was definitely not an original unit used in building up a sequence, but a number connecting two systems or scales of measurement. From this it was promoted to the status of a numerical rank: the Sumerian ranks rise by powers of 60, with decimal ranks inserted as intermediate steps within the broad sexagesimal intervals:

60^0	10	60^1	10×60^1	60^2	10×60^2	60^3
aš	*u*	*geš*	*geš-u*	*šar*	*šar-u*	*šar-gal* (= *šar-geš*)

This progression is obviously the work of the "scientific" priests. *Šar*, 60^2, is clearly a limit of counting. As so often happened in the history of number words, here too the next rank $60^3 = 60^2 \times 60$, which according to the computational principle ought to be called *šar-geš*, as in column 4 in the table of numerals above, was called *šar-gal*, the "great *šar*"; this is a word formation like our own "thousand" (see pp. 132 and 47). A noteworthy feature is the recurrence of this "great" formation at 60 and again at 60^2.

In forming larger numbers, the number following the numerical rank multiplies the latter: *šar-u* = $60^2 \times 10 = 3600 \times 10 = 36,000$. But so far nothing at all is known about the nature and origin of combinations of numbers of different ranks; *geš-u* is 60'10 = 600, but what is 70 (*u-geš*?), or 72?

Šar is clearly a limit of counting (row C in the table on p. 164) in both number words and numerals. Row D in the same table is obviously formed with *šar*: first came the 10-multiple *šar-u*, $60^2 \times 10$, then the 60-multiple *šar-geš*, $60^2 \times 60$ (circle with 10-symbol), and finally *šar-gal*, $60^3 \times 60$, the "great" *šar* (circle with 60-symbol), which did not appear in the old form but does in the new cuneiform numerals. Whereas the symbols in rows A and B are the oldest, the evolved numerals, those in row C, represent the first natural evolution of these written numerals and the symbols in row D an artificial expansion; thus we have a conceptual succession of gradations similar to the Roman numerals and to our own sequence of number words (see p. 144).

Cuneiform writing came into general use soon afterwards. The old slender cylindrical stylus now became wedge-shaped, and like an elongated right triangle in cross section. If one perpendicular of the stylus was held vertically and the other horizontally, the impression made by the stylus was a vertical "wedge." If the stylus was turned so that the two mutually perpendicular sides were inclined, the result was an oblique angle or hook, as one can readily see from the illustration of the clay tablet (see Fig. 36). From then on the

Fig. 36 A Babylonian clay tablet with cuneiform numerals in positional order, ca. 1800 B.C. It can be seen that each number is made up of two basic symbols, the wedge (for 1) and the hook (for 10), and that every "digit" is a combination (or arrangement) of these. In the middle of the second row from the bottom we can read the numbers 29 and 31, and to their right, 53 and 49. In the last column at the extreme right, the numbers from 1 to 13 (and 14, 15) are written from the top to the bottom. Size: 13 × 9 cm.
From: O. Neugebauer and A. Sachs, *Mathematical Cuneiform Texts* (American Oriental Series 29, New Haven 1945).

Babylonian numbers were written as combinations of wedges and hooks (column 4 in table on p. 164): unity was a single wedge and 10 a hook (Aa, b); the circles in column 3 were now approximated by 4 cuneiform wedges (as in column 4).

And now something happened that was unique in the history of numerals: the distinction between the large and the small symbols (which was made in rows A and B in the table on p. 164) was abandoned, and it was from this abandonment of the characteristic symbols — not from any particular intellectual deliberation — that the positional method of writing numerals arose!

If instead of the cuneiform characters Aa and Bc, we were to write X and x according to the Babylonian system, then XX xxx would clearly be $2 \times 60 + 3 = 123$. Now if we no longer distinguish between the capital X and the lower-case x, but make them all the same size, like the small wedges in Bc, then the symbol xx xxx is now governed entirely by the law of succession of magnitudes: the xx on the left is the 60-multiple of the xxx on the right. Yet the succession of ranks remains undefined, because there is no symbol for the absence of a rank, no middle or final zero; thus our number $2 \times 60 + 3$ could just as well be $2 \times 60^2 + 3$ (because there is no middle zero) or $2 \times 60^2 + 3 \times 60$ (no terminal zero). Hence we have an undefined system of positional numerals that observes the law of succession of magnitudes but not of the order of magnitudes, which then have to be deduced from the context. It was not until quite late (in the 6th century B.C.) that a sign for a gap or missing symbol first appeared within a Babylonian numeral, and still later in astronomical tables at the end of a numeral as well.

Written in positional cuneiform symbols, the number

758 is   and   3750 (see Fig. 34) are:

$12 \times 60 + 30 \times 10 + 8 =$ $1 \times 60^2 + 2 \times 60 + 3 \times 10 =$
$12, 38$ $1, 2, 30.$

With this new system of numerals, in which all numbers could be written with only two different symbols, the Babylonians achieved great skill in computations, and along with it a high level of mathematical development. Indeed their problems and solutions, especially in the theory of equations, are still regarded with wonder and admiration today. The Babylonians made use of multiplication tables for their computations, as the following illustration and its exact transcription show (see Fig. 37).

As the rank of the positional number decreases in this sliding succession of magnitudes — we shall see that it inclines less strongly toward its intellectual origin than the Indian — it directly begets the sexagesimal fractions: the succession glides over the ranks, from whole numbers right into fractions, without stopping. In mathematical notation (see p. 56):

$$\cdots + a_2 \times 60^2 + a_1 \times 60^1 + a_0 + a_{-1} \times 60^{-1} + a_{-2} \times 60^{-2} \cdots$$

But the most significant feature is that calculations involving fractions do not differ in the slightest degree from calculations with whole numbers (in great contrast to the duodecimal Roman fractions we described earlier), and this is what gave sexagesimal fractions

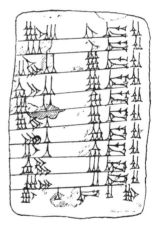

Fig. 37 Babylonian multiplication table for products from 18 × 1 to 18 × 60, from the temple library at Nippur, about 1350 B.C.

1st row: 18 (times) 1 (=) 18
2nd row: 2   36
. . . .
last row: 11   3, 18 —
that is, $3 \times 60 + 18 = 198$. The reverse side of the tablet shows 18 × 12 through 18 × 20, followed by 18 × 30, × 40 (× 50 is omitted), × 60. Intermediate products like 18 × 37 are arrived at by combining 18 × 30 + 18 × 7. For the transcription see II. V. Hilprecht, *Mathematical, Meteorological and Chronological Tablets from the Temple Library of Nippur.* Philadelphia 1906.

their present worldwide currency. The Greeks firmly established the succession of magnitudes in the following terms:

prôton (and) *deúteron hexēkostón* (*méros*),
"first (and) second sixtieth (part)."

The Alexandrian astronomer Ptolemy used sexagesimal fractions in his astronomical work entitled *Megálē Sýntaxis* ("The Great Compilation"), which the Arabs later passed on to the West as the *Almagest* (< *al-mégistē*, "the greatest"). Until the time of Copernicus, this was the fundamental textbook of astronomy. Latin works (translated from the Arabic) presented the sexagesimal fractions verbally as

(*pars*) *minuta prima* (*et*) *secunda,*
"first (and) second diminished (part),"

both of which still live on in our "minutes" and "seconds."

Thus not only the number 60, but also 360, came into our own culture. This was most probably occasioned by the sun's complete circuit through the heavens in roughly 360 days, from which we have the number of parts into which any circle is divided. This reflection of the celestial numbers in the terrestrial was the work of the Babylonian priests. The solar day was divided into twelve two-hour periods in analogy with the twelve-fold division of the zodiac. Moreover, the seven known heavenly bodies were used not only for classifying and naming metals, colors, and parts of the body, but also for the cyclical ordering of days (i.e., the seven days of the week are named after the planets). This arrangement rests essentially on astrological beliefs, which in every culture gave rise to astronomy and thereafter continued to live on beside it.

After the conquests of Alexander the Great, Chaldean astronomers streamed westward into the ancient Mediterranean lands and acquainted those peoples with their own culture, in which the number 12, as well as 60 with its multiples 300 and 360, were so important. Many traces remain in the Bible among other places, as in the 12 tribes of Israel, the 12 apostles and the $12^2 \times 10^3$ "sealed" servants of God of the Apocalypse (Revelations 7:4, and 14:1 & 3).

Around 2500 B.C. the Sumerian state was overthrown politically by the Semitic Akkadians who came out of the desert, but the Sumerian culture overwhelmed that of the conquerors. In addition to the Sumerian numerals, the Akkadians also took over cuneiform writing and adapted it to their own very different language. They called the number 60 *šuššu* (cf. the Hebrew word, p. 115), which the Greeks altered to *sóssos* and which shows a great resemblance to the German word *Schock* ["shock"]. From 600 came the word *ner*, Greek *néros*, and from the Sumerian 60^2 came the Greek *sáros*. Quite understandably, these borrowed words were soon eroded to indefinite round numbers, just as in the case of the Greeks and Romans, among whom *sescenti*, 600 became a round number for the same reason.

Another of the many mystical connections of the number 60 can be seen in the Platonic number:

60^4 years = 1 "world year" = 360 × 36,000 years = 360 "world days" (in each of 36,000 years).
1 human life span = 100 years = 36,000 days.

Thus one day in a man's lifetime is the same as one year in a "world day." The human life span is the 360th part of a world life span or 1 degree of the circular orbit of the universe, if we equate days and years.

THE BREAK AFTER SIXTY IN THE GERMANIC NUMBER SEQUENCE

After the foregoing excursion, in which we explored the numbers 12 and 60 and the reasons for their prominence, we must go back once more to the main subject of our discussion and ask ourselves: Is this break after 60 in the Indo-European number sequence (for example, Gothic *six tigjus — sibuntehund*) due to Babylonian influence?

The reader may answer hastily: of course! If so, he fails to remember that the migration of certain aspects of Babylonian culture into the Mediterranean world, such as the Babylonian measures (time, angles, and weights) and their round numbers, took place in the last few centuries before the birth of Christ and thus in the full light of recorded history, whereas the Indo-European number sequences go all the way back to the murky realms of prehistory, long before 3000 B.C. Hence the chronological overlap of the Mesopotamian and Mediterranean cultures is subject to question. There is also serious doubt about their geographical contact: the Sumerians inhabited the southern part of Mesopotamia, while the Indo-European peoples in that period lived very far from the Tigris and the Euphrates.

Then what about the derivation of the "shock" — the relationship between the Greek *sóssos* and the Babylonian *šiššu*? If the former was really derived from the latter, then this happened in the Greek language long after the Indo-European number sequences had already been formed. But the most telling objection to this seeming bit of evidence is that even if the "shock" does have a Babylonian origin, as is widely believed, it is still unthinkable that the mere use of this one measure of quantity, which was no more frequent than the *Mandel* or the *Stiege*, could have altered the very deepest phonetic roots of an Indo-European number word. And not just the word for 60, but also the words for all the tens following 60! Since people quite frequently count beyond sixty, 70, 80, and 90 should also have acquired new names, if the hypothesis were valid. But it is definitely out of the question. The Babylonian number rank 60 was rooted in a very different conceptual soil from that of the Indo-European number words after 60.

If the break at 60 is not due to Babylonian influence, then how did it come about? Here I should like to offer a suggestion which has already been made by others (see M. de Villiers, *The Numeral Words*, London 1923). The change in Indo-European number sequences from 70 on is focused in every case on the syllable -*hund* (or *ant*-; see p. 94). And this syllable is the numerical rank "hundred," whose effect extends back as far as 70. We have established numerous instances of this sort of reverse influence of a particular numerical rank which at first stood out prominently in the advancing number sequence, and then sought to be tied down into the preceding portion. This is an event in the building of the number sequence which can be readily understood in the light of early man's conception of numbers (see p. 46). The Germanic fashion of overcounting and counting by halves, which we discussed at length, created the number word for 95, for example, by reaching back from the numerical rank 100 into the numbers preceding it (see p. 76).

But why the backward influence from 100 to exactly 70? The break in the sequence of tens mirrors a similar break in the sequence of units. The latter can be seen in Latin, for example, in which the words for 7, 8, and 9 are formed with the ending *-em* that has been taken from *decem*, 10: *novem*, *octo* (a dual form for *octem*!), *septem* (see table on p. 92). Whether this conceals an erstwhile counting back from 10, which occurs in other number sequences (such as the Finnish; see p. 75) where 9 = 1 subtracted from 10, 8 = 2 from 10, and 7 = 3 from 10, is a possibility that may require further study. Comparative philology alone can decide whether such suggested derivations for 9 as the Greek *en-ap-ken* > *ennefen* > *ennea* and Latin (*u*)*nu-*(*a*)*p-hem* > *novem* have any validity.

Now let us summarize: the reverse influence of a numerical rank upon the number words preceding it (as from 10 back to 7), as well as the reflection of a particular feature from one order of magnitude (in this case, the units) to another (the tens) — a typical example of this is the break in the Slavic number sequences after 4, which is repeated after 40, 400, and 4000 (see p. 24) — these are the results of natural forces, so to speak, which are manifested in the structure of a number sequence and thus may quite readily have affected other Indo-European number sequences as well. For these reasons the non-Babylonian explanation seems far more natural than the influence of an alien number sequence, which, though in itself quite possible, could at the very most have inserted a single foreign number word into the Indo-European sequence (just as the Aryan *sata*, 100, came into the Finnish language), but would never have had the power to alter all the subsequent number words as well.

Even if the Babylonian sexagesimal gradation is not the direct answer to our question, however, it still has deepened our understanding of the problem, quite apart from any further insight into the cultural history of numbers which we may have gained from our digression into Babylonia.

Hidden Number Words

In the course of our discussion thus far we have looked into the meanings of number words, we have tried to make out the original images that underlie them, and we have found that with the passage of time the images of the units, in particular, have suffered so much attrition that we could barely make out the few distinct traces they left.

Now we come to hidden number words! These are words to which a number word has contributed a supporting image, which sometimes appears openly, as in the German word *Zwie-licht* ("twi-light"), but may at times be fully concealed, as in *Samt*, "velvet." Words of the latter type are especially noteworthy. From the philologist's point of view it is particularly fascinating to observe how cleverly and ingeniously man has used a number to indicate the essential character of some object which he wished to name and to which no one now would attribute any numerical quality. We shall confine our examples to the most familiar languages, but the reader is urged to refer to others, especially Chinese.

ONE

The word *simple* < Latin *sim-plex*, "once-folded" (*plicare*, "to fold") is one of a large group of words whose original stem is the Indo-European root *sem*; we have already listed many of these derivatives (see p. 146). Here we can add the German word *Sint-flut* ("flood, deluge"), which in Old High German was *sinvluot*, wherein the Germanic root *sin* means "always, all-embracing"; it is a deluge which inundates everything, and has nothing whatever to do with *Sünde* ("sin"). In the same class is the German suffix *-sam* ("*als Kaiser Rotbart lobesam*"), which as an independent word once meant "of like quality"(compare the Latin *similis*, "like, similar," p. 146).

The German word for zero, *Null*, contains the other Indo-European one-form, *oinós* > Latin *unus*. The Latin diminutive ending equivalent to "-ling," or the German *-chen* is *-ulus*: *mus*, "mouse" — *musc-ulus*, "little mouse." Thus from *unulus* > *ullus*, "little one, one-ling, any at all." The Indo-European negative syllable *n* combined with this to form the Latin word *n-ullus*, "none, not any." The numeral 0 obtained its name "null" because in the medieval view it was "no (numeral)," *nulla* (*figura*); its history will be discussed later. Another relative of "one" is the Latin *non*, "not" < Old Latin *ne-oinom*, "not one," as is the German *nein* < *n-ein(s)*. Other members of the same tribe are the English words *on-ly* < *one-like* and *al-one* < *all-one*. Moreover the Italian and French words for "boar" are *singolare* and *sanglier* respectively, derived from the same idea of one-ness, perhaps because the wild boar lives or ought to live alone.

The German word *Eimer*, "bucket" < Middle High German *eimbar*, Old High German *ein-bar*, amber, is borrowed from the Latin *amphora*, whose diminutive form *amporla* is still in current use as the word "ampoule." The Greek word *am(phi)phoreús* (< *ampho*, "both" and *pherein*, "to carry") referred to an earthenware jug with two handles. In popular colloquial speech *Ein-henkel*, "one-handled jug," was derived from *eim-bar* (from the Germanic *beran*, "to bear, to carry"; compare the words *fruchtbar*, "fruitful" and *Bahre*, "barrow, bier"). Analogously, the *Zuber*, "tub," is a two-handled vessel.

The "ace," the playing card with the value of one, comes, of course, from the word *as* in Roman fractions. There is also a large family of words derived from the Latin *pris — prior — primus*, "early" — "front" — "foremost, first," among which we shall mention only the *Primel*, "primrose," the French *prin-temps*, "spring" ("early time") and *prince*, from the Latin *prin-ceps*, "he who takes the first place" (Latin *capere*, "to take, seize").

We shall conclude, appropriately, with the "migraine," which is related not to "one" but to "half": Greek (*he*)*mi-kranía*, "half-skull (pain)."

TWO

The word "two" can either join ("twin," "twain") or divide (German *Zwist*, "dispute, discord"). Man experienced both quite early in his development. Thus there is a vast family of related words in which the number word "two" is latent in various forms: German *zwi*, Norse *tvi*, Latin *bis, dis-, vi-*, and *ambo*, and Greek *dís, día*, and *ámphō*. The reader may recall our comments about the number Two at the beginning of the number sequence (see p. 12).

From the Old High German word for "two" come the German *zwisk*, "every two," whence *in zwisken > inzwischen, zwischen*, literally "in the middle of two" and thus "in between," and the English *betwixt* and *between*. *Twisko*, the "two-sexed," is the epithet applied in the Germanic sagas of the gods to the Earth-born ancestor who was able to produce a son without having a wife. We find another derivation of "two" in the Icelandic *tuenner*, "every two"; *tuennern* thus means something like "to make two, to double," which is the modern German word *zwirnen*, "to twist, entwine." *Zwirn*, which is *twine* in English, is hence a doubling, a string made up of two strands joined together, as in the English word *twist*; in German the word *Twist* also means a certain kind of cotton yarn. The same word *Zwist* in the Oberlausitz region refers to a cord made up of doubled strands. In the same category are the words *Zwillich*, "drill cloth," and *Zwilch*, a "two-thread cloth," for the Old High German *zwi-lich* is an imitation of the Latin *bi-lix*, "two-threaded," from *licium*, "thread." A verbal cousin of *Zwilch* is *Drilch*, "drill, denim, canvas, a cord or cloth made up of triple thread" (see p. 176).

The word *Zwist* also occurs with the meaning of "strife, discord." The explanation for this lies in the Old Norse word *tvistra*, "to divide" and thus "to break apart into two"; we shall shortly have more to say about this class of words.

The word *Zweifel*, "doubt," is clearly an offspring of *Zwei*, "two," not only in German but also in numerous other languages: for example, Old High German *zwivo*, Old Norse *tyja*; Greek *doiế* (cf. "double"). The Gothic *twei-fls* expresses exactly the same idea as the Greek *di-plóos*, "twofold" — *zwie-spältig*, "split, forked, divided," expresses its meaning clearly. The concept of "dubious" is even more picturesquely expressed in the Old High German Book of Gospels of Otfried:

thiz selba uuas ihm untar zuein — "the same was to him between two."

The very same turn of speech still lives on in the Italian expression *stare intra due*, "to stand between two" meaning "to doubt." In Latin the word *du-bius* consists of the elements *du-*, "two" and *bius* < Indo-European *bhu*, "to be" (> Latin *fu-i*, "I have been"). The intensified Latin form of the same word, *dubitare*, has given rise to the French *douter* and the English *doubt*. Similarly in the Greek *en-doi-ázein*, "(to be) between two" we can distinctly make out the "two"-root, just as we can see the number word *kettö*, 2, in the Hungarian *ketelkedik*, "to doubt."

The German word *Zweig*, "twig," has the meaning of "fork," in Middle High German *zwisel*; a place known as *Zwisel* in the Bavarian Forest is named after a fork in the road. *Geweih*, "horns, antlers," is verbally a *Ge-(z)weig*, a "forked," in which the dropping of the *-z-* goes back to the two form *vi*, which is in the same class as the Gothic words *wit* and *weis* (pp. 13 & 150). Besides these we have the group of related words *zwicken* ("pinch; gripe; worry"), *zwacken* ("nip; tease"), and *zwecken* ("to pine, tack"); cf. the word *Zweck*, "goal, purpose," literally "that which is between two fingers or pincers." The *Zweck* was also a "pinched-off" piece of leather placed at the center of a target to serve as a "mark" or bull's-eye.

The numerical meaning underlying the words "twilight," "twin," and *Zwiespalt* and *Zwietracht* (both meaning "discord, dissension") is obvious. But one would probably not expect to find a "two" in the German word *Zuber*, "tub"; yet in its Old High German ancestor *zwi-bar* we see it as the "two-handled" cousin of the *Eimer*, the "jug" which has only "one handle" (see p. 171).

Both Greek and Latin have given a whole series of words to German and English as foreign loan words, whose connection with "two" is thus sometimes hard to see. Who would think of it, for example, in "diploma" and "diplomat"? The Greek *di-plóos* means "twofold, double" and is related to Latin *duplus* > French *double*, which German has borrowed as *doppel* and English as *double*. The Greek *diplóma* is "folded, doubled." The Roman emperors often granted certain rights and privileges to veteran soldiers upon their discharge from service. These were originally inscribed on the inner surfaces of *two* bronze tablets which were held together by rings through one of their long sides, like a loose-leaf notebook. The two tablets were then closed against each other and a wire was drawn through two holes drilled in the center, twisted together several times, and the two ends held with a wax seal. This was a "diploma," an imperial writ specifying certain rights and privileges. Today a "diplomat" still presents his "diploma," his written credentials, to the foreign government to which he is accredited.

The Roman writing tablets that could be similarly folded together were called a *diptychon*, and a "diptych" in art is a painting on two panels hinged at the center; a *tri-ptychon*, "triptych" is a three-paneled altar painting whose two outer wings fold over to cover the center panel (from the Greek word *ptýx*, "fold").

In the word "duel" we can recognize the Latin *du-ellum*, a "fight between two antagonists," which was an older form of the later *bellum*, "war." Here too we see the transition from *duis* to *bis*, "twice," with which the Italians still call for an encore in the theater and concert hall.

And so we come to the large and fascinating family of *bis*-words in the Romance languages. This group has two kinds of members — those in which *bis* has the purely numerical meaning of "two," and those in which it distorts or even imparts a pejorative connotation to the main root word. A good example of the first kind is the French "pushcart," *brouette* < *bis-rouette*, a "two-wheeler"; it is beautifully camouflaged in the German word *Protze*, the two-wheeled artillery caisson or guncart, from the Italian *bis-rozzo* < *ruota*, "wheel." The French *bi-cyclette* and English *bi-cycle*, on the other hand, are quite obviously two-wheeled conveyances. The *Zwieback*, Italian *bis-cotta*, French and English *biscuit*, German *Biskuit*, is "twice-baked" bread; in French a thick, strong broth is thus called a *bisque*, "twice" or "thoroughly cooked," and curiously this is also the term for a penalty point in tennis. A person who squints is known in Spanish as *bis-ojo* and in French as *bigle* (from the Latin *oculus*, "eye"), meaning "two-eyed" or "double-sighted."

Why is the intercalary day (February 29) known in the Romance languages, for instance in France, as *bissexte* and the leap year as *bissextil*, Italian *bisestile*, "two times sixth"? The last month of the old Roman year was February (see p. 25). Every four years an intercalary day was inserted in February, after the 24th, and came to be known as the "second 24th" day of February. In the Roman manner of designating the days of the calendar, this was "the sixth day before the Calends of March" (that is, before March 1, the beginning of the new year), *dies sextus ante Kalendas Martias*; hence the intercalary day was *bi-sextus*, the "twice sixth" day before the Calends. From ancient times this was popularly regarded as an unlucky day, and the French word *bissetre* still means "misfortune."

In Germany the *Bise* is a colored braid on uniforms, like the stripe along the trouser seam. The French *bis-er* is *bis* expanded to a verb, which literally means only "to do something twice" but now has the meaning of "to put on another color," especially black as in mourning; one would naturally not see this nuance in the word itself. Similar derivations from the Latin *bini*, "every two," are the Italian *binario*, "rail, track" and the French *bin-ocle* (< *oculus*, Latin "eye"), a "double glass" like binoculars or opera glasses.

Now some examples of the second *bis*-group: *bis*- can also reduce the meaning or make it pejorative. In French, it is, in this case, readily altered to *bes*, as we have already seen in the words *bissac*, "double-sack" and *besace*, "begging or beggar's sack" (see p. 14). Something which is no longer completely sour becomes *bes-aigre*, "sour-ish"; similarly in Italian *bis-lungo* is "longish," *bis-tondo* is "roundish" and if one does not really sing (*cantare*) one hums (*bis-cantare*). Whoever is "filthy as a pig" in Italian is *bis-unto* (< Latin *unctus*, "anointed"). An "oversight" in French is *be-vue* (< *voir*, "to see"); a "blinking or fluttering before the eyes" is *ber-lue*, "bad light," from the Latin *lux*, "light," whence comes the French *bluette* < *be-luette*, a "spark" which is a "side-light" and is not part of the main light. The French *ba-fouiller* is "to babble, to talk nonsense" (*fouiller*, "to dig"), and *bis-quer* is "to become vexed."

In German this change in a word's meaning is beautifully exemplified in the *Bilanz* and *Balance*, "balance, equilibrium." Both derive from the same words *bis* and the Latin *lanx*, "scale, steelyard." The first of the above German words, which became a part of commercial language by way of the Italian *bi-lancio*, means a "balance" between assets and liabilities and hence something that has achieved equilibrium and come to rest, whereas *Balance* refers to suspended objects which are not yet in equilibrium but are constantly and restlessly seeking it; for similar reasons the French call the balance-wheel in a watch the "restless."

The Latin word *bis* was derived from an old two-form *dụis*, which also gave rise to the common prefix *dis-*. These two syllables impart a doubling or a pejorative sense to a great number of words that are also used in German and English: to "de-fame" is "to bring into dis-repute" (from Latin *fama*, "name, reputation, report"); "di-stance" means separation; "dif-ference" is a "carrying apart, partition" (from Latin *ferre*, "to carry"); what is "dif-ficult" is "not easy" (from Latin *facilis*, "easy"). The antithesis of *con-cordia*, "a heart (and soul)," is *dis-cordia*, "discord, dissension" (from Latin *cor*, "heart"). The corresponding French prefix is *des-*, as in *dès-espérer*, "to des-pair"; in Italian this is often shortened before a consonant to *-s* as in *s-cordanza*, "bad temper, ill humor" or in *s-fumare*, "to make dim, obscure" (from *fumare*, "to smoke"; cf. the word *sfumato*, literally "smoke-filled, smoky," used as a technical term in the fine arts to describe the soft, blurred outlines of figures as in Leonardo da Vinci's paintings).

To "set something apart" in Latin is *dis-putare*, whence our word "dispute"; *putare*, "to cut," may possibly have arrived by way of the "notched stick" at its more common meaning of "to think, consider, estimate" (see Tally Sticks, pp. 223 ff.). One who must "comprehend something through and through," or *dis-cipere* (from Latin *capere*, "to grasp"), is a *dis-cipulus*, a "disciple" or "pupil." "Dis-cipline" is then "instruction, learning" and by transference also "order." Of the many more examples that belong to this group of words, we shall cite only "di-rector," the "ruler," whose prefix echoes the old Roman maxim *divide et impera*, "rule by dividing into two parties." And *di-vid-ere* may well be the strongest word for "make two of," for it contains both the "two"-roots *dis* and *vi(d)*.

This *vi-* < Indo-European (*d*)*ụi*, "two," which we already met in the Latin *vi-ginti*, 20 (see p. 150), is formed from the Indo-European *ụidh*, "to sunder," a word that recurs in the German *Witwe* and English "widow," a woman who has been "sundered" from her husband, and in the German *Waise*, the orphan who has been "sundered" from its parents, whose opposite is the "in-divid-ual" who is "undivided."

In "dialogue" and "dialect" we have the Greek form *diá* as in *diá-logos*, "speech between two" and *diá-lektos*, which is properly a conversation in which two persons take part and has by transference come to be a "dialect" in the modern sense of a specific form of collo-quial speech. The number of words formed with *diá-* in German and English is very large; we shall mention only *dia-gnosis*, "thorough knowledge" (from Greek *gignóskein*, "to know, recognize") and *dilemma*, "dilemma," a "pinch between two horns" (< Greek *lambánein*, "to take"). Greek words compounded with *dícha*, "doubled," may also have pejorative meanings, as in *dichó-nous*, "two-minded" and hence "equivocal, ambivalent, false, faithless." We may end this great family with the example of the Devil, Greek *diá-bolos*, he who "throws between" and thereby causes discord (from the Greek *bállein*, "to throw"); linguistically, the (German) *Teufel* and the (English) *Devil* came from the Greek through the Gothic *diabulus*. Moreover, the German *Daus* and English *Deuce* in dice games was originally the Two, and comes from the Latin *diavolus*, "two-throw."

There is one more "two"-form to be mentioned, which is also com-monly encountered in foreign loan-words in German and English; this is the Greek *ámphó* and Latin *ambo*, "both." An "amphi-theater" is a "double theater"; in the history of Roman architecture this was in fact made up of two semicircular banks of seats, as in the ancient theater with a stage, to make a round building without a stage but with an enclosed arena ("sand" in Latin), as in the Colos-seum in Rome, where circuses and gladiatorial contests were housed instead of dramatic performances. We have already mentioned the *ampoule* in connection with the *Eimer* or "one-handled jug" (see p. 171). As a prefix *amb-* means "from both sides": in Latin *amb-igere* is "to pull from both sides" and thus "to doubt, to dispute" (cf. English word "ambiguous"); *amb-ire* means "to go back and forth," so that *ambition* was originally an ardent or zealous moving about.

Let us look back: What a huge multitude of words man created from the number "two"! This is understandable: it is the first step past "one," from the I to the You and beyond into the world at large, where all that is one is perpetually divided into two, and the divided are joined into one.

THREE

Drillich, *Drilch*, or *Drell*, "drill," a triple-stranded material used for making army jackets, can be immediately recognized, on the analogy of *Zwillich*, as a Germanization of the Latin *tri-licium*, "three-threaded." The word *Trense*, "braid, twist," has the same origin, for the Latin *trini-care* (< *trini*, "each three") literally means "to three" and here "to braid from three strands," as with hair; in Spanish it gave rise to the verb *trenzar*, whence *trenza*, "plait, braid, cable." This word was brought to the Netherlands by the Spanish army of occupation, whence it came in the 16th century to Germany as *trensse*, a light bridle with a snaffle-bit (in contrast to the Hungarian *Kandare*, a heavy bridle with a curb-rein). The French *tresse*, Italian *treccia*, from the same root, which was thus originally a "braided" edging, was brought to Germany in the 18th century along with French fashions in clothing.

A *Dreidraht* [literally a "three-cord"] in German is a slow, dawdling, tiresome person, a bore. *Drahteln* in the Swabian dialect is "to delay, dawdle" and may very well have taken its meaning from *drehen*, "to twist, turn" and *Draht*, "wire, cord." *Dreidraht* then is probably an alliterative and cumulative doubling, formed all the more easily here because the number three serves to intensify the meaning of the word.

The foreign loan-word *trivial*, "trivial," in German goes back to the Latin noun *trivium* (< *tri-via*, "three-roads"), which in Roman times and well into the Middle Ages referred to the three basic disciplines of the curriculum: grammar, rhetoric, and dialectics. Once the student has mastered these, he could embark upon the *quadrivium*, the "four paths" of arithmetic, geometry, astronomy, and music. Together these made up the "seven liberal arts." Hence *trivial* acquired its meaning of "commonplace, ordinary, unimportant, and uninteresting." The famous Trevi Fountain in Rome, which is named after its three great streams of water, used to be called the Fontana Trivia.

The word "tribute" goes far back into linguistic history. The modern word is derived from the Latin *tri-bus* < Indo-European *tri-bhu-s*, from *-bhus*, "to be" (see p. 173) and originally meant a "third or third part" and then a "district" or "community," because Rome arose from the three Italic communities or tribes — the Romans, the Sabines, and the Albans. Thereafter *tribus* came to mean a group of related people, a clan or *tribe*, and then a tax or customs district, so that *tributum* is a "payment, tribute." A "tribune" was originally an official in charge of such a district; the raised seat from which he dispensed justice and made decisions was the "tribunal." The German word *Tribüne* (< French *tribune*) then retained only its meaning of "elevated, raised" and became the *Bühne*, or stage of a theater.

This very class of words is a significant example of the wide range of meanings a language can embrace with one simple number word. An intriguing instance of the same thing is *testament*, a word in which "three" is so well concealed that even an experienced philologist may have trouble recognizing it. "Witness" in Latin is *testis* < *ter-stis*, *tri-sto*, "to be a third party." The witness stands by as a third person, as in the illustrative anecdote about pearl-trading related in the next volume, in which such third persons are excluded by using finger-gestures under a cloth (see chapter on Finger Counting, pp. 199 ff.). "Can anyone else (a third person) hear us?" we ask when we want to discuss a private matter. One rap, a second rap, and still another rap by a third person standing by still seals a bid at a German cattle auction. In Russian the corespondent in a divorce suit on the grounds of adultery is called *tretij*, "the third party." There was an Old Latin (Oscan) word *tristaamentud* > *testamentum*, "being the third." The "last will and testament" is thus a bequest to one's heirs in the presence of a third person as witness.

It may also be worth noting that the month of May in Old Saxon was once named *trimilki*, because during that season a cow could be milked three times a day.

Why was a certain kind of small, old-fashioned pocket pistol called a *terzerol*? Allegedly because of the male hunting falcon, which was about a third smaller than the female; Italian *terzuolo* < Latin *tertius*, "third." Originally people saw in the English word *travel* only the strain and effort it entails, which is still expressed in the French word *travail*, "work," or the English "travail." *Qual*, "torment," originated in the Latin *tri-palium*, an instrument of torture which had three stakes, and still lives on in French expressions such as the *travail d'enfant*, "birth pains," and *travaille de la pierre*, "suffering from (gall-) stones."

The *Kümmelblättchen*, "caraway leaf," is actually a foreign loan, although it looks very German. It refers to a game played by card sharpers, in which three of the cards are covered. But where is the "three" hidden in this word — in the *kümmel*? Yes, because this is not a true German root, but a misunderstood assimilation of *gimel*, the third letter of the Hebrew alphabet, which we met earlier (p. 121). The Hebrews, like the Greeks, used their letters as numerals, so that *alef* was 1, *bē* 2, and *gimel* 3.

FOUR

To begin with, the square has always been expressed by the number word "four." The basic form of this is *quadratus*, which Albrecht Dürer turned into the German *Vierung*; today the word is used only in church architecture, for the square formed by the intersection of the long and short arms of a Roman basilica. Thus we come to the large clan of foreign loan-words in German which begin with *Quadr-*, above all *Quadrat*, "square," and *Quader*, "parallelepiped," from the Latin *quadrus*, "four cornered" and *quadrare*, "to square" (after *quattuor*, 4). The *Quadrant*, "quadrant," usually has the meaning of "a fourth part, a quarter"; we think of the Roman coin, the *quadrans* or quarter *as* (see Fig. 32 on p. 158), and in mathematics it refers to a quarter of a full circle; but the French "clock face," *cadran*, which is now generally circular, takes its name from the days when sundials had square faces. Then we have the fractional form with *t*: *Quartal*, a "quarter year" and *Quartier*, a "quarter (of a city)," which is a standard example of the semantic fading of number words: neither in German nor in other languages does it have any connection with an actual division into four parts. In Venice such a quarter or district is called a *sestiere*, literally a "sixth" of the city. The true meaning lives on, however, in the Dutch expression *een kwartier over sesse*, "a quarter past six." The *Quarte* and *Quarta* ("fourth, quarter"), are derived from the Latin ordinal form *quartus*, "the fourth"; the "quartet" is the same with the addition of the Italian diminutive ending -*etto*. The *r*-less word *Quatember* is derived from *quattuor tempora*, the "four seasons"; in the Roman

Catholic Church this is the name for the three holy days that introduce each quarter year.

The Italian *quadro* is a painting, properly speaking a square panel,
while the same word *cadre* in French means "frame." In French
military terminology, especially in the form *carrée*, it also has the
meaning of the "core" of a troop of soldiers, which it took on from
the Italian *squadra*, "battle square" (< Latin *ex quadra*). In the 16th
century the augmentative form *squadrone* (see p. 134) also went
into the French language as *escadron*, from which the word "squadron" was formed. *Geschwader*, "squadron" (or airplanes or ships)
is a good Germanization of the same word *Schwadron*, a "squadron"
(of cavalry). The same origin might be attributed to *schwadronieren*,
"to boast, to cut open," but this word may also have come from the
Middle High German *swatern*, "to gossip." The basic word in its
unaugmented form is also present in the English *squad*, a small detachment of a dozen or so soldiers. The Italian *quadrigla* is a "four-
cornered" troop of soldiers or horsemen, whose French form
quadrille, a dance of four couples, clearly betrays the numerical
idea behind it.

The *Kartaune* < Medieval Latin *quartana* is a "quarter"-cannon
because it shot 25-pound instead of 100-pound cannonballs. The
Kaserne ("garrison"), which since the time of Louis XIV has been
the name for a very large barracks or group of barracks in which
soldiers were housed, was originally merely a shelter for "four"
soldiers on guard duty (from the Latin *quaterna*, "every four").
The same Latin word in the Middle Ages referred to a pile of paper
consisting of four large sheets, one on top of another, which were
folded in the middle and bound together, thus forming 16 pages —
the "bend" which still governs the strength of a book. *Quaternio*
became *caterna*, *quayer*, and finally French *cahier* ("notebook")
and English "quire," which in Italian is still called *quaderno*.

The "diamond" on playing cards, which is *Karo* in German, was
originally a French word: Latin *quadrum*, diminutive *quadrellum* >
Old French *quarrel, carrel* > *carreau* (Latin *-l* > French *-au*, as in
pellis > *peau*, "hide"); thus *kar(r)iert* means "checkered" or broken
up into small squares.

When rocks are broken up, they are also shaped into "square" or
quadratic form, so that the pit where this is done is a "quarry"
in English. "Dice" are among the most common examples of a small
"four-cornered" body. Thus we find them in Roman mosaics as
tesserae, giving rise to the English word *tesselated* (from the Greek
word *téssera*, "four"; see p. 104). Another word of Greek origin
is *trapezium*, which in German and English is a four-sided geometrical
figure, while the "trapeze" is a device used in acrobatics: the Greek
trápeza is actually *(te)trápeza*, a "four-foot" or table with four legs
(< *péza*, "foot" and *tétra*, short form of *téttares*, "four"). (In
Modern Greek *trapézi* is a "table" and *trapeza* a "bank," revealing
the latter's origin on the money changer's table — Trans.)

The Italian word for "to cut into four parts, to quarter" is *squartare*; for "to hoard, to be stingy" Italian has the amusing expression
squartare lo zero, "to divide zero into four." The Italian hangman

is called, among other things, *squartatore* from the old custom of quartering a man's corpse after hanging him.

People tend to think of the square not only as something with four corners, but also as something regular, something orderly. Thus a "square man," in Greek *tetrágonos anér*, is a good, a righteous man, and in Italian the expression *essere fuor di squadra* means "to be unsuitable" (literally: "outside the square"); the English "it does not square with me" is similar. The French *tête carrée*, however, is a stubborn, obstinate "squarehead," whereas the Italian *testa quadrata* also has the meaning of a capacious and hence a "clever, intelligent head."

Now for a Germanic example: the British quarter-penny was called the "farthing" < Anglo-Saxon *feorth-ing*, a "small fourth." A small "quarter" and thus a little piece of cake is a *farle* for short, a fourth of a keg (as of butter) is a "firkin" in English, and a quarter of a grain measure, the Scottish bushel, is a *firlot*. If we add to these the "fourth" of Romance origin, the "quarter" and the "quadroon" or "quarter-blood," the racially mixed person who has one white and one mulatto parent (so that his blood is thought to be "¼ black"), we have a very interesting "quarter"-series in which the number word changes form according to the object it specifies — a phenomenon we have learned to recognize as a sign of early formation.

Let us conclude with some examples of the way in which a pure number word can become the name of an object. In the Celtic group of languages the Irish *cethir*, "four," has also come to mean "cattle" (= "four-footed"). Frenchmen use the word *quatrain* (and also *sixain*) for the goldfinch, because this bird has 4 (or 6) white spots on its tail. Similarly in Modern Greek a rose is *trianta-phyllo*, the "thirty-leaved" flower, and the nightingale in Finnish is *sata-kieli*, the "hundred-voiced" bird.

FIVE

During the 17th century, the English acquired a taste for a refreshing drink from the East Indies, which was made with five ingredients: arrak, sugar, lemon juice, seasonings, and water. To this beverage they gave the name "punch," after the Sanskrit Hindi number word *panča*, "five."

The "quintessence" is the *quinta essentia*, the "fifth being" and hence that which is "essential." According to Aristotle there were four "essences" (elements), fire, water, air, and earth; the fifth essence was the ether, the all-embracing Divinity.

The "quint" is the name for the highest string, the E-string, on the violin. Yet this is not the fifth, but the fourth string on the modern instrument; it retains the name from the old viol, which had five strings.

The French *quinte* is the jurisdictional boundary of a city with a periphery of about 5 miles; similar explanations make clear the underlying meanings of the Spanish *quinta*, "country house," *quinteria*, "tenant farm," and *quintar*, to muster all the inhabitants of a city up to the city limits for military service. There is a peculiar French expression *la quinte de toux* for a fit of coughing, from the superstitious belief that this recurs every "fifth" hour. And the French word *quinte* for "feint" and then "fancy" comes from fencing jargon, in which the "quinte" is the "fifth" thrust — three *quintes* with three different meanings. On the other hand, the French hundredweight, the *quintal*, has nothing whatever to do with "five" and is derived from Arabic (see p. 186).

The German word *Quentchen*, the "dram" or "little fifth," was originally a fifth of a ten-gram or half-ounce weight, and later became a quarter.

In the Bavarian and Austrian dialects Thursday, *Donnerstag*, is also sometimes called *Pfinztag*, whose first syllable reflects a Gothic *pinte* < Greek *pémptē*, "fifth"; thus it is the "fifth day" of the week, counting from Sunday. *Pfinztag* must have migrated with the Goths, who were Arian Christians, up the Danube River to Bavaria, for it shows the High German phonetic shift from *p* to *pf* (see table on p. 127, column II line 5); had it come from the Roman Catholic Church, it would have arrived later, and in the form *Pinztag*. This is a very impressive example of how the age and origin of a word can be revealed in its form. Thursday is *Pémptē* in Modern Greek, in which most of the days of the week have numerical names: Sunday is *Kyriaké*, ("the Day of the Lord"), Monday *Deutéra* ("Second Day"), Tuesday *Trítē* ("Third"), Wednesday *Tetártē* ("Fourth"), Thursday *Pémptē* ("Fifth"), and then Friday *Paraskeué* ("Day of Prepara tion" for) Saturday, *Sábbato(n)* (the "Sabbath").

SIX

Here we encounter the semester, for example, from the Latin *sex-menstris*, a "six-month" period, but also a word whose connection with "six" one would never suspect — *Samt*, "velvet." The Middle High German word *samat* was borrowed around the year 1200 from the Old French *samit*, which came by way of the Italian *sciamito* from Medieval Latin *(e)xamitum* < Greek *(he)xámiton*, from *héx*, "six" and *mítos*, "thread." In Byzantium this was a cloth woven from "six threads" and thus the finer cousin of the *Zwilch*, "canvas" and *Drilch*, "drill," which we discussed before (see pp. 172–176). In Italy the material was made with shorn upright threads and was known as velvet.

The *siesta* is the Spanish "sixth (hour)" < Latin *sexta (hora)*, which was midday in the old 12-hour day running from sunrise to sunset; in southern lands, of course, it has come to be the hour at which people take their midday naps.

An old German grain measure was the *Sechter* or *Sester*, from the Latin *sextarius*, a "sixth" (of the Roman measure of volume, the *congius*). From the same Latin word comes the *Ster*, which is still used in Southern Germany to indicate an amount of cut wood equal to 1 cubic meter. The word is also known in Italy as the bushel, which took the forms *(se)stario, staio,* and *satiolo,* in Italian, and *setier* in French.

In Russia waiters used to be addressed somewhat contemptuously as *Shestochka*, "little six," after the rather insignificant six-card in card games (from the Russian *šestj*, "six"). We have already mentioned the old Roman leap year, the *bisextil*, in our discussion of the word "two" (see p. 174).

SEVEN

The mysterious and superstitious significance of the number seven originated in the number of "planets" — bodies which did not have fixed places in the heavens but were "wandering" stars: the Sun, the Moon, Venus, Jupiter, Mars, Mercury, and Saturn — known at the time of the Babylonian astrologers. They were regarded as messengers of the gods, and terrestrial things like the metals, colors, precious stones, parts of the body, and later on the days of the week were ordered according to their pattern. This led to our present seven-day, planetary week whose days are named after the pagan gods; Latin designations such as *Dies Solis*, "Day of the Sun," etc., probably reflect Egyptian influence. The Jews, who brought the Babylonian seven-day week to Rome in the first century A.D., had other names for the days (see below).

The colloquial German expression *die böse Sieben*, "the evil seven," for a vicious-tempered wife is very remotely related to its object in that the seven-card in a certain medieval card game could take all the others, even face cards with the images of the Pope and the Emperor. The image on the seven-card was originally the Devil's, and later came to be that of a woman.

Septen-trion(al) is the French word for "North" ("northern"), and the Italians say *settentrione*, after the seven threshing oxen (in the constellation of the Great Bear) that the "ox-driver" Boötes constantly drives in a circle around the north pole of the heavens (from the Latin *tero, trivi*, "to grind, thresh"). Grain was threshed in the old days by driving oxen or other animals round and round in a circle over the grain spread out on the threshing-floor.

Our numerically named months Septem-ber, Octo-ber, Novem-ber, and Decem-ber also begin with seven, from the Latin *september* (*mensis*), the "seventh (month)," and so on. The discrepancy that September today is actually the ninth month, and hence should properly be named "November," stems from the old Roman way of counting the months of the year, which we discussed earlier (p. 25).

The English *sennight* < *seven nights* reflects the old Germanic habit of counting the nights instead of the days: *Weihnacht* ("Christmas"), *Fastnacht* ("Shrove Tuesday"), the English *fortnight* (14 days), and the German expression "8 days" for a week are all remnants of this old custom (see p. 157). The word *Woche*, "week" is common to the Germanic languages: Bishop Wulfila used the Gothic word *wiko*, "sequence to which we come" for the Christian time interval of seven days; the Old High German *wohha* is related to the word *Wechsel*, "change." The Greek word for week, in contrast, is *hebdomáda*, a "seven-ness"; correspondingly in medieval Latin it was *septimana*, whence came the Italian *settimana* and French *semaine*, but the word "weekly" is *ebdomario* in Italian and *hebdomadaire* in French, from which comes the French word for a weekly periodical.

Apart from the Sabbath, the opening day of the week, the Jews named the days of the week by ordinal numbers from the "First" to the "Sixth" day after the Sabbath. The early Christian church adopted this manner of ordering the days of the week, which was used also by the Greeks and the Goths, with the Sabbath being the holy or feast day, Sunday the first and Friday the sixth day of the week. Thus, as we have seen, Thursday, as the fifth day after the Sabbath, came to be called *Pfinztag*. Later on the Church shifted the holy day from the Sabbath to Sunday and then counted Monday as *feria prima*, continuing through Friday, *feria quinta*, followed by *sabbatum*; the Roman *feria* was a "day of celebration on which workmen and tradespeople rested" (> Italian *fiera*, French *foire*, "fair"); in the Church calendar each day is the holy or name day of a particular saint.

From the standpoint of cultural history, it is extremely interesting that the names for the days of the week were more-or-less permanently adopted by various nations; for example, the Germanic and Romance peoples generally did not keep the numerical names at all, so that in German only *Mittwoch*, "Wednesday," and *Samstag*, "Saturday," succeeded in replacing the names of the heathen gods Mercury and Saturn. Two less commonly known examples, which also reveal the borrowing of foreign number words, are the Russian and the Hungarian days of the week (see tables on pp. 98 and 113):

	Russian	Hungarian
Sunday	*Voskresenje*, "Resurrection"	*Vasarnap*, "Market-day"
Monday	*Ponedjelnik*, "not yet work-day"*	*Het-fö*, "seven-head"
Tuesday	*Vtornik*, "the second day"	*Kedd*, from *Kettö*, 2
Wednesday	*Sreda*, "middle day"	*Szerda*, "middle" (Russian borrowing)
Thursday	*Četverg*, "the fourth day"	*Csütörtök*, "the fourth" (Slavic)
Friday	*Pjatnitsa*, "the fifth day"	*Pentek*, "the fifth" (Slavic)
Saturday	*Subbota*, "the Sabbath"	*Szombat*, "Sabbath" (Greek)

* *Editor's note:* This is Menninger's translation. The Russian word means something like "the day after the end of the week."

NINE

The "nones" are the prayers offered at the ninth hour, Latin *hora nona*, or 3 o'clock in the afternoon according to the old twelve-hour day (cf. "siesta," p. 181); this is also the origin of the English words *noon, forenoon,* and *afternoon*. On a different plane, prostitutes in ancient Rome were called, among other things, *nonariae* because they did not open the doors of their houses until the ninth hour.

TEN

The tithe (*Zehnte* in German) was a tax based on an ancient Jewish form which was adopted by the Christian Church and later by the Roman authorities; it was a tenth part of the harvest and was collected in a "tithe barn." Since this tax was prevalent everywhere, we find the same formation in the English word *tithe* < Anglo-Saxon *teotha*, Old High German *zehanto*, and French *dîme* from the

Latin *decima* (*pars*). *La Dîme* was the title given by the Dutchman Michael Stevin in 1585 to his textbook on decimal fractions; the same title in Flemish was *De Thiende*. In ancient Greece a *dekátē* or tenth part of any booty was set aside as an offering to the gods.

The word *decimate* goes back to the Roman custom of having only every tenth man pay a penalty instead of extracting it from the whole body of men. Thus if a Roman legion mutinied, every tenth legionnaire was put to death. The medieval saying, *Clericus clericum non decimat*, "a churchman does not decimate a churchman," is the equivalent of the German proverb, *Eine Krähe hackt der anderen kein Auge aus*, "one crow does not peck out the eye of another."

Similarly from Roman military language came the words *Dekan* in German and *dean* in English: in a Roman army camp a *decanus* was a "corporal" in charge of a tent containing ten soldiers. In the Middle Ages the term was elevated to refer to the head of a university faculty or an ecclesiastical official. The ambassador with the greatest length of service among those accredited to a government is known by the French title *doyen*, the dean of the diplomatic corps in a particular capital.

Boccaccio's *Decamerone* (Decameron) was so titled because it is a collection of tales told during a period of "10 days" (Greek *déka*, 10, and *hēméra*, "day"). The German word *Decher*, "dicker," was originally the term used in the fur trade for a bundle of 10 pelts. This designation goes back to the days when the Germanic tribes had to pay a *decuria*, a "tithe," of hides as tribute to the Romans (*c* > *ch* as a result of phonetic shift; see table on p. 127, column I, 1).

From the Latin *deni*, "every ten" came the *denarius*, the "penny" in the Carolingian coinage system (Italian *denario*, "money"; French *denier*) whose initial letter *d* became the abbreviated symbol for both the English *penny* and the German *Pfennig* (in the flowery form ₰). But the Persian *dinar*, which belongs to the same family of related words, was the name of a gold coin. "Penny" goods are called *derrata* from which are derived *denarata* in Italian and *denrée* in French.

In many countries the army drew freely on number words and even numerical ranks for the titles of various officers and commanders of units. The "corporal" or leader of 10 men in the Roman army was a *dec-urio*, just as the *cent-urio* or commander of a "century" derived his title from the Latin word for 100. In Hungarian a corporal is called *tiz-edes* (10), a captain *szaz-ados* (100), and a colonel *ezer-des* (1000), and the Turkish language and army had the exact equivalents in *on-baši*, *yüz-baši*, and *bin-baši* (*baši*, "head"); other military ranks based on the numbers of men commanded were the Greek *hekatónt-archos*, Hindi *śata-patis*, Gothic *hunda-faþs*, and the Old High German *zehanzo-herostos*, the former "100-lord" or captain.

TWENTY

Who would have thought that "hussar" has anything to do with twenty? But the Hungarian word *husz* is 20, and *hus-ar* means "twenty-worth." In the 16th century the Hungarian King Mathias

Corvinus decreed that each group of twenty households in a town was to furnish one soldier, mounted and fully equipped, a "hussar," for the army.

FORTY

Among the Semitic peoples the number 40 has been a prominent and commonly used round number (as in the tale of "Ali Baba and the Forty Thieves" in the *Thousand and One Nights*). It occurs very frequently in the Bible: Moses waited 40 days for God to appear on Mt. Sinai; Jesus lived for 40 days in the wilderness; and in the Deluge "the rain was upon the earth forty days and forty nights" (Genesis 7:12). The Christian (*Dominica*) *quadragesima*, properly speaking the "40th Sunday" but actually the sixth, comes 40 days before Good Friday. From the Latin number word were formed the Italian *quaresima* > French *carême*, "Lent." In Greece Lent is still called *sarakosté* < (*tès*) *sarakostós*, the period "of 40" days.

Yet a similar derivation for the Russian *sorok*, 40, from the Greek (*tès*) *sarak(ostés)* is impossible on both phonetic and historical grounds. The word *sorok* is lacking, for example, in Church Slavonic which was strongly and directly influenced by Greek. The Russian number word was probably derived instead from the Old Norse *serk(r)*, "pelt"; this found its way into Russian popular speech through the fur trade between the Slavs and the Germanic Norsemen, who counted 40 furs to 1 *timbr* (*Zimmer*, "room") and 5 *timbr* to 1 *serkr* (cf. *Decher* and *dicker*, p. 184; for *serkr* see the German *Bär-serker* ["berserker"], "the wild warrior in the bearskin").

The quarantine period, during which arriving ships were held for observation in case there were any diseases aboard, was decreed for the first time in 14th-century Venice as *quaranta giorni*, "40 days," in reference to the Biblical number 40. Both the practice (the method of protection against the importation of diseases) and the word were brought to Germany in the 17th century by the French (*quarantaine*, "aggregate of 40 units").

FIFTY

The "fiftieth (day)" after Easter was called *pentēkosté* in Greek and was brought into Church Latin as a foreign loan-word: Italian *pentecosta*, French *pentecôte*, English *Pentecost*. The Greek word came into German through the Christianized Goths along the Danube. The Gothic *paintekuste* became Old Saxon *pincoston* and Dutch *pinksteren* and, after undergoing the Second or High German Phonetic Shift (see Table on p. 127), appeared in Modern German as *Pfingsten*. For its analogue in cultural history, see *Pfinztag* (p. 181). These linguistic migrations suggest that the Germanic peoples became familiar with Christianity in Southeastern Europe sometime around the 5th or 6th century, as indicated by the two Germanic phonetic shifts.

In England the same holiday has been known from the earliest days as *Whitsunday*, or "White Sunday," because Pentecost was a favored day for baptism and the newly baptized church members dressed in white. The German *Weisser Sonntag* for Roman Catholics is the

Feast of the First Communion, which comes eight days after Easter. Easter, however, is the old heathen word for the feast in honor of the goddess of spring, Ostara, whose name is preserved in both English and German (*Ostern*). The Catholic Church also took over the name *Pas-ha* with the Jewish Passover festival whence the Italian *pasqua* and French *Pâque*. The Jewish festival, of course, commemorates the Exodus of the Israelites from Egypt (Hebrew *pesach* = "deliverance").

SEVENTY

Septuaginta (Latin "seventy") is the name of the Septuagint, the Greek translation of the Old Testament, which was purportedly the work of 72 learned Jews in Egypt around 200 B.C. The Jews who lived in Egypt at that time spoke only Greek.

HUNDRED

The Latin *centum* appears in the now extinct German words *Zentgraf*, "hundred-count," and *Zentgericht*, "hundred-law": the Medieval Latin word *centa* (shortened from *centena*) was, in the legal language of the Franks, a settlement known as a "hundred"; later it came to be the fourth part of a *Gau*, a German province.

The "centner" was derived from the Middle High German *zentenaere*, from the Medieval Latin *centenarium* (*pondus*) a "100-(weight)," which is precisely translated by the English word *hundredweight*. The Dutch word *centenaar* clearly preserves the original Latin form. This weight had a remarkable history in the French language, in which it is called the *quintal*; it is *quintale* in both Italian and Spanish. The word was drived from the Latin *quintus*, "fifth," but has no discernible connection with the actual number 5. What happened was that the Latin word *centenarium* became *kintar* in Moorish Spain, and the Romance words for the hundredweight arose from this, as a result of extensive trade with the Arabs.

In ancient Greece the "hecatomb" < Greek *hekatón*, 100 + *bous*, "cattle," was a sacrificial offering of 100 cattle such as Pythagoras, among others, presented in gratitude to the gods after discovering his famous theorem.

THOUSAND

Roman distances were measured in *mille passuum*, "1000 (military) double steps or paces" of 5 feet each, and thus about 1.5 kilometers; from the Latin *milia*, treated as feminine singular, comes our word "mile." The German word *Kohlenmeiler* ("charcoal pile") is derived from the Latin *miliarius*, "containing 1000 pieces," and has been used in the Eastern Alpine region to designate the amount of wood in a layered pile of logs.

Thousand is often used as a round number in German: *Tausendschön* ("daisy"), *Tausend-fuss* ("centipede, millipede"). Thus it may seem remarkable that in other languages, other numbers have had their meanings inflated in exactly the same way: in Modern Greek, for example, the daisy and the centipede have names which literally mean "five-beautiful" and "forty-foot" respectively. Since numbers that are not numerical ranks become semantically inflated mainly for mystical reasons, their interpretation is no longer our concern. We shall merely mention the flower called the milfoil or yarrow, which is *Schafgarbe* in German, because its botanical name *mille-folium* was Germanized not to "1000-leaf" but, probably because of the very similar sound of the Greek word *meló-phyllon*, to "sheep-leaf" (*Schaf-blatt*).

We shall conclude appropriately with the artificial decimal system of measures that was devised in France at the time of the French Revolution and officially adopted in Germany in 1872. These measures use the Latin names of the numerical ranks to form the fractions of the unit measure and the Greek names for its multiples. To indicate length, for example, we have the following measures:

milli , centi , deci meter < meter > (deka-, hecto-), kilo-meter.

NUMBER

Now we have finally come to the end of our discussion and interpretation of number words. But where does the German word *Zahl* ("number") itself come from? In Germanic we find a *tal*, which appears for example in the Gothic *talzian*, "to teach," and in Old Norse as a "report, narration, story." Thus we have the English word "tale," as well as "to tell," and "to talk." In the King James version, Psalm 147:4 reads:

"He telleth the number of the stars; he calleth them all by their names."

The English expression *all told*, which means "all in all" and literally "all counted," uses the word in this sense, and a *teller* in both England and the U.S.A. is a cashier in a bank; in the British Parliament he is the voice or vote "counter." The Middle High German word *tal* likewise has both meanings, "number" and "speech, discourse." Hence it follows that among the Germanic peoples *tal* was a much-used expression related to their domestic and agricultural economy and the calculation of its yield ("accounts"). If we trace this word back through its Indo-European relatives, we come up, surprisingly, against the Latin words *dolare*, "to hew, cut, chop," *dolium*, "tub, cask, drum," *dolere*, "to grieve, feel pain," and *delere*, "to destroy," all descended from the root *del*, "to cleave, hew, cut." Thus we can understand the original Germanic word *talo* as a "notch" carved into a stick for the purpose of counting. The old German measure of length, the *Zoll*, which was originally a small notched stick, also belongs to this class of words.

But this word does not occur outside the Germanic languages. Instead, we have the Greek *a-ri-thmós*, "number," *né-ri-tos*, "numberless, innumerable," and then the Old Saxon and Old High German *rim*, "series, number," and the Modern German word *Reim* ("rhyme") — all these, through their common element -*ri*-, betray their origin in the Indo-European root *rei*- or *re*-, "to count, to arrange in order" (whence our word *ri-te* meaning "holy order"; we have already come across it in the word *Hund-ert*, see p. 130).

The Latin *numerus* (> French *nombre*) has give us the German *Nummer* and the English "number"; this goes back to a root *nem-*, "to count, order, arrange," which appears in the Greek words *némō*, "I divide" and *nómos*, "law," and the Latin *nummus*, "coin."

With this we approach the end of our study. Our hidden number words have offered us some unusually clear glimpses of the workings of early man's mind and of his verbal sureness of aim. But first and foremost, they have shown us the enormous and unsuspected importance of number words in his life and environment. He did not use his "one — two — three" just to count his sheep and cattle. To early man, number words were far more meaningful. Each had its own individual features, each was imbued with visual significance and had all the vitality of a young word still rooted in the everyday life of ordinary people.

Now that we have really come to the end of our wanderings through the realm of spoken numbers, let us pause for a while and look back over the road we have traveled. Although we may have had some notion, when we set out, that the number sequence was "made up" for the express purpose of ordering men's affairs, we have now come to recognize its most fundamental and important quality, the evolution of the number sequence.

The Evolution of the
Number Sequence

The number sequence was not created or "made," it did not spring more or less fully formed from the mind of a single man of genius — it grew up and evolved slowly and randomly with man himself and his various languages. Like a frail plant, it grew and budded timidly, going from "I — you" to "one — two" and then on to three and four, which was the first of the early limits of counting. These limits became articulations in the number sequence, like the joints in a cornstalk. We found more of them at the ranks 10, 100, and 1000.

Yet the vigesimal group made up of man's fingers and toes also left frequent visible traces. Limits of counting, groupings, and changes in word form are all evidence of the slow evolutionary growth of the number sequence, as are the specifications of words meaning "many," and foreign loan-words.

We saw the surprising number and variety of "number towers" which individual peoples erected, although we had assumed at the beginning that there was only one such tower of numbers — our own — and that all the others differed only in being expressed in a different language. We saw number sequences made up of gradations of 10 and 20, continuous and interrupted sequences, cumbersome sequences which pile up layer upon layer, and fast-climbing sequences in which the pure pleasure of building up the series almost gave wings to the numbers and their meanings. This multiplicity is profoundly impressive, for it is manifested not only in the actual, practical, day-to-day existence of various peoples, but also in their intellectual and spiritual lives and in their mental creations. Although the number sequence is conceptually unique and stands alone, the many different attempts and tendencies that surround it constitute an absorbing intellectual drama which leads up to its ultimate denouement — the number sequence as we know it today. This unfolding drama casts an especially strong spell when we view it against the panorama of world history.

Number words and history! The interweaving of our number sequence with the Indo-European family of languages has enabled us to look far back into prehistory, where only shadowy outlines are to be seen, and where we learned with astonishment that many now distinct and separate peoples at one time spoke the same language. This mother tongue first built up a number sequence out of gradations of 10. We have also seen that the daughter languages became separated from their mother tongue some "thousands" of years ago, and that some of these enjoyed a rapid and rich development while others held on with difficulty to their secondary positions, and were forced to borrow from their sisters when they did not have the strength to evolve for themselves.

Then let us recall the abundance of events that have taken place in the bright light of recorded history. To mention only a very few: the Norsemen brought the Germanic vigesimal gradation to France. Decayed and hidden number words in the English language bear witness to an earlier Celtic occupation of the British Isles. Christianity with its Latin number words and habits of counting displaced the

Old Norse custom of "over-counting." The same Norse over-count testifies to the contiguity of the Germanic peoples with the Finns and reveals the traces of a once voluminous Scandinavian trade that extended into Central Russia as far as the Urals; this is curiously evidenced by the Russian number word *sorok*, 40, which the Norse fur-traders left behind like a foundling, in the Russian number sequence.

FOREIGN LOAN WORDS

These are among the best witnesses to past contacts and relationships between various nations. The most magnificent example of all is provided by the Hindu-Arabic numerals, which we too borrowed and which are now commonly used all over the world, exemplifying the most significant symbol of mankind's universality today.

Spoken number words offer an abundance of examples of language borrowing one or more foreign numerical ranks because it had not yet formed its own. For instance, Finnish (*sata*, 100), Hungarian (*szas*, 100), and probably also Slavic (*suto*, 100) have taken over the second decimal rank from one of the Aryan tongues (*satam*); and the Hungarian number word *tis*, 10, very likely has the same origin (*dasa*). There is no doubt that the Finno-Ugric peoples at one period lived within the general sphere of influence of the Persian language, probably in Eastern Russia (on this, see also p. 111). The Finnish number sequence from 11 through 19 follows the Germanic pattern.

Foreign borrowings have been even more numerous in the case of the third decimal rank. The Germanic word for this (*þusundi*, "thousand") was borrowed by the Lithuanians (*tukstantis*) and the Slavs (*tysešta*), and from the latter in turn by the Finns (*tuhata*). The Christian church brought the Greek word *chilioi* into Bulgarian and Serbian as *hilijada*, where it stands alongside the Germanic form. Similarly the Latin *mille* migrated into Celtic (*mile*) and Basque (*mil*). The Hungarian language adopted a foreign word for "thousand," *ezer* < Persian *hazar*. This Persian word is very widespread: we find it in Armenian, in the languages of the Caucasus region, in Crimean Gothic, and even in Hindustani, into which it was introduced along with many other Persian and Arabic words, because Persian was the official court language of the Moslem princes of Northern India. The Slavs in the Ukraine have not one but three words for 1000: the Germanic *tausent*, the Hungarian-Iranian *jezero*, and the Slavic *tysesta*.

The borrowings of still higher number words are even more extensive — we need only think of the almost universal word *million*; forms of this word, in addition, have often changed their numerical value, as with the Persian *läk*, 10^4 < Sanskrit *laksa*, 10^5 or the Persian *kurur*, 5×10^5 < Hindustani *karor*, 10^7. And we should also mention the borrowing of Chinese number words by the Japanese before we go on to examine the Gypsies' number sequence.

What sort of people are the Gypsies? Hungarians? When they first appeared in Western Europe around the beginning of the 15th century, they themselves gave people to understand that they were descendants of the ancient Egyptian race; for this reason the English still call them (*e*)*gypsies* and the Spaniards *gitanos*. But let us take a look at their number sequence:

1 *jek*	6 *šob*	20 *biš*	70 *efta-var-deš*	7 × 10
2 *dui*	7 *efta*	30 *driganta*	80 *ochto-var-deš*	8 × 10
3 *trin*	8 *ochto*	40 *dui-var-biš* 2 × 20	90 *ena-var-deš*	9 × 10
4 *štar*	9 *ena*	50 *jek paššel* ½′100	100 *šel, jek-šel*	
5 *panč*	10 *deš*	60 *tri-var-biš* 3 × 20	1000 *deš-šel, jekzeros* or *tisikos*	

Egyptian? Hardly. This is an obvious miscellany of Indo-European number words. Wherever they may have come from initially, since time immemorial the gypsies have been wandering in Indo-European lands. The first units of their number sequence suggest that their initial home may have been somewhere in the Indo-Iranian linguistic region.

More specifically: the gypsy number words for 1 ... 6, and 10 come from the Indo-Iranian group; their 20, *biš*, also goes back to the Persian *bist* < Avestic *visaiti*, Sanskrit *vimśati*; their 100, *šel*, is closely related to the Afghanian *sil* < Iranian *satem* (wherein $t > 1$); their 7, 8, and 9 are Greek and their 30 is Roman in origin.

As the number sequence is further built up, these gypsy words continue to show a variety of patterns: their 20, 40, and 60 are old 20-groups; the intervening steps 30, 70, and 90 were formed from 10 and inserted later, drawing the 80 out of the vigesimal category into their own. 100 as an old grouping exerted a reverse influence on 50, which is ½ × 100. The word *deš-šel*, 10′100, shows that 1000 is an old limit of counting; but along with this indigenous gypsy formation there are also synonyms borrowed from Iranian (*jek-(e)zeros*) and Slavic-Germanic (*tisikos*).

Here we shall end our voyage through the cultural history of spoken numbers, with the extraordinary insights it has given us into the developing human mind and into the lives of primitive peoples and their languages!

From this fertile soil the number sequence grew naturally, much like a plant. It hesitantly tried the first steps forward, to two and then to four, then stood still, then pressed on further as far as ten, where it gathered itself and pushed boldly up to a hundred and then a thousand, and there stood still again, always keeping pace with commercial and social evolution. Then came the point at which its natural growth ceased, from which it no longer evolved unconsciously; now man deliberately drove it ahead over every hurdle and obstacle, past million and billion up to infinity, far beyond the scale or grasp of his ordinary life. Now the popular roots of the number sequence died out: it became an "abstract" sequence, an extraordinary and ingenious, but purely "technical," intellectual tool.

This artificial extension of the number sequence was the verbal analogue of the world-wide currency which the Hindu numerals, "our" numerals, have attained. These two events — the upward growth of the spoken number sequence to infinity and the world-wide spread of the Hindu-Arabic numerals — symbolically express the dominant position which numbers achieved in man's conceptual environment and have occupied ever since. It is significant that both these events occurred in the Western world and at the threshold of modern history, around the year 1500.

Although both events have the same symbolical importance, they are quite different in that the verbal expansion was inward, so to speak, and (until it reached the artificial word *million* and its successors) remained closely tied to a particular language, whereas the written numerals moved from nation to nation, from the European West to the Asiatic Far East and from the "civilized" world of white men to the lands inhabited by still primitive peoples. Number words and numerals: from the very beginning, writing has been the wandering sister of the stay-at-home spoken language. Even "our own" system of writing came from outside our own linguistic region.

This brings us to the concepts and ideas associated with written numerals. Numbers, of course, do not just trip lightly over the tongue — both economic and social necessity have forced men to write them down and to use them for keeping records and for making computations. Today we possess a complete sequence of numerals and have developed procedures for computation. But we shall find, just as with spoken numbers, that written numerals did not suddenly appear fully formed against a blank historical background; they too evolved through many different forms devised and used by individual cultures. We shall see that people, before they could conveniently make computations and carry out operations on paper with written numerals, made ingenious use of the abacus and the counting board. The ability to calculate solely with written numbers was provided for the first time by the Indian numerals.

Now, in the second section, let us turn to the cultural history of written numerals and the computations made with them.

WRITTEN NUMERALS
AND COMPUTATIONS

Preface

The next section of our cultural history of numbers deals with those written numerals and computations that have been used by ordinary people in simple domestic calculations. The main subject of this section is the development of the numerals and computations used in Western Europe. From time to time, however, some light may be shed on this development by occasional glimpses of the numerals and computational operations of other peoples, whose cultures may be either closely or distantly related to our Western civilization.

The Indian numerals which we use today in writing and computation will be discussed before we take up the Roman numerals and counting boards. The history and influence on written computations of the now forgotten counting board or abacus will be the subject of a thorough treatment, more exhaustive than any other published thus far, to my knowledge. It is not only an absorbing segment of the history of numbers and mathematics, but also and above all it forms the background for the fascinating events which accompanied the introduction of Indian place-value notation into the Western world. We come upon this seldom-trod path in the midst of the intellectual revolution from which Western man emerged to "take the world into his hands." Numbers have been one of the most important means by which he learned to master his environment.

The previous section dealt with number words — the number sequence and spoken numbers; this section concerns itself with numerals. Each section, representing spoken numbers or written numerals, can stand alone, although many themes are common to both and many developments and events will not become fully clear until light has been cast on them from both directions.

Introduction

When we speak of Roman or Egyptian or even our own numerals, we almost always mean the "official" numerals in which a state's records are kept and which all individuals are thus obliged to know. Most people are completely unaware of any other kinds of written numbers.

Yet historically every "official" system of numerals is preceded by some set of primitive number symbols, signs which individuals or associations or villages have used to "write" down numbers, in the form of notches carved in a piece of wood or of knots tied into a string cord, or in some other popular manner. A study of these pre-"official" recorded numbers will acquaint us with a multitude of now generally forgotten things that are still alive in language and custom, and above all will provide valuable insights into the "early" concepts which underlie or antedate written representations of numbers. These insights are important, moreover, because even the formal numerals frequently embody early forms which can be understood only with a knowledge of various folkways. A surprising amount of light is thrown on Roman numerals, for example, by notched tally sticks.

To "write" a number is to make it visible and record it permanently, in contrast to speaking it. Thus finger counting, the representation of numbers by various signs made with the hands and fingers, makes numbers visible but nevertheless evanescent, in a sort of intermediate stage between spoken and written numbers. These "finger numbers" are not merely child's play: they had great importance and significance in earlier times, as they still have today in many parts of the world.

These two small bits of information about finger counting and primitive number symbols already suggest that a people's formal written numerals do not weigh as heavily in the scales of culture as spoken numbers do. The written numerals are usually foreign imports; only seldom, as in the case of Egyptian and Chinese, do they grow out of the same native soil as the spoken numbers.

We should suspect from this fact that letters have also been used as number symbols. The Goths wrote alphabetical numerals in imitation of the Greeks, and so did virtually all peoples within the sphere of influence of the (Semitic) alphabetical form of writing. For many centuries they were the numerals used by Greek mathematicians.

These letter numerals and the so-called "early" numerals, whose systems were based only on the laws of ordering and grouping, were used merely to record numbers; they were not well suited to computations. For their numerical calculations people used the counting board or abacus, whose now forgotten history we shall also look into. We shall see its use in the world of the ancient Greeks and Romans, we shall find special forms of the abacus still in use in China, Japan, Russia, and even in our own culture, and then we shall turn our attention to the medieval counting boards used in Western Europe: to their different forms, to their remarkable ancillaries, the *calculi*, *apices*, and other counters, and finally to the main procedures used in making computations on the abacus.

Thus we shall have gained some insight into the way people during the Middle Ages were working with numbers at about the time the "new" Indian numerals began to penetrate into Europe by way of Italy, around 1500.

Where did the Indian numerals come from and how did they develop? What was new about them? By what route were they brought into the lands of the western Mediterranean and from there to northern Europe? The answers to these questions will unfold a vast drama of civilization extending from India through the Arab world into the West, from the quiet monastic cells of the early Middle Ages to the progressive offices of the Italian and German merchants and beyond, up to the threshold of modern times. We shall take part in this fascinating pageant of intellectual history, and thus understand how and why the Indian numerals came to be the only written numbers used by all the major nations of the world. These, and only these, enabled man to extend his capacity for calculations to undreamed-of heights. For better or for worse, they have subjected the world to the tyranny of numbers.

After this broad view we shall finally focus our attention on our narrower subject, spoken numbers and written numerals, this time in the Far East, in China and Japan. The culture of the Chinese and the Japanese is so unique and so isolated from the rest of the world that it is useful in two ways: it serves both as a comprehensive review of the main threads by which we guided ourselves through the complicated cultural history of numbers, and also as a counterpart, an analogue to our own world of numbers, from which we can learn what is extraneous or peculiar to our own numbers and what is unique and intrinsic to numbers and numerals themselves.

The common threads which we have followed through different times and civilizations will thus remind us that the common spirit of mankind first once began to evolve a single set of numbers, and then struck out along different paths represented by various separate cultures, until it finally arrived at the most highly perfected system, that of the Indian numerals. As these spread out over the whole inhabited world and came to be universally used, pushing aside all other numerals, they have become a symbol of the essential unity of mankind, which is manifested not at the outset, but only at the end of a long course of evolution.

Finger Counting

Finger Counting

Nunc mihi iam credas fieri quod posse negatur:
octo tenes manibus, si me monstrante magistro
sublatis septem reliqui tibi sex remanebunt.

"Now you shall believe what you would deny could be done: In your hands you hold eight, as my teacher once taught; Take away seven, and six still remain."

A Roman riddle, which was never solved during the Middle Ages (see p. 204)

Language uses words to capture numbers. But words are ephemeral. From time immemorial, mankind has tried to find some way of making words and numbers permanent. The answer to this problem was writing.

But to fix spoken words by means of a pictorial or phonetic form of writing is a very difficult and laborious task, which only a few peoples in the world have succeeded in accomplishing. Even we, for example, though we speak a language of our own, write it with symbols that we did not invent. For "our" letters are actually Roman, and the development of alphabetical writing goes back further than the Romans, the Greeks, and the Phoenicians, all the way back to the ancient Egyptians.

With numerals, however, the story is a little different. A truly native system of writing numbers which all members of the culture had to learn and use, and which was thus an "official" set of numerals, was also developed by only a small number of peoples; the numerals used by the vast majority of cultures have been foreign imports. The digits which we commonly use today originated in India; before these, we wrote numbers in the Roman fashion. But before and quite often along with such "official" numerals, people from the earliest times have always devised their own primitive ways of recording numbers, by cutting notches, or scratching strokes, or tying knots, and, they used these means exclusively for carrying on the business of their households, or their trades, or perhaps their villages: we shall discuss these primitive numerical symbols in the following chapter.

But first we will consider still another remarkable predecessor of written numerals: finger counting, numbers formed on the fingers, which is the subject of the present chapter.

The Romans had a method of representing the numbers from 1 to 10,000 on the fingers of both hands — a form of "finger-writing." We shall call these generally overlooked arrangements of numbers "finger counting," to differentiate them from the much earlier finger-gestures of primitive cultures, which very rarely went beyond the "ten" that can be counted on the fingers of both hands.

Finger counting came to Western Europe as part of the heritage of classical antiquity, and was very commonly used during the Middle Ages; thereafter it was replaced by the Indian numerals and has now been forgotten completely. Today such "finger numbers," in somewhat altered form, are used only by Arab and Indian traders in the Middle East.

The technique of finger counting seems to have been passed on mostly by word of mouth. No textbook describing it has been preserved from Roman times. This very fact may be evidence of its widespread use among common or illiterate people, for things which do not require special teachers or schools are generally not written down.

THE VENERABLE BEDE AND HIS FINGER COUNTING
One man did write such a textbook, however. This was the English Benedictine monk who is known to history as *Baeda Venerabilis*,

the Venerable Bede, one of the greatest and most important scholars of the early Middle Ages. He died in A.D. 735 (see Fig. 247). Posterity has rightly considered him "worthy to be honored."

In those times, when the barbarian migrations over the European continent were gradually coming to an end and the Carolingian empire of Charlemagne was beginning to take form, the Church sent its emissaries, especially the Benedictine monks, to Ireland and England to lay the foundations for a Christian culture. In an age when all books were copied by hand, the monasteries were the sole refuge of learning.

This early medieval learning, apart from purely ecclesiastical and religious matters, amounted to nothing more than collecting and passing on the inheritance of Roman civilization. With all the industry and diligence which only an unworldly monk could bring to the task, this heritage was brought together from all available sources and thus handed on to the later Middle Ages: *Semper aut discere aut docere aut scribere dulce habui*, "I have spent my life pleasantly in learning and teaching and writing," said the Venerable Bede, who never once left the English monastery in which he lived. We have him to thank for the only complete record of finger counting in existence, *De computo vel loquela digitorum*, "On Calculating and Speaking with the Fingers." This forms the introduction to his work on chronology, *De temporum ratione*. To the monks of the Middle Ages this "computation of the times" meant primarily calculating the highly variable date of Easter Sunday. Spring begins on March 21. In 325 the Council of Nicaea officially decreed that Easter should be the first Sunday after the first full moon of spring, so that it would never under any circumstances fall on the same day as the Jewish feast of Passover, which is celebrated on the eve of the first full moon of spring.

Easter Sunday, in turn, established the times of still other variable Church holidays: Pentecost or Whitsuntide, the day of the Ascension of Christ, the feast of Corpus Christi, which was celebrated from 1247 on, and the feast of the Trinity, after 1334. Thus in every monastery the *computus paschalis* or "calculation of Easter" had to be performed by several monks.

Bede's flexus digitorum ("*finger-bending*"). Now we shall follow Bede's own account: first in translation with brief explanatory comments and then in the original Latin text (see p. 206). As he goes along, the reader is urged to form these numerical gestures with his own fingers. The illustrated finger-counting tables (Figs. 38 and 41) give a rough idea of how these were formed, but only a rough idea, since they are not all clear and some of them are wrong. But if the reader "spells out" these numbers on his own hands, not only will the original sources come alive for him, but he will realize their true significance. Otherwise he might well mistake these finger numbers for a form of idle amusement.

"Before we begin, with God's help, to speak of chronology and its calculation," begins the Venerable Bede, "we deem it necessary first briefly to show the very necessary and ready skill in finger counting:

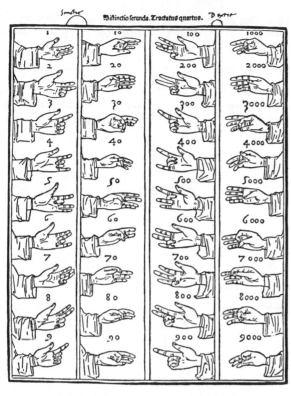

left hand   right hand

Fig. 38 Finger counting from the *Summa de Arithmetica* of the Italian mathematician Luca Pacioli, the first important mathematical work to be printed,
which was published in Venice in 1494. In contrast to the Venerable Bede's
account, the hundreds and thousands are made on the right hand.

1: If you wish to say "one," you must bend the little finger of your left
   hand and place its tip on the palm;
2: for "two" lay down the ring finger next to it;
3: for "three" also the middle finger;
4: for "four" you must again raise the little finger;
5: for "five" also the ring finger;
6: for "six" you must extend the middle finger, and then the ring finger,
   which is called the *medicus*, alone remains bent down upon the palm.

This finger was so named in ancient times because it was believed
that a vein runs directly to it from the heart. Since 6 was a "perfect"
and hence a sacred number, it was deemed worthy above all the
others to bear the ring, and thus was made on the "ring" finger. A
"perfect" number is exactly equal to the sum of its divisors — apart
from the number itself, of course. The divisors of 6 are 1, 2, and 3,
the sum of which is 6. The number 12 is not "perfect," since its
divisors 1, 2, 3, 4, and 6 add up to 16. The next three perfect numbers
are 28, 496, and 8128.

7: for "seven" extend all the fingers and bend only the little finger over
   the wrist.

In contrast to 1 (as well as 2 and 3), in making 7 (as well as 8 and 9)
the fingers were bent not at the middle knuckle but at the bottom

joint, so that they extended far over the palm or the fleshy pad of the thumb. Pacioli failed to make this distinction (see Fig. 38; it is made correctly in Fig. 42 however).

8: for "eight" lay the ring finger down next to it (see Fig. 42).

Here we have the solution to the riddle quoted at the head of this chapter. The answer lies in finger counting: if we form the number 8 on the hand and then remove the 7, what remains is the finger-gesture for 6!

9: for "nine" place the "impudent" (middle) finger next to it (see Fig. 42).

It is important to note that only the last three fingers of the left hand are used to form the units. Now we can see how all the numbers through 9999 could be indicated just by the fingers of both hands (the names of the fingers below are represented by their initial letters):

left hand		right hand	
units	tens	hundreds	thousands
L — R — M	I — T	T — I	M — R — L

In this arrangement the person making the numbers worked backward, whereas the person opposite or facing him read the numbers in ascending order from right to left, thousands, hundreds — tens, units, as we write them today.

Actually the Venerable Bede also showed them in the same order. Pacioli's description of fingers counting, written some eight centuries after Bede's, only interchanges the hundreds and thousands; the second known description, by Jacob Leupold (see Fig. 41), adheres to Bede's:

If you wish to say 10, you must place the nail of the little finger over the middle of the thumb.

For 20, place the tip of the thumb between the index and the middle fingers.

For 30, join the nails of the index finger and thumb in a loving curve ("in a tender embrace").

For 40, place the thumb beside or on top of the index finger and extend them both.

For 50, bend the tip of the thumb inward (toward the palm) like a Greek Γ (gamma).

For 60, lay the tip of the index finger over the thumb bent as above. (Thus the illustration in Fig. 38 is incorrect: the thumb should be bent toward the palm.)

For 70, place the thumb within the bent index finger so that the thumbnail touches the middle knuckle of the index finger (Fig. 38 is correct for this).

Or: place the tip of the thumb in the knuckle of the index finger and place the latter over it.

For 80, "fill" the index finger, bent as above, with the extended thumb, so that the latter's nail thumbnail touches the index finger (with its upper side).

By "fill" Bede means that the thumb is to "close" the curve of the index finger in such a way that the thumbnail touches the nail of the index finger.

For 90, lay the index finger on the base of the thumb.

Or bend the tip of the index finger back over its own base and place the thumb over it.

Figures 38, 41, and 44 do not clearly illustrate all the prescribed gestures; some of them are wrong. Jacob Leupold is obviously interpreting Bede's description to the extent of his own understanding (Fig. 41). Bede's depiction is consistent with the well written and much later Arabic sources, so that we can be fairly sure that our translation of Bede and the explanatory comments correctly reflect the true ancient finger counting (see p. 214).

Bede continues:

Up to now you have used the left hand, but you shall make 100 on the right hand as you did 10 on the left." (Pacioli interchanges the hundreds and the thousands)
Similarly all the remaining hundreds to 900.
And 1000 on the right hand as 1 is formed on the left.
And so on up to 9000.

Now the reader himself should form the numbers 21, 75, 206, and 5327 on his own fingers. He will be surprised to find that underlying these finger gestures is a positional or place-value system: the four ranks of numbers — the units, tens, hundreds, and thousands — are made with four different groups of fingers! The gestures here play the part of digits, of which there are not 9, however, but 2×9. The reason for this lies in the nature of the system, in which first 3 and then 2 fingers are available to form each rank of numbers. "Zero" was indicated by the normal relaxed position of the fingers.

Thus the place-value concept took form even in ancient times, when the Indian system of numerals — the one we use today — was not yet developed, and it was used in the early Middle Ages, when the Indian numerals had not yet been brought to the West. We shall meet the positional concept once more on the abacus, again quite independently of our modern place-value notation. The Hindu-Arabic numerals were therefore not the first in history to embody the concept of a place-value notation.

Digiti and articuli ("fingers" and "joints"). In medieval books on numbers and mathematics, all numbers were divided, after the manner of Boëthius (who lived in the 6th century A.D.) into three classes: the nine *digiti* or "fingers" from 1 to 9, the *articuli* or "joints" meaning all the numbers divisible by 10, such as 20, 700, 850, etc., and the *numeri compositi* or "combined numbers" consisting of both previous classes, like 23 and 857 (see Figs. 162 and 227).

Where did these remarkable names come from? There is no doubt that they were derived from finger counting, in which the nine units were formed with the three "whole" fingers and the tens (to 90) with the index finger and thumb together (except for 50) and in such a manner that these two always touched at the "joints." These names, to be sure, did not appear in finger counting itself, but were technical

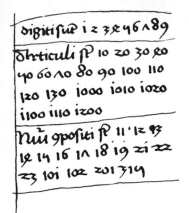

Fig. 39 The old classification of the natural numbers into *digiti, articuli,* and *numeri compositi,* from a medieval manuscript of the 14th century. Bayrische Staatsbibliothek, Munich.

Fig. 40 2000 shown on the fingers; an illustration from a copy of Bede's work on finger counting, dated 1140. From D. E. Smith, *History of Mathematics,* Boston, 1928.

terms connected with the abacus, on which they served to designate the number of places in multiplication.

Example: we wish to multiply 40 × 70, which means placing 4 tens × 7 tens on the abacus; since 4 × 7 = 28, the medieval rule reads:

decenum per X si multiplicaveris dabis digitis C et articulis mille,

"give to the digitus 8 the value 100 and to the articulus 2 the value 1000" (result: 2800).

In the Salem Algorism, a 12th century manuscript on arithmetic which we shall deal with extensively (see Fig. 232), 35 and 67 are to be added. When learning to add the units, school children say for 7 + 5 = 12, "put down 2 and carry 1"; pupils in medieval monastery schools learned:

scribe digitum 2, transfer articulum 1,

"write down the digitus 2, carry over the articulus 1."

Today the units are still called *digits* in England and *doigts* in France.

The words of Bede's original text will help the reader who knows Latin to visualize these curious finger gestures, and at the same time give him a sample of the style of medieval learned writing (from 2 on, *quum dicis* is abbreviated to *q. d.*):

1: *Quum ergo dicis unum, minimum in laeva digitum inflectens, in medium palmae artum infiges.*

2: *q. d. duo, secundum a minimo flexum ibidem impones.*

3: *q. d. tria, tertium similiter afflectes.*

4: *q. d. quatuor, ibidem minimum levabis.*

5: *q. d. quinque, secundum a minimo similiter eriges.*

6: *q. d. sex, tertium nihilominus elevabis, medio duntaxat solo, qui Medicus appellatur, in medium palmae fixo.*

7: *q. d. septem, minimum solum, caeteris interim levabis, super palmae radicem pones.*

8: *Juxta quem, q. d. octo, medicum,*

9: *q. d. novem, impudicum e regione compones.*

10: *q. d. decem, unguem indicis in medio figes artu pollicis.*

20: *q. d. viginti, summitatem pollicis inter medios indicis et impudici artus inmittes.*

30: *q. d. triginta, ungues indicis et pollicis blando conjuges amplexu.*

40: *q. d. quadraginta, interiora pollicis lateri vel dorso indicis superduces, ambobus duntaxat erectis.*

50: *q. d. quinquaginta, pollicem exteriore artu instar graecae literae Γ curvatum, and palmam inclinabis.*

60: *q. d. sexaginta, pollicem, ut supra curvatum, indice circumflexo diligenter a fronte praecinges.*

70: *q. d. septuaginta, indicem, ut supra, circumflexum pollice immisso superimplebis, ungue duntaxat illius erecto trans medium indicis artum.*

80: *q. d. octoginta, indicem, ut supra, circumflexum, pollice in longum tenso implebis, ungue vindelicet illius in medium indicis artum infixo.*

90: *q. d. nonaginta, indicis inflexi unguem radici pollicis infiges. Hactenus in laeva.*

100: *Centum vero in dextera, quemadmodum in laeva facies.*

200: *Ducenta in dextera, quemadmodum Viginti in laeva.*
300: *Trecenta in dextera, quemadmodum Triginta in laeva.*
    Eodem modo et cetera usque ad DCCCC.
1000: *Item mille in dextera quemadmodum unum in laeva.*
2000: *Duo millia* – – – *Duo* – –
3000: *Tria millia* – – – *Tria* – –
    Et cetera usque ad novem millia.

Bede goes on to show how the numbers from 10,000 to 1 million are formed using finger gestures which sound artificial and were probably rarely if ever used. Perhaps Bede himself extended the limit of counting here. Leupold's 18th-century mathematical compilation depicts these higher numbers (see Fig. 41):

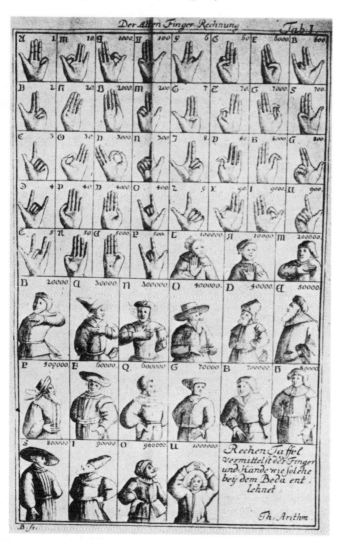

Fig. 41 Bede's finger counting illustrated in Jacob Leupold's *Theatrum Arithmetico-Geometricum*, published in 1727 — a thousand years after Bede.

10,000:   *Porro quum dicis decem millia, laevam medio pectori supinam appones, digitis tantum ad collum erectis.*

20,000:   *Viginti millia quum dicis, eandem pectori expansam late superpones.*

30,000:   *Triginta millia quum dicis, eadem prona, sed erecta pollicem cartilagini medii pectoris inmittes.*

40,000:   *Quadraginta millia quum dicis, eandem in umbilico erectam supinabis.*

50,000:   *Quinquaginta millia quum dicis, eiusdem pronae, sed erectae, pollicem umbilico impones.*

60,000:   *Sexaginta millia quum dicis, eadem prona femur laevum desuper comprehendes.*

70,000:   *Septuaginta millia quum dicis, eandem supinam femori superpones.*

80,000:   *Octoginta millia quum dicis, eandem pronam femori superpones.*

90,000:   *Nonaginta millia quum dicis, eadem lumbos apprehendes, pollice ad inguina verso.*

100,000:   *At vero centum millia, et ducenta millia, et cetera usque ad DCCCC millia, eodem quo diximus ordine in dextera corporis parte complebis. Decies autem et centena millia, quum dicis, ambas sibi manus insertis invicem digitis implicabis.*

10,000:   If you wish to say ten thousand, place the back of your left hand against your chest with the fingers extended, and point to the neck.

20,000:   place the left hand with the fingers spread out over the chest (shown differently in Fig. 41).

30,000:   point with the thumb of the extended hand to the cartilage in the middle of the chest.

40,000:   you must place the back of the hand over the navel.

50,000:   point with the thumb of the extended hand to the navel.

60,000:   grasp the left thigh from above.

70,000:   place the back of the left hand over the left thigh.

80,000:   place the palm of the hand over the thigh.

90,000:   place the left hand over the small of the back, the thumb pointing toward the groin (and thus forward).

Then for 100,000 through 900,000 do the same on the right side of the body in the same order. For 10 times 100,000 (which is 1 million), place both hands together with the fingers interlaced.

So much for the Venerable Bede. Now that he has introduced us to finger counting, we shall have something to say later on about its significance (see p. 212).

FINGER COUNTING IN ANTIQUITY

As we follow our present path, we come to a very important milestone of cultural history, St. Jerome's translation of the Bible into Latin. Jerome died in A.D. 420. This Church father retains a place in popular affection as a result of the legendary story about the lion and the thorn which he removed from its paw. Albrecht Dürer's famous picture shows the Saint in his little house, bent over his writing desk and completely absorbed in his translation of the Bible into Latin.

Jerome's *Vulgata*, the "Vulgate" or "popular" Bible, is still used by the Roman Catholic Church as its official Holy Writ.

St. Jerome also wrote commentaries on the Bible, including one on Matthew 13:8 which tells the Parable of the Sower:

But others fell into good earth, and brought forth fruit, some an hundredfold, some sixtyfold, some thirtyfold.

Jerome's comment on this passage is so curious that it is worth quoting here in the original Latin:

Centesimus, et sexagesimus, et tricesimus fructus, quanquam de una terra et de uno semente nascitur, tamen multum differt in numero. Triginta referuntur ad nuptias: nam et ipsa digitorum conjunctio, quasi molli osculo se complectens et foederans, maritum pingit et conjugem.

Sexaginta ad viduas: eo quod in angustia et tribulatione sint positae, unde et in superiore digito deprimuntur, quantoque major est difficultas expertae quondam voluptatis illecebris abstinere tanto majus et praemium.

Porro centesimus numerus, quaeso diligenter Lector attende, a sinistra transfertur ad dextram, et iisdem quidem digitis, sed non eadem manu, quibus in laeva manu nuptiae significantur et viduae, circulum faciens exprimit virginitatis coronam.

The hundredfold and the sixtyfold and the thirtyfold fruit, though they may have sprung from the same soil and the same seed, nevertheless differ greatly in their number. Thirty is a symbol of marriage: for this manner of placing the fingers, which are joined and interwoven as in a sweet embrace, represents both man and wife.

Sixty stands for widowhood, because the widow endures trouble and tribulation that weigh on her, just as (the thumb) is pressed down by the index finger which lies above it (in forming the number sixty). But the more difficult it is to abstain from the pleasures that were once enjoyed, the greater will be the reward.

But now attend carefully, dear Reader: the number hundred is transferred from the left hand to the right, where it is formed with the same fingers but not on the same hand as the hand which symbolized marriage and widowhood. The circle now formed on the right hand stands for the crown of virgin purity.

These interpretations of St. Jerome's have very graphically preserved the finger gestures for the numbers in question (see Fig. 44). They also bear witness to the popularity of finger counting in the fourth century, since the Church father has used their images to make explanations and draw analogies which would be completely incomprehensible to anyone without a knowledge of finger counting.

Moreover, today in Naples (and perhaps not only there) the finger position for the number 30 still has the same symbolic meaning as in Jerome's commentary: *molli osculo se complectens*, a "tender embrace." And a Roman writer tells us that the same gesture, but made on the right hand, served as an appeal to Venus!

Universal familiarity with finger counting, at least in late Roman times, was also demonstrated by another Church father, St. Augustine, who died in A.D. 430 when he was Bishop of the town of Hippo in North Africa, and who also provided us with much evidence about finger counting. One passage from his writings shows that not only numbers, but also calculations, were made on the fingers. He is explaining Verse 21:11 of the Gospel according to St. John, in which Simon Peter cast his net at Christ's command and brought up 153 fishes. St. Augustine tries to see a mystical significance in this number: it is made up, says he, of all the numbers from 1 through 17; but 17 in turn is composed of 10, the number of the Ten Commandments, and 7, the number of the Holy Spirit, and is thus particularly significant. Then Augustine continues:

> *Apud vos numerate: sic computate. Decem et septem faciunt centum quinquaginta tres: si vero computes ab uno usque ad decem et septem et addas numeros omnes — unum, duo, tria: sicut unum et duo et tres faciunt sex; sex, quattuor et quinque faciunt quindecim: sic pervenes usque ad decem et septem, portans in digitis centum quinquaginta tres.*

Count for yourselves and figure as follows: 10 and 7 make 153, for if you count from one to seventeen and add together all the numbers — 1, 2, 3: 1 and 2 and 3 are 6; 6 and 4 and 5 are 15; you finally arrive at 17 and carry 153 on your fingers.

Thus as one counts up to 17, the fingers constantly join together the previous numbers until they finally "carry" 153. This example is very important, for it reveals the main purpose of finger counting: to record temporarily the intermediate sums in mental arithmetic: $1 + 2 = 3$, $+ 3 = 6$, $+ 4 = 10$, $+ 5 = 15$ and so on.

Macrobius, a Roman contemporary of Augustine in his own writings, drew heavily on older works, relates that

> Many people regard Janus as the god of the Sun; thus he is often shown in statues as forming the number 300 with his right hand and the number 65 with his left (*manu dextera trecentorum, et sinistra sexaginta quinque numerum retinens*). This symbolizes the days of the year, which is the Sun's chief creation.

This passage is taken almost word for word from Pliny the Elder. Pliny is referring to the statue of Janus in Rome, which was shown forming the number 365 on its fingers as an attribute identifying him as the god of time and of the year (*digitis figuratis CCCLXV dierum nota*). With Pliny, who died in A.D. 79, we arrive in ancient Rome where, as he informs us, there was a statue in the forum of Janus, the two-headed god of all beginnings and endings ("January"), which showed him representing the 365 days of the year on his fingers. There could scarcely be better evidence of the universal familiarity with finger counting in Rome than this passage, which describes it casually as nothing at all out of the ordinary.

The Roman satirist Juvenal (died A.D. 130) also betrays his knowledge of finger counting when he says of Nestor, the oldest and wisest of the Greek princes at the Siege of Troy:

> *Felix nimirum qui tot per saecula mortem*
> *distulit atque suos iam dextera computat annos,*

Happy is he who so many times over the years has cheated death
And now reckons his age on the right hand.

For the fingers of the right hand were used to form the hundreds!

Another confirmation is provided by Quintilian (1st century A.D.), the famous teacher who was celebrated even in the Middle Ages, and who said: "The uneducated man betrays himself less by his fear of the answers than by mistakes in computation because of the uncertain and unclear numbers he forms on his fingers."

And with Firmicus Maternus, who wrote a textbook of astrology in A.D. 340, we look into a schoolroom:

Vides ut primos discentes computos digitos tarda agitatione deflectant,
Do you see how awkwardly beginners in computation bend their fingers?

The finger numbers from 1 through 15 have been preserved on some very rare Roman *tesserae* (counters); of these, two beautiful ivory specimens showing the numbers VIIII and VIII formed with the fingers, are illustrated here (Fig. 42). They come from the collection of Lord Hamilton, the English ambassador to Naples during the Napoleonic Wars. Since such *tesserae* have been found only with numbers up to 15, they were probably counters in some kind of game.

Fig. 42 Two Roman counters showing the numbers VIIII and VIII formed by finger gestures. Ivory, 2.9 cm in diameter, 2 and 4 mm thick. Probably 1st century A.D. British Museum, London.

More evidence of the similarity between individual numbers in Roman finger counting and those described by the Venerable Bede could easily be cited, but let us turn our attention for a while to finger counting in ancient Greece. Here the evidence is extremely scanty, but the 5th-century B.C. comic dramatist Aristophanes does say once in *The Wasps*: first the income of the city of Athens is to be counted, and then the expenditures of the judicial bench, "not indirectly by using counters, but right on the hand":

μὲν λόγισαι φαύλως μὴ ψήφοις ἀλλ'ἀπὸ χειρός.

This passage is as vague as the expression *pempázein*, which refers to nothing more specific than a numerical grouping of five stemming directly from the hand (see p. 41).

But why is Greek testimony so scarce? The Greek evidence comes from early writings; the Roman from writers who lived during and after the time of Christ. The Roman period is historically connected with our own, and its character has been preserved to the smallest details. We are probably not far wrong if we attribute the advancing use of finger counting to the expansion of the Roman Empire.

Many barbarian tribes, speaking a multitude of foreign languages, lived outside the Empire but within the sphere of Roman culture and trade. As the commerce of Rome expanded, the merchant's needs also grew: the numbers with which he worked became larger and so did the computations he had to master. This is where the counting board or abacus was useful. Intermediate computations could be recorded in some way by a slave, as St. Augustine shows in a very informative example. In the absence of paper, the numbers were "written down" on the fingers at once and thus made visible. There was always a division of labor between the computer and the recorder in a counting house, as we see very clearly from medieval illustrations: one man calls out the sums and amounts, another makes calculations, and a third writes (see Fig. 193 and compare with the picture of the Roman gravestone, Fig. 138).

The use of finger counting was also stimulated from the outside, by real problems that arose in trading with peoples who spoke other languages. For numbers formed on the fingers could be universally understood by merchants, even without speaking the language of their barbarian partners.

One even senses that finger counting became a kind of technical jargon of commerce. Were the numbers formed by the fingers the source of the number words that grew up out of the earliest stages of mankind's development? Of course not. Did they spring from a single creative mind, as a sort of offspring of learning? No, although there was something artificial about them. They had elements of both: a custom that grew up anonymously to meet the needs of ordinary life, and a deliberately thought-out answer to a particular problem; they were a sort of commercial language in themselves, like the pidgin (= "business") English used in the Far East and the South Pacific.

This interpretation finds support in the custom of

FINGER COUNTING IN ARABIC AND EAST AFRICAN COMMERCE TRADE

In the seaports and market places of the Red Sea, Arabia, and East Africa, merchants have evolved a finger language that is understood in every market of every country in the region. Buyers and sellers come to terms underneath a cloth, a fold of garment or a strip of muslin from a turban, by touching the fingers of each other's hand and thus bargaining in complete privacy. Our illustration shows two Indian pearl traders settling the price of a pearl under a handkerchief (see Fig. 43).

The privacy thus achieved is one of the main purposes of this commercial finger language, for in these countries all transactions are concluded in the open market place. If idle bystanders and others are not to be let in on the price, seller and buyer must come to an understanding both silently and secretly. It also has the advantage

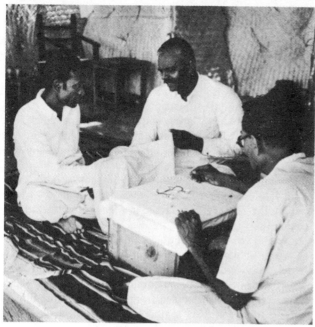

Fig. 43 A pearl merchant of Southern India bargaining with a buyer by means of finger gestures made under a cloth. This rare picture (taken in 1956) by the Indian photographer Nimal Abeyawardene was very kindly lent to the author by Th. Martens of Munich.

of allowing traders to deal with each other directly, without letting a possibly untrustworthy interpreter come between them. Not only the European, Indian, Arab, and Persian merchants, but also those from the interior of the continents, the Abyssinians, Somalis, and Bedouins, all understand the finger language of the shores of the Indian Ocean.

One of the rules of Arab and East African commerce is the following: if the buyer's hand touches the seller's extended index finger, this means either 1, 10, or 100. Both parties to the transaction are agreed about the numerical rank of the price; if not, they first specify the coin in which they are bargaining. This resembles our own habit of saying, for example, 5, 6, or 12 when we actually mean 5000, 6000, or 12,000 dollars.

Correspondingly, if the buyer touches the seller's 2 (or 3 or 4) first fingers, this means, 2, 20, or 200 (3, 30, or 300; 4, 40, or 400); touching the whole hand then means 5, 50, or 500. The little finger alone means 6 (60 or 600), the ring finger alone 7, the middle finger alone 8, the bent index finger 9 and the thumb 10 (or 100 or 1000). If the buyer strokes the index finger from the middle joint knuckle toward the tip, he is saying $-\frac{1}{2}$ or "one half off"; if he strokes the index finger toward the knuckle, this means $+\frac{1}{2}$. The fractions $(-)$ $+\frac{1}{4}$ and $(-)+\frac{1}{8}$ are similarly indicated; these exhaust the "vocabulary" of this Oriental language of trade. In simple terms:

2500 = 2 × the thumb and 1 × the whole hand grasped;
 4½ = 4 fingers touched and the index finger stroked toward the middle knuckle;
  76 = ring finger and then middle finger touched (a succession of numerical ranks!).

This finger language is "spoken" very fluently and rapidly, and mistakes are virtually never made.

An agreed and mutually understood finger language is used today, not between people who speak different languages, but in the world's largest cattle market in Chicago. Like that of the Middle East, this custom reveals something about the essence and purpose of finger counting, even though the Roman method is not used. But we also have some late evidence that our Roman finger numbers continued to be used for a very long time in the Arabic world of the Mediterranean.

In the year 1340 the Byzantine scholar Nicholas Rhabdas of Smyrna wrote his Greek "Treatise on the Finger Measures" (*Ekphrásis toû Daktylikoû Métrou*), which is almost the same as the Venerable Bede's, even to the smallest details. The former could not have been paraphrased or copied from the latter — the differences in place and time, some six centuries later, weigh against such an idea, as do other considerations. Even the Arabs knew the Roman finger numbers, as Arabic and Persian documents from the 14th century show. All their gestures are the same as Bede's, except that the right and left hands are reversed, with the units beginning on the right hand in accordance with Arabic script which is read from right to left. When praying, say these sources, the believer should place his right hand on his leg as if he were forming the number 53 (thus the index finger is the only one extended).

How is the bowstring grasped when shooting with a bow and arrow? With the tips of the index finger and the thumb, "as the number 30 is formed."

One Arab poet very ingeniously mocked another, Khalid by name, who had become rich: "Khalid went with 90 and came back with 30" — he came back poorer, apparently. But if the reader forms these numbers on his fingers, he will see that for 90 the thumb and index finger are pressed tightly together to form a "thin," "lean" figure, whereas for 30 they form a full, "rich" circle.

We may conclude with the Arabic counterpart to Juvenal's reference to Nestor, two lines by a poet with the charming name of Sheref-ed-din Ali al Jezdi:

Were I to recount all the wonders of the world,
I should count as many as the numbers which the left hand forms.

ROMAN FINGER COUNTING IN THE WEST

In the summer of 822 I undertook, with Tatto's guidance, to study arithmetic. He began by explaining to us the books of the Consul Manlius Boëthius concerning the various kinds and classes and meanings of numbers. Then we learned to count on the fingers, from the books which Bede wrote on this subject.

Thus wrote Walafried Strabo, who in the year 842 was Abbot of the Monastery of Reichenau on the Bodensee. We infer from his account that finger counting continued to be in use in the West as well. But here it had moved from the market place into the scholar's study. Finger numbers were no longer merely the universal language of commerce; now they began to serve the higher art of mathematical computation. This, indeed, was the role we found them playing in the Venerable Bede's writings (see p. 202).

Now we are in a position to understand the words of Brother Berthold of Regensburg (1220–1272), one of the most celebrated and popular preachers of the Middle Ages:

"man zalte in der alten ê an den vingern. des kunnet ir ungelerten liute niht, wan es ist der gelerten vil, di es niht kunnent. man zelt also, so sin sehzic wirt, so leit man den dumen in die linken hant."

In former times people counted on their fingers. But unlettered people could not do this, and even among the educated there were many who could not (count on their fingers). Counting was done thus: if the number was sixty, for example, one began with the thumb in the left hand.

Here Berthold himself erred, for the number he describes is not 60, but 50.

The first secular book, or at any rate the first printed one, in which we again come across finger counting, is another scholarly mathematical compendium, *Summa de Arithmetica, Geometria Proportioni et Proportionalita* by the Italian Luca Pacioli, which was published in 1494; the illustration of Fig. 38 is taken from this book. That finger numbers were adopted by a mathematician, who actually chose them because of their suitability for his purposes (unlike the monks, who merely described them because they were handed down to them), shows the important position that finger counting attained in medieval mathematics. *Abacus atque vetustissima veterum Latinorum per*

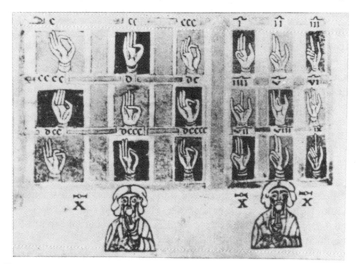

Fig. 44 Finger numbers from a 13th-century Spanish codex. The numbers from 100 through 900 are shown on the left, and those from 1000 through 9000 on the right; below are the gestures for 10,000 and 20,000. These hands with their slender, elongated fingers are drawn in the Byzantine style.

digitos manusque numerandi consuetudo — "The Abacus (= counting board) and the Age-old Custom of the Ancient Latins of Counting on the Hands and Fingers": such was the title of the book published by the German writer Aventin in Nuremberg in the year 1522. Here we see the connection between finger counting and the abacus. We shall learn the techniques of making computations on the abacus later; but finger counting here clearly serves the purpose of recording temporarily the intermediate results of these computations. The Italian mathematician Leonardo of Pisa (1180–1250),* whose important contribution to mathematics are still highly regarded, testifies to this. At one point he refers to finger counting by the expression *servare* — that is, "to keep, store" — meaning what our schoolchildren mean when they say "carry." And he goes on to say, very clearly,

ponens semper in manibus numeros ex divisione euntes,

constantly keeping in his hands the numbers which result from the division.

But the most remarkable thing of all is that Leonardo here wants to use finger counting not in conjunction with the abacus, but with new Indian numerals, whose chief proponent he had become in the West. The relevant passage, which is of the greatest importance for the history of mathematical computations, reads as follows:

Predictis figuris Yndorum earumque gradibus secundum materiam superius descriptam cum frequenti usu hene cognitis, opportet eos qui arte abbaci uti voluerint, ut subtiliores et ingeniores appareant, scire computum per figuram manuum, secundum magistrorum abbaci usum antiquitus sapientissime inuentum.

If these numerals of the Indians and their place-value notation are to be thoroughly mastered through constant practice, it behooves those who would become adept and expert in the art of computation to learn to count on the fingers, which the masters of computation according to the old manner once found to be invaluable.

Leonardo's words *ars abbaci*, moreover, refer to the art of computation in general and not to the counting board alone (see p. 425).

Our final witness for the use of finger counting in computations is the "royal" passage concerning the number 73 formed on the fingers, for it specifically mentions the *abacistae*, those "learned men" who knew how to make computations on the *abacus* or counting board. The author of this passage was the Hohenstaufen Emperor Frederick II (died 1250), who constantly studied this absorbing subject while struggling with the Pope for mastery over the civilized world. He never ceased to be a friend and patron of learning, of the arts — and of hunting. The last of these was so dear to him that he even wrote a famous book about falconry: *De arte venandi cum avibus* — "On the Art of Hunting with Birds." In this book Frederick explains how the falcon is held by a skilled huntsman:

Manum vero non plicet interius neque exterius, sed in rectitudine brachii teneat, conjugendo pollicem extensum indici, et replicet indicem ad extremitatem pollicis, et erit modus, secundum quem abacistae tenent septuaginta

---

* *Editor's note:* also known as Leonardo Pisano or Fibonacci.

cum manu, et alii digiti eiusdem manus replicentur in palmam sub illis duobus digitis, ut firmius sustenentur, ad similitudinem tenentis numerum ternarium, et sic ex replicatione indicis super pollicem, et trium digitorum in palma sub illis, teteat manum ad formam abacistae tenentis septuaginta tria.

The hand is held facing neither inward nor outward, but extended in the same direction that the arm is extended. The index finger is then laid over the extended thumb and bent forward over the end segment of the thumb: this is just the way in which the masters of computation (*abacistae*) form the number seventy with their fingers. The other fingers of the same hand are bent over the palm beneath these two fingers (the index finger and thumb), to support the latter, just as the number three is formed. Thus the index finger is bent over the thumb and the three other fingers beneath it, in the manner of a master of computation forming the number seventy-three."

The reader may recall the intriguing Arabic counterpart of this passage, which states that the bowstring is to be drawn with the fingers held so as to form the number thirty.

In the 16th century, however, as the Indian numerals came to be increasingly deeply rooted in the West as a result of printing and because of their concise and convenient forms which at last made purely written computations possible thus assuring victory over the counting board and the Roman numerals, the days of finger counting were coming to an end. Although Fig. 41 is taken from an 18th-century book, this was one of the very last works to describe finger counting, and it was even then mentioned merely as a curiosity and no longer treated as a useful skill. Soon thereafter no one knew anything about finger numbers: they finally disappeared forever from the history of civilization.

There are still a few places where finger counting lives on today or where it has only lately died out; it was used by peasants in the Auvergne, in Walachia, in Bessarabia, and by gypsies in Serbia. But even these people used it not in the old Roman manner, but as a means of computing.

COMPUTATIONS ON THE FINGERS

Up to now we have regarded finger numbers as a form of numerals, which in a sense served just to record numbers temporarily. But the fingers can also be used as a calculating machine, so to speak, with the numbers to be operated on "put in" and the final result read off after a simple intermediate calculation. Finger counting was used in this manner for the small multiplication table from 5×5 to 9×9 and for the large multiplication table from 10×10 to 15×15. This was the method:

Let us multiply 6×8, for example. Hold the hands extended upright and on each of them form the excess over 5 in both numbers ($6 = 5 + 1$; $8 = 5 + 3$), so that one finger is bent on the left hand and three on the right. Count the bent fingers: $3 + 1 = 4$; these will be the tens of the final result: 40. Multiply the standing fingers: $4 \times 2 = 8$; these will be the units of the final result. Answer: 48. Thus there is no need to memorize the multiplication table beyond 5×5.

To multiply 13 × 14 as an example from the large multiplication table, the procedure is as follows: bend the fingers on each hand to represent the excess over 10 — thus three fingers are bent on the left and four on the right hand. Their sum, 3 + 4 = 7, gives the tens: 70; their product, 3 × 4 = 12, gives the units; 70 + 12 = 82; this is the excess over 100, and the final answer is thus 182. Here too one need know the multiplication table no further than 5 × 5.

Actual computation on the fingers is surprisingly effective, especially for simple people with little or no schooling. Remember that we ourselves would be utterly unable to make calculations had we not learned the multiplication table. Today pupils memorize it at the very beginning of their schooling, but in the Middle Ages this was such a difficult matter that people employed special tables called *mensae Pythagoricae*, from which such products as 6 × 8 could be read off, just as we might consult a table to find the product of 16 × 18 (Fig. 45).

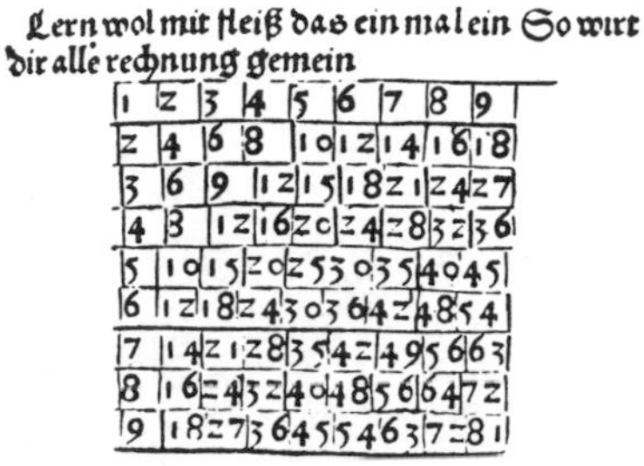
Lern wol mit fleiß das ein mal ein So wirt dir alle rechnung gemein

1	2	3	4	5	6	7	8	9
2	4	6	8	10	12	14	16	18
3	6	9	12	15	18	21	24	27
4	8	12	16	20	24	28	32	36
5	10	15	20	25	30	35	40	45
6	12	18	24	30	36	42	48	54
7	14	21	28	35	42	49	56	63
8	16	24	32	40	48	56	64	72
9	18	27	36	45	54	63	72	81

Fig. 45 Multiplication table from an arithmetic book of the 16th century. Compare the early medieval multiplication tables (Figs. 113 and 162).

One gets the impression that each new such "labor-saving device" in mathematical computation was greeted with joy. Perhaps this was also how the method of finger-multiplication was received; at any rate, Leonardo of Pisa writes:

. . . *multiplicationes in manibus addiscendo semper utantur colligere, ut animus pariter cum manibus in additionibus et multiplicationibus quorumlibet numerorum expeditior fiat,*

. . . multiplication with the fingers must be practiced constantly, so that the mind like the hands becomes more adept at adding and multiplying various numbers. (Compare the passage quoted from St. Augustine, p. 210).

Other medieval and early modern writers, however, although they wrote about finger counting, say nothing about computations using the fingers; perhaps they were not too frequently used.

"Complementary operations" were, of course, quite readily employed in the Middle Ages, as we shall see when we come to the abacus. Instead of actual numbers, such as 6 × 8 for example, medieval arithmeticians worked with their "complements" within some rank, in this case 4 and 2 up to 10. In many textbooks on arithmetic we find not multiplication tables as we know them, but a description of the following kind of operation:

6 4 6 × 8; subtract the 10-complement of one number from that of
8 2 the other: 6 − 2 = 8 − 4 = 4; after the remainder write the
4 8 product of the complements 2 × 4; answer: 48.

Was this peculiar method of multiplication inherited from the Romans, or was it an independent popular invention of the Middle Ages? That is hard to tell. As evidence in favor of the former hypothesis we may cite its use among common people in the Auvergne and Wallachia: both these regions were ruled and influenced for many centuries by Rome. One could also point to the excess over 5 that is inherent in Roman numerals: 6 = VI, 7 = VII, 8 = VIII, 9 = VIIII. But that is all. Only more thorough and accurate anthropological research in areas where "early" computations are still or were recently used can resolve this question one way or the other.

Moreover it must be kept in mind that for a primitive computer the multiplication table itself represents a significant advance, and that it can always be adjusted by addition. We need only think of the Egyptians, who in general could perform such multiplications as $25 \times 43 = 1075$ merely by multiplying by ten and by doubling the factor 43 and then adding:

/ 1	43	
10	430	
/ 20	860	The numbers in the lines marked by a slash are added
2	86	together.
/ 4	172	
25	1075	

SOME FORMS OF FINGER COUNTING USED BY PEOPLES OF OTHER CULTURES

In the Egyptian ell, which contained 28 "fingers," we have an example of the individual units of a measure being counted in the middle: I, II, III, (symbols for) 4, 5, 6, 7 (Fig. 46). But the Egyptian

Fig. 46 Egyptian system of measures with the numbers 4 through 7 "counted on the fingers." Drawing of the first half of the scale (with the beginning, I, omitted). Original about 25 cm long.

4, 5, 6, 7 are clearly finger numbers: 4 is a pictograph of a hand with the thumb closed over the palm, 5 a hand with thumb extended away from the hand, 6 a hand with thumb extended and fingers bent inward (to augment the 5 formed on the other hand), and 7 symbolized in some manner that is not quite clear. Quite remarkably, right next to these we find the "true" Egyptian number symbols, such as !!! ∩ = 16, etc. (see Fig. 8); the symbol ⇔ relates to fractions of the ell. The "early" finger numbers were still so tightly interwoven with the ell, a measure that arose from the ordinary lives of common people, that the inconsistency in this system of numerals was not felt at all. In addition, the ancient Egyptian symbol for 10,000, a pointing finger, also very likely goes back to some ancient form of finger counting (see Figs. 8, 9, and 10).

Ancient China offers another instructive example. The Chinese represented the numbers 1, 2, and 3 by those numbers of strokes respectively; the strokes can readily be interpreted as numbers once formed on the fingers. There is nothing special about that. But the Chinese symbol for 4 is a square with two small vertical lines drawn within it (see Figs. 47 and 265). Yet this was originally the picture of a hand (without the thumb). Ancient Chinese coins show, in addition

Fig. 47 Chinese numerals for 4, 8, and 9; these were originally representations of numbers formed on the fingers. These were found on ancient coins (cf. Figs. 48 and 280–281); for their modern forms, see Fig. 264.

Fig. 48 Chinese coin bearing the inscription: "Series 19." The number 10′9 is on the left side of the coin. This is a rare piece dating from the beginning of the Han Dynasty, around 200 B.C.
From the collection of R. Schlösser, Hannover.

Fig. 49 Statue of Buddha with sacred finger gesture. From the highest terrace of the Borobodur temple in Java. Ninth century A.D.

to the four simple vertical strokes, two perpendicular horizontal strokes joining them; these were clearly suggested by the form of the human hand (see Fig. 279). From this the symbol gradually was eroded to its modern form.

The Chinese word *pa* means "to split, cleave, cut," and the Chinese number word for 8 is also *pa*. The old symbol for this number consists of two "separated" signs which may represent a single gesture for "to split." This would mean that the similarity between the two words was extended to the written characters, which may well have happened, as shown by other examples. But could this not also be a picture of "two hands," 4 + 4 (see Fig. 47)?

At any rate, it is quite certain that the modern character for 9 is derived from the picture of a shaking hand, which is a number gesture. This appears on an ancient coin, on the left side of which, next to the square hole in the center, is the number 19, referring to the series in which this coin was minted (see Fig. 48). In Southern China a sudden movement of the hand to the right ear still stands for the number 9.

But what was still clearly recognizable in ancient Egypt was transformed in China by the masters of calligraphy into a matter of good style. Here, however, we have one of the rare examples which reveal the way in which "early" finger numbers were incorporated into the "mature" system of numerals.

Finger counting? At first the reader may well have looked on it with great suspicion: What could it be but a form of trivial amusement, once it did more than represent the numbers 1 through 10? Now we have seen that numbers counted variously on the fingers have always existed and are still in use today. The hands "speak," the fingers form the individual "letters." For the deaf and dumb, this is the only means of communication. But over and above these practical methods are the "sacred" finger gestures with which we shall conclude this chapter.

In Italy, Byzantine mosaics predominated in ecclesiastical art well into the 12th century. The image figure of Christ looms up in the vault of the apse (as for example in the church at Cefalù in Sicily), bending the ring finger and the thumb of his right hand in a solemn gesture to form a circle. A deep symbolism undoubtedly underlies this "Christian" gesture, as it also does the arm and finger movements of the Siamese temple dancers and the *mudras* of the priests on the island of Bali.

The Indian Buddha indicates a level of spiritual achievement (see Fig. 49) by the gesture he forms with his fingers. These finger gestures are as numberless as the various stages on the road to perfection; just as in Romanesque and Gothic churches every column capital is different from every other, no two gestures made by the often hundreds of Buddha images in an Indian temple are the same. These gestures of Buddha's do not represent numbers, of course, but they do show that "finger writing" has played a leading role in other cultures as well. Moreover, Indian Buddhist writings divide the art of computation into three stages or levels: counting on the fingers, which is called *mudra*, mental arithmetic, and higher computations.

Folk Symbols for Numbers

Tally Sticks

"The notched tally stick itself testifies to the intelligence of our ancestors. No invention is simpler and yet more significant than this."
J. Möser, *Patriotische Phantasien*, 1776

Numbers formed with the fingers vanish in an instant. How was a merchant to keep his accounts, or even to make his calculations, with these alone? To represent the next number he must first destroy the previous one. Finger counting shares the visual quality of writing, but it differs in its impermanence which it has in common with number words. This is what gives numbers counted on the fingers their intermediate position between spoken numbers and written numerals.

EARLY WRITING AND READING

Today we are so accustomed to reading and writing that we can scarcely imagine what it was like when most if not all people could do neither. The letters of the alphabet are the very foundation stones of our education. What is the first thing we learn in school? "Dot your i's and cross your t's" — in other words, how to write properly! It is of profound significance that one of the greatest and most difficult achievements of a culture has now become the most elementary prerequisite of an education.

Suppose someone were to do away with all numerals, even the Roman? Then we would have some idea of the state of written numbers in the early Middle Ages. To be sure, Roman numerals found their way into the monastic cloisters along with other aspects of the culture inherited from ancient Rome; of course, Roman numerals were later brought by the students from these monastic schools into the world of literate people, of merchants, of scribes in chanceries. But the farmer, who never went to a monastery school, also had to keep track of his crops and livestock and harvests. The peasant could also be a creditor or a debtor. Where did he get the numerals to reckon with?

From no one but himself! He developed his own number signs, which were adequate for his own ordinary dealings, and which no one but himself could read and understand. Who, for instance, without being told how, could read the symbols on the tally sticks of Fig. 58 or on the magnificent Swiss notched stick of Fig. 52, with the carving of a cow at the top? Who knows what numbers are signified by the notches on the war club from the Fiji Islands (Fig. 6)? Perhaps a member of the tribe which carved the club could still read them, just as many Swiss can identify a particular hamlet or valley from the carvings on a notched stick. But their significance is always personal and valid only within a limited area: these are folk-signs, number symbols devised by simple people themselves to meet their everyday needs. They are not learned and taught in school as part of the general culture. In their level of development, of course, these are primitive numerals, since they naturally draw on the simplest laws of written numerals, ordering and grouping (pp. 39 ff.).

The primitive nature of these folk symbols is also revealed in the way by which they are made: they are carved or cut or scratched. In

[223]

the history of language, this is the earliest forerunner of the art of writing.

The word *schreiben* came into the German language from the Romans: Old High German *scriban* < Latin *scribere*. But the Latin word goes together with the Greek *skaripháomai*, which, like the Greek word for "write," *gráphein*, originally meant "to scrape, scratch." Yet the Germanic peoples did not originally learn to inscribe symbols from the Romans, for they had their own word for this: the modern German *ritzen* ("to scratch"), Anglo-Saxon *writan*. This word replaced the Latin loan word in the English *write*, and is preserved in Modern German in the word *um-reissen* ("to sketch, outline"); a *Grund-riss* in German is a line drawing. In his *Germania* (Book X) Tacitus, the Roman historian, describes the customs of the earliest Teutons.

Virgam frugiferae arbori in surculos amputant eosque notis quibusdam discretos super candidam vestem temere ac fortuito spargunt. Mox si publice consultetur, sacerdos civitatis, si privatim, ipse pater familiae precatus deos caelumque suspiciens ter singulos tollit, sublatos secundum impressam ante notam interpretatur.

From a sapling the Teutons cut off a twig, cut this up into small pieces, put certain marks on the sticks and then toss them at random upon a white cloth. If anyone wishes to know what the fates hold in store, the priest of the community, praying to the gods like the father of a household, his face raised toward the heavens, three times picks up a stick and interprets it according to the marks which have been carved upon it beforehand.

From this old Germanic custom of casting lots derives the modern German expression *Buchstaben lesen* ("to read letters") (= *auflesen*, "to pick up the lots"), which in English is *read* (from *raten* ["to guess, solve"]) the "secrets" hidden in the symbols carved on the sticks (Old Norse *run* > "rune"). These, of course, were not the ancestors of the letters we use today.

TALLY STICKS WERE UNIVERSAL

Paper, which was a Chinese invention, first came to Germany in the 14th century. In the beginning it was too expensive for common use; like parchment in the monasteries, it was used only for the most important documents. Moreover great skill was required for writing on it. The common people's "paper" was wood, its writing instrument was a stylus or sharp knife and its letters were notches or grooves. For this reason, we find some form of tally used everywhere.

Even the various European names for these tallies provide some very informative glimpses into the cultural history of early writing and computation. In Middle and Low German there were *Kerbholz* ("notched stick"), *Dagstock* and *Knüppel* ("cudgel"); in Bavaria and the Tyrol people used the *Span* ("chip"), the *Kärm* and the *Raitholz* (= *Rechenholz*, "reckoning" or "tally stick"). The German Swiss had *Tessele* (< Latin and Italian *tessera*, "dice," then "four-cornered tablets" and finally "mark, stamp"; cf. p. 179) as well as the *Alpscheit* (literally, "Alp log") and the *Beile*, the latter word probably derived from the Medieval Latin *pagella*, "scale," from which the German words *Pegel* ("water gauge") and *peilen* ("to take soundings, measure the depth of water") and the English word *pail* have also most likely descended.

Austria and Vienna have their *Robitsch* or *Robasch*, a borrowing of the Slavic word *rovaš*, "notched stick." "*Es ist mir in die Rabuse gegangen*" ("I've had to go into notches"), a Bohemian would say regretfully when he had to borrow money. This peculiar word comes from the Slavic *rubatj*, Russian *rubitj*, "to cut, notch," to which the Russian ruble (*rublj*) is also related; the ruble was originally a piece of silver, about the thickness of a finger, cut off from a long bar of silver. These Slavic words, along with the Old High German cognate words *ruaba*, "number, count" and *ruabōn*, "to count, reckon," betray the connection between "cutting, notching" and "counting, reckoning," which crops up time and time again. It is most obvious in the English word *score* (< Old Saxon *sceran*, "to shear, cut"), which in addition to "cut" and "keep count" has also taken on the meaning of the number word for "twenty" (see p. 50); it is no less evident in the original Germanic word *talo*, the "notch in a piece of wood" which also means "number" (see p. 187), in the Serbian *broj*, "number" < *britj*, "to cut" and, to take an example from a primitive culture, in the Bantu word *vala*, which means "to notch" but also "to reckon, to count."

In Sweden there was the *karvstock* and in Holland the *kerf*. The Romans used the word *talea*, a "cut twig," for a "staff" or "stave." From this were formed the Medieval Latin *talare*, "to cut," which became Italian *tagliare*, Spanish *entallar* and French *tailler* (French *tailleur*, English "tailor"). Thus the notched stick was called a *taglia* or *tessera* in Italy, a *tarja* in Spain, a *taille* in France and a *tally* in England.

This category, finally, also includes the words for "book," or *codex*, as the medieval manuscript volume was called (for instance, the *Codex argenteus*, see p. 259). The book takes its name from the *Buche*, or "beech wood," from which the wooden writing tablets were originally fashioned, just as the Latin *liber* is named after the "bark," and the Greek *biblos* after the papyrus, on which it was written. The material on which the writing was done thus gave its name to the written product. The Latin *caudex*, too, was the "log of wood" that was cut up into *tabulae* or *tabellae*. Such wooden tablets were then tied together at one side into a bundle, and the resulting bundle was called the "beeches" < Gothic *bokos*; the Middle High German expression *lesen an den buochen* meant "to read the wood tablets" — the singular form of the word *Buch* not appearing until later. Shortly thereafter the herdsmen of Graubünden tallied up the amount of milk they got from their cows in wooden "books" of this kind: each "book" consisted of a wooden writing tablet enclosed between two hinged wooden covers (see Fig. 50). Such wooden "books" were the earliest ancestors of the modern book; but even earlier they were bundles of notched sticks linked together at one end, of which we can find examples from Switzerland, Russia, and China (see Figs. 61, 64, and 75). The Romans later covered the wooden writing tablet with a layer of wax upon which they

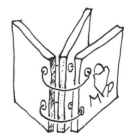

Fig. 50 Wooden book of the Graubünden herdsmen.

inscribed letters with a stylus; this became the Roman writing tablet (cf. also the "diploma," p. 173). We find the same curious connection between wood and computation again in the Roman habit of using the word "codex" to refer to the "book" in which they kept their dispatches in successive order — *nomina in codicem digerit.*

But the most striking bridge between "cutting" and "computing" is made up of the Latin words *putare, imputare, deputare,* and *computare. Putare* means literally "to cut" (cf. *amputare,* which Tacitus uses in the same sense as the modern English word in the passage quoted on p. 224); *imputare* is "to cut in, indent," so that "to incise a notch for someone" takes on the figurative meaning of "to assign a debt to someone"; *deputare* is then the exact opposite of "indent" with its figurative meaning of "impute," since whatever had been tallied up to someone's account was later "cut away" from the notched stick after the debt was paid or the obligation discharged. *Computare* then clearly became "to compute, to reckon," as shown by the word *computus,* which in the early Middle Ages meant a "record of time." The word *putare* thence acquired its standard meaning of "to reckon, think, consider, believe."

Now after all this, what does a tally stick actually have to do with computation? We shall find the answer to this question in the following section.

VARIETIES OF TALLY STICKS

The Simple Number Stick

A contemporary of Pieter Brueghel's (died 1569) has an anecdote about the famous Flemish painter:

While he was in Antwerp, he lived with a young girl whom he would have married if she had not had the unfortunate habit of constantly telling lies. He made a pact with her, that for each lie she told he would cut a notch in a piece of wood; he took a good long stick, and when the stick was filled with notches, there could be no longer any question of marriage; this happened before very long.

One could scarcely find a better example of the meaning of the German phrase, *etwas auf dem Kerbholz haben* (literally "to have something on the notched stick" = "to have a skeleton in the closet").

From the standpoint of writing numbers, what exactly did the astute Pieter Brueghel do? Every time this daughter of Eve bent the truth beyond the breaking point, he took his knife and tallied one up to her account: for each lie he cut one notch, in one-to-one succession, so that one lie = one notch, again one lie = one notch and once more, one lie = one notch, ..., just as the Fiji Islander did with his club (see Fig. 6, p. 39). The notches were thus a supplementary quantity whose number value, because of its abstract, "colorless" nature, was used to represent the lies, with all the emotional burden they carried. A notched piece of wood of this kind is a simple *number stick.* We can identify it from the one-to-one succession of the notches — their ordering, in other words — which we now recognize as one of the earliest and most basic laws of the number sequence.

Number sticks have been used in every period and by all peoples.

Just as the Fiji Islander notched the handle of his club, the woodsman counts up the bundles of brushwood he has produced and, in a very different sort of occupation, the vineyard worker uses his vine-cutting knife to mark on his stave the number of basketfuls of grapes he has delivered. Many primitive peoples keep a record of the passage of time on their so-called calendar sticks in just the same manner.

The inhabitants of the Nicobar Islands in the Indian Ocean often have to count the numbers of coconuts they have picked. Since a plain number stick would never suffice for this purpose, they cut themselves a rod half a meter long from a bamboo stalk and split one of its ends into a number of lengths, which remain connected like the brush of an old-fashioned broom. Then the number of coconuts is marked by notches carved in the split lengths at the end (see Fig. 51).

There is no more impressive evidence of the universal use of simple number sticks in all periods of history than the illustration of the two carved bones of Fig. 51, with their notched edges. There seems to be very little difference between them, but they are separated by thousands of years: the bone at left was notched by a prehistoric man, but the one on the right by a Swiss farmer in our own time.

Number sticks have the force of legal documents when the number of days worked or the amount of goods tallied up on them serve as a record for compensation; they then become a sort of check, or "commercial paper." In ordinary life they once played a role whose importance we can scarcely imagine today. Tradesmen, innkeepers, bakers, and smiths kept their accounts on tally sticks, as they do today in ledgers, and they often bound their flat notched sticks together to form something like a book (see Figs. 61 and 64).

When in Schiller's play, Piccolomini's health was being toasted in Wallenstein's camp, during the Thirty Years' War, the barmaid presented a bottle to the company with the words:

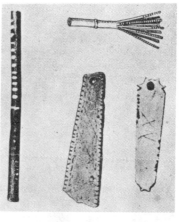

Fig. 51 Number sticks. *Left:* an English brushwood cutter's tally stick (Germanic Ethnological Collection, Berlin); *top:* a number stick of split bamboo used to count coconuts; *bottom left:* a prehistoric; and *bottom right:* a more recent notched bone, both from the Museum of Ethnology in Basel.

Das kommt nicht aufs Kerbholz. Ich geb es gern.
Gute Verrichtung, meine Herrn!

"This doesn't go on the tally; I give it free.
Success in battle may you see!"

An Italian saying goes:

Come e bello vivere in questo regno:
si mangia, si beve e si segna
tutto su un pezzo di legna!

How pleasant it is to live in this land:
We eat, we drink and we just tally
It all up on a piece of wood!

But the innkeeper just laughs and says: "In this inn, unfortunately, there is no chalk and no tally stick." In other words, no one is allowed to eat and drink on credit.

When a person chatters on endlessly without having anything to say, the Germans say that he *redet aufs Kerbholz,* "talks on the notched stick." *De kerfstock loopt to hoog,* complains the Dutchman, "The bill is getting too high." And if he wants to send away someone who doesn't pay his bills, then *is de kerfstock ijzeren* ("turned into iron"): no more notches can be carved in the tally stick.

Thus the notched stick appears in figures of speech as a "debt" or a "bill." From an 18th-century writer comes this fine passage: "When I look up and think back, I see a great tally stick, which reminds me that I have sinned against you a thousand times more."

How common bookkeeping on tally sticks was, even in such a progressive commercial town as Frankfurt-am-Main, is shown by this quotation from the Weavers' statutes of the beginning of the 14th century:

Item so will auch die gemeyn geselschaft, dass alle rechenmeyster nit sollenn die kerben einem andern leihen, das eynem wirt zusteet.

It shall likewise be the rule for the entire guild that no master calculator shall lend to one person the tallies which are owed to another.

The use of tally sticks as bills or records of obligations can be traced all the way back to the time of the early Germanic tribes. A feature of the common law of the Franks and the Alemanni was the *festuca*, a rod that was exchanged between the parties to a transaction in matters of law, especially those involving promises of payment. According to Salic law, the debtor had to turn over to his creditor his *festuca*, on which his own identification symbol and the sum involved in numbers were marked (as in Fig. 52). In this *fides facta per festucam*, this "promise made by the *festuca*," the rod in question was nothing more than a notched stick on which the debt was tallied. The custom was continued into the Middle Ages, when a purchase was sealed "*mit hant und mit Halmes*, "with hand and stick," as was then the law and the custom.

Once the debt was paid or the obligation otherwise discharged, either the tally stick was burned or else the sum was cut off — that is, the creditor cut the notches off the stick, "he made the tally (actually the wood) smooth"; this is shown beautifully by the old Russian tax rods, which as a result of repeated addition and removal of notches finally became very thin (see Figs. 61 and 62). The same was true of the Romans' *tabula rasa*, the "erased (smoothed-out) tablet"; this expression has by transference come to mean "an old debt cancelled" and hence "a new leaf." On the wax-covered writing tablets which the later Romans used, the writing was erased merely by rubbing it over with a finger or the flat end of the stylus.

Fig. 52 *Stiala de latg*, a tally stick carved by a cowherd to record the amount of milk produced by the whole village. The one on the left is "cut off," i.e., the account is settled. From Bündner Oberland, Switzerland. 12 and 15 cm long.

Now we can understand the lines in which Fischart, the most satirical portrayer of his own times, ridicules a mendicant nun:

Sie will kurzum Gott nichts schuldig pleiben,
sondern will das Kerbholz rein und glatt abkerben.

In short, she wants to owe nothing to God;
but wants the tally stick clean and smooth.

Martin Luther, addressing a friend to whom he had not written for a very long time, began his letter thus: "I must cut off the tally stick, for I have not answered your letter for a long time."

Each installment paid on a debt was also recorded in this manner, as a municipal document dating from the year 1588 indicates: "A woman who is a debtor is to pay 12 gulden per year, until the whole debt has been paid; and each time this payment is to be cut off from her tally stick."

A sale of goods in 1453 is documented by a court judgment:

... und haben beide gebeden das gericht, das is soliche gut dem Becker abesnide von siner kerben und Clase Riess zusniden an sine kerbe.

... and both /parties/ have requested this judgment, /to the effect that/ Becker shall remove these goods from his tally and Clase Riess add them to his own.

It is not clear from this citation whether the amount of the debt was being transferred to a lease of the goods or whether this was a transfer from one household to another. But the words do show very clearly that tally sticks played an important role in the lives of peasants.

The whole significance of tally sticks and people's familiarity with them from the very oldest times is summed up in a nutshell by Skirnir's Song in the Norse Edda, in which the "scratched" magic spell is broken by erasing the incised mark (*thurs*, "scratch," was the name of the evil rune þ, which always brought bad luck; see Fig. 95, column 12, line 9):

I scratch a *thurs*   and then three runes:
Lust, sorrow   and pangs of love;
I scratch them out   as I scratched them in,
When they were needed.

The other kinds of tally sticks we shall now examine show how significant and practical these wooden numerical records were to our immediate ancestors.

Stialas de Latg

In these fascinating popular "milk sticks" of the Tavetsch Valley in the Bündner Oberland district of Switzerland, we shall find a kind of tally stick used to keep the "books" of a coöperative enterprise. Thus their numerical value and legal validity is extended from the individual to a whole village community.

From a piece of alder or ash wood the Alpine herdsman each day would carve a 5- to 8-sided rod, some 15 to 20 cm long, which he colors with red chalk; thus the light-colored uncolored carving stands out quite vividly (see Fig. 52). Then for every peasant who owns some cows in the common herd, an individual symbol is carved on the rod (in this case, two symbols on each face). Below each such symbol runs an incised line, the individual peasant's "account," on which the cowherd tallies off the amount of milk yielded by his cows. And when the herdsman adds a special ornamental carving, like the reclining cow in our illustration, these *stialas* are no longer primitive records, but a gay and colorful form of folk art.

Swiss cattle spend the whole summer at pasture in the high Alpine meadows. A particular herd may consist of cattle belonging to a number of different owners, let us say seventeen. The milk which the cows give every day must be processed and delivered at once. Since the mountain peaks in the Tavetsch area are close to the villages, each day a farmer, A or B or C, etc., according to an accepted and fixed order, goes to the mountain pasture and makes cheese, using the milk not only from his own but from all the cows in the herd. Obviously records of the amounts of milk used have to be kept; this is the chief herdsman's duty. Below the individual symbols of each of the 16 farmers who did not make cheese on a particular day, or all the farmers except A, notches are cut for the amount of milk from his cows that was used that day:

one notch crossing the long groove means 10 *Crennen* (= $10 \times \frac{5}{4}$ pounds),
one cut (without removing the wood) means 5 *Crennen*,
one notch in the bottom right edge means 1 *Cr.*,
one cut in the same edge without removing the wood means $\frac{1}{2}$ *Cr.*,

smaller fractions are not counted. Thus farmer R, for example, has "lent" $4 \times 10 + 4$ *Crennen* of milk to farmer A on a given day. Finally farmer A, who made the cheese that day, carves his own personal sign on the surface of the butt end, which is the bottom surface as the *stialas* are shown in our illustration, and takes them home with him as a record for the milk he has borrowed from the other farmers in his village.

After the cycle has been run through and each of the seventeen farmers, A, B, C, etc. has made cheese on one day, they meet on a Sunday after church service to "combine the milk" — that is, to settle accounts. Each brings along his *stiala*. The amounts which farmer A has recorded on his own *stiala* are what he owes; the amounts tallied up to A's account on the other *stialas* are what the others owe him.

With this simple form of bookkeeping there is no possibility of cheating: no one would add notches to his *stiala* because that would increase the amount he owes; and no notches can be removed, because this would immediately show up on the red wood. If A on his day in the Alpine pasture used 60 *Crennen* of B's milk, and B on his day used 90 *Crennen* of milk from A's cows, B owes A the difference of 30 *Crennen*. A cuts away or "strikes out" his 60 *Crennen*, since he owes nothing to B, and writes down on a piece of paper that he still has 30 *Crennen* more coming from B. The differences are either paid then and there or else carried over into the next round.

We may well admire the clear and simple method of recording these very complicated debit and credit transactions. On these *stialas*, moreover, for the first time we find number symbols with different values — for $\frac{1}{2}$, 1, 5, and 10 units. They are not really numerals like 1, 2, and 3, for they do not follow in order; they are actually signs for groupings: one notch stands for a "group" of two cuts.

The *stialas*, which continued in use well into the 20th century, were the records of transactions among the many "shareholders" of an Alpine community. Generally, however, there are only two parties to a transaction: buyer and seller, or creditor and debtor. Both must keep records of the obligation undertaken and discharged. But receipts and promissory notes call for paper and writing, which simple people neither understand nor like. Then how can the transaction be recorded on a tally stick so that no one can cheat? This was accomplished by the split or

Double Tally Stick

A long piece of wood is cut lengthwise almost to the end; the part with the large end is the "stock" (the main stick) and the split-off portion is the "inset" (the piece laid on the main stock). In Vienna these were called *Manderl* and *Weiberl* ("little man" and "little woman") respectively.

In the case of the beautifully made Finnish double tally sticks, stock and inset are identical and are held together by wooden pegs (see Fig. 53).

Fig. 53 Finnish tally stick to record units of work done. Size 25 × 2 × 1.5 cm.
Kansallis Museum, Helsinki.

When a payment or delivery is either made or received, the debtor inserts his inset in the stock, which the creditor generally keeps, and notches are cut into or removed from both pieces at once. Then both parties take back their own pieces and keep them until the final settlement. In this marvelously simple fashion the "double bookkeeping" makes any cheating impossible. These "countersigned" records are called *contretaille* and *tacca* (notch) *di contrasegno* in French and Italian, respectively.

Until quite recently a three-part tally stick was used by the snow-removal men in Vienna (see Fig. 54). The middle portion with the head was kept by the driver, one of the side pieces by the foreman at the loading platform and the other by the foreman at the unloading platform; thus there was a record in triplicate. Since both foreman had as many different side pieces as there were drivers, one for each, they would hang them together on a string around their necks. All three parts of a tally stick were marked with the same number, here 174.

Here, too, no cheating was possible. For this reason the double or triple tally sticks always had the force of legal documents.

And if anyone is not good at settling his accounts by writing and reading, he shall be satisfied with crudely made tallies or tickets: Then if one party brings to court a tally or ticket as evidence of his debt, and the other party produces the corresponding or matching ticket or wooden tally, and they are found to be the same, credence shall be given to them and the amounts stated on them shall be acknowledged,

says the Basel statute book as late as 1719. Even Napoleon's book of statutes, the *Code civile* of 1804, which was in force in a number of German provinces as well as in France, still had the provision that

Les tailles correlatives à leurs échantillons font foi entre les personnes qui sont dans l'usage de constater ainsi les fournitures qu'elles font ou recoivent en détail.

Fig. 54 Tally stick in three parts, used by snow-removal men in Vienna. 30 × 4 cm.
State Museum of German Ethnology, Berlin.

The tally sticks (= insets) which match their stocks have the force of contracts between persons who are accustomed to declare in this manner the individual deliveries they have made or received.

The absolute security against deception which is the special advantage of stock and inset was transferred, as early as the 10th century, from tally sticks to written documents:

Am Tage St. Andreä anno 1594 *sind zwey gleichlautende, einer Handschrift aufeinander ausgeschnittene Briefe, denen jede Parthey einen zu sich genommen, ausgefertiget worden, [weshalb man] nicht nötig gehabt, dergleichen Kontrakte zu unterschreiben, sondern es haben solche, wenn sie ordentlich aufeinander gepasset, eben die Kraft des Beweises gehabt, wie denen ausgeschnittenen Kerbhölzern beigeleget wird.*

On St. Andrew's Day in 1594 two identical letters were prepared, one written above the other and the two cut apart, of which each party took one copy; [wherefore] it was not necessary to sign the contracts, but if they matched one another properly, they had the force of legal documents, just like the matching portions of a carved wooden tally stick when they are fitted together.

These were the notched tickets mentioned in the Basel statute just quoted; they were also known as notched letters, split papers, or split tickets. In Medieval Latin they were called *cartae partitae* or *dentatae*, "divided" or "toothed papers" (see Fig. 55).

Fig. 55 Notched tickets: these documents written out twice and then, after the manner of the double tally sticks, cut apart along a zigzag or wavy line. Only an exactly matching half was accepted as true and correct. Kunsthistorisches Museum, Vienna.

The contract was written out in full two or more times on a sheet of paper or parchment, letters were drawn in between the two or more copies of the text, and the parts were then cut apart through the letters along a jagged zigzag or wavy line The wavy cut has been preserved here and there in the coupons of stock and bond certificates or other commercial paper, at the end of the booklet from which the coupons or cheques are detached. A trace of the old usage also persists in the English word *charter*, meaning to "rent or hire a ship or boat," for the *charter party*, the contract between the merchant and the ship operator, is literally the old Latin *carta partita*, a "paper cut apart";

in the Hanseatic League of medieval European trading towns the German term *Zertepapier* and the French term *chartepapier* were used. The English *indenture* preserves the serrated or denticulated edge of the notched paper in its very name (< Medieval Latin *indentare*, "to cut teeth into"; *dens*, "tooth").

The check as a certificate of demand for payment goes back to the notched tally sticks. The English Royal Treasury, as we shall see in the Exchequer tallies, kept its records of income, such as taxes, and expenditures on notched tally sticks (see p. 236). But the Exchequer also issued the stocks of double notched sticks or *tallies* (perhaps, for instance, with a £20 notch) to its citizens as certificates of payment. The holder could then go to the Royal Treasury, which held the inset or matching piece, and if the stock and inset agreed, the stock would be redeemed in money. In English *to check* still means to compare an original document or piece of writing against a copy to see if the copy is correct. Hence the written certificate or money order, which was to be presented for remittance and checked against its security, later came to be called a "check" or "cheque"; we shall see later (p. 347) how this word is related to "chess" and "checkerboard."

A *check* is thus generally an "identification mark," just as the Greek word *sýmbolon* originally meant a "distinctive mark" (< Greek *sym-bállein*, "to throw together") — that is, a broken shard, for the most part with writing on it, which fitted the piece from which it had been broken. Another such mark of identification or recommendation was the Roman *tessera hospitalis*, the "guest-countersign."

In addition to this verbal evidence for the long survival of the notched tally stick, there is also the very interesting and rare instance of a written testimony: the word for "contract" in Chinese is symbolized by two characters at the top, one for a tally stick (stick with notches) and one for a knife, and another at the bottom which means "large" (see Fig. 56). A "contract" or "agreement" in Chinese is thus literally a "large tally stick"!

Popular speech has naturally also taken note of the peculiar quality of the double tally stick. "To keep a tally with someone" was the German commercial jargon for a business relationship; in a figurative sense it then came to mean to be close to someone in a personal sense, just as German country people say, "*Sie hat's mit ihm*" ("She's in love with him" — literally: "She has it (a tally) with him").

Fig. 56 "Contract": Chinese character made up of a tally stick and a knife; the character at bottom means "large."

The English language is very rich in expressions of this kind, because in England the notched stick (the *tally*) itself played a dominant role in the state's finances until well into the 19th century (see the *Exchequer tallies*, p. 236). *To keep tally with somebody* is the equivalent of the German expression just mentioned. *They were tallies for each other*, i.e. "they were like double tally sticks to each other," by transference means "one was the spitting image of the other." From this came the use of *tally* in the sense of "counterpart" and of the verb *to tally* with the meaning "to match or fit together": *the account does not tally* means "the calculation is not correct." The reader can figure out for himself why *to live tally* means "to live in sin" and just what a *tally-wife* is.

The *tallyman* or "junk-dealer" got his name from his habit of carving notches in a stick. He is the proprietor of a *tallyshop* where second-hand goods are bought on credit and paid for in installments; the Frenchman says *acheter à la taille* when he means "to buy on credit." Thus in the *tally* the original wooden notched stick has become paper, so to speak.

Special Kinds of Tally Sticks

The notched number stick and the double tally stick, which we have just discussed, may well appear to have exhausted the forms of notched sticks used to record numbers. Certainly these were by far the most common. But especially in some remote valleys of Switzerland tally records of such peculiar kinds have been preserved almost to the present day that we must look at at least three examples.

An *Alpscheit* (literally "Alpine wood-billet") from the Lötschental district (Fig. 57) was a villager's certificate showing that he was

Fig. 57 An *Alpscheit* or "Alpine billet" with insets, dating from 1752, on which the pasture rights of the peasants in a village are recorded. It is triangular in section, 1.30 m long and 9 cm wide, weighs more than 3 kg and has more than 70 cut-outs into which corresponding inset pieces were fitted.

entitled to let his cattle graze on the communal pasture. Small pieces were cut out at regular intervals from each of the three surfaces. The peasants kept these inset pieces, or *Beitesseln*, as evidence of their individual pasture rights, for the notches and cuts carved into them, when they matched those on the cut-outs, were the proof of their particular "cow" rights:

one long groove is 1 cow right,
one short groove is $\frac{1}{2}$ cow right,
one long cut is $\frac{1}{4}$ cow right,
one short cut is $\frac{1}{8}$ cow right.

One cow was considered the equivalent of 10 sheep, in terms of pasture rights, but one had to have $\frac{1}{8}$ cow right to be allowed to pasture one sheep. The peasant would keep his small inset pieces in a small, ornately carved box, while the cowherd would keep all his "Alpine billets" strung together on a cord.

Here we see a special use of the notched tally stick: the inset piece which was fitted into one of its grooves was not a promissory note but a certificate of rights that gave its owner a share or a fraction of a share in the community pasture. Yet it was not the exact equivalent of a stock certificate, because it did not entitle the peasant to fractional ownership, but merely to proportional use of the pasture. One cow right in a good mountain pasture was considered to be worth 1000 Swiss francs. Thus these tiny carved bits of wood had the same effect as securities or bonds.

In contrast to the cow rights, however, the *Kapitaltesseln* ("capital tallies") of Visperterminen, Switzerland, were the exact equivalents of bonds. Quite apart from their economic importance, they are all the more interesting from our standpoint because of their remarkable numerals. The community had funds which it would lend out to individual farmers. As a promissory note for his debt, the borrower gave the municipal government a tally stick with his personal mark carved on it and the amount of his debt marked on the opposite side. These tallies were strung on a cord through the holes in their ends (see Fig. 58) and kept until the debt was paid off.

Fig. 58 Capital tally sticks: on the obverse side are carved notches showing the debt owed by a peasant; on the reverse is his personal symbol. From Visperterminen, Switzerland.

The Swiss water tallies are also of interest because of their numerical symbols (see Fig. 59). In the area of Wallis, Switzerland, the aqueducts leading down from the mountains play an important role in the local economy, because the arable land becomes very dry in summer. Thus the streams of glacial water are carefully collected and led through gullies and ravines, often over long distances, to the cultivated fields where the crops are grown. These glacial streams are the property of the community which may, for example, have 34 such aqueducts. The farmers in the area must buy the rights to share this summer water. Like the pasture rights, the water rights are recorded and certified by carved tally sticks.

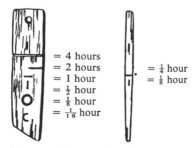

= 4 hours
= 2 hours
= 1 hour
= ½ hour
= ¼ hour
= 1/16 hour

= ¼ hour
= ⅛ hour

Fig. 59 Water-tallies from Wallis, Switzerland.

Ethnologists also include among tally sticks the so-called "messenger staffs" or "turn tallies" on which are carved the personal symbols, for example, of the men who take turns in ringing the church bells. Whoever has the stick has the duty that time around. We shall not describe or illustrate these, however, since the notches carved on them have no numerical significance.

Now that we have briefly seen a few examples of number symbols carved in wood, let us take a look at the most outstanding example of the tally stick.

THE BRITISH EXCHEQUER TALLIES

These are unique in cultural history. In the Exchequer tallies the ancient number stick of obscure and common origin achieved its highest development: it became an official government record.

Some years ago, when repairs were being made in Westminster Abbey, several hundred "exchequer tallies," or notched sticks used by the British Royal Treasury, were found together with documents and the remains of the leather sacks in which they had probably been kept (see Fig. 60). These dated from the 13th century.

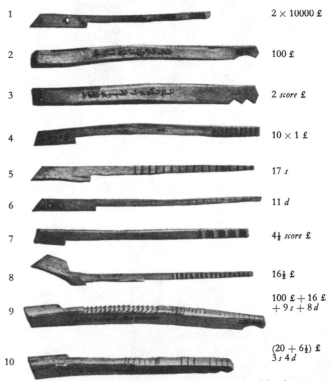

1	2×10000 £
2	100 £
3	2 *score* £
4	10×1 £
5	17 *s*
6	11 *d*
7	$4\frac{1}{2}$ *score* £
8	$16\frac{1}{2}$ £
9	100 £ + 16 £ + 9 *s* + 8 *d*
10	$(20 + 6\frac{1}{2})$ £ 3 *s* 4 *d*

Fig. 60 Exchequer tallies from the 13th century. On these sticks the treasury officials carved the amounts of tax owed and paid. These continued in use until 1826; the method of carving the notches remained the same until that year. In the case of Nos. 2 and 3 the notches are in the bottom surface, and the head of the tally points to the rear.
Society of Antiquaries, London.

Ever since the 12th century the English Royal Treasury had kept its records in books and on tally sticks, by methods that remained virtually unchanged until the 1820's. Thus from *tally* was formed the old English word for "tax," *tailage* or *tallage*, which resembles the related French word *taille*.

The Court of Exchequer, which was the state tribunal of audit, consisted of several departments. Here we are concerned with the central office, where the sheriffs, who administered the shires or provinces of England, settled their accounts with the Crown. The table in this office was covered by a counting cloth with a square checkerboard pattern, from which the whole court of audit took its name (see p. 347). Over this table the sheriff reported to the Treasurer, item by item, and item by item the Calculator laid out the amounts in *calculi* or counters on the chequered cloth and thus arrived at the final sum. Then the Cutter carved a *tally* to record the amount paid or owed to the Crown. The whole procedure was observed and audited by members of the court and higher officials. The use of tally sticks and counters enabled everyone to understand without having to know how to read and write.

This is how such an account was typically settled: if the official in charge of a district owed the Crown £100 annually in taxes and duties, he would make the first payment at Easter, let us say £40. In testimony of this a *tally* for £40 would be carved, of which the official retained the *stock* (or *stipe*) as a receipt for his payment and the court of audit kept the inset piece (the *foil*) for its own records. The name of the payer and the nature and amount of the payment were incised on both pieces. Then on Michaelmas-day (September 29) the official had to make good the entire annual levy of £100. To do this, he would submit his *stock* for the amount of £40 and pay the rest. The *stock* was very carefully compared (*checked*; see p. 233) with the *foil* in the Treasury's possession; if they agreed, the amount was written up to the payer's account. If any deception or cheating was detected, the official would be arrested on the spot. The payments made by the county officials were entered in the receipt book, and when their obligations were discharged their accounts for the year were closed with the phrase: *et quietus est* or *and he is quit* (of his debt).

As early as the year 1300 the *tallies* were issued by the Royal Bank to be used as a medium of exchange in England. For example, King Edward I's steward once received a *tally* as a money order in place of a payment owed by one of the inhabitants of London. From this man, who himself was a debtor of the Royal Treasury, the steward could demand and receive the sum of money marked on the tally. In this manner the state avoided the burdensome collection of debts owed to it and at the same time satisfied its own creditors: a cashless system of exchange, using notched tally sticks as cheques. From the middle of the 14th century on this system was in constant use, and did not begin to decline until the rise of banking institutions in the 18th century. But this use of tally sticks is no longer to be seen anywhere.

The banking term "dividend" probably also goes back to the English tally, which from sometime in the 13th century was also frequently called *tallia dividenda,* or simply *dividenda,* "a stick to be divided," especially in using the insets (foils) in the King's possession to make purchases for the Royal Court. In return for his goods, the tradesman was given a *dividenda* which he would later redeem at the Treasury, just as today a bond-holder, in his "coupons," has a certain share in the earnings of the enterprise which issued the bond. After the 18th century in England the "dividend" referred not to the tally or coupon but to the actual share of the earnings, and from there it acquired its present general commercial usage.

Because of their official nature, the carving of the notches in the tallies had to be uniformly regulated. Fortunately we are adequately informed about this and about the procedures of the court of audit by the *Dialogus de Scaccario* ("Dialogue Concerning the Chessboard"), written in 1186 by the then Royal Treasurer Richard, Bishop of London. The rules prescribed here, and above all the *tallies* themselves, remained in force for many centuries. In 1782, to be sure, it was commanded that tallies be no longer issued, but they remained valid until 1826. Then in 1834, when the state's official collection of cancelled *tallies* was being destroyed, a vast number of them were burned with such excessive zeal in the furnaces below the Houses of Parliament that the Parliament buildings themselves went up in flames.

Now, what does the *Dialogus* have to say about the "cutting of the tallies" — *de incisione talearum*?

In summo ponunt m. li. (librae) sic ut incisio eius spissitudinis palme capax sit . . .

The notch for £1000 is placed at the end and is as large as the hand is wide (see No. 1 in Fig. 23);
for £100 the notch is as large as the thickness of a thumb, and to distinguish it from that for £1000 it is not straight but curved (see No. 2);
for £20 it is as large as the thickness of the little finger (No. 3);
for £1 it has the breadth of a ripe barleycorn (No. 4);
for 1 shilling it is smaller but still large enough to be seen as a notch (No. 5); whereas
for 1 penny only a cut is made, with no wood being removed (No. 6);
for a half of any of these units a notch or cut half the length is carved: one cut slantwise, one perpendicular to the edge (Nos. 7, 8).

This prescribed succession of magnitudes was to be strictly observed: the largest number is always on the outside and the others follow in order. Examples 9 and 10 in Fig. 60 show how simply and easily the tally cutter avoided any possibility of confusing the larger notches with each other: The highest units were carved into the bottom surface (like the notch for £100 in No. 9), but the smaller ones at the top from left to right, so that the lowest ranks stand above the highest, from which they are very clearly differentiated!

Now we can make an observation both revealing and fruitful for the cultural history of numbers. On the tallies we see that the intermediate rank between £100 and £1 was not £10, as we might have expected, but £20: — we meet the old vigesimal grouping again in the *scores* see p. 49)! *A score of pounds* is £20. But what is a *score*? The old Saxon word *sceran* is similar to the English word "shear," to cut, cf. German *Schere*, scissors. Thus a *score* is something which has been cut or carved — a notch! This was incised on a tally stick whenever a group of 20 was to be counted (such as 20 sheep), and from this notch the vigesimal grouping acquired its English name *score*. There is a Finno-Ugric counterpart: The Lapp word *tseke*, "notch," likewise came to be the number word for 10 (see p. 113).

Now we clearly understand the significance of the English expressions *to run into scores*, meaning "to go into debt," and *on score*, meaning "on credit" (on the tally stick, without paying). Previously we saw many English expressions in which *score* was used with the meaning of 20 (pp. 49 ff.); but we can also find many passages in Shakespeare in which the word *score* still has its original meaning of "notch" and "tally stick":

Our forefathers had no other books than the score and the tally.
(King Henry VI, Part Two, Act 4, Scene 7.)

Then from Scene 2 in the same act we have:

I thank you, good people. There shall be no money, all shall eat and drink on my score.

Of Macbeth it is said at the end of the tragedy (Act 5, Scene 7) that

He parted well and paid his score.

The English language shows another interesting development. If someone lent a sum of money to the Bank of England, a *tally* was carved to record the transaction, of which the Bank retained the inset piece or *foil* (from the Latin word *folium*, "leaf") and the creditor received the *stock*. Thus he became a *stock-holder* and possessed a *bank-stock*, which had the very same worth as paper money issued by a government. From this custom comes the modern expression "stocks" for "shares in a business enterprise" or "paper money."

Notched tally sticks used to keep official records of state finances! After this rare case in a country with a high level of culture, we may conclude with a simple "early" example such as could once be found virtually everywhere. This is a Russian "tax book" for a Cheremysian village. The Cheremysians lived along the banks of the Middle Volga, and spoke a language belonging to the Finno-Ugric group.

For each of the 13 households in the village the tax collector had a tally stick. On the stick he carved an identifying symbol at the head and the number of inhabitants (see Fig. 61; half a notch for a child), and the amount of tax due annually at the end. Then the amount paid was "cut off." From Fig. 62 one can see clearly that this was done repeatedly and that the tax collector did not make a new tally stick every year.

Now let us glance back over the ground we have covered: what have we learned from the tally sticks? For one thing, they represent the oldest form of bookkeeping. They do not simply show the amounts involved but also record the transaction between buyer and seller or creditor and debtor, without giving either party any opportunity for cheating, even for such highly complicated exchanges as those

Fig. 61 A Russian "tax book" from the Middle Volga area. For each household there is a tally stick on which are carved an identifying symbol, the number of inhabitants and the amount of tax to be paid. One tally stick is about 17 × 2 × 1.5 cm in size.
Kansallis Museum, Helsinki.

involved in the milk-and-cheese coöperatives of the high Alpine meadows. They were account books in which the finances of a household or a village community were recorded in a generally very sensible fashion, and here and there they even rose to the status of "legal tender" or "share certificates" or medium of exchange. Their close connection with the economic life of mankind, which remained unchanged from the very earliest times until just recently, makes the notched tally sticks awesomely impressive as documents.

We finally have the secret of the "earliest" writing which was born of the common people. As uninitiates, we cannot read it. But we do know one thing: It was a method of writing numbers, and it consisted of primitive numerals.

Now let us examine the "numerals" or number signs themselves.

Fig. 62 A tally stick from the Russian "tax roll" of Fig. 61. The paid debts were "cut off" and new sums incised in the same spot.

THE NUMBERS ON THE TALLY STICKS

The simplest tally sticks merely bear rows of notch after notch; they form the amount to be counted by means of the supplementary quantity of notches carved. The rows of notches are organized by groups. Systems of written numbers formed according to this rule, like the Egyptian or the Roman numerals, we shall call "early" numerals or row numerals (see Fig. 8).

Number Notches

On the tally sticks we see groupings of notches requiring some sort of symbol different from the simple unit notches. At first we may be inclined to think that a notch is a notch. Yet the tally sticks we have seen thus far show a great variety of different kinds of notches (see Fig. 63): notches carved transversely and perpendicularly into the

Fig. 63 Various forms of notches.

wood, incisions, half-notches and half-incisions, slanting notches and incisions, notches and incisions cut into the edges of the stick, round notches, beveled notches, straight and round notches carved in the middle of the wood surface, etc.

But all these various special forms of notches were avoided in the English Exchequer tallies. They used only simple notches, differentiated from each other by size and position, and thus surprisingly arrived at a consistent and regular system of measures.

Quite often we have seen notches in the form of an X or a V (see Figs. 53, 58, 59, and 64). Anyone not familiar with tally sticks might think that these were Roman numerals transferred to the wooden sticks. Although this did happen here and there, did the peasants in some remote mountain village really have to wait until Roman numerals were introduced before they could carve X and V. Aren't these shapes actually the easiest ones to whittle with wood and knife? To ask such a question is to answer it. One hardly needs to see or refer to prehistoric notched bones with such notches (see Fig. 51) to realize that the other commonly used Roman numerals L, C, M, or ∞ rarely if ever occur on tally sticks, and that one likewise does not see back-counted formations like IV or IX. Carpenters also form the numerals I, V, and X with their axes when they number the beams and timbers they have fashioned.

Now let us ask the same question the other way around: Are the Roman numerals, at least I, V, and X, actually forms of notches cut into tally sticks? I believe that further facts will confirm that this is so, apart from mere similarity of form and the fact that such notches are very easy to carve.

At the beginning, the Roman numerals observed the basic and simple laws of ordering and grouping, like the notches in tally sticks. X is a group-form. How did this come about? By a crossing of the usual symbol for unity. We shall see this simple device over and over: crossing two signs makes a group-form. In the Indian Kharosthi numerals X = 4 is clearly the symbol for a grouping (see Fig. 12). Excellent examples are again provided by the Swiss tally sticks, such as the number 42 incised on the second stick from the left in Fig. 58.

Fig. 64 A bundle of Alpine number billets, small flat sticks some 20 cm long on which are carved the cow-rights to which their owner is entitled; the owner's name or symbol is on the reverse side. The most ornate of these sticks, the one at the extreme right showing the number 122, gives the total. This bundle, from Saanen in the Canton of Bern, is dated 1778.

For the most part a crossing of straight-line symbols means 10, so that if we wish to give a name to it, we can call it a "ten-stroke." A striking example is to be seen on the chief tally stick of Fig. 64, which shows the total, where even the V notch is crossed (in the number 122; see *b* in Fig. 65). Here we see that the unit notch which stands by itself is also crossed through into an X notch; the number 122 could also have been formed as in *c* in the same illustration.

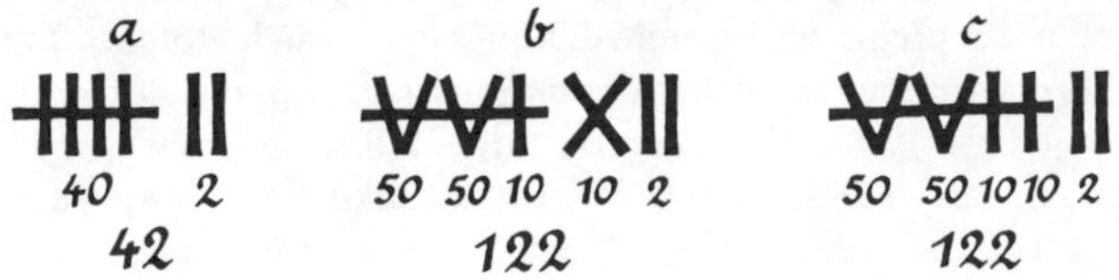

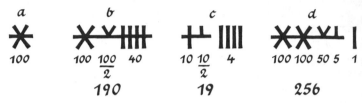

Fig. 65 The ten-stroke in notches on Swiss tally sticks.

A double crossing then indicates the group 10 × 10, or 100 (see *a* in Fig. 66). Thus we find even such numbers as 190 and 256 on tally sticks (see Fig. 58).

Fig. 66 Double crosses and half-signs formed by notches carved in Swiss tally sticks.

These notched numbers carved in tally sticks from a remote Alpine valley suggest a possible explanation for the extraordinary and hard-to-trace Roman numeral L for 50. The rules governing the formation of numbers on tally sticks are of a very early date, for the amounts recorded by them are ordered in a series and grouped. But if the quantity makes up only half a grouping, only half the symbol for the particular group is carved! For example, the V symbol on the ten-stroke is half an X and hence means half of 100, or 50. This is very readily seen on the right-hand tally stick in Fig. 58 in the numbers 19 and 256 (see also Fig. 66, *c* and *d*); in the case of the last number the half unit-notch on the ten-stroke stands for 5.

A study of tally sticks reveals two very important things about early written numerals:

The crossing of a carved symbol most frequently means a grouping, usually 10. (From the "crossed" Roman numeral X comes the Latin word *decussare*, "to cross through" from *decussis*, "the number X," from *decu-* for *decem* 10 used in compounds).

Half the symbol stands for half its normal value (see the Roman symbols for fractions, p. 161).

Now since they have played an important role in our culture for hundreds of years, let us again examine the

ROMAN NUMERALS

primarily the symbols I, V, X, L, C, D, and M. If the unit I is crossed it produces the symbol X for 10, half of which, V, has the value 5. In Fig. 67 the Etruscan coin at the left shows the bottom half of the ten-symbol, Λ, whereas in Rome the upper half, V, was commonly used (see Fig. 66).

But what about L for 50? This numeral has nothing to do with the letter "L," which it coincidentally resembles. Our illustration shows

Fig. 67
ROMAN COINS (top)
Left: with the value of 60 (sesterces), 215 B.C.;
right: with the series number 70 of the minting (90 B.C.). This is one of the so-called "serrated coins" (*nummus serratus*) which Tacitus says the Germanic tribes prized especially highly (see p. 358).
ETRUSCAN COINS (bottom)
With the symbols Λ and X; the coin on the right bears the Gorgon's head (5th century B.C.).
Landesmuseum, Darmstadt.

two Roman coins, on which an arrow-shaped symbol and an up-side-down T can be recognized. Moreover, an inscription carved on a Roman milestone contains the numbers 51 and 74 (see Fig. 71, lines 4, 5, and 6). Now the development is quite clear: The two oblique strokes on the arrow-shaped symbol were incised as a curve or a horizontal straight line, until finally this numeral came to be written just like the letter L.

What was the origin of the initial arrow-shaped form? It may be a half-symbol, like the V of X, in this case the upper half of a double or twice-crossed numeral I, such as we have seen on wooden tally sticks (Figs. 58, 59, and 66). Thus it may have developed from a deeply rooted early form (see Fig. 68).

Another possible explanation is that it is an old form of the Greek letter Ψ. The Romans took their alphabet from the Etruscans, and they in turn from the Greeks (see columns 7 and 9 in Fig. 95 on p. 265). Since they did not make use of the letter Ψ, this was free to be used as a numeral. Thus the Roman numeral for 50 is thought by some to have developed from an arrow-shaped variant of the Greek Ψ (Fig. 95, p. 265, line 26).

Fig. 68 The Roman 50-numeral L as a half-symbol for 100.

Correspondingly, this hypothesis (which is Mommsen's) sees the Roman numerals for 100 and 1000 as originating in similarly "free" Greek letters of the Greek alphabet. Since the Etruscans had replaced the aspirated Θ and Φ by the unaspirated T and P, the former two were also available for use as numerals Actually, there was an old Φ-form, a circle with two diameters at right angles to each other, which resembles the Roman numeral for 1000 (compare Figs. 69 and 70 with Fig. 10). The Θ supposedly developed into the Roman numeral C (as shown in Fig. 70).

This transformation does not seem too convincing, however, because the use of foreign letters looks too "learned," especially these particular letters, because their respective numerical values of 9 and 500 which they had had in Greece now became 100 and 1000 in Roman civilization (see Fig. 95, p. 265, lines 9 and 24). Couldn't the Romans themselves or their ancestors have invented symbols of their own?

Fig. 69 Coin of the ancient city of Pherai in Thessaly, with the initial letter *phi* in the old form (instead of φ), which the Romans may have adopted as their numeral for 1000 (cf. Fig. 10). 4th century B.C.; 1.8 cm in diameter. Staatliche Münzsammlung, Munich.

The symbols on the small tablet which the computer holds in his left hand in the Etruscan cameo illustrated in Fig. 134 are certainly numerals. We see here a few familiar forms, and also a circle with an oblique cross within it. This may have been the old sign for 1000. The cross may well first have right itself to assume a vertical-and-horizontal position (*b* in Fig. 70) and then dropped the horizontal bar (*c*); then it would have strongly resembled the Roman numeral for 1000. This derivation is quite surprisingly supported by an analysis of the symbol (*d*): D for 500 is doubtless the right half of the symbol; this leaves the left half, C, as the numeral for 100! This last step cannot be documented, yet the "breakdown" does find some confirmation in the 500-symbol on a Roman milestone inscription on which it still bears the horizontal bar of the original full form (see Fig. 71, fourth line from bottom; cf. *d* in Fig. 70).

Was the Etruscan symbol, the circle with an oblique cross, an independently invented form or was it in turn derived from the Greek

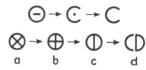

a b c d

Fig. 70 Possible origins of the Roman numeral C for 100:
top: from the Greek letter Θ;
bottom: from a supposed Etruscan numeral for 100 or 1000 (cf. Fig. 97). If the second hypothesis is accepted, then *c* could be the Roman symbol for 1000 (as in Fig. 10, p. 44). Its left "half," *d*, may then have become the Roman 100, while the right "half" D became the Roman numeral for 500.

Fig. 71 Inscription on a Roman milestone on the Via Popilia in Lucania. This shows the old form of the numeral for 50 (lines 4, 5, and 6 from the top) and a peculiar form for 500 (line 4 from the bottom). The inscription is 74 cm wide, with letters 3 cm high. Around 130 B.C.
Museo dell Civiltà Romana, Rome.

The inscription reads: "I built the road from Regium to Capua and erected all the bridges, milestones, and postal stations on it. From here to Novceria the distance is 51 miles, to Capua 84, to Muranum 74, to Consentia 123, to Valentia 180, to the statue on the seashore 237 and from Capua to Regium in all 321 miles. Likewise as Praetor in Sicily I tracked down Italic fugitive slaves and returned 917 persons (to their masters). Moreover I was the first to establish that on public lands hersdmen must yield to farmers. At this place I built the forum and other public buildings."

theta Θ? At the present time there is no way of answering this question.

At any rate, neither of these two hypothetical derivations is inconsistent with the idea that the arrow-shaped 50 form was originally half of a horizontally crossed $\times$ standing for 100. This "early" numeral could well have disappeared later on. Or perhaps, instead of being crossed with a horizontal stroke, which would have required a considerable degree of accuracy in carving, a curve or parenthesis was placed next to the $\times$, so that ($\times$ stood for 100 and ($\times$) for 1000. Then the Etruscan symbol would have had a completely indigenous, "early" derivation, as we have seen in the case of the tally sticks. Even the unique form ∞ for 1000, which appears in the inscription of Fig. 73, could then easily have been derived directly from ($\times$).

Now let us get on with our analysis of the Roman numerals themselves.

The numeral for 100 was later written as C, which (quite by coincidence!) was also the initial letter of the Latin number word *centum*, 100, just as the numeral for 1000 developed into the form M, with which (again, purely by coincidence) the number word *mille* began, while the symbol for 50 came to be standardized as L, which does *not* correspond to any Latin number word. By contrast, in the early Greek number symbols from time immemorial the initial letter of the name for a numerical rank served as the numeral for that number: for example, *Δ* was the numeral for 10 because it was the first letter of the Greek word *ΔEKA*, "ten" (see Fig. 97). This method of writing numbers was very common in the Middle Ages (as, for instance, the Indian numeral 2 became identified with Z probably because of the word *z-wei*, "two"; see Fig. 118) but never appeared at all in Roman times. For 1000 the Romans could also use the curious form ∞, which ever since the English mathematician Wallis proposed it in 1655 has been accepted as the mathematical symbol for infinity (see Fig. 73).

All the other Roman numerals are quite clear: D for 500 is half the symbol for 1000. The higher numerals, those for 10,000 and 100,000, are artificial extensions of the already available symbol for 1000; this is more evidence that 1000 was an old limit of counting. The numerals for 5000 and 50,000 were at one time the halves of the symbols for 10,000 and 100,000 (see Fig. 72, also Fig. 31).

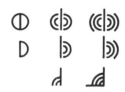

Fig. 72 The Roman half-symbols for
500   5000   50,000
These are the "halves" of
1000   10,000   100,000
shown in the top row. At the bottom are the transformations that appear on Roman denarii (see Fig. 31).

"Caesar sesterces 100 millions"

Fig. 73 The Roman numeral for 100 million here consists of the symbol ∞ placed within a frame, which itself originated from an extension of the numeral for 100,000 (see Fig. 10). From an inscription dating from the year A.D. 36. Letters 12 mm high.
Scavi di Ostia (Excavations), Rome.

And what about the million, the ⊠ or X standing beneath a gate? The gate is nothing more than a rectilinear version of the curved sign for 100,000, which we have seen already on the *Columna rostrata* (Fig. 10). It is an abbreviation of *centena milia*, the last of the verbal numerical ranks for which there was a full name. By exact analogy to the verbal expression *decies c. m.*, "ten times a hundred thousand," the numeral was written as an X beneath a curve (see p. 44, where we spoke of this numeral in a different connection).

In medieval manuscripts a numeral often had its value multiplied by a thousand by the addition of a superior bar, for example $\overline{\text{II}}$ for MM, 2000 (see Fig. 44); this was not common in Roman texts, however, where for the most part it served only to identify the symbol as a numeral, i.e., to distinguish it from the letter of the same form (see p. 281).

Now let us summarize. The Roman number symbols did not evolve continuously and smoothly any more than did the Roman number words. They began with very ancient forms on tally sticks, certainly up to X and perhaps up to 100. However the symbol for 1000 may have originated, whether from a foreign letter or an indigenous carved form, it was at any rate treated as an early numeral: it was halved to produce a numeral of half its value and by the same token multiplied 10 and 100 times by *decussatio*. In this manner the limit of counting was pushed forward as far as 100,000. Then when the breakthrough came in calculations involving millions, the limit remained at the numeral for a million, which was no longer ordered in series but counted and thus built up in gradations.

Thus the Roman numerals are the "sisters" of the Latin spoken numbers. Like the latter, the former were not invented all at once by one person, but were evolved gradually by the people who wrote them.

The Chinese Han Sticks

If we closely examine inscribed Chinese wooden sticks (Fig. 74) from the time of the Han Dynasty (200 B.C. to A.D. 200), we come across the "ten-stroke" not only at 10, but also at 20, 30, and 40, just like the insets on the tally stick from one Alpine village (see Figs. 58 and 64).

Fig. 74 The ten-stroke on Chinese Han sticks:

10   20   30   40   the same, fused together.

This ancient method of writing 20 and 30 has been preserved even to our own time as a special form in China and even more in Japan. The interesting feature is this: a single Chinese character could also have only a single one-syllable pronunciation. The usual Chinese way of writing 20 (2 × 10) has two characters, and is thus pronounced "*erh-shih*" "two-ten." But the old special form of 20 has the special monosyllabic name, *nien*; the Japanese, in contrast, follow the rule in saying *ni-ju*. The same applies to 30 (Chinese *sa* instead of *san-shih*; see p. 450). Hence, there are two Chinese number words for 20, each corresponding to a numeral.

These Han sticks are among the very oldest documents that record events of ordinary or commercial life. They were found in a guard post of a part of the Great Wall of China which projected far into the wasteland, where the Gobi Desert meets the eastern spur of the Tarimbek Mountains. The characters written on them deal with the duties and the life of the border troops; the ones illustrated here (see Fig. 75) record the payments made for baked bricks used in building the Great Wall (1, 2) and the distances traveled in the course of these operations (4, 5, 6).

The Han sticks are important not only because of the historical events they record but especially because they represent a significant intermediate stage between the tally stick and written numerals. Are these slender pieces of wood tally sticks? Naturally; and like all tally sticks they are marked and read from top to bottom. The later numerals for the tens, however, have not yet replaced the older

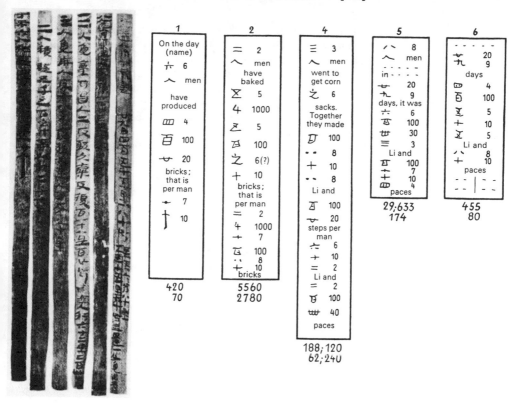

1		2		4		5		6	
On the day (name)		=	2	=	3	/、	8	·· ··	20
六	6	八	men have baked	人	men went to get corn	人	men	九	9 days
人	men			乙	6 sacks. Together they made	in ·· ··	20	四	4
have produced		乂	5			九	9 days, it was	百	100
		4	1000	刀	100	六	6	乂	5
皿	4	乙	5	··	8	西	100	十	10
百	100	乃	100	十	10	卅	30	乂	5
↻	20	乙	6(?)	··	8 Li and	=	3 Li and	↗	8 Li and
bricks; that is per man		十	10 bricks; that is per man	乃	100	亞	100	十	10 paces
·	7	=	2	↻	20 steps per man	丗	7	··	
十	10	4	1000	六	6	丗	10	· ·| ·	
		··	7	十	10	皿	4		
		乃	100	=	2 Li and	paces			
		··	8	=	2				
		十	10 bricks	乃	100				
				卌	40 paces				
420		5560		188; 120		29; 633		455	
70		2780		62; 240		174		80	

Fig. 75 Wooden number sticks from the time of the Han Dynasty, with early Chinese characters. These were found on the Great Wall and are about 2000 years old. Each stick is about 40 × 2 cm in size; translations (except for 3) are at right; *Li* is "mile."

method of writing with notches crossed through, except that here the brush replaces the knife.

The memory of the tally sticks remained alive for thousands of years. Even today, the Chinese write their characters in vertical columns and always only on one side of even the thickest paper. What is the Chinese word for "book"? A bundle of sticks. Thus the Chinese written character for "book" is a cluster of tally sticks strung together (Fig. 76), and appears to be an Asiatic relative of the Russian "tax book" and the Swiss "Alp book" (see Figs. 61 and 64). The word "law" in Chinese is then very interestingly conveyed by a Chinese character representing a "book" of tally sticks solemnly elevated on a table (see bottom character of the figure).

Now, back to the notches themselves.

Notched Numbers Tied to Specific Measures

These are definitely numerals — that much is certain. But they are not abstract numerals, for the numbers they represent are tied to the objects numbered. The unit notch which means "5 Crennen of milk" on one tally stick has the value of "4 hours of water" on another and of "1 franc" on a third. Thus the specific measure is present as a silent partner in these early number symbols.

Fig. 76 "Book" and "law" in Chinese characters.
Top: "book" (new form, with old form below), a bundle of tally sticks as in Fig. 61;
bottom: "law," a bundle of tally sticks lying on a table.

The notched tally sticks of a Wallachian shepherd are beautiful examples of the way in which a numeral is still tied to a particular object or objects. From the first, second, and third peasants, let us say, the chief herdsman receives a certain number of sheep for which he is responsible and records these numbers on a long tally stick. The peasants in turn record the number of sheep they have handed over on their own tally sticks. Now the herdsman is obligated to remit certain amounts of cheese to the sheep owners. The amounts of cheese are recorded by the shepherd on the other side of his tally stick, using quite different number symbols for this purpose. In the case of written numerals, this is the exact counterpart of the classes of spoken numbers: The numerals depend on the nature of the objects whose quantities they represent (see p. 31).

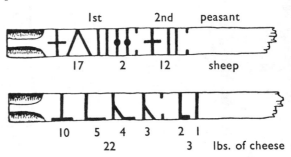

Fig. 77 Shepherd's tally sticks on which the kinds of notches used to record amounts of cheese differ from those that represent peasants and sheep.

Thus we have found the same formations and the same early laws in both spoken number words and written or carved numerals. Just as number words were liberated by stages from the objects they counted, the early numerals were first freed from their association with the wooden tally stick, then the forms of the notches were written instead of being carved, and here and there the resulting numerals also were removed from their content and became "abstract" numerals that could be applied to any or all objects to be counted. But some did not extend their domain beyond the village or the rural district; they remained valid only for peasants and their farm economy. These were the peasant numerals.

Peasant Numerals

Gottfried Keller, in his well-known novel, *Der Grüne Heinrich*, tells the following story:

In the house opposite there was a dark, open hall completely filled with various kinds of junk.... Far in the background there sat a heavy-set, aged woman dressed in old-fashioned clothes.... She could read printed letters only with great difficulty, and could neither read nor write Arabic numerals; instead, her whole mathematical arsenal consisted of a Roman One, a Five, a Ten, and a Hundred. As she had absorbed these four numerals in her early youth, in some faraway and forgotten land, handed down to her after being used for a thousand years, she manipulated them with remarkable dexterity. She kept no books and had no written accounts, yet she was quite able at all times to keep track of her whole trade, which was often carried on in several thousand very small transactions, rapidly using a piece of chalk to cover a table top with large columns of these four numerals. When she had posted all these small amounts in this manner, from memory she would simply arrive at their sum by erasing one row after another with her moistened finger as quickly as she had written it down, at the same time posting the results at the side. Thus new and smaller groups of numbers appeared whose meaning and purpose no one but herself knew, since they invariably consisted of only the same four bare figures which to outsiders looked like some kind of ancient magic writing. She was never able to carry out the same operation with pencil or pen, nor with a slate-pencil on a slate, because she not only needed all the space on a table-top, but also because she could only draw her large symbols with a piece of soft chalk.

We could scarcely find a better representative to demonstrate peasant numbers than this magnificent old woman junk-dealer of Keller's. What does she need any newfangled numerals for? They are just as mysterious and magical to her as her own are to others; but to her own clear and simple mind her peasant numerals are also perfectly clear and simple. Whether or not they were actually Roman numerals is less important than the fact that she chose these particular forms and clung to them steadfastly. L and D and M did not appear in her calculations any more than they do in the following examples:

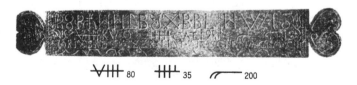

Fig. 78 An Alpine tally stick on which the herdsman has recorded in peasant numerals the numbers of different animals (cows, oxen, goats) he has pastured in the Alpine meadows during a summer. At the top are the initials of the owners' names and the year 1813; below the herdsman has written "Jöri Bregetzer made this *Scheita* ("billet")". Length 35 cm. Museum für Schweizer Volkskunde, Basel.

The Alpine tally stick carved by the herdsman Jöri Bregetzer in the year 1813 shows the numbers of various animals in his charge during that summer (see Fig. 78). According to his account, these were 80 cows, 35 *Galtiere* (oxen) and 200 goats (cf. Fig. 66).

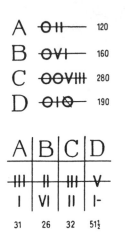

Fig 79 "Bar-figures," developed from the tally sticks used in Switzerland.

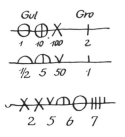

Fig. 81 Peasant numerals from Carniola. The numerals, which are different for gulden and groschen, were written "on the line." The ten-stroke appears in 10 and 5 groschen; half-symbols in the middle; at the bottom the sum of 256 gulden and 7 groschen, is written on the line, which to prevent cheating is terminated at the extreme left by a small hook.

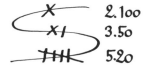

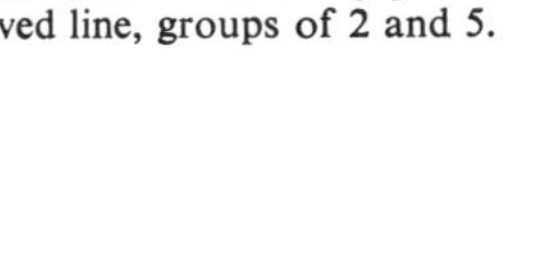

Fig. 82 Peasant numerals used in scoring the game called *Jassen*. Place-value indicated by position on curved line, groups of 2 and 5.

These numbers are still carved into the wood, but this Alpine billet is more than just a tally stick. The numerals on it have now been separated from their objects and are used to count anything. This development of the numerals on tally sticks can still be seen, even in our own times, in Switzerland. Here each shareholder in an Alpine meadow in a particular district had a long, flat wooden billet on which was recorded the amount he contributed to the communal cheese-making. From this *Beilen-Rechnung*, or hatchet-figures (see p. 224), developed the written *Stangenrechnung* or "bar figures," in which the *Stange*, the "bar" is our ten-stroke (see Fig. 79).

The diagrams can be understood without further explanation. At the right are recorded the amounts of milk contributed during the day by the community members A, B, C, and D; the daily amounts are then combined on the bars at left. Here, too, we see an instance of back-counting in the writing of these numerals.

Time and again the ten-stroke crops up in combination with the old tally numbers (see Fig. 80):

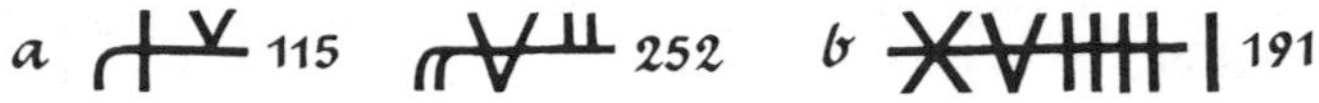

Fig. 80 Peasant numerals with ten-stroke: (*a*) from the Prätigau; (*b*) from the Wallis district.

The symbols for money in use in Carniola well into the 19th century (see Fig. 81) are of interest because of their great antiquity. Here we see the writing "on the line" (at the top in the Figure at left), which to make cheating impossible has a small hook at the beginning (bottom left), the half-symbols for half-values (middle) and the ten-stroke which combines the individual single-unit gulden into groups of ten. (In the figure, *Gul* = gulden and *Gro* = groschen).

When even the denominations of money and the groups are to be differentiated by special numerals, as is done in *Jassen*, a card game played by peasants in the south of Baden, this is done very cleverly by keeping them apart by a line (see Fig. 82). This line is drawn in a large S-shape and thus provides three separate segments, for the 100-, 50-, and 20-groups, the only ones used in scoring this game; on the first two segments the numerals are grouped by twos and on the last by fives.

Until the middle of the last century people in the Canton Uri were taught to add up such sums as 457 + 60, for example, in peasant numerals, as shown in Fig. 83. In this operation numbers were written down, erased and others written down again, just as the old woman did in Gottfried Keller's book (see p. 249).

The "line" or "stroke" was originally an image of the old tally stick itself. In several places in Switzerland, as grain and potatoes were delivered in sacks a tally was kept by making marks on a vertical line, as illustrated in Fig. 84; every group of five strokes was made alternately on the right and left of the vertical line, so that the total (which in our example is 13) could be read off more easily. Hence the German expressions, "to have someone on the line," meaning to have a debt someone owes marked up on a line (= tally stick), as well as "to put someone up on the board," meaning that from then on all his sins will be marked down or chalked up to him exactly.

Our series of examples ends with a Styrian peasant calendar of the year 1398, from which the month of September is reproduced here (see Fig. 85). The forms of the written numerals obviously resemble the notches carved on tally sticks, so that they look rather like Norse runes. Again we see the numbers written "on the line" along with the clear groups of 5 and 10. Even if calendar sticks were no longer preserved today, ethnologists and anthropologists would have no reason to doubt that these early peasant numerals were derived from the tally and only later became independent of it.

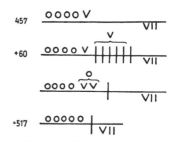

Fig. 83 The addition 457 + 60 = 517, as performed in peasant numerals.

Fig. 84 Counting "on the line."

Fig. 85 Farmer's calendar from Styria, with peasant numerals which clearly recall the notches carved on tally sticks. The letters stand for the days of the week. The numbers 1 through 19 refer to the nineteen-year lunar cycle, in which every year has one number, the so-called "golden number." The bottom row reads:

2 10 18 5 15 4 12 1 9 ... 16 5 13.

With this we come to the end of our discussion of tally sticks, which have given us a fascinating glimpse into the simple business dealings of unlettered people in earlier days. The tally stick was the peasant's and the tradesman's and the merchant's account book; on it, with knife on wood, they learned to carve the primitive numerals which were sufficient to manage their own affairs, and which they took with them and kept on using even as they were learning to "write" properly. But for the cultural history of numbers it is important to understand that the early rules which govern the arrangement of spoken numbers apply to written numerals as well. It could even be argued, in fact, that the simplest forms of written numerals preceded spoken number words and that these laws of ordering and grouping were transferred from the numerals to the number words. We see, moreover, that these rules are universal: we find them used all over the world, even in China, in the symbols that make up an "abstract" system of numerals. We have been able to arrive at this understanding only by traveling over the long path we have followed thus far in the present volume, since today tally sticks have disappeared completely in our civilization and have been replaced by the sophisticated form of writing now in use. But we must admit that even primitive people can indeed "read and write."

Knots Used as Numerals

Before concluding our consideration of folk symbols, we must look at just one more curious form of primitive numerals, or recorded numbers, one that has no trace of writing in it: the knots. Knots have of course, never evolved into "written" numerals.

The custom of counting both days and objects by tying knots is found all over the world. The Tibetan prayer-strings and the rosary are both forms of number-strings on which the prescribed numbers of pious exercises is recorded in knots. King Darius of Persia gave his subjects a cord with 60 knots tied into it when he set off on his attempt to conquer ancient Greece; each day they were to untie one more knot, and if he had not returned by the time the last knot was untied, they were no longer to wait for him (see p. 153). Since such number-strings are the exact equivalent of the number or tally stick, there is no need to say much more about them. But there have been a few curious forms of number-strings with knots with which the reader is probably not familiar and which go beyond the mere lining up of one knot after another (see Fig. 86).

In the Ryukyu Islands in the Pacific Ocean, between Japan and Taiwan, workmen braid strands of straw or reeds with fringes in various different forms to indicate the wages they have earned. Each form of fringe signifies a particular unit of value, so that together they make up a kind of place-value notation. A free end stands for one unit and a knot for five units.

The knotted cords of Peru, called *quipus*, the only form of "writing" known, played a very important role in the Inca Empire because they recorded all the official Inca transactions concerning the land and subjects of the empire.

From a main strand (which was some 50 cm long) hang the often variously colored strings (each about 40 cm long) into which the knots have been tied (see Fig. 87). One of these strings would be used to show the number of sheep, for example, another for goats, and a third for lambs or kids, in the same manner as Bolivian herdsmen today still keep track of their flocks. The knots themselves indicate the numbers.

Three different kinds of knots were tied on the *quipu*: single (overhand) knots (No. 1 in Fig. 88), double or figure-of-eight knots (No. 2), and slip knots with 2 to 9 loops (No. 3). As the photograph (Fig. 87) also shows, these knots are not distributed at random over the strings but are arranged in a decimal gradation so that the hundreds are closest to the main strand at the top, the tens follow below in a second horizontal row, and the units are tied into the ends of the strings at the bottom; the units alone may be represented by a double or figure-of-eight knot (for 1) or a slip-knot (for 2 to 9). Thus 235 is "written" with 2 knots in the top row, then 3 single knots in the middle and finally a slip-knot with 5 loops at the bottom end.

If the reader already suspected that numbers underlie these knots, his suspicion will have been strengthened by the slip knots with various loops and by the obvious order in which the knots are arranged, and it will have been changed to certainty by the "head

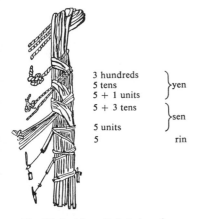

3 hundreds	
5 tens	yen
5 + 1 units	
5 + 3 tens	
5 units	sen
5	rin

Fig. 86 A string of plaited reeds on whose fringes workmen of the Ryukyu Islands have recorded the wages owed to them. Here they amount to 356 yen, 85 sen, and 5 rin.

Fig. 87 Peruvian *quipu*. The knots are tied in order of numerical rank, with the units at the end; it can be seen clearly that the knots representing each rank are not just spaced at random, but that they run horizontally across the *quipu*, each row at a particular height. In this example the ends have been destroyed. About 40 cm long.
Linden Museum, Stuttgart.

strings" marked by the letter K in Fig. 88. These are threaded through the loops at the top of a number of strings, and the knots tied in them give the sums of the numbers shown by the knots in the strings.

The most striking feature of the *quipus* is their close similarity to the Exchequer tallies: in Peru, as in England, a very early and primitive form of numerals achieved official recognition and was used to keep the state's financial records. The only difference between them was that without exception all accounts of the Inca Empire were kept in these *quipus*, because apart from these knots no other system of written numerals was known. In every Inca settlement there were four official *quipu* keepers, known as *camayocs*, who tied the knots in

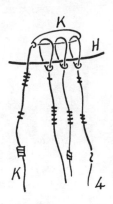
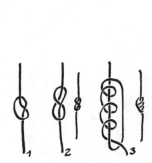

Fig. 88 The various knots in a Peruvian *quipu*: 1: single or overhand knot; 2: double or figure-of-eight knot; 3: a kind of slip-knot with three loops; 4: the head string, marked K in this illustration, which is run through the loops at the head of the other three strings showing the numbers 150, 42, and 231, bears their sum, which is 423.

these strings and submitted them to the central government in Cuzco. There is no doubt that this perhaps intentionally obscure manner of recording numbers, which only initiates could read, was a strong support for the monarchical absolutism of the Inca ruler. An example of the opposite situation is provided by ancient Athens, where the government of the city-state was obliged to reveal its records openly to the citizens so that every official was subject to "democratic" criticism (see p. 269) — dictatorship and democracy in bookkeeping!

At first glance, it seems puzzling that the Incas could also "write down" their history, their laws, and their agreements contracts on *quipus*. Garcilaso de la Vega, son of a Spaniard and an Inca princess, gives a very revealing account of the reception of the Spanish ambassador by the Inca:

Among the common people and the nobility who attended the Inca in the audience hall there were two official historians who recorded Hernando de Soto's message and the Inca's reply in knots.

How this was done is revealed to some extent in a second passage in which Garcilaso complains of a poor translator:

His translation was not good and not accurate; this was unintentional, of course, since he did not understand the meaning of what he had to translate. Instead of the three-in-one and single God he read three gods plus one make four (!), adding the numbers so as to make the expression intelligible to himself.

This quotation shows how things and events which are not numerical could be translated into numbers as an aid to memory in handing them down by word of mouth: a primitive form of quantification of non-numerical data, quite familiar in the modern world, where it is carried out to the most minute details by data-processing machines.

These people, the "scribes" or ayutas, formed a high-ranking stratum of Peruvian society at the time of the Spanish conquest. We really do the same thing, if we want to consider the knots we tie in a string or a handkerchief as a reminder.

Is the *chimpu* of the Bolivian and Peruvian Indians descended from the *quipu*? Almost certainly, for it too records numbers (such as 4456, for example in Fig. 89) on strings in a form of place-value notation, but in a different and surprisingly sensible manner, using fruit seeds strung like beads. The strings are first knotted together at the top. Then each numerical rank, thousands, hundreds, tens, and units, is indicated by the appropriate number of seeds. The four seeds representing the number of thousands are threaded on four strands, four more for the hundreds on three strands, five seeds for the tens on two strands, and finally six seeds for the units on just one strand. Then the four strands are knotted together at the bottom (u – – – – – u′ in the Figure) and the number 4456 is now "written down" and can be stored away.

The Chinese too at one time used knots to record numbers. Lao-tse, the Chinese philosopher who lived in the 5th century B.C., urged his compatriots of Tao-te-king to go back to the simple ways of doing things: "Let people again tie knots in cords, and these will serve as writing."

After seeing the *quipus*, we can understand to some extent what Lao-tse meant by this admonition and what kind of attitude he was recommending. A paper war in knots is inconceivable. But wherever writing has been made too easy, it will rage, as our civilization has learned since the invention of the typewriter.

Some scholars suspect the peculiar fashion, characteristic of the Indian Devanagari script, of arranging the letters symmetrically on each side of a vertical line (see Fig. 243), may well be a vestige of an old *quipu*-like form of "knot-writing."

Even German folkways offer an interesting example of "knot-writing" in the *Miller's Knots*, used by millers in their transactions with bakers up to the beginning of the twentieth century (see Fig. 90).

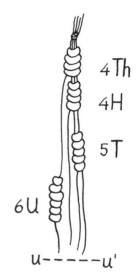

Fig. 89 The *chimpu* of the Peruvian and Bolivian Indians, a descendant of the *quipu*. This one shows the number 4456.

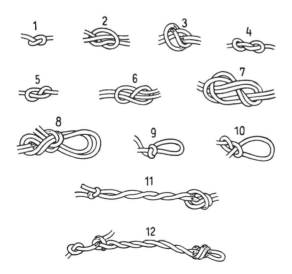

Fig. 90 Miller's knots used to indicate amounts and kinds of flour.

The particular knots illustrated here were used in the province of Baden.

The miller had to "write down" somehow the amounts and the kinds of flour and meal contained in the sacks he delivered. For this purpose he would use the draw strings that tied the mouths of the sacks. The quantities and measures were indicated by knots (1–7 in the Figure) and the kinds of meal or flour by loops or tufts (8–12). The miller's measure of flour was the *Sester*, an earlier measure of volume containing 10 *Mässel* (for the origin of the word, see p. 181).

1 *Mässel* = one simple overhand knot   (1)
2 *Mässel* = the same knot with a strand drawn through it   (2)
       or tied in the bight of the draw string   (3)
5 *Mässel* = ½ *Sester*, and
10 *Mässel* = 1 *Sester*, both represented by special knots   (4, 5)
2 *Sester* = the same 1-*Sester* knot tied in the bight or with a
       strand drawn through it   (6)
6 *Sester*   likewise represented by a special knot   (7)

Here for the first time we see that different numbers can be represented not only by series of the same knot, but also by special individual knot-symbols, which thus become knot-numerals: 1, 2, 5, 10, 20, and 60 measures. The intermediate numbers are then generally formed by combinations: 8½ *Sester* = 6 + 2 + ½ *Sester*.

The manner of indicating the kinds of meal and flour delivered by the miller, for instance "hog-mash, rye, barley, 1st and 2nd grade wheat-flour," can be seen from the illustration (8–12). Barley rye, for example, was represented by a loop tied into a *Sester*-knot.

Like the tally sticks, these ingenious miller's knots are striking evidence of the inventiveness of simple, unlettered people.

Alphabetical Numerals

The Gothic Numerals

"Let him that hath understanding count the number of the beast: for it is the number of a man; and his number is Six hundred threescore and six."
Revelation 13:18

Notches, knots, and peasant numerals have only a limited, or even purely personal, validity. For this reason their elevation to the status of government records in the case of the British Exchequer tallies and the Peruvian *quipus* was exceptional — a historical curiosity, so to say.

During the Middle Ages the "official" method of writing numbers in Western Europe was the Roman numerals. These eventually made their way north via the monasteries. Here they found no opponent of equal strength, for the popular numerals could make no claim to universal or widespread dominance, and they were, moreover, quite compatible with the Roman. Nowhere north of the countries bordering the Mediterranean was there a generally used system of written numerals, nor any universal form of writing, for that matter. Thus the Roman numerals continued to predominate in northern Europe even after the fall of the civilization which invented them; thanks to their "early," primitive nature they met with such ready reception that they were never felt to be alien, and in the 16th century they were stubbornly defended as "*de düdesche tall*" (the German numbers), as against the Hindu place-value notation that was then invading Europe.

The Greek numerals, on the other hand, never came to be deeply rooted among the Germanic Goths.

In their southeastward migration to the shores of the Black Sea and the lower reaches of the Danube, the Goths entered the sphere of Greek culture. The Visigoth Bishop Ulfilas, who died in A.D. 381, translated the Bible for the benefit of his Christian congregations. This magnificent monument of Germanic linguistic history is best preserved in the famous *Codex argenteus* (of which 187 out of 330 folios still exist). This manuscript, with its letters of silver on purple-colored parchment, which after long wanderings has finally found a permanent home at Upsala in Sweden, was written around A.D. 500 in Italy, most likely by an East Goth copyist.

Ulfilas, a highly educated man who knew both Greek and Latin, invented this Gothic alphabet expressly for his translation of the Bible (see Fig. 91). For this purpose he chose 17 Greek and 3 Latin (h, r, s) letters, as well as 7 Germanic runes (those for j, u, f, o, and perhaps also q, hw, and þ); the last of these may have been derived from the Greek *thēta*, θ, by removing the horizontal stroke in the center and placing it vertically at the side. Two letters of Greek origin, ϛ and ↑, did not correspond to any Gothic phonetic value, but served only as the numerals for 90 and 900, respectively. With this we come upon the remarkable fact with which we are concerned here:

The 27 letters of the Gothic alphabet served simultaneously as numerals.

Ulfilas arranged the Gothic number words according to his Greek model (see p. 270). The 27 = 3 × 9 different Gothic letters are used to represent the group of nine units, the group of nine tens, and

Numerical value	Gothic	Derived from	Phonetic value	Numerical value	Gothic	Derived from	Phonetic value	Numerical value	Gothic	Derived from	Phonetic value
1	A	λ	a	10	ı ï	I	i	100	R	R.	r
2	B	B	b	20	K	K	k	200	S	S.	s
3	Γ	Γ	g	30	λ	λ	l	300	T	T	t
4	d	Δ	d	40	M	M	m	400	Y	Y	w
5	E	E	e	50	N	N	n	500	F	F:	f
6	u	q:	q	60	G	Ç:	j	600	X	X	x
7	z	Z	z	70	n	n:	u	700	Θ	Θ:	hw
8	h	h.	h·	80	π	π	p	800	Q	Q:	o
9	φ	Ψ:	þ	90	Ч	Ç	—	900	↑	↑	—

Fig. 91 The Gothic alphabet consists of 17 Greek and 3 Latin (.) letters and 7 Germanic runes (:). Bishop Ulfilas' Gothic Bible, the *Codex argenteus* (see Figs. 92 and 93) was written in this alphabet.

the group of nine hundreds. Thus, for example, we read the parable of the sower (Mark 4:8)

"... *jah wahs jando. jah bar ain .l.*
jah ain .j. jah ain .r. jah qaþ: sae ..."

[And other fell on good ground, and did yield fruit that sprang up and increased;] and brought forth, some thirty, and some sixty, and some an hundred. [And he said unto them]...

The *Codex argenteus* is written in uncial letters with the words run together and without punctuation, except that periods are used to mark the ends of sentences. When letters serve as numerals, they are covered by a horizontal stroke and occasionally set off by periods at

Fig. 92 The numerals x̄ ⵣ̄ R̄ for 30, 60, and 100, from the Gothic Bible (folio 294). In this photograph the silver letters appear white against the dark background of the purple parchment; thus the negative, with black letters against a light background (as shown in Fig. 93), is easier to read.

each side. The Gothic alphabetical numerals are used especially to indicate the verse numbers in the margin (see Fig. 93). References are also made to the corresponding Bible passages within the decorative arches at the bottom of the page. In addition, some number words, especially the thousands, are also written out in full (see Figs. 27 and 28). But the Goths made no computations with these alphabetical numerals, nor did people use them to keep their ordinary, daily accounts. Of the Goths' own early written numbers we know absolutely nothing.

With these Gothic alphabetical numerals we have touched upon a subject which very seldom appears in the cultural history of numbers, but which is nevertheless of the very greatest significance. Let us now turn our attention to this.

Letters and Numbers

Fig. 93 Left margin of Folio 125 of the Gothic Bible, showing verses No. 156, 157, 158, and 159 and references to the same verses (Gospel According to St. John) in the arch at the bottom. Folio size 25 × 30 cm, letters 0.5 cm high, text itself measures 17 × 14 cm.

Numbers seek form as words and as symbols. In essence, however, letters are the embodiment of words, not of numbers. Yet letters also come into contact with numbers, both indirectly through the medium of number words and directly in themselves, as we have just shown in the Gothic alphabetical numerals.

Letters and numbers meet in number words in two different ways. The first and most frequently encountered is the writing of numbers as number words. This is easily understandable when no computations are to be made and there are not many numbers. We have seen from our study of spoken numbers that every culture began by expressing its numbers as words. But even in their books on arithmetic the Arabs, for example, although they both knew and used the Indian numerals, also wrote out their numbers in words. Here we may recall the Indian symbolic numbers (see p. 120). The Chinese ideographic numerals are the only known instance in history in which the written number word has been the same as the numeral (see pp. 458; also p. 53).

The second way in which letters and numbers come together indirectly is in the use of abbreviations for the written number words as numerals. In the early Greek system of writing numbers, for example, the numerals were the initial letters, like the *Δ* of *ΔΕΚΑ* = 10. Another and less familiar example of the same thing is the Arabic *siyaq* script, which is still used in Persia. This script uses the abbreviated and stereotyped written number words as numerals in a peculiar form of place-value notation (see p. 276).

Now for the direct contact between letters and numbers. One of our most priceless possessions is the established sequence of the alphabet with its twenty-odd letters in their fixed order. One need only recall the great convenience of our dictionaries and directories, which the Chinese with their ideographic characters do not enjoy. The fixed order of the letters in the alphabet may have originated in astrological concepts. Perhaps it goes back to the 30 or so constellations through which the moon passes on its travels. At any rate, numbers are indissolubly linked to the fixed sequence of the alphabet: the first letter stands for the first number, the next letter for the next number, and so forth.

The first to recognize and take advantage of this numbering quality of the letters were the Greeks, who did not invent the alphabet themselves but took it over from the Phoenicians. It was also the Greeks who realized the three possible ways in which letters can be associated with numbers:

1. The continuous one-to-one sequence in which the numbers 1 through 24 were represented by the letters *A* through *Ω* respectively. In this manner the Alexandrian scholars numbered the books and lines of Homer's epic poems and the ancient Greek masons numbered their blocks of marble. If numbers higher than 24 were called for, then *AA* = 25, *AB* = 26, *AΓ* = 27 and so on were used. Even today we still occasionally use this system of numbering, although we rarely if ever go beyond the first few letters of the alphabet. A

similar method of numbering was used by the Indian mathematician
Aryabhata.

2. The graduated succession is that which we saw in the Gothic alphabetical numerals. The alphabet, since it had fewer letters, was expanded
to 27 = 3 × 9, so that each group of 9 letters would represent the
units, the tens, and the hundreds. How the thousands were formed
in this system we shall discuss later, when we shall also see that the
Greek mathematicians (even Archimedes and Diophantus) used
these alphabetical numerals for their computations. This system of
symbolizing numbers was retained in Greek culture (Byzantium)
until it was gradually replaced during the 14th century by the
"Indian" place-value method.

3. The place-value method. Here the first nine Greek letters, α
through ϑ, served as "numerals" in the Indian manner. Along with
a zero sign, these letters were the exact equivalents of the Indian
numerals 1 through 9 plus 0 (see p. 273). This method is still in
use today in a Javanese and a few Southern Indian number systems.

The Greeks both thought of and used all three of these methods of
representing numbers by letters. Thus we have a remarkable record
of the numerical use of letters all the way from the beginning of
Classical Greek to the end of Byzantine Greek culture.

But what was the origin of the alphabet, which played so important a role in the development of numerals? We shall briefly
sketch the history of the alphabet as a background to discussing its
use by the Greeks to represent numbers. Then we shall add to this a
survey of the other forms in which the connection between letters
and numbers has been manifested.

HISTORY OF THE ALPHABET

One of the greatest wonders of the history of human culture has
been the world-wide spread of the 22 symbols which the Phoenicians,
a minor trading people, devised to represent the consonants of their
Semitic language. For not only did the Hebrew and Arabic alphabets,
which are used to write these other Semitic languages, derive from
the Phoenician, but the entire groups of Turkish-Mongolian and
Persian-Indian (by way of Aramaic) and above all the European
alphabets (by way of Greek) as well (see Fig. 94).

Trade and Writing: The Phoenicians were merchants who sailed the
Mediterranean Sea and traveled through the lands bordering on it;
and it was their later successors, the Arab traders, who carried the
Indian numerals from the Orient to the West!

The alphabetical form of writing is thought to have developed
approximately as follows: at the outset every form of writing (such
as the Egyptian, the Babylonian, or the Chinese) was made up of
picture-symbols for words (like the picture of a head standing for
"head"). Then when the image of the head was replaced by the
initial sound h, this had become a letter.

An alphabet, in which each sound is, generally speaking, represented
by one symbol, is the last and highest stage in the development of
writing, which usually runs from the ideographic through the
syllabic to the phonetic form. The alphabetical form of writing follows

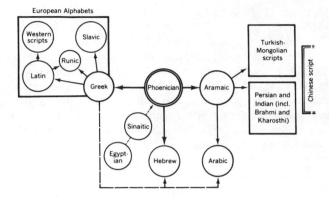

Fig. 94 The origin and spread of the Phoenician alphabetical writing. The dashed line (running down from the Greek alphabet) shows the direction in which the Greek alphabetical numerals migrated.

a major intellectual principle: All the innumerable words that make up the language are resolved into a limited number of some 20 to 30 individual phonetic symbols. This principle originated in ancient Egypt, whence it was passed on by the so-called Sinaitic inscriptions, and probably also through the influence of Crete and Cyprus, to be finally perfected by the Phoenicians. It was in Phoenicia that the Egyptian hieroglyphic image of "house" was streamlined down to a "letter" which merely sounded the initial sound *b* of the Semitic word *beth*, "house."

In column 1 of the table of Fig. 95 are some of the documentable Egyptian hieroglyphs; column 2 gives their Phoenician letter-forms and column 6 their Hebrew names (derived from the Phoenician). The reader is urged to examine the table so that this segment of the history of writing may come to life for him; he will also better understand the numerals that have been derived from letters of the alphabet.

The Phoenician form of writing consisted of 22 symbols which — and this is a peculiarity of all Semitic alphabets — stood for the consonants only (see p. 116). Around the 11th century B.C. the Greeks became acquainted with the Phoenician alphabet and adopted it for themselves. In so doing, they took over the already established sequence of the names, forms, and phonetic values of the signs (compare columns 7, 8, and 10 with columns 4 and 6 in Fig. 95).

Thereupon the Greeks made two important changes: to adapt the alphabet to their own language, they altered some of the letters to vowels (*alpha, epsilon, iota, omicron, upsilon,* and *omega*), and they added the three letters *phi* φ, *chi* χ and *psi* ψ. The Ionian-Milesian alphabet — for there were many early Greek scripts — was officially adopted in Athens (with the forms shown in column 7 of the table). All western alphabets are derived from this.

Thus we come to the third important achievement of the Greeks: the letters have numerical values according to their fixed order in the alphabet (column 9). This tying of the letters to numbers which

Line	EGYPTIAN	PHOENICIAN	HEBREW 3 Letter	4 Phonetic value	5 Number-word	6 Name	GREEK 7 Letter	8 Phonetic value	9 Number-word	10 Name	LATIN 11	RUNES 12 Letter	13 Number value
1	𓃻	𐤀	א	’	1	alef 'Rind'	Αα	a	1	álpha	A	ᚠ	4
2	𓉐	𐤁	ב	b	2	beth 'Haus'	Ββ	b	2	bétha	B	ᛒ	18
3		𐤂	ג	g	3	gimel 'Kamel'	Γγ	g	3	gámma	C	–	–
4	𓂧	𐤃	ד	d	4	daleth 'Tür'	Δδ	d	4	délta	D	ᛗ	24
5		𐤄	ה	h	5	he	Εε	e	5	è-psilón	E	ᛗ,1	19,13
6	Y	𐤅	ו	w	6	waw 'Nagel'	Ϝς	–	6	waù	F	ᚢ	1
7		𐤆	ז	z	7	zajin 'Waffe'	Ζζ	z	7	zéta	(G)	ᚷ	·7
8	𓉔	𐤇	ח	h	8	heth	Ηη	ä	8	éta	H	ᚺ	9
9	𓎛	𐤈	ט	t	9	teth	Θϑ	th	9	théta	100?	ᚦ	3
10	𓂝	𐤉	י	j	10	jod 'Hand'	Ιι	i	10	jóta	I	1,ᛋ	11,12
11	𓂭	𐤊	כ	k	20	kaf 'offene Hand'	Κκ	k	20	káppa	K	ᚲ	6
12		𐤋	ל	l	30	lamed	Λλ	l	30	lámbda	L	ᚱ	21
13	〰	𐤌	מ	m	40	mem 'Wasser'	Μμ	m	40	mŷ	M	ᛗ	20
14		𐤍	נ	n	50	nun 'Fisch Schlange'	Νν	n	50	nŷ	N	ᛏ	10
15		𐤎	ס	s	60	samek	Ξξ	x	60	xî	–	–	–
16	👁	𐤏	ע	‘	70	ayin 'Auge'	Οο	o	70	ò-mikrón	O	ᛇ	23
17		𐤐	פ	p	80	pe 'Mund'	Ππ	p	80	pî	P	ᛈ	14
18		𐤑	צ	s	90	sade	–	–	–	–	–	–	–
19		𐤒	ק	q	100	qof	Ϙϙ	–	90	kóppa	Q	–	–
20	🐒	𐤓	ר	r	200	reš 'Kopf'	Ρϱ	r	100	rhô	R	ᚱ	5
21	𐤔	𐤔	ש	š	300	šin 'Zahn'	Σσ	s	200	sigma	S	ᛋ	16
22	✗+	𐤕	ת	t	400	tau 'Zeichen'	Ττ	t	300	taû	T	ᛏ	17
23			ך	-k	(500)	(kaf)	Υυ	ü	400	ŷ-psilón	V	u: ᚾ	2
24			ם	-m	(600)	(mem)	Φφ	ph	500	phî	1000	w: ᛈ	8
25			ן	-n	(700)	(nun)	Χχ	ch	600	chî	X	ng: ᛜ	22
26			ף	-f	(800)	(fe)	ψΥ	ps	700	psî	50?	-z: ᛉ	15
27			ץ	-s	(900)	(sade)	Ωω	ō	800	ô-méga	–	–	–
28							↑ⲟↇ	–	900	sampî	–	–	–

Fig. 95 The development of alphabetical writing, from the Egyptian (column 1) through the Phoenician (2) to the Hebrew (3), Greek (7), Latin (11), and Germanic (12) alphabets. The Greeks gave their letters numerical values (9), and the Hebrews (5) adopted a similar pattern. Column 13 shows the numerical values assigned to the Runic characters.

the Greeks invented then was handed back to the Semitic donors of the letters themselves: Hebrew adopted the alphabetical numerals (column 5), and they went from there to Syriac and thence to Arabic. Two coins (*shekels*) (Fig. 96) dating from the time of the First Uprising of the Jews (A.D. 66–70) show, above the cup in the center, the old Hebrew/Phoenician signs for 2 and 4, meaning "in the second (and fourth) years of the Uprising." With the aid of Fig. 95 we can

Fig. 96 Hebrew coins. Above the cup they show the alphabetical numerals for 2 and 4 (1st century A.D.). State Numismatic Collection, Munich.

very easily decipher the inscription, which runs from right to left, as *šql yšr'l*, "shekel of Israel." The Hebrew letters shown in column 3 of the table are the so-called "square writing"; with the addition of the terminal forms *-k, -m, -n, -p* (or *-f*) and *-s* the sequence of numerals, which once ran only as far as 400, was extended to 900.

An elderly Jew once told the present writer that in his youth he used to hear the Jewish cattle dealers trading as follows: "*mem shuk* (40 marks), *gimel shuk* (3 marks), *lamed gimel* (33)" and so forth.

The connection between letters and numbers gave rise among Judaic and early Christian scholars, as well as the Greeks, to *gematria*. This word is very likely a corruption of the Greek *geometria*, which the Jews wrote as *gmtr*, and referred to the study of numerology in general. Since each of the letters in a word had a numerical value, the word itself also had its own individual number. Thus "Amen," Greek $\alpha\mu\eta\nu$, $= 1 + 40 + 8 + 50 = 99$; hence this word at the end of a Greek prayer would often be written as $\kappa\vartheta$, 99.

The Jews, so as not to take the name of the Lord in vain, would write the number 15 not as *yh*, but as *tw* ($= 9 + 6$), for *yh* are the initial letters of the name "Jehovah" (*yhwh = Yahweh*). For similarly "sacred" reasons the Irish would avoid the Roman numeral X $=$ 10 on account of the name $XPI\Sigma TO\Sigma$, and the Chinese Boxers during their rebellion against the foreigners ($=$ Christians) removed the symbol $+$ for 10 from their coins because of its fancied resemblance to the "Christian" letter X (see Fig. 264). Because the Greek word for death, $\Theta ANATO\Sigma$, begins with a *theta* $\Theta = 9$, the Greeks instead of the number 9 would write the circumlocution $8 + 1$ or $4 + 5$; there is a counterpart to this superstitious custom in Japan (see p. 453).

Persons or things the letters of whose names had the same total numerical values were also thought to be mystically related. In the Middle Ages, for instance, people would "calculate" the outcome of a dual or individual combat from the arbitrary sums of the letters or digits in the combatants' names; the one with the larger sum would be the "predicted" winner. If the numerical value of Siegfried's name was 238 and the conventional sum of this was 4, and if the numerical value of Hagen's name was 65 and the conventional sum of this was 2, then in a duel between Siegfried and Hagen, gematria predicted that Hagen was sure to be vanquished. By the reverse process, names were often thought to underlie numbers: the most famous example of this is in the Book of Revelations (13:18):

Let him that hath understanding count the number of the beast: for it is the number of a man; and his number is Six hundred threescore and six — *kaì ho arithmòs autoû hexakósioi hexékonta héx.*

There can be no doubt that a name is latent here. In early Christian times this number from the Apocalypse was deciphered, among many others, as *Neron kaisar,* "Caesar Nero" — an interpretation that is quite possible if the Hebrew writing of this name, *nrụn qsr,* is transcribed into Greek and then "encoded":

nun	*reš*	*waw*	*nun*	*qof*	*samek*	*reš*	
50	200	6	50	100	60	200	= 666.

The "numbering of letters" and the "lettering of numbers" was the art of *isopsephy,* as this sort of toying with the numerical values of various words was called (from the Greek words *ísos,* "equal," and *pséphos,* "pebble, counter, number") in all seriousness. As late as the 16th century the German mathematician Michael Stifel (died 1567), who is highly regarded by posterity, valued his "word calculations," as he called them, above his more important work in mathematics.

Now we shall turn to the numerals used by the Greeks: how did ordinary people, not to mention the great Greek mathematicians, write numbers, and how did they make their computations?

The Two Greek Sets of Numerals

The Greeks had two different systems of numerals, an early one that arranged the numbers in order and grouped them like the Roman numerals (these will thus be called here the Greek "row numerals"), and a later erudite system of alphabetical numerals which first appeared in the 5th century B.C. but was not adopted as the official system of numerals in Athens until the 1st century B.C.

The Greek row numerals had individual symbols for the numerical ranks 1, 10, 100, 1000, and 10,000; these (except for 1) were the initial letters of the corresponding number words. They could be grouped by fives by using the initial letter *Π* of the word *pénte*, 5, as shown in Fig. 97.

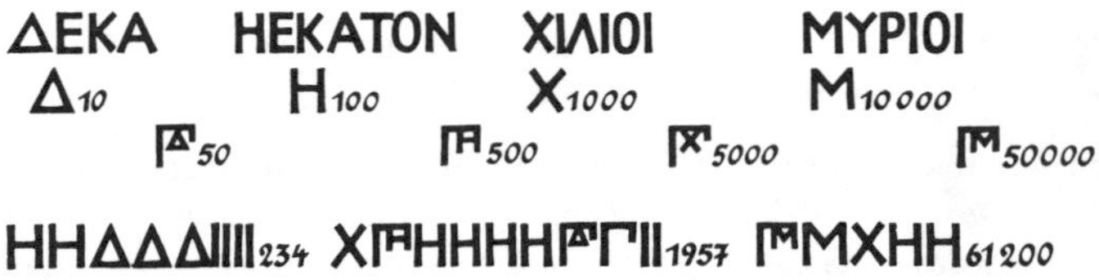

ΔEKA   HEKATON   XIΛIOI   MYPIOI

Δ*10*   H*100*   X*1000*   M*10000*

Π*50*   Π*500*   Π*5000*   Π*50000*

HHΔΔΔIIII*234*   XΓHHHHHΠΓII*1957*   ΠMXHH*61200*

Fig. 97 Early Greek numerals (row numerals). Like the Roman numerals, these are patterned on a decimal 10-grouping interrupted by a quinary 5-grouping. The units are represented by vertical strokes. The 5-grouping appears in the initial letter of the word *ΠÉNTE*, "five." This illustration shows the higher ranks, whose numerals are the initial letters of the number words (see also Fig. 128).

This system of written numbers thus made use only of the symbols for the old decimal grouping, which is evidently combined with a quinary 5-grouping.

These early Greek numerals have been given the very unfortunate name of "Herodian numerals" — unfortunate, because the grammarian Herodian lived around A.D. 200 in Byzantium, and thus more than five centuries after the appearance of the numerals named for him. Moreover Herodian mentioned them only once in passing. He was in no sense their inventor. Thus calling the early Greek "row numerals" Herodian has as little justification as naming the Hindu numerals we now commonly use after Adam Riese.

These Greek numerals were used in Attic inscriptions from the middle of the 5th century until well into the 1st century B.C., chiefly in the public tribute lists and the accounts which the financial officials of the city state presented to the citizens. The inscriptions of the first category listed the amounts paid annually by subjects who owed taxes to the Athenian treasury. Figure 98 shows a fragment of a square marble block, 1 × 0.4 × 3.6 m in size, on which the numerals themselves are about 4 cm high; at the top left is the name of the tribute payer.

In these tribute lists the numbers always refer to payments of money; in this case the numbers after the semicolons are obols and the others drachmas. The monetary scale was 1 talent (T) = 60 minae, 1 mina = 100 drachmas (⊢), 1 drachma = 60 obols (|); in addition, the coin called the *stater* (Σ) was worth 4 drachmas.

Fig. 98 Athenian tribute list with early row numerals 4 cm high. Fragment of a square stele, almost 4 m high; 5th century B.C.
The inscription reads:
(*AP*)*O THRAIKES PH*(*OROS*) — "Payments from Thrace":
skiathioi 66 *drachmai* 4 *oboloi;* then
500 50 8;2 16;4 600 25.

The method of indicating the coin denomination was quite interesting (see Fig. 99). Either the coin was not specified, in which case it was understood to be drachmas, or it was placed on the left side of the number (top row in figure), or it took the place of the units in the numeral (middle row), or else it was incorporated into the number symbol itself (as in the bottom row).

Fig. 99. Greek statements of amounts of money: From top to bottom: 12 drachmas — 12 staters — 12 drachmas 3 obols — 2 talents 105 talents

In this method of specifying the coin in which an amount of money was stated, we have a beautiful documentation of the early stage of development in which numbers were still tied to the things they counted: "2 talents," "2 staters," or "2 obols" all have different symbols for the number 2.

Fig. 100. Report of the treasurer of Athens giving the city-state's income and expenditures in the year 415 B.C. At the end is the total (*kephálaion*) of 327 talents. This inscription measures 50 × 40 cm. British Museum, London.

We can read the total amount on an inscription in which the Athenian treasurer in 415 B.C. made a public accounting of his four-year term of office (Fig. 100):

KEPHALAION AN(alómatos toû epì tès) ARCHES
"head-number" (= total, Greek *kephalé*, "head"), which in this
treasurer's term of office came to 327 talents.

Such were the numerals used by the city of Athens. The Boeotians, on the other hand, for 100 would write the first two letters *HE(katón)*, and for 500 would place a *Π(énte)* in front (see Fig. 101). For 1000 they used the arrow-form of the letter *chi* (instead of *X* — see Fig. 95, column 7, row 25, and Fig. 133 on p. 304).

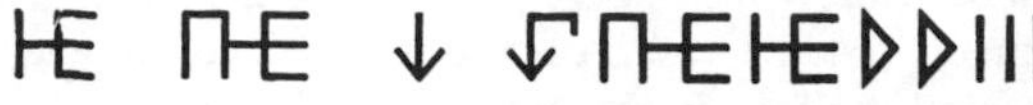
100
500
1000
5623

Fig. 101 Boeotian numerals.

Although we see very clearly that the Greeks used these signs to write numbers, it is very hard to understand how they made computations with them. And in fact, as we shall see presently, they did not; they used counting boards instead. Using this system of written numerals, Archimedes would have been unable to compute the value of π or bracket it, by means of a 96-sided polygon, between 3-10/70 and 3-10/71. For such purposes numbers would have to be understood and visualized much more precisely and organized on some basis other than just the awkward grouping. These needs were met by

The Greek Alphabetical Numerals

As the table of Fig. 95 shows, the twenty-seven letters of the Greek alphabet (including the special forms) were arranged for use as numerals as follows:

Units	1	2	3	4	5	6	7	8	9
	Α	*Β*	*Γ*	*Δ*	*Ε*	*Ϝ*	*Ζ*	*Η*	*Θ*
	α	β	γ	δ	ε	ϛ	ζ	η	ϑ
Tens	*Ι*	*Κ*	*Λ*	*Μ*	*Ν*	*Ξ*	*Ο*	*Π*	*Ϙ*
	ι	κ	λ	μ	ν	ξ	ο	π	ϙ
Hundreds	*Ρ*	*Σ*	*Τ*	*Υ*	*Φ*	*Χ*	*Ψ*	*Ω*	ϡ
	ρ	σ	τ	υ	φ	χ	ψ	ω	ϡ
Thousands	͵α	͵β	͵γ	͵δ	͵ε	͵ϛ	͵ζ	͵η	͵ϑ

For this purpose the Greeks added to their normal alphabet three Semitic letters for which they had no sounds (or no longer had them by the 5th century B.C.). These (see column 8 of Fig. 95) were: for 6 the *digámma* Ϝ with its later form ϛ, which was also called *st-ígma* because it served as the abbreviation for *-st-*; for 90 the *koppa* ϙ; and for 900 the *sampî* ϡ, probably so named for the Semitic *sade* that comes after *pi*, through which the Greek alphabetical numerals from 90 upward displaced the Semitic numerals (compare column 5 with column 8; *hōs án* π, "like π"). Bishop Ulfilas adopted the arrow form of the *sampî*, 900, for use in his Gothic Bible (see Fig. 91).

Examples: PIA is 111, written in descending order of magnitude, but sometimes also in ascending order as AIP and more rarely without any order as IAP. In manuscripts the alphabetical numerals were topped with a short horizontal stroke, like $\bar{\varepsilon}$, 5, and $\overline{\sigma\lambda\delta}$, 234, or placed between periods or lines of dots ·*E*· or ⁝*E*⁝ or even sometimes turned on end like ⁝m⁝. A prime to the right of the numeral meant an ordinal number, such as ε', "the fifth," or a unit fraction, $\frac{1}{5}$. The thousands were written just like the units, but with a short vertical stroke at the bottom left: $_{,}\varepsilon$, 5000 or $_{,}\alpha\omega\nu\varsigma$, 1856.

The next numerical rank, the *myrioi* = 10,000, could be written in three different ways:

The first used the symbol M from the early row numerals, but with multiples of *myrioi* no longer shown by repeating the numeral; thus 30,000 was not MMM but was written as in a place-value notation (see Fig. 102):

(1)   (2)   (3)

Fig. 102 The myriads were not repeated in the Greek alphabetical numerals, but their number indicated by placing the symbol for the appropriate unit in front as in 1*a* or on top (1*b* above). Here $\Gamma = 3$:

(1) 30,000   (2) 32,000   (3) 30,002

The second method used a dot after the myriads instead of the letter M, as did Diophantus of Alexandria (3rd century B.C.).

$1507284 = 150'7284 \ \rho\nu.,_{,}\zeta\sigma\pi\delta$ and $_{,}\alpha\tau\lambda\alpha.,\varepsilon\sigma\iota\delta$ 1331 5214;

The third method used two dots placed over the myriads, as in later Greek manuscripts:

$50,000 = 50000 \ \ddot{\varepsilon}$ or $50,000,000 = 5000$ myriads $_{,}\ddot{\varepsilon}$.

Now for the advantages and disadvantages of these Greek alphabetical numerals. Their advantage over the old row numerals can hardly be missed; they represent an enormous simplification, since they use only one sign for each unit (rank). Thus in inscriptions we generally find alphabetical numerals used wherever there was little room, for example to indicate numbers on sacred offerings, like the

Fig. 103 Greek coins with alphabetical numerals. The first two are from Alexandria, with the numerals A = 1 and IB = 12, the years of the respective rulers' reigns (about 250 B.C.); the third coin, at right, comes from Byzantium and shows the year (right to left) $ZI\Phi = 517$ of the so-called Pontinian Era (which started in 297 B.C.); in other words, A.D. 220. The L on the coin at left indicates that the number following is a year.
State Numismatic Collection, Munich.

ΨΝΔ, 754, and the *ΣΟΓ*, 293, on two Greek vases from the 4th century B.C., which are among the oldest preserved documents with these numerals. In Alexandria they first appeared on coins about a century after Alexander the Great, and in this way came to be commonly used. Figure 103 shows two such Alexandrian coins and one from Byzantium.

On a sundial dating from the 9th century A.D. the hours are numbered in alphabetical numerals (see Fig. 104), and today in Athens there is still a *Γ*-September Street.

Fig. 104 Sundial with Greek alphabetical numerals, from a church in Boeotia, 9th century A.D.
Photograph by E. Delp.

And now another adantage of these alphabetical numerals, in fact their most important advantage: With these numerals it was possible at long last to make computations in writing, without having to use an abacus.

To us, however, who are thoroughly familiar with the Indian system of numerals, this very quality may seem to be questionable because of the lack of graphic identity of the units, tens, and hundreds. Thus $4 \times 10 = 40$ in the Indian system becomes $\delta \times \iota = \nu$ in the Greek alphabetical numerals, and the (to us) almost identical computation $4 \times 100 = 400$ is $\delta \times \rho = \mu$. The multiplication table and the addition table become obscured by the plethora of relationships.

Nevertheless Archimedes and Diphantus made their computations with Greek alphabetical numerals. But how?

At first glance this looks more difficult than it really is. To begin with: people read and memorized tables not as $\delta \times \iota = \mu$, but in terms of number words, "four times ten equals forty." Thus the ear could discern the similarity of $4 \times 100 = 400$ even if the eye could not. Moreover prepared multiplication tables were commonly used, as an arithmetician of the 14th century, Nicholas Rhabdas Artavazdos of Smyrna, has testified. The following multiplication table shows the beginning of this system, for $2 \times 1 = 2$, $2 \times 2 = 4$, $2 \times 3 = 6$, etc., and correspondingly for $20 \times 10 = 200$ and $200 \times 100 = 20,000$:

β	α	β
	β	δ
	γ	ς

κ	ι	σ
	κ	υ
	λ	χ

σ	ρ	$\ddot{\beta}$
	σ	$\overset{\circ}{\delta}$
	τ	$\ddot{\varsigma}$

. . . .

$2 \times 1 = 2$ $20 \times 10 = 200$ $200 \times 100 = 20{,}000$
    2   4   20   400   200   40{,}000

. . . .

For example, let us calculate 25×43, which we have already done according to the Egyptian method (see p. 219)

κ	ε		25	
μ	γ		43	
ω	ξ		800	60
σ	$\iota\varepsilon$		200	15
,α	$o\varepsilon$		1000	75 = 1075

The Greeks began with the highest rank: $20 \times 40 = 800$; then they went on to $20 \times 3 = 60$; then $5 \times 40 = 200$; and finally $5 \times 3 = 15$; final answer: 1075. Each intermediate multiplication, such as 20×40, was broken down into a computational problem, $2 \times 4 = 8$, (the so-called *pythménes*, "root problem," from *pythmén* = "root, stem, base"), and the gradational or rank problem of placing the resulting 8 in the correct order of magnitude. This computation in two steps corresponds to the procedure on a modern slide rule, except of course that finding the right order of magnitude was harder for the Greeks than for us, because the numerals κ and μ have no zeros, as do 20 and 40, from which the numerical rank of the final answer 800 can be easily read. For the old rule for finding the numerical rank, which was derived from the counting board, see p. 316.

The study of practical computations, which the Greeks termed "logistics" in contrast to "scientific arithmetic," developed and flourished in Alexandria, the capital city of Hellenistic learning and culture, and were used from then on throughout the Hellenistic world and the Byzantine Empire. As the Indian place-value notation penetrated deeper and deeper in the West (from the 15th century on), it gradually displaced computations with alphabetical numerals. But not completely. The two systems of numerals finally reached an "equilibrium": the new numerals provided a place-value system in which *the Greek alphabetical numerals for the units were used as* "*Indian numerals.*" The zero, which had the form of a dot according to the Eastern Arabic fashion, was added to the units α to ϑ:

$\alpha\vartheta\varepsilon\varsigma$ 1956 β. 20 β.. 200 $\beta.\beta$ 202.

A Greek manuscript of a 15th-century textbook on arithmetic gives the following two-stage problem in computation involving fractions (as an easier method of using alphabetical numerals):

$12\frac{2}{7} \times 15$; rearranged as $86/7 \times 15$; without denominators $86 \times 15 = 430 + 860 = 1290$.

Transcription:

$\alpha\varepsilon$	$\alpha\beta\,\dfrac{\beta}{\zeta}$
	$\eta\,\varsigma$
	$\alpha\,\varepsilon$
	$\delta\,\gamma\,\cdot$
	$\eta\,\varsigma$
	$\alpha\,\beta\,\vartheta\,\cdot$

Translation:

15	$12\dfrac{2}{7}$
	86
	15
	430
	86
	1290

Fig. 105 A Greek computation carried out after the "Indian" manner. From a 15th-century manuscript. The digits are the Greek letters *alpha* through *theta* and the zeros are represented by dots. At right, next to the problem in fractions, is the division $1290 \div 56 = 23-1/28$.
Courtesy of Prof. K. Vogel, Munich.

Some Other Connections Between Letters and Numbers

The *Katapaya system* of Southern India arranged the 34 consonants of the Sanskrit alphabet to form the ten numerals 1 through 9 plus 0. Some digits thereby acquired three, and others four, different phonetic values; 1, for example, could be *k-t-p-y*, whence the name "Katapaya." Zero has only the sounds *nj* and *n*. In general, each consonant was sounded together with *a*, although it could also be combined with the other vowels *e i o u* without changing its numerical value. Thus the number 111 could be expressed, to use examples from German, not only by the word *Kette* but also by the word *Paket*. The Indian astronomers used to call the lunar cycle *anantapura* = n-n-t-p-r = 00612; this number, reading from right to left, gave the number of minutes of the half-month (15 × 24 × 60 = 21,600). This simple example shows the method and the advantage of designating a number by a word which itself has some quite different meaning, to be used as a secret means of communication, as a mnemonic device, or perhaps merely to clothe computations in poetic dress, as we saw earlier with the Indian verbal number symbols (see p. 120). The impression one gets from these Indian numbers, especially if we also recall the Indian number towers, is of the great joy that the Indians found in playing with numbers and their easy familiarity with place-value notation. Unlike the Greek alphabetical numerals, this Katapaya system of course had no practical significance.

The ancient *Irish Ogham characters*, which are found on inscriptions dating from the 4th century A.D. but are certainly older, are just the reverse of the Katapaya: here numerals are used as letters! The vowels are represented by up to five dots and the consonants by up to five perpendicular or slanting strokes on one side or the other of a long vertical line (see Figs. 106 and 107). The corner of a grave stele might serve as the vertical line on which the Ogham characters were aligned. These curious number symbols made "on the line" are strongly reminiscent of tally sticks. Northerners, it seems, generally loved secret writing.

The *old Germanic runes*, originally 24 in number, were probably derived from the old Northern Italic alphabets some time during the first three centuries A.D. The once variable forms became fixed by being carved into wood; all the horizontal strokes were drawn at an angle, as in the runic A and F, so that they would not run parallel to the grain of the wood, and the curved lines were either made straight or broken, as in O (see Fig. 95, column 12, rows 1, 6, and 16). There were two Germanic peculiarities: the old "sacred" order of the letters was changed (from this alphabet were formed the positional numbers shown in column 13), and the old names of the letters were changed to symbolic names like *f-e*, "possession, property"; *u-ruz*, "aurochs"; *þ-urs*, "giant"; *a-nsuz*, "god"; *r-aido*, "ride"; *k-enaz*, "torch." This leads us to the original point of the runes: these letters were (initially) not intended or used for writing, but for casting lots and telling the future, and were thus "written" (Old Saxon *writan*) or "inscribed" (Latin *scribere*) on small wooden

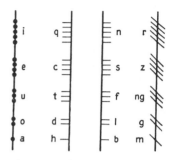

Fig. 106 Irish Ogham characters, expressing letters in the form of primitive numerals.

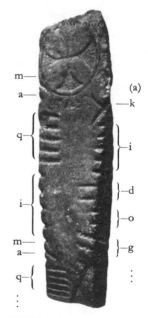

Fig. 107 Irish gravestone from Aglish, with Ogham characters, around A.D. 300. These are read from bottom to top beginning on the right and then continuing at left from top to bottom (beginning and end are lost; X is sounded as *k*):
"... *godika maqimaq*..."
"(grave of Lu)dudex, son of *maq*..."
Size: 88 × 25 × 5 cm.
Irish National Museum, Dublin.

[275]

sticks, as Tacitus describes in the passage from his Germania, already quoted (see p. 224; Old Norse *run* = "secret" or "counsel").

The important thing for us here, however, is that the runes had no numerical values. On the other hand, they were often replaced by numerals as a form of secret code. The 24 = 3 × 8 runes were divided into three groups of eight each (or "genders"), quite possibly because the old Germanic tribes likewise divided the rim of the heavens into eight parts. The number 8 was sacred. Thus the rune A was the fourth rune in the first gender. The two ordinal numbers (4 and 1) were written on a stick, as the memorial tablet from Rök, Sweden (Fig. 108), shows. This tablet bears the longest known inscription in runic characters: a passage in which a father curses the enemies of his slain son.

The secret runes are to be read clockwise beginning at the top left: 3 − 2, 1 − 5, then 2 − 2, 2 − 3, and 3 − 5, 3 − 2 — that is, *ol ni-röþr*, "as an old man of 90 years." Thus the first and third genders have been interchanged. After about A.D. 800 the runic characters, which were originally used by all the Germanic tribes, appear only in Scandinavia. Runic inscriptions, which are quite rare, continued to be carved until some time in the 14th century.

Finally, let us deal with the *Persian siyaq* script, which we have already encountered earlier (see p. 74; from Arabic *siyaq* = "order"). Persian is an Indo-European language, a descendant of Avestan. It was once written in cuneiform characters borrowed from the Babylonians. Since the Arab conquest of Persia in the 7th century, the Arabic alphabet has reigned there along with Islam, and with it many Arabic words have entered into the old Persian language.

As for the Persian numerals, the Arabs wrote out the number words for a long time even in mathematical writings. The *siyaq* script which made numerals out of abbreviated or eroded number words, probably goes back to this time (Fig. 109). Because they employ some

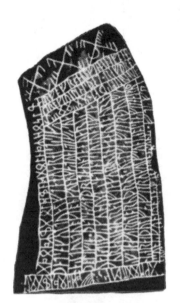

Fig. 108 Runic stone from Rök, Sweden, around A.D. 850 At the top are secret runes replaced by number symbols. More than 2 m high; now in the National Park, Stockholm.

| 5 | 50 | 5000 | 500 | 5005 |

Fig. 109 Persian *siyaq* numerals. The two parallel strokes in the number at the right stand for missing digits (zeros). This script is read from right to left; 5, represented by a loop, appears at the beginning of each of these numerals.

features of a place-value notation together with special signs for higher numbers, the *siyaq* numerals represent a unique intermediate stage between numerals and fully written-out number words.

To mention only one feature: the units (5), the tens (50), and the thousands (5000) all have the same beginning (5) but are differentiated by their terminal symbols; 500, however, is written differently, although the sign for 5 can be discerned in it as well. Another very interesting trait is the presence of a sign denoting the absence of a digit, but its use was not absolutely mandatory; its form however resembles that of the ancient Babylonian symbol for a missing digit (see Fig. 228). In writing a number, the units and the tens were transposed, as we have already seen (see p. 74).

The *siyaq* numerals were used for many centuries in Persian and Turkish financial and official documents. They had the advantage of not being familiar to everyone, and also of being hard to alter by anyone who might wish to falsify the accounts. In the bazaars of Iran they are still used by the old merchants for keeping their own records, but always only in terms of amounts of money and always expressing

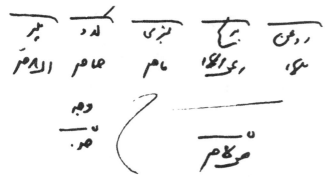

Fig. 110 Computation written out in *siyaq* characters.

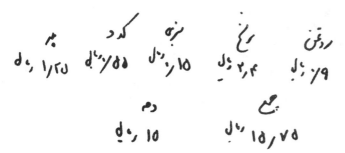

Fig. 111 The same computation as Fig. 110, written in Eastern Arabic numerals. Figures 110 and 111 were written in a "bilingual" manner, by a Persian. The numerical ranks are in the top row, and the numbers beneath.

Siyaq:	1250	550	2400	900;	(below) 50 (?)	(1)5750
East Arabic:	1.25	0.55	2.4	0.9	15	15.75

Kindly provided by H. Horst, Teheran.

the new *rial* currency in terms of the old *dinar* currency (1 *rial* = 1000 old *dinars*, whence the difference between the two sets of numbers in Figs. 110 and 111). The *siyaq* characters are used only for writing down the numbers; computations are made on the Russian abacus, the *ščët*, which in Persian has been and is still today called *tshoke* (see Figs. 148 and 149).

In the last century the *siyaq* characters were officially replaced by the Eastern Arabic style Indian numerals, i.e., the oriental forms of the numerals we ourselves use. Today the Iranian merchant must make out his tax declaration in the modern Indian-Arabic numerals (see Fig. 234).

In addition to the *siyaq* and the Indian numerals, however, Iranians still make use of the Greek alphabetical numerals, which the Arabs first learned in Syria. They arranged the numerals in Hebrew fashion according to their *abujdad* (so named after the first four letters, *a–b–j–d*), and for centuries used them for writing the figures in their astronomical tables. Today in Iran they are used only to number the pages of a book's preface, just as we commonly use the Roman numerals for the same purpose.

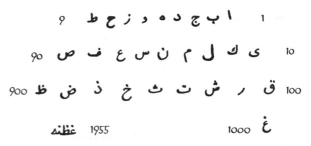

Fig. 112 The Arabic alphabetical numerals. These are arranged in the same order as in Hebrew (see Fig. 95, column 3). The numbers were written right to left, in the opposite direction from the digits in the Indian-Arabic numerals (Fig. 234). In the number 1955 we see the four numerals 5′50′900′1000 combined.

Now let us look back over our material: we have seen the co-existence of three systems of numerals that not only represent three different intellectual stages of development, but also indicate the three different ways in which numbers can be associated with letters. We have penetrated deeply into the fabric out of which time has woven the tapestry of cultural and ethnic history. We have — to mention only the main strands — seen the alphabet of the Phoenicians and the alphabetical numerals which the Greeks derived from them, and we have recognized the Arabs' custom of writing out number words and the Indians' method of writing numerals. Finally, one more important point: The *siyaq* characters as a peculiar intermediate link between number words and numerals suggest that the Indian numerals (from 4 on) may have started out as abbreviations of the corresponding number words.

With this insight we leave the alphabetical numerals behind. Gothic, Greek, Indian, and Persian numerals — truly a rich harvest has been reaped from the 22 seeds of the first Phoenician alphabet. Now we must return from the Orient to our western world.

The alphabetical numerals did not spread from the Goths to the other Germanic tribes in the West. This was probably because they were not a popular, but a "learned" set of numerals that might have been suited for use in the old and highly developed culture of Byzantium, with its army of scribes, but not by young and primitive peoples who were in many cases settling down for the very first time after centuries of nomadic wandering.

But these Teutonic tribes were able to comprehend the primitive system of the Roman numerals which they adopted without hesitation and treated as their own.

The "German" Roman Numerals

Roman Numerals in Cursive Form

Ich habe diss rechenbüchlein, dem Leyen zu gutt vnnd nutz, dem die Ziffernzale am ersten zu lernen schwer, durch die gewöhnlich teutsch Zal geordnet.

"To make this little book of computations agreeable and useful to its reader, who will find numerals hard to learn at first, I have used the common German numbers throughout."

A 16th-century arithmetician

Over the course of time many different variants of the old Roman numerals came into use in the West. The Romans themselves, for instance, never wrote 1000 as M or 2000 as MM, but as (I) and (I)(I), or at the most occasionally as IIM, using the M as an abbreviation for the number word *mille* (see Figs. 10 and 135). In the Middle Ages, however, M became quite common as a symbol for 1000: MMCXII = 2112.

Just as often, however, the M was replaced by the 1000-stroke above the number, which was known as the *titulus* in the Middle Ages and as *vinculum* in classical Latin, and which the later Romans also used occasionally; one inscription has $\overline{\text{VI}}$XXXIX (= 6039) and likewise IↄↄↄXXXVIIII; see also Fig. 44. This 1000-stroke above the numeral is to be sharply differentiated from the stroke which merely distinguishes numerals from letters and likewise appears in classical Roman times: $\overline{\text{V}}$ = 5; $\overline{\text{IIIVIR}}$ = *triumvir*. This gave rise to our habit of enclosing Roman numerals between horizontal strokes: $\overline{\underline{\text{III}}}$, $\overline{\underline{\text{V}}}$, $\overline{\underline{\text{X}}}$.

The Roman numeral for 1 million was an X within a frame: $\boxed{\text{X}}$; the reason for the X has already been shown (p. 244).

As the art of writing spread, and came into increasing use by merchants and managers of various kinds of enterprises, the abstract but letter-like Roman numerals were written more and more as letters. Thus X became an *x* and V a *v* or, as often happened in the Middle Ages, a *u*; thus whoever "made an X for a U" tallied up double the amount, 10 instead of 5, and thus by extension figuratively cheated someone. After the so-called "uncial" script using only capital letters developed into the medieval form of writing with lower-case letters, a number such as 763 would then also be written as *dcclxiij*. And in the "Gothic" script which became common in Germany as well as the "German" script that developed from it, the numbers 19 and 7 came to look like ɼiɼ and ʋij (see Fig. 114). To prevent falsification of a number, the last digit *i* was always lengthened and written as *j* (see Fig. 122).

The page from the Rüsselsheim book of accounts reproduced in Fig. 114, which dates from the middle of the 16th century, is an excellent example of the medieval "German" Roman numerals in their fully cursive form. Observe the fluency of the whole writing, especially in the number 19 in the fourth line from the bottom, which was required of bookkeepers in writing numbers (see p. 427). Compare also the beautiful examples illustrated in Figs. 121–123.

The transformation of the Roman numerals into letters of the Latin alphabet then also made possible the fascinating riddles in which a Latin verse would be concealed within the numerals of the date of an occurrence or vice versa (the so-called "chronograms" or "date riddles"):

LVtetIa Mater natos sVos DeVoraVIt,
"Mother Lutetia has devoured her own children."

This cryptic saying refers to the massacre of St. Bartholomew's

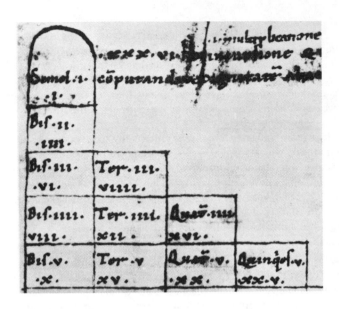

Fig. 113 The beginning of two multiplication tables from 13th-century monastic manuscripts: *semel* once, *bis* twice, *ter* 3 times, etc. Both are still fully "Roman." Above the second is written: *et ad scientiam utilissima*, "... and especially useful for learning ...." The resemblance to multiplication tables of other peoples and times is interesting (see Index).
State Library, Munich.

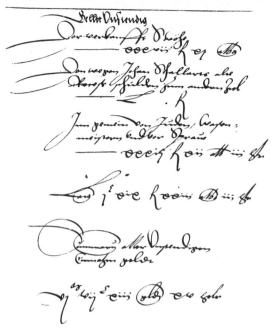

Fig. 114 "German" Roman numerals, as used in a book of receipts by an official of Rüsselsheim, 1554.
Hessian State Archive, Darmstadt.
The page reads:

Gelltt unstending.

Vor verkaufft Strohe————————————————————37 *Gulden (f)* 11 *Albus*
Von wegen Johan Schellerts alte Regesschulden zum andern
Ziel ————————————————————————————————50 *Gulden*
Inn gemein von Juden, Wasenmeistern und vor Spraw
————————————————————————————32 *Gulden* 12 *Albus* 3 *Heller*

Latus 1ᶜ 19 *f* 23 *alb* 3 *hlr*

Summarum aller unstendigen Einnahm geldt

6ᴹ 7ᶜ 13 *Gulden* 15 *Heller*.

"Moneys received and paid.

"From sale of straw————————————————36 *Gulden* (f) 11 *Albus*
"Transfer of Johan Schellert's old debt to another purpose————50 Gulden
"Total received from Jews, harnessmakers and for hay
————————————————————————————32 Gulden 12 Albus 3 Heller

"In all, 1ᶜ 19 f 23 alb 3 hlr

"Totals of all variable moneys received

6ᴹ 7ᶜ 13 (= 6713) Gulden 15 Heller."

Night — Lutetia Parisiorum was the old name for the capital of the Celtic tribe known as the Parisi — by incorporating the Roman numerals for the year 1572 in which the Huguenots were slaughtered in Paris.

MEDIEVAL NUMBERS

Sebastian Franck (1499–1542) wrote in his *Spiegel und bildtnis des gantzen erdbodens* ("Mirror and Image of the Whole World"):

Zimmet kumpt von Zailon .CC. vn LX teutscher meil von Calicut weyter gelegen. Die Nägelin kummen von Meluza/für Calicut hinausgelegen vij. c. vnd XL deutscher meyl.

Cinnamon comes from Ceylon, which lies .CC. and LX (= 260) German miles beyond Calicut. Cloves are brought from the Moluccas, which are vij. c. and XL (= 740) German miles from Calicut.

What strikes us here? The first time Franck writes 200 as CC in the old manner; but a few lines further down he writes 700 as vij. c. and not as DCC. Thus the unit numeral VII appears as the number in front of the rank C, to which the seven C-groups have been condensed. We have seen exactly the same thing happen with spoken numbers (see p. 45). With this change the Roman numerals have embarked on the path to a place-value notation. Yet no positional principle was ever fully formulated in terms of Roman numerals, because medieval usage placed numerals in front of the rank symbols only for the numbers 100 and up but clung to the serial principle for the tens below (see Fig. 114). No one took the mental step of realizing what had actually been done with what appeared as mere abbreviations.

A further obstacle to the development of the primitive Roman numerals into a "mature" form of numerals is revealed in the way the number 123456789 was written in the year 1792:

C	MM	C	M	C	
i	xxiij	iiij	lvj	vij	lxxxix
1	23	4	56	7	89

This is the same difficulty inherent in pronouncing or spelling out large numbers: the problem of breaking down the numerical ranks that come after the thousands. Scribes were led into the habit of again always using C and M instead of the higher ranks and thus destroying the original clarity of the system. MM, for example, stood for million, 1000 × 1000 (see p. 143).

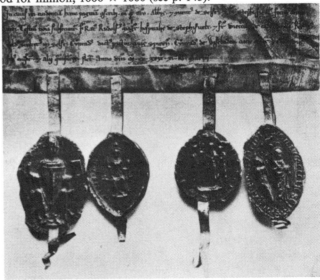

Fig. 115 A document written on vellum in the year 1229, with seals. In the last line the year appears in ordinal form in Latin: *Anno d(o)m(ini) millesimo ducentesimo*.. abbreviated to the then common form *m°cc°*, etc. Hessian State Archive, Darmstadt.

Now for some more examples, whose instructive features the reader will not fail to observe:

$\bar{c}\cdot\overline{lxiiij}\cdot ccc\cdot i$	164301	time: 1120
II·DCCC·XIIII	2814	1220
cIɔ·Iɔ·Ic	1599	1600
ⅭↃ D C X L	1640	1640
IIII milia·ccc·L·VI	4356	13th century
CCCM	300 000	1550

In the Rüsselsheim book of accounts the sum of 1859 Gulden is written sometimes as 1^M viij^c lix Gulden and sometimes as xviii^c lix Gulden.

In Papal Bulls 20 was written as $\overset{x}{X}$ and 8 might appear as $\overset{=}{V}$. Furthermore,

XV.Cet:II   1502
XIIII^c ain jar   1401

and Köbel (Fig. 117) wrote

MVI^{c}XII   1612.

One arithmetician distinguished *CM* (= 900) from C^M (= 100,000). There are vigesimal groupings:

IIII$_{xx}$ et huit   88
VII.XX.VII $7 \times 20 + 7 = 147$
XIIII.XX.XVI   296
mil.IIII^c IIII^{xx} et V   1485.

In one church there is inscribed the date: I·V^c·V 1505; whoever inscribed this number was evidently already familiar with the new place-value notation; it is interesting that he gets around the zero by using the lower-case c.

The fraction $\frac{1}{2}$, which appears almost exclusively in money calculations, was thought of as half of unity and was thus written and printed as a 1 with a stroke through the middle (see Fig. 116), as in the

12$^1/_2$ 76$^1/_2$ 4$^1/_2$ 9$^1/_2$

Fig. 116 Medieval handwriting, showing various forms of the fraction $\frac{1}{2}$.

table of silver weights in the Bamberg book of computations (see Fig. 189). The letter *S* was used in medieval manuscripts as an abbreviation for the Latin word *semis*, "half," superimposed on V and X and then read as "five less half" (4$\frac{1}{2}$) or "ten less half" (9$\frac{1}{2}$). (We have already discussed these peculiar word-forms, p. 77.)

Truly a wonderful variety of modes of organization — sometimes just sequence and grouping, then the use of digits as coefficients bedeüt biß figur der selben tayl aine .

I
IIII *Dieße figur ist vñ bedeüt ain fiertel von ainem gantzen/also mag man auch ain fünfftail/ayn sechstail/ain sybentail oder zwai sechstail 2c. vnd alle ander brüch beschreiben/Als* $\frac{I}{V} \mid \frac{I}{VI} \mid \frac{I}{VII} \mid \frac{II}{VI}$ *2c.*

VI
VIII *Diß sein Sechs achtail/das sein sechstail der acht ain gantz machen .*

IX
XI *Diß Figur bezaigt ann newn ayilfftail das seyn IX tail/der XI.ain gantz machen .*

XX
XXXI *Diß Figur bezaichet/zwentzig ainundreyßigt tail /das sein zwentzigt tail .der aines undreissigt ain gantz machen .*

IIᶜ
IIIIᶜ.LX *Diß sein zwaihundert tail/der Sierhundert vnd sechzigt ain gantz machen .*

Fig. 117 Fractions in "German" Roman numerals, from Köbel's *Rechenbüchlein* (Short Book of Arithmetic), 1514.

crops up, and at still other times we see the old vigesimal grouping and the physically truncated whole used to represent the "half" — but always, despite all the many differences in form, easily understood.

The very multiplicity of such number-letter connections suggests the indigenous, popular nature of these numerals. Anyone could work with them. This also applies to the calendars, genuinely popular books, which up to 1500 were for the most part written with the "German numbers" (unless they used the old peasant numerals; see Fig. 85). And in the year 1514 the arithmetician Köbel put out a very popular and often consulted short book of arithmetic computations, which was printed entirely, even including the fractions, in Roman numerals: *Ich habe disz rechenbüchlein, dem Leyen zu gutt vnnd nutz, (dem die Ziffernzale am ersten zu lernen schwer) durch die gewönlich teutsch Zal geordnet* ("To make this little book of computations agreeable and useful to its reader, who will find numerals hard to learn at first, I have used the common German numbers throughout").

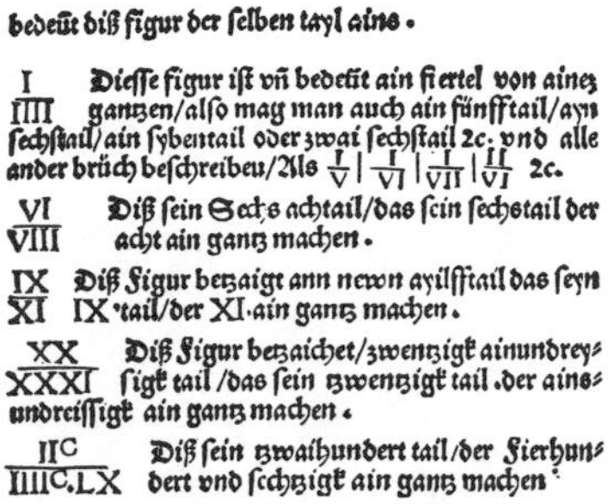

Even as late as the middle of the 18th century the French spoke of Roman numerals as *chiffres des Finances, dont les Financiers se servent ordinairement*, the "numerals of the (state) accounts, which financial officials ordinarily use."

Besides their clear and easily understood organization, the visual quality of the individual Roman numerals also contributed to their ready adoption by the common people. Just as the Germans now say of a man who is forty years old, that he has *genullt* (zeroed) four times, in the Middle Franconian dialect the expression was that he has *gekreuzelt* (crossed) four times.

Es war das concilium gehalt
zu Costnitz, das geschah als man zahlt
ein Ringgen, 4 Rosseisen dabei,
ein Schlaifen, Haggen und ein Ay,

The council was held
in Costnitz; that happened when we counted
a ring, 4 horseshoes too,
a ploughshare, a hook, and an egg,

reads a rhymed monastic chronicle of about 1500. The *Ringge* (= *Rinke*) is a buckle, the ploughshare means actually the iron tip of the plough, the horseshoes and the hook are self-explanatory, and the egg clearly means "one": the result is ⓄCCCCXVI, which is thus 1416, the date of the Council of Constance.

Date riddles of this kind first appeared around the end of the 14th century and continued until well into the 16th. The destruction of Basel in the year 1356 by plague and fire is recorded in the jingle:

Ein ringge und sin dorn
Trü rossyssin verkorn
Ein zimmeraxst und der gelten Zahl
Da (ver)fyel Basel überal.

A ring and its thorn,	Ⓞ
Three horseshoes chosen,	CCC
A carpenter's axe and the gelten number,	L VI
That's when Basel came to an end.	

Gelten here refers to jugs or pitchers, of which there were six at the Wedding of Cana.

THE GRADUAL PENETRATION OF THE NEW INDIAN NUMERALS

The fact that the Roman numerals were so deeply rooted in the customs and affections of the people at first made it exceedingly difficult for the new Indian numerals, the "figure numbers," to replace the old familiar Roman numerals. We shall learn later that in northern Europe the Indian numerals first began to be used by ordinary people about 1500. This date, the change from the fifteenth to the sixteenth century, is the great intellectual watershed of modern history, the time when all the new movements generally came to the fore. A variety of developments were also preparing the ground for the common adoption of the new system of numerals. Trade and commerce were expanding rapidly, and the merchant's control over his transactions had to grow accordingly. Money was replacing barter; the merchant princes in Lübeck or Nuremberg could no longer simply carve tally sticks: now they increasingly had to keep proper accounts, write numbers and make computations. The need for knowledge was universal. Now it was the towns and cities with their recently acquired prosperity which supported and required education, not just the cloistered monasteries. Schooling became an important issue, which was also aggravated by the problems of the Reformation and was thenceforward an urgent question. To be sure, the mental framework of the Middle Ages still strongly asserted itself and continued to live on for a long time to come; Humanism respected the fresh, new way of life of the cities and, instead of stepping out boldly as a pacemaker, tended to follow in the footsteps of foreign fashion. But above all printing extended and spread all these new trends: now almost anyone could learn and master things that had formerly been accessible to but a few.

At the turn of the sixteenth century all these various intellectual ingredients, the old and the new, the wished-for and the known, both properly comprehended and misunderstood, were boiling vigorously in the same pot. The following numerals may serve to symbolize this confusion and multiplicity:

M·CCCC·8II	1482	With zero: I·0·VIII·IX	1089
1·5·IIII	1504	IVOII	1502
CC2	202	ICC00	1200
15×5	1515	I·II·τ·τ	1200
Cδ	104 (!)	15000·30	15030

On his Erasmus Altar at Louvain the painter Dirk Bouts placed the number MCCCC4XVII. In one book of that period, the pages were numbered as follows: 100, 100–1, 100–2, . . ., 200–4, 200–5, etc.

Let us see if we can characterize more precisely the period in which the new numerals were introduced. The stonecutter carving the year of Zeugwart Hallenburg's wife's death on his gravestone outside the Frauenkirche in Munich in the form M·DC·Z4 showed that he was about a hundred years behind the times as far as using the new numerals was concerned (see Fig. 118). On the other hand, his

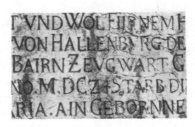

Fig. 118 M·DCZ4 — that is, 1624 — carved on a tombstone (shown at top) outside the Frauenkirche in Munich.

compatriot and fellow guild member, who produced the beautiful knights' grave shown in Fig. 119 and carved the year of one knight's death in the form *m·cccc·lxxxij* at its head, put the date when his brother died on the left side of the tombstone in the new form, .94. — in the same breath, so to speak. The time when these two knights died and when knighthood died out generally was one of transition for it was also the epoch in which the merchant flourished.

And in the merchants' account books the new way of writing numbers took hold slowly but inexorably, at least where it was actually used. Unfortunately the books kept by the House of Fugger in Augsburg go back only as far as 1494; thus we must look instead at the account books of the city of Augsburg, which provide an almost unbroken record of the town's incomes and expenditures from the year 1320 on. These examples from the Augsburg municipal archives at the same time offer some valuable glimpses into the history of record-keeping, bookkeeping, and coinage (see Figs. 120–124).

THE ACCOUNT BOOKS OF THE IMPERIAL FREE CITY OF AUGSBURG

In the beginning both the transactions and the contracts were written in Latin (Fig. 120); later it was only the amount involved, invariably incorporated into the running text, until one day it was "displayed" separately (Fig. 121). The new numerals at first were confined to quoting the year (Fig. 123) until they finally dared tackle the running amounts. But even then the scribe protects himself by writing the money amounts in Roman numerals within the text before displaying them in Indian numerals (Fig. 124). This was in A.D. 1470. After this the ice was broken, although it still took more than half a century before the amounts were written entirely in the new Indian numerals. In the books kept by the progressive and sophisticated Fugger family, in contrast, as early as 1494 (the Fugger books are extant from then on) we find amounts recorded exclusively in the new numerals. In the year 1533 Anton Fugger had an inventory of all his property

Fig. 119 Tombstone of the two knights Ludwig and Hans von Paulsdorf. The first of these brothers died *Anno d(o)m(ini) mccclxxxij* (1482) *am pfinztag vor vith*, "in the Year of Our Lord 1482 on the day of Pentecost" (date carved at the top); the second died *darnach im .94. am freitag vor liechtmessen,* "afterwards, in the year '94, on the Friday before Candlemas." Bavarian National Museum, Munich.

prepared, and in this we can already see the elements of "modern" bookkeeping (Fig. 125).

A remark or two is in order on the currency and the methods of indicating denominations of money:

The amounts were stated in pounds (*lib, lb*) of 240 pence each (*denarius*, abbreviated *d* with a flourish) as well as in shilling (*solidus, s* with a flourish) pennies, that is shillings of 12 pence each, and shilling heller, that is shillings of 12 heller (*h*) each. Besides these ordinary "short" shillings of 12 pence each, there was also the "long" Bavarian shilling of 30 pence, with eight "long" shillings to the pound. The shilling and the pound, which incidentally was originally a weight of around 367 grams according to Carolingian law, were not actual coins, but only accounting or computational pounds and shillings — larger abstract units into which the minted penny coins were grouped. Another coin was the Rhenish gulden, abbreviated *rf* (both letters with flourishes), which like the pound initially contained 240 pence but by about 1475, in Augsburg at least, was worth only 168 pence. (See pp. 343 ff. for a further discussion of medieval European money.) The joint use of the pound and the gulden side by side is also to be seen in the Dinkelsbühl counting tables and the counting cloths of Munich (see Figs. 178 and 180). Because of the varying standards of currency it is sometimes very difficult to convert the amounts of money recorded at various times into modern currency.

1. In the year 1320 the Augsburg city clerk put down everything in his elegant handwriting in Latin in the oldest municipal account books, called *Baumeisterbüchern* ("master builders' books") because they deal principally with payments to master builders (the Roman units are reproduced here as *i* and *j*):

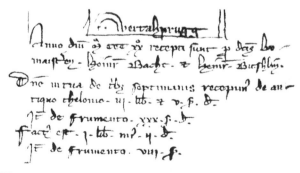

Fig. 120

Wertachprugg. Anno Domini M°CCC XX° recepti sunt pro dictis bumaisteren Heinricus Bache et Heinricus Bitschlin.

   Domine in tua (anima) de tribus septimanis recepimus de antiquo
   thelonio — iii lib et v.s.d.
   item de frumento — xxx s.d.
   factus est i lib minus ii d.

"Wertachbrücke (city tower). In the Year of Our Lord 1320 receipts for the said builders Heinrich Bache and Heinrich Bitschlin were:
God, in thy (soul). [This is the introductory formula which was repeated at the beginning of each new entry or section, and which is the ancestor of the phrase *Mit Gott* that appears in German account books of past generations.]

In three weeks we have received from the old customs collector —
3 pounds and 5 shilling pennies (= 3 × 240 + 5 × 12 = 780 pence).
In addition, for cereals — 30 shilling pennies.
This gives 1 pound less 2 pennies." [At that time 1 shilling was worth around
8 denarii, for the reckoning here is clearly based on the "long" shilling of
30 pence.]

NOTE: The word *minus*, abbreviated here, was in later times represented only
by a short horizontal stroke at the top. When this stroke was written alone, it
evolved into our present minus sign.

2. In the year 1400 the transactions were written in German, but
the sums in Roman numerals and clearly set off:

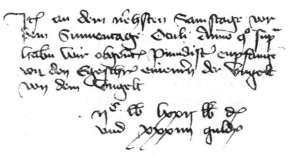

Fig. 121

Item an dem nächsten Samstag vor dem Sunnentag Oculi anno quo supra haben
wir den obengenannten Pumaister empfangen. Von den eingeschriebenen einemern
*   der Ungelt. Von dem Ungelt*
*   ii^c lb lxxij lb d*
*   vnd xxxiiij guld.*

"Besides, on the next Saturday before the Sunday of Oculi in the aforementioned year we received hired the aforenamed master builder. Of the
recorded moneys received (customs, duties, taxes, etc.). Moneys are
272 pound pennies
and 34 gulden."

The official of Rüsselsheim some hundred and fifty years later still
kept his accounts in much the same manner (see Fig. 114).

3. In the year 1410 we are given the sum of all amounts received,
which the city clerk wrote down majestically and with flourishes
(Fig. 122). Here we see the thousands and the hundreds marked off

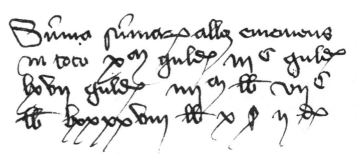

Fig. 122

Suma Sumarum alles einemens in toto
x^M guld iij^c guld lxvij guld
iiij^M lb vij^c lb lxxxxviij lb x s ij d.

"The full sum of all receipts in toto is
10367 gulden and 4798 pounds 10 shillings 2 pence."

from the groups of tens and units, and written with the unit designations — in other words, with C used to indicate rank, as in 3C, and not just in sequence, as CCC. This separation of the hundreds and thousands was still taught by Adam Riese for reading numbers (see p. 142). Another surprising feature is the appearance side by side of gulden and pounds in the final sum, without any corresponding conversion. From this method of writing it becomes clear that such sums could have been arrived at only through computations on the counting board.

4. In the year 1430 the amounts of money were still given only in Roman numerals, but the date itself in Indian numerals stands above the entry (Fig. 123).

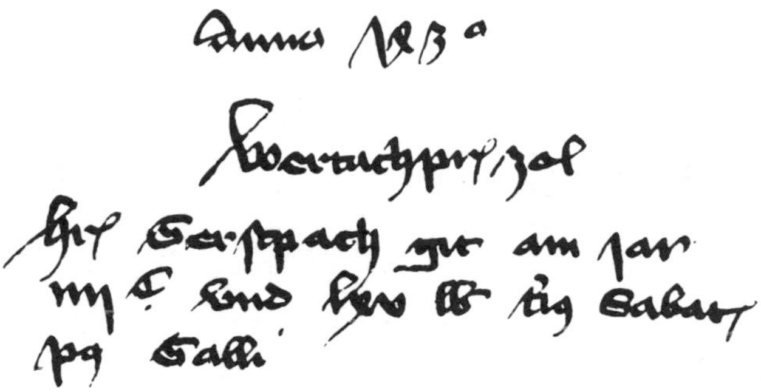

Fig. 123 *Anno* 1430. *Wertachbrücken-Zoll. Herr Geistpach git ain jar* 400 *und* 70 *Pfund tertius sabat bis Galli.*

"In the year 1430. Toll at the Wertachbrücken gate. For one year Herr Geistpach hands over 400 and 70 pounds on three Saturdays (that is, in three installments)."

5. In the year 1470 there is an entry that refers to the *kaiserlich Steur* ("Imperial taxes"). The transaction is recorded in Roman numerals and then the amount is repeated in the new Indian numerals (Fig. 124, top).

6. In the year 1500 the amounts hitherto also written in Roman numerals (for security) were now commonly recorded in Indian numerals at the side, each monetary denomination in its own column (Fig. 124, bottom).

7. In the year 1533 Anton Fugger, then one of the wealthiest merchants in all of Europe, made an inventory of all his possessions. It is worth taking a brief look at the way his accounts were drawn up and also reading the total of one section and the first few entries in his exchange book, in gulden, shillings, and heller (Fig. 125).

I shall leave the reader to search out other interesting points for himself, such as the manner in which the abbreviation of *libra*, "pound," developed from *lib* through ~~*lib*~~ and *lb* into the present symbol for pound, £ or ₷ or how *x* by infinite repetition as in Fig. 122 ultimately became ʒ:

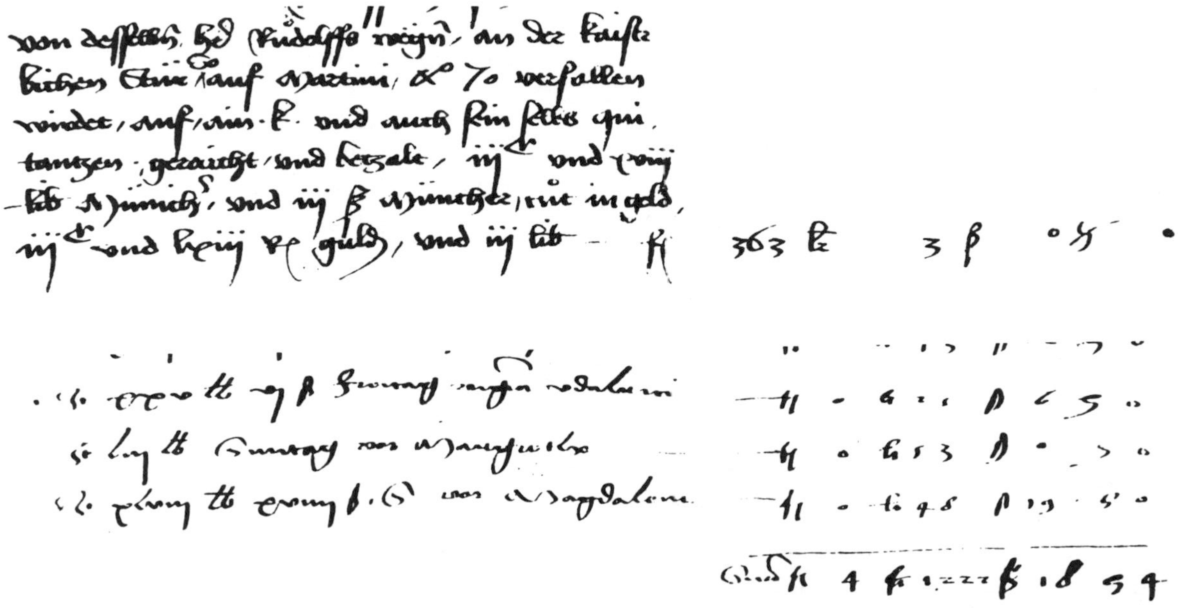

Fig. 124 Entries of 1470 and 1500 in the Augsburg ledgers.

Top:
*von desselben Herrn Rudolfs wegen an der kaiserlichen Stür, so auf Martini anno 70 verfallen wirdet, auf ain k. vnd auch sein selbs quitantzen geraicht und bezahlt iii*ᶜ *und xviij lib Münnicher und iij s Münchener, tut in Gold iii*ᶜ *und lxiii r. guld, vnd iii lib — f 363 lb 3 s 0 h 0.*

"For the same Herr Rudolf, an Imperial receipt was drawn for his payment of the Imperial tax which fell due on St. Martin's Day of the year '70, and a receipt was also given to Herr Rudolf, in the amount of iiiᶜ and xviij lib Munich and iij shillings Munich, which is in gold —

Bottom:

iiiᶜ and lxiii r. guld, and iii lib — f 363 lb 3 s 0 h 0."

item xxv lb vj s freitag vigilia uldalrici	. . . .	f	0 lb	25	s	6	h	0
item liij lb sundag vor Margareten		f	0 lb	53	s	0	h	0
item xlviij lb xviij s samstag vor Magdalenen		f	0 lb	48	s	19	h	0
Summa		f	4 lb	1222	s	18	h	4

"then 25 pounds 6 shillings on the eve of St. Ulrich's Day
then 53 pounds on the Sunday before St. Margaret's
then 48 pounds 19 shillings on the Saturday before St. Magdalen's"

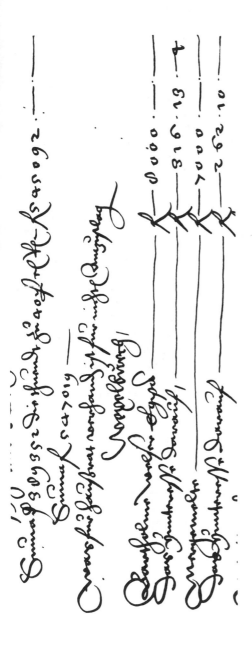

Fig. 125 An excerpt from Anton Fugger's inventory of his total worth, which he had made in 1533. In his *Wexelbuech* (exchange book) the name Welser appears in the first item; this was the name of another rich merchant of Augsburg. "The interest thereon" refers to amounts of interest due on loans made by Fugger.
Reproduced with kind permission of the Fürstlich Fuggersche Foundation, Augsburg.

Suma Dukaten 389 352 die thuendt zu 40 Kreizer
rheinische Gulden   f 545092 . – . –
Suma f 547019 ———
Was für Haussrat vorhanden ist für nichts anzuschlagen
         Wexelbuech.

Bertholme Welser cgpa (compagnia) ———————— f 8000 . – . –
Das ynteresse darauf ——————————— f 316 . 13 . 4
Mergemelte ———————————————— f 7000 . – . –
Das ynteresse darauf ——————————— f 262 . 10 . –
"Total ducats (*d* with a flourish) 389 352, each of 40 crowns
Rhenish gulden   f 545092 . – . –
Total f 547019 ———
Amounts on hand for household expenses are not to be included in total worth
         Exchange book.

Bartholomew Welser Co. —————————— f 8000 . – . –
The interest therefrom ———————————— f 316 . 13 . 4
Items variously mentioned ——————————— f 7000 . – . –
The interest therefrom ———————————— f 262 . 10 . –

[293]

WRITTEN NUMERALS AND COMPUTATIONS

By now the reader may be wondering why so much fuss is made over the mere fact that a new set of numerals replaced an old one. Whoever asks such a question shows that he does not understand the essence of the Indian system of numerals.

The primitive rules of ordering and grouping have now been superseded. At our present stage we can scarcely imagine the extreme difficulty of abandoning the highly satisfactory visual quality of the number groupings in order to adopt the more sophisticated principle of place-value. At first glance one is likely to notice that the new system is easier to write — but this was true only for those who were already familiar with it. Arithmetic texts of the 16th century go to unbelievable lengths, both in verbal exposition and in tables *der Verglychung tudtscher vnd ciferzal* ("for comparing the German numbers and the cipher numerals"), to make the new numerals and their use clear to people; we shall revert to this again later (see pp. 334 and 433).

"Numbering" — that is, the mere writing of the numbers — was in itself a leading subject in schools and books, and even in catechisms, which are among the oldest books of popular instruction. In a typical catechism printed in 1525 in Wittenberg, for example, we find, along with the *teyn Bade Gades* ("Ten Commandments of God"), the *vade unse* (the Lord's Prayer), the *Döpe* (Baptism) and the *Bycht* (Confession) also a section on *de düdesche tall mit den cifern* (the German numbers and the new ciphers).

But though at first glance one merely notices the greater brevity brought about by the new numerals, a second glance lets one see a little deeper: with the new digits we can now for the first time make computations!

With this remark we finally realize that writing numerals and making computations are two entirely different things; up to now we have generally had nothing to say about computations, although we quite thoroughly discussed spoken numbers and written numerals. But didn't people make calculations with Roman numerals? No, they did not! The fairly simple multiplication

	325 × 47	in Roman numerals	CCCXXV · XLVII
	2275		MMCCLXXV
	13000		ⅲ MMM
	15275		ⅲ ⅅCCLXXV

looks clearly impossible to the uninitiated reader.

The how did people in the Middle Ages make the necessary computations — not to mention the Romans themselves, who developed the system of numerals used throughout the Middle Ages? This question about computations, and thus about the real purpose of written numbers, brings us to a fascinating subject of cultural history which has now largely been forgotten — the counting board or abacus.

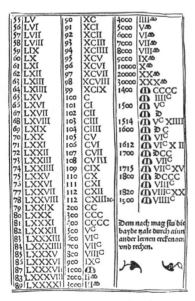

Fig. 126 A table for learning the new system of numerals, from Köbel's book of arithmetic, 1524.

The Abacus

The Nature of the Counting Board

When an aborigine wants to add 6 coconuts to the 17 he already has and then give the total number, how does he arrive at the answer? Just as we do when we recite the alphabet: he counts 17 coconuts and then counts 6 more. He does not "compute" at all; he merely keeps on counting until he reaches 23. Yet in a sense he *has* made a calculation by combining two numbers into one. Then he can replace the coconuts by a supplementary quantity of pebbles or sticks, as the Wedda tribesman did (see p. 33). Thus he has physically "grasped" his answer, or even "written" it, if you will. Arab traders supposedly often carry important numbers about with them, in the form of such "pebble writing," in a small satchel. If we disregard their different face values, the coins we use are also supplementary quantities that give us the numbers we need, and hence are numerals of a sort: take 743, for instance. We do not carry this number about with us in the form of seven hundred forty three separate pennies — rather we group or combine it into a five-dollar bill, two one-dollar bills, four dimes and only three copper pennies. Why? Because, apart from other obvious advantages, by grouping it in this manner we can read the number more easily.

How are computations made in primitive system of numerals that does not use these "pebble numbers"? Let us begin by seeing how a peasant adds 60 and 457 in three separate steps (see Fig. 83). As a first step, he writes down the initial number 457 in his own symbols: he then puts down the number 60 quite independently of the first number; last of all he groups together the two numbers, first in V-groupings and ultimately in O-groupings, so that he can finally read off the answer quite easily. This example shows step by step what happened all at once in putting down the initial number: the peasant eliminated all the rows successively and replaced them with groupings, just as Gottfried Keller's old woman junk-dealer did (p. 249).

Peasant numbers are essentially no different from Roman numerals. But was the same procedure followed in the counting house of a great merchant: writing down and calculating the sums in chalk on a tablet or board in Roman numerals, and then erasing them and entering only the final results in the account book? Of course not. In making their computations, merchants, government officials, and counting house clerks abandoned the Roman numerals and once again calculated with pebbles — or counters — no longer just by lining them up in rows as the native did, however, but on a counting board.

We can understand why: For one thing, the counting board or abacus helps to break numbers down into groups so that they can be handled more easily; then too, it makes the numbers movable so that they can be combined in various ways without constant erasing and rewriting. How does it do that? A large number like 6238 could be broken down into groups represented by different kinds of pebbles, large and small, or of various colors. But this is still not the way it is done: the counters are given their grouping values by their positions. Thus a board or tablet is divided into strips or columns, one each for the units, the tens, the hundreds, and the

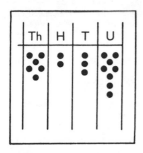

Fig. 127 Basic form of the counting board, showing the number 6238.

thousands, into which the number 6238 is readily broken down and "written" by means of the counters (Fig. 127). Thus the nine or less counters that appear in any one column can be visualized without any trouble at all.

Is this really how it was done in the Middle Ages? The question is quite pertinent, for this would mean that a mature form of complete place-value notation was used in computation, whereas in actual writing people clung blindly and stubbornly to the primitive rules of ordering and grouping numbers.

But it was indeed just so: in computation people used a mature place-value system with gradations based on 10; in writing the old numerals were retained. The two went along hand in hand for many centuries. The price paid for the "popular" simplicity of the Roman numerals was their uselessness in computation, but this disadvantage was again nicely circumvented by using the counting board with its more advanced gradational arrangement. The mutually complementary use of the numerals and the counting board thus created a fully adequate and convenient tool for simple computation, which people were therefore extremely reluctant to part with. For what is in fact the essence of the new Indian number system? That it combines the two aspects, writing numbers and making computations, into one single procedure, by extending the more advanced principle of the counting board — its place-value notation — to the numerals as well. The separation between computation and the writing of numbers was never realized during the Middle Ages; people wanted utility, not intellectual or spiritual perfection. It never occurred to anyone even to try to take the step which the Indians had taken. Cultural history shows over and over again that regardless of how close a subject matter may be to its ultimate and perfect expression, only the intellectual readiness of a culture to take this step will ensure its development to complete maturity. A striking example of this statement is the place-value notation we have been talking about: not only did Medieval Europe possess it for many centuries, but it was thoroughly familiar to people even in antiquity — on the counting board.

Now let us take up the history of the abacus.

The Counting Board in
Ancient Civilizations

Little is known about the manner in which computations were made by ancient peoples — not only the Babylonians, the Egyptians, and the Indians, but also the Greeks and the Romans. The problems and questions in Egyptian papyri offer some intriguing glimpses into the mathematical thought of Pharaonic times, but the actual operations by which the Egyptians found or attempted to find the solutions must be laboriously deduced from the rules, and can often be only guessed at. We know the numerals used by all of these ancient cultures. These support the hypothesis that computations were performed on counting boards. But whenever the written, pictorial, or archaeological documents are lacking, we find ourselves groping in the dark.

And such documents are, unfortunately, missing throughout the ancient world except for Greece and Rome; only a few actual counting boards and representations or descriptions of counting boards used by the Greeks and Romans are still preserved. No wonder, then, that these are especially treasured as rare and precious documents of cultural history. The following pages will discuss and illustrate a few rare specimens, some perhaps for the first time.

THE SALAMIS TABLET
The only ancient Greek counting board that has been preserved is a tablet of white marble found about the middle of the last century on the island of Salamis (Fig. 128). Its exact date is not known.

Two groups of parallel lines have been chiseled into this tablet, one group of eleven lines crossed by a perpendicular line through their middle, and another group of five shorter ones at some distance from the first group. Along the two long sides of the tablet and across one of the shorter sides are letters, which can be identified as early Greek numerals and also as denominations of ancient coins (cf. Fig. 99).

T ᚠX ᒪH ᒪ△ ᒣF ICTX

Talent = 6000	1000	100 Drachmas	10	1	1 $\frac{1}{2}$ $\frac{1}{4}$ $\frac{1}{8}$ Obol

The three symbols at the right are those for the half, the quarter, and the eighth obol, which were respectively called the *hemiobólion* (represented by the half-symbol C because in Boeotia O signified a whole obol), the *tetartemórion* (T for "quarter-piece") and the *chalkós* (X, "copper" or "ore"); thus the last two are not to be confused with the symbols X for *chílioi* and T for *tálanton*, which stand at the left end of the row of numerals. When separated from their currency values, the fractions of the drachma become abstract fractions like the Roman:

1 drachma = 6 obols = 12 half = 24 quarter = 48 eighth obols.

These symbols for numerical values testify that the tablet was actually used for computations, very likely in government financial

[299]

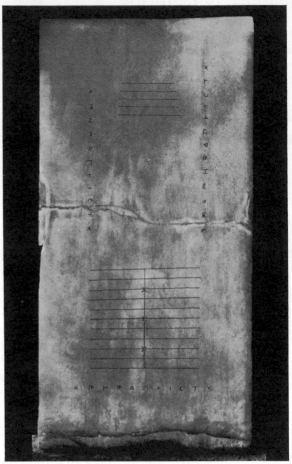

Fig. 128 The Salamis Tablet, the only large ancient Greek counting board preserved today. Size: 149 × 75 × 4.5 cm (7.5 cm thick at the edges). National Museum, Athens.

offices, with whose public accounting we have already become familiar (see Fig. 100). Then if we read the following passage from the historian Polybius (2nd century B.C.), we shall see that the Salamic Tablet is essentially like the counting board already described in general terms.

The courtiers who surround kings are exactly like counters on the lines of a counting board, for, depending on the will of the reckoner, they may be valued either at no more than a mere *chalkós*, or else at a whole talent! (See also the quotation from Aristophanes on p. 211).

The *chalkós* and the *talent*, of course, are the lowest and the highest values on our tablet, at the extreme right and left ends of the scale. The counters took their values in computation from these: if they were placed under the symbol for the *chalkós* they had the value of an eighth of an obol, but if they were moved to the space beneath the talent sign, their value was increased 300,000 fold; a more graphic example of place-value notation could scarcely be found.

As early as the end of the seventh century B.C. Solon, the wise law-giver of Athens, compared a tyrant's favorite to a counter whose worth depended entirely on the whim of the person who pushed it from one place to another. This simile lived on through the centuries, as long as counting boards were in use: we shall meet it once again, this time in 18th-century disguise.

The Greeks called the counting board or table *abákion*, and the Romans *abacus*; the Greek word *ábax* means "round platter" or "stemless cup" and thus also a "table without legs." It is unlikely that this word was derived from the Semitic *abq* (dust), although the Semitic peoples also, especially later on, had similar sand-covered tablets which they used for drawing figures and making computations (see p. 331). *Abacus* was also the name for the ornate Roman display tables with costly inlays of marble or silver, for the smooth wall panels covered with plaster of marble-dust and for the top plate in the capital of Doric columns. In ancient times this word thus had the general meaning of a flat table or board. The Roman name *abacus* was used for the counting board throughout the Middle Ages. Greeks today still say of someone who is resourceful *ksérei tòn ávakon*, "he knows his abacus."

The counters were called *pséphoi* (pebbles). Since the word *pséphízein* (literally, "to pebble") is also the general word for "to compute" or "to calculate," the Greek language itself provides suggestive evidence as to where and how computations generally were made. As late as the 14th century the Byzantine scholar Maximus Planudes entitled his textbook on computations with the Indian numerals *Pséphophoría kat' Indôus*, "Pebble-placing in the Indian Fashion," showing how deeply rooted the ancient expression was in the abacus.

In addition to placing them in the columns of the counting board, the Greeks also arranged counters independently to form "geometrical numbers" — triangles, right angles, squares, etc. These "pebble-numbers" (counter-numbers), which turn up later in the history of mathematics as "figurative numbers," thus represent the oldest association of numbers with geometrical shapes. With the aid of these pebble-numbers the Greeks discovered quite a few important laws of number theory, for example the rule that the difference between two consecutive square numbers is always an odd number, the "corner" (*gnómon* in Greek) that is added to the previous square in a series of squares, as shown in the margin.

On the other hand, no information has been preserved concerning the actual operations performed on the abacus itself; nevertheless we can infer these with a high degree of probability from the general arrangement of the Salamis Tablet, from the remaining fragments of other Greek counting boards with columns drawn or incised on them, and above all from the medieval counting boards, about which we are plentifully and adequately informed.

To use the Salamis Tablet the reckoner would stand facing it on the right-hand long side. His *pséphoi* (counters) would be piled in the middle between the two groups of parallel lines. The group of lines at the left end of the board as he faced it were used for the integers,

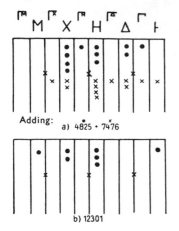

Adding:
a) 4825 + 7476

b) 12301

Fig. 129 4825 + 7476 = 12301 on the Salamis Tablet.

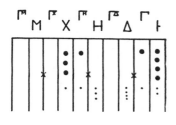

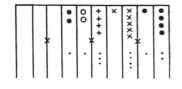

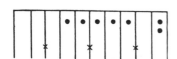

Fig. 130 3509 − 1847 = 1662 on the Salamis Tablet. *Top:* putting down 3509 (●) and 1847 (.); *middle:* changing 1 thousand into two five-hundreds (○), one five-hundred into five hundreds (+) and one hundred into ten tens (×); *bottom:* answer 1662.

the group at the right end for fractions. The small crosses indicated the columns: the one at the right distinguished the units and the one at the left marked off the talents, or the myriads if the computation did not involve money.

Let us say, for example, that the quantity 7476 is to be added to 4825. First the reckoner forms the number 4825 (here indicated by the symbol ●) with pebbles in the appropriate columns, thereby observing the rule of quinary grouping which is always followed in each even column (the 2nd, 4th, etc.), as shown in Fig. 129. Then he simply "gives" the second number 7476 (indicated by ×) to the first, as expressed precisely by the Greek technical term *syntithénai* and the Latin *addere* (from *ad-dare*), which became our word *addition*. The first step of this procedure thus consists of "writing" or putting down the initial number, the second step of addition proper, and the third step of grouping, in which 5 units make up one 5-group and two of the latter a 10-group, or one unit in the next higher column. After the calculation has been "purified," as this grouping was called in the Middle Ages (the Romans called it *purgatio rationis*), the reckoner can then read off the final number, the answer 12301. The result may also have been placed over the intermediate numbers of the computation as a "head number" (perhaps above the middle line), as suggested by its Greek name *kephálaion* (from *kephalé*, "head"; see Fig. 100) and its Latin name *summa* (from *summus*, "the highest"). Thus the very same steps which we saw in peasant calculations are also essential to the reckoning table.

Subtraction, for example 3509 − 1847, is performed in the same graphic manner. The first number is put down (represened by ● in Fig. 130) and then the second number is "taken away" from the first: Greek *aphaireîn* and Latin *subtrahere*, "to draw away," whence comes our own word *subtract*. In this process the second number must be either remembered or recorded elsewhere. On the Salamis Tablet this may have been done in one of the rows of numbers at the side, or arranged so that the first number appeared above and the second below the middle line.

It makes no difference with which column the subtraction begins. One can start with the highest rank just as well as with any of the others. The important thing is that one must be able to "take away"; in other words, the number of *pséphoi* of the first number in a column may not be smaller than that of the second number. If this happens, the ranks must be "changed" to lower ranks. Contrary to the procedure in addition, the numbers are "de-grouped," one ten being converted to two fives and one five into five units in the right-hand column. Then the "taking away" can be done quite easily, and the "rest" of the *pséphoi* that remain on the counting board are the answer, in this case 1662.

What was the purpose of the rows of number symbols carved along three sides of the Salamis Tablet? Evidently they served as a "supplementary" record of numbers to be placed on the abacus. In multiplying $874 + \frac{5}{6} + \frac{1}{12} + \frac{1}{24} + \frac{1}{48}$ by 93, the calculator had to keep these initial numbers constantly before his eyes (see Fig. 131). He now had to use the columns to build up the answer little by little (an example

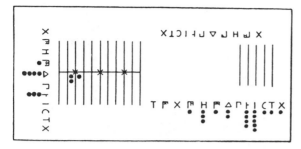

Fig. 131 $(874 + \frac{5}{6} + \frac{1}{12} + \frac{1}{24} + \frac{1}{48}) \times 93$;— that is, 874 drachmas 5 obols and 1 half, 1 quarter and 1 eighth obol, times 93, multiplied on the Salamis Tablet.

of the way multiplication was performed on the Roman abacus will be explained on p. 316). Perhaps in these operations a second reckoner stood on the left side of the table and made calculations on his side of the middle line. As counterparts of this in the Middle Ages, we have the Dinkelsbühl reckoning table and the counting board of the "three gentlemen of Basel" (see Figs. 177 and 178).

Besides the Salamis Tablet and the passage quoted from Polybius, we also have, by chance, a third witness, in the form of a picture of such a reckoner at work, on the famous

DARIUS VASE

This magnificent ceremonial vase in the red-figured Attic style, which commemorates the Persian Wars, was found in an ancient grave near Canosa (the ancient Canusium) in Apulia and is now in the Museo Nationale in Naples (Fig. 132). Southern Italy, including Apulia, was part of the realm of Greek culture in ancient times.

In the top row of the three rings of figures, Zeus and the other gods take Hellas, menaced by the Persians, under their protection. In the middle row, the Persian King Darius is seated on his throne, and a Persian standing in front of him seems to be warning him against undertaking his proposed expedition against the Greeks. Arranged about Darius are figures symbolizing the members of his court and officials of his government, including, in the bottom row, the revealing representation of Darius's royal treasurer (see Fig. 133).

A man dressed in Persian costume stands in front of the table with a sack of coins. He is bringing in the tribute paid by the various conquered peoples, who are represented by the individual figures kneeling behind him. The Persian treasurer sits at a small table on which are the signs MϞ (Boeotian for × 1000), HΔΓ and O (Boeotian for obol), ⟨ (or C, the sign for the half-obol) and T, which has the same meaning as on the Salamis Tablet (see p. 299). A striking feature is the lack of the five-group, for the symbol Γ here clearly stands for the units. Thus apart from the fractions, only the decimal system is used. In contrast to the Salamis Tablet, the counters or pebbles here lie, without inscribed columns, directly beneath the symbols for the numerical ranks. Possibly the pebble seen above

Fig. 132 The Darius Vase with its depiction of a calculator at his reckoning table (bottom row), one of the only two known ancient representations (see Fig. 134 for the other). Height 1.30 m, greatest circumference 2 m. Probably 4th century B.C.
Museo Nationale, Naples.

Fig. 133 The treasurer on the Darius Vase, seated at his reckoning table, which has counters scattered upon it.

(below in this picture, which is upside down) the H stands for the five-group, after the Roman custom (see p. 317). Thus the number 1731–4/6 can be read on the Darius Vase; of course this whole representation is symbolic.

In his left hand the reckoner holds a diptych or double tablet, on which we read the word *talanta H*, "100 talents." On this tablet are posted the amounts which he successively finds on the abacus. Thus a living representation of computation in ancient times has been preserved through sheer good luck.

THE ETRUSCAN CAMEO

There is a surprising similarity between the Darius Vase and a tiny but exquisitely carved cameo of Etruscan origin (Fig. 134). On this gem, in almost exactly the same position and attitude as the treasurer on the Darius Vase, a man is seated at a small, three-legged table on which the counters are represented by three minute spheres. Like the reckoner on the vase, he holds a tablet in his left hand, this time with Etruscan numerals, which leave no doubt as to the man's activity. Perhaps the tablet alludes to the double rows of the signs for the numbers 1, 5, 10, 100, and certainly also 1000, which the reckoning table also has. The symbols ∧ and × are familiar to us from the Etruscan coins we have already seen (see Fig. 67).

The similarity between the representations on two works of art of such divergent origins, which were also certainly made at different times, bears witness to the universality of the reckoner, whether as merchant or as money-changer. Even his names in different languages, Greek *trapezítēs* and Latin *mensarius*, "tabl-er" (from the Latin *mensa*, "table") show unequivocally that computations were made on tables, just as the present "bank-er" ultimately derived his name from the medieval "counting bank" which was his tool (see Fig. 250).

Fig. 134 Etruscan cameo, showing a reckoner at his table. 1.5 cm high. Cabinet des Médailles, Paris.

With the Etruscan cameo we have entered the world of the Romans.

THE ROMAN HAND ABACUS

The only two specimens still preserved show that the Romans arranged the counting board very cleverly for the purpose of making smaller computations. One of the two is in the great collection known as the Cabinet des Médailles in Paris (see Fig. 135, which is a photograph of a plaster cast); the other, which once belonged to the famous

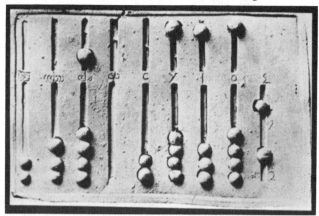

Fig. 135 Roman hand abacus. Plaster cast of the specimen in the Cabinet des Médailles, Paris. Almost full size. Between the two rows of grooves are the Roman number symbols for

1 million 10   1   1 $\frac{1}{2}$ $\frac{1}{4}$ $\frac{1}{3}$

         Integers   Unciae

Jesuit Athanasius Kirchner (died 1680), is now in the Museo delle Terme in Rome. To emphasize the "hand" size of this Roman abacus, the writer photographed this specimen held in his own hand (Fig. 136). The Augsburg scholar Marcus Welser also possessed a Roman hand abacus which is now lost, but which he fortunately illustrated and described precisely (in his *Opera historica et philologica*, 1682); Welser's abacus was the same as the other two, except that it had three small grooves at the right end (instead of only one). On the basis of these three, we can give a reliable account of the Roman hand abacus:

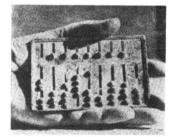

Fig. 136 A real hand abacus, the second preserved specimen, now in the Museo delle Terme in Rome.

On a small bronze tablet are carved 8 long grooves (*alveoli*) with the same number of short grooves at the top; at the right end is another long groove (three more on Welser's abacus) without the short groove above. On the raised surface between the two rows of grooves are the Roman symbols for the whole numbers, for the *unciae* and along the extreme right groove for the half, quarter and third uncia (called *semuncia*, *sicilius* and *duella* in Latin; see p. 161). In these grooves run little spheres (called *claviculi*, "little nails"), four in each long groove (5 in the case of the *unciae*) and one in each short groove above.

And how were the computations carried out? These dealt only with the simplest problems, for the most part probably the addition of two or more numbers. We shall run through a computation with integers; the reader will find the essential principles of operations with fractions on p. 158. To "insert" the first number 5328, the counters are pushed toward the raised bridge near the center. The rows above the bridge are the five-groups. Thus to form the 5, in the short groove at the top, the counter is pushed downward; to form the next digit three counters are pushed up in the next unit groove, as the diagram shows; the reader can follow the rest of the procedure himself.

Now, if a second number is to be added to the first, it cannot be placed separately on the hand abacus as it is with the free counters on the reckoning table, and the answer arrived at by combining the two. Here the combining must take place simultaneously — that is, the second number, which the computer keeps in his head or writes down elsewhere, must be immediately blended step by step with the first.

We now have only the two above-mentioned actual hand abaci, but we also have an invaluable Roman pictorial representation of a computer at work (see Fig. 138). In the left corner of the sculptured

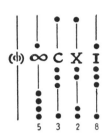

Fig. 137 5328 on the hand abacus.

Fig. 138 A Roman *calculator* (from *calculi*, "counters") with his hand abacus, calculating at the "dictation" of his master. From a 1st century A.D. Roman gravestone in the Museo Capitolino, Rome.

gravestone of a merchant stands the calculator of the household with his portable abacus, which is here portrayed symbolically. His stance and posture speak for themselves and are very indicative of the proceedings: the master "dictates" the amounts, the *calculator* adds them up. Every orderly head of a household kept a *codex accepti et dispensi*, a "book of receipts and disbursements," and in the *tablinum* of his Roman house he had a special counting room.

THE HAND ABACUS IN ASIA
The small Roman reckoning instrument did not take hold in the West. This is understandable, for it presupposed a skill in working with numbers which only the merchant would generally possess. On the other

hand, it penetrated to a surprising degree through all of the Far East, especially China and Japan, where it is now universally used.

Thus we shall interrupt our passage through antiquity at this point, to examine the Asiatic counterpart of the Roman abacus with its attached counters or buttons (*claviculi*), before we go on to discuss the operations procedures with loose counters (*calculi*).

The Japanese Soroban

This is an exact analogue of the Roman hand abacus. Within a rectangular frame, on each thin rod (Japanese *keta*, "row, wire") slide 6 small beveled beads, the sixth being separated from the other five at the bottom by a long strip perpendicular to the vertical wires or rods. The differences between the Japanese *soroban* and the Roman hand abacus are the following (see Fig. 139): instead of four counters at

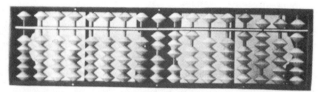

Fig. 139 A Japanese *soroban* with 17 wires or rods (*keta*), typical of those which may be bought anywhere in Japan today. The number 231 is shown in the middle, and 1956 at the right. Size: 21 × 5.5 cm. This is a specimen in the present author's possession, through the kindness of Nobuko Yokota, Tokyo.

the bottom there are generally five; the vertical rods or wires have no numerical symbols, which are unspecified; the *soroban* has more vertical columns, as many as seventeen; and it has no columns especially set aside for fractions. They do have in common, however, the quinary and decimal grouping of the counters which run along fixed rods (or grooves). The Japanese might well wish, "If I only had one more *keta* salary!" when he dreams of multiples of ten decimals. On the *soroban*, as on the Roman hand abacus, only the counters pushed against the bridge or separating bar have value; in Fig. 139 the numbers 231 and 1956 are formed on the soroban (see also Fig. 141).

The *soroban* was introduced into Japan from China, probably in the 16th century. The origin of its name is unexplained. In the 1870's it was almost replaced by written computations with the "western" (that is, Indian) numerals, which had by then come into universal use throughout the world, but since about 1930 it has again been in great demand as a result of the growth of Japanese industry and trade. The Japanese chambers of commerce and industry offer yearly examinations and competitions in computations on the *soroban* in which, to name only one year, in 1942 about 40,000 participants competed for the prize.

As he writes, the author has before him a Japanese textbook, in its 4th edition of 1954, which provides instruction "using the latest methods in the mastery of the *soroban*, from the basic elements to the highest levels of proficiency." This book has been "registered and

Fig. 140 A Japanese calculating on the *soroban*, adding up amounts taken from a ledger.
Photo by Wolfram Müller.

accepted by the Soroban League of the Chamber of Industry and Commerce." It clearly and thoroughly teaches the four fundamental operations of arithmetic, using many examples and practice exercises, and every aspect is thought out and presented fully, down to the positions of the fingers in moving the beads (see Fig. 141).

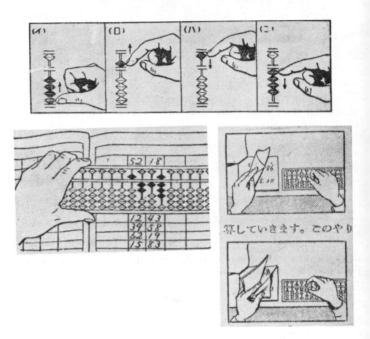

Fig. 141 Excerpts from a Japanese *soroban* textbook published in 1954. *Top:* showing the proper technique for moving the counters along a *keta*. *Bottom left:* placing the number 5218 on the *soroban*; observe that numbers today are written in Indian numerals, but that computations are still (generally) made on the *soroban*. *Bottom right:* how the merchant adds the posted amounts together. The left hand turns the pages of the account book, while the right hand calculates. Compare the similar representations of reckoners at work in other cultures and at other times (Figs. 133, 134, 151, and 195).

There is an art even to pushing the counters. On the author's *soroban*, which is the type generally used in commerce, one *keta* with its beads is 5 cm long. This, however, includes the cross-bar, so that there is only a bare 4 mm of room left to move a counter in! And two adjacent *ketas* are only about 12 mm apart, while the sharp edges of the counters on them are separated by only 1 mm. Moreover the counters slide along the *ketas* almost completely without friction — all these circumstances call for a degree of dexterity and precision of movement which our European fingers, at least at the beginning, simply do not possess. The first few attempts to use the *soroban* strongly resemble the proverbial bull in the china shop, which breaks more dishes with its rear end than with its head: the beginner always finds himself moving more counters than he intends to.

All four operations of arithmetic are taught, especially multiplication and division. The many *ketas* or wires permit all the numbers in a computation to be recorded, as for example, both the factors 27 × 6 and also the result 162. As always on an abacus, the rules of place value are extremely important. To make these clear by an example (see Fig. 142): The factor 6 is placed on the left, and 27 under

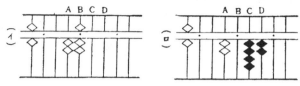

Fig. 142 27 × 6 = 162 on the *soroban*; from the *soroban* textbook.

ketas A and B. The multiplication 6 × 7 = 42 is performed and placed on *keta* D, and the number 7 is thereby removed from *keta* B. Then follows 6 × 2(0) = 12(0), which is immediately combined with 4 into 16; final answer: 162.

To carry out this procedure, the reckoner must therefore know the multiplication table, the rules of place value and the proper method of instantaneously combining numbers — in other words, he must constantly work problems in his head. For this reason great importance is attached to elementary education in Japan, including the stress, surprising to us, on computations on the *soroban*. When the Americans occupied Japan in 1945, they at first ridiculed the *soroban* for its supposed backwardness. Naturally they had to demonstrate their own "progressive" methods, and so they organized a calculating match in Tokyo, which was observed by 3000 spectators. This competition, which was in many respects one between two very different cultures, had a surprising outcome.

The contest matched 22-year-old Kiyoshi Matsuzaki, a Japanese Communications Ministry clerk with seven years' special abacus lessons, against 22-year-old Pvt. Thomas Ian Wood of Deering, Mo., an Army finance clerk with four years' experience on modern machines. Matsuzaki, who flipped the wooden beads with such lightning dexterity that he was immediately nicknamed "The Hands," used an ordinary Japanese *soroban*, selling for about 25 cents before the war. Wood's electric machine cost $700.

The abacus won the addition event — columns of four- to six-digit figures — taking all six heats and finishing one of them more than a full minute ahead of Wood. The abacus also won in subtraction. Wood staged a rally in multiplication, since abacus multiplication requires many hand motions; but Matsuzaki was out in front again in division and in the final composite problem. "The Hands" also made fewer mistakes.

One reason why Matsuzaki won is that like all abacus veterans, he does the simplest arithmetic in his head, pegging the results on the abacus and going on from there."*

In addition, the Japanese, who today are taught the *soroban* in elementary and secondary school, also learn to make computations with our familiar Indian numerals, just as our own bookkeepers do who work calculating machines in offices, but with the important

*Quoted directly from "Hands Down!" in *Reader's Digest*, Vol. 50, March, 1947, p. 47.

difference that the latter no longer need to do mental arithmetic, so that they become increasingly rusty at it and their skill declines to the level of the mere physical operation of the machine, whereas their Japanese colleagues always stay in practice and thus constantly improve their skill in arithmetic. The Japanese can make mistakes in computation on his *soroban*, to be sure, while the machine makes only "typographical" errors; this is an advantage of the machine, of course, but also a large drawback, as suggested by the constant complaints about the mangled computations produced by clerks without adequate skill.

The Chinese Suan Pan

This device (literally, "reckoning board") has two beads on the separated 5-group wires instead of the single one on the Japanese *soroban* and the Roman hand abacus (see Fig. 143). This improvement makes addition a little easier; one can first form $6 + 8 = 14$ on one wire or rod and then add them together, whereas on the *soroban* they must be combined immediately. Like our own slide rules, some are almost a foot and a half long, while others may be no larger than a matchbox. The formation of the Chinese character for *suan*, "to calculate," is interesting: two hands holding a reckoning board made of bamboo (Figs. 144 and 145). As in Japan, children in China are instructed in the use of the *suan pan*.

The *suan pan* is the parent of the *soroban*, which, as we have already seen, was first introduced into Japan in the 16th century. Today both are indispensable aids to computation in shops and offices all over the Far East. The shopkeeper can add up several numbers on the abacus generally faster than we can by writing them down. And just as an experienced clerk in a western bank depresses the keys of his adding machine without looking at them, the Chinese and the Japanese play on their *suan pan* and *soroban*, pushing the counters back and forth with a light "click-click" instead of the rattling noise made by the machine. The long use and popularity of the Oriental abacus are strikingly documented by our two illustrations of Figs. 146 and 147. A charming woodcut from a Japanese book of the 18th century, showing a rich merchant at work in his counting room with his assistants, and a photograph, taken in 1957, of the interior of a crowded department store in Peking, where the *suan pan* still lies ready on the sales counter.

From these descendants, which are still in the full flower of life, let us turn back and take a final look at their ancestor, the Roman hand abacus. This too was a calculating instrument which merchants carried with them and kept ready for use in a fold of their togas. Its significance may be judged from the fact that it was and still is retained in China and Japan along with the highly developed Oriental written numbers. How much more must the Romans, with their cumbersome numerals, have relied on it! To be sure, it lacked the flexibility of its Far Eastern descendants, it was tightly bound to the symbols carved over each groove, and it was quite clumsy, especially in connection with the Roman fractions. Thus the scope and level of the computations that could be made on it were more limited. It is not impossible, neverthe-

Fig. 143 Chinese *suan pan* showing the numbers 10 and 1872. Size: 6 × 45 cm — thus this is a pocket abacus (in contrast to the *suan pan* shown in Fig. 147)
Ethnographic Museum, Frankfurt/Main.

Fig. 144 *Suan*, "to calculate." This Chinese character also stands for the *suan pan* and the Japanese *soroban*.

卄   𥫗   目

Fig. 145 Meaning of Fig. 144: Two "hands" holding a "reckoning board" made of "bamboo."

Fig. 146 A Japanese merchant, writing down the sums produced for him by his two *soroban* calculators. Woodcut from an 18th-century Japanese book (cf. Fig. 256, p. 435).

Fig. 147 Scene in a Peking department store. On every counter lies the sales clerk's *suan pan*. Photograph taken in 1957 by H. Pabel.

less, that the development of the Far Eastern forms of the abacus was stimulated by the Roman hand abacus. The *suan pan* did not appear in China before the 12th century. Consider how far eastward the Roman Empire extended; consider that during the Han Dynasty (200 B.C. to A.D. 200) there was a lively trade between China and the Roman Empire in silk, iron, and hides, as a result of which some

20 million sesterces left the land each year (according to Pliny). Moreover connections between the Mediterranean and the Orient can be traced as far back as the 5th century B.C. The Chinese word *ser*, for "silk," was adopted by the Greeks (*tò sērikón*), who called the inhabitants of the Middle Kingdom the "Serians." These ties, continued for a long time, were renewed by the conquests of Alexander the Great and then were drawn even closer by the Arab merchants. When Kublai Khan established the gigantic Mongol Empire with his conquest of China, European merchants, craftsmen, clergy, and scholars streamed into his realm. Germans and Frenchmen held high official positions at Kublai Khan's court. Marco Polo was his trusted advisor. Venice, which was likewise at her zenith, even maintained a commercial colony there.

Thus there were quite enough living bridges over which the Roman hand abacus, or the idea of it at least, could have been carried to the Far East. Evidence against this hypothesis, of course, may be the fact that the counters on the Oriental abacus do not run in grooves but are strung on rods or strings; this device is surely derived from the knotted strings which have become familiar to us in many forms in Asia (such as the prayer strings adorned with beads, and others).

And the tortuous paths along which these cultural influences may travel may be discerned from the fact that the Far Eastern calculating device migrated back to us in the West.

The Russian Ščët

There is still another Asiatic form of the abacus with perpendicular wires on each of which ten beads are strung, thus having no quinary grouping. On these the fifth and sixth beads are colored differently from the rest, to make counting easier. These are the Russian *ščëty* (Russian *ščët*, "counting, reckoning," nominative plural *ščëty*). Whether or not these go back directly or indirectly to a Chinese

Fig. 148 A Russian estate owner at his desk, with his *ščët* at his right hand, always ready for use. From a photograph taken in 1909.

ancestor, they were (and still are) at any rate universally used in Russia by officials, by shopkeepers, and by people in general.

Chance has placed in the author's hands an old picture of a Russian estate owner, who always had his *ščët* ready at hand on his desk, used for computing everything, really everything, that could possibly be computed (Fig. 148). The photograph is especially significant because it was not intended especially to show the *ščët*.

Our second illustration shows another *ščët* — this one from Persia (Fig. 149). This is a way of saying that it was also used there, as in Turkey, by merchants and tradespeople, and is still in use by older people. Of all the forms of the abacus, this one has been the most "popular," the one most readily accepted by common and primitive peoples. A person who needs one can easily make it for himself, it is easy to use in theory, and it requires no previous schooling or long practice as does the *soroban*. Of course, its "capacity" is also more limited.

We can see how computations are made on a "home-made" *ščët* which a refugee from East Prussia presented to the author (Fig. 150). The wires themselves have positional value. Since it was regularly used to add up sums of money (rubles), the fourth wire from the bottom had a smaller number of beads, generally four (for quarters), and then served as a decimal point to separate the two denominations, the rubles from the kopeks. Above it, in increasing order upward, are bars for the units, tens, hundreds and so on. The lowest horizontal bar, also with four beads, could be used for calculating the fractions $\frac{1}{4}$, $\frac{1}{2}$, $\frac{3}{4}$ (see Fig. 149).

Whoever has become adept in the use of the *ščët* is very reluctant to abandon it for any other device. After giving a talk about the abacus on the radio, the author received a letter from a lady, a displaced person from one of the Baltic countries, saying that she had had a *ščët* made for herself and was now working as a bookkeeper in a large Frankfurt business establishment, using the *ščët* alone. He found her actually at work with her small and modest calculating "machine," surrounded by her "electric" fellow-workers, quite content and under no constraint, for she was in no wise deficient in her computations (Fig. 151). She was adding up item after item from a long series of numbers, rapidly and accurately moving the beads with her middle finger and adding each new number to the running subtotal (and thus at once "combining" them).

Anyone who calculates on the *ščët* soon comes to recognize the value of specially marking out the two middle beads in the row of ten. There is no doubt that this small calculating device is one of those simple but highly significant inventions of man which (like the wheel, for example) have enabled him to overcome hitherto impossible obstacles.

The three very rare photographs (Figs. 152, 153, and 154) — they were all taken prior to World War II, but the author has been unable to find out by whom — show the *ščëty* in use by various "ordinary people." But the significance of the *ščët* goes far beyond this. It shows graphically what difficulty "primitive" man had in coping with numbers, and how numbers grew up slowly, keeping pace with a

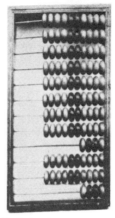

Fig. 149 A Russian *ščët*, from Persia. 20 × 13 cm. Photograph by H. Horst, Tehran.

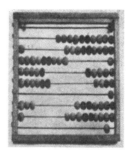

Fig. 150 Russian *ščët*, showing the number 1956.31.

Fig. 151 A displaced person from the Baltic countries employed in a large Frankfurt office, calculating on her trusty *ščët* amid her electric colleagues. Photograph by the author.

Fig. 152 The Russian *ščët* in a village school.

Fig. 153 *Ščët* used by herdsmen in the Altai Mountains.

Fig. 154 *Ščët* used on an estate, where the bookkeeper is tallying up a peasant's time and wages.

people's culture. This slow growth or evolution, which we have so often had to reconstruct from spoken numbers and written numerals, presents itself here symbolically right before our eyes.

Perhaps when the reader first set eyes on a picture of a *ščĕt*, he was struck by the thought, "Why this is just like the small abacus on our baby playpens" (Fig. 155). And in this he is, perhaps to his own surprise, absolutely right: "our" abacus or calculating device is a child of the Russian one. Its history is an excellent example of the way chance can affect the migration of a cultural feature. During his invasion of Russian in 1812, Napoleon had in his army an engineer with the rank of lieutenant, Poncelet, whose name will be familiar to knowledgeable readers as the founder of projective geometry. During the French retreat Poncelet was captured by the Russians and brought to Saratov, on the Volga. While there, living among simple people, he became so impressed with the excellence of the *ščĕt* as a device for teaching children that, upon returning to France, he introduced it into all the schools in the city of Metz. From there the *boullier* (from the French word *boule*, "sphere, ball") spread all over France and Germany, and even America. Thus it may well be said that the Roman hand abacus has come down to us by way of a long detour through Asia.

Fig. 155 Little Johnny and his "calculating machine."

Today the abacus is used in western culture, if at all, only as a sort of visual introduction to the techniques of arithmetic computation. In this respect it follows a view expressed long ago by Adam Riese (see p. 431): Place values were to be assigned to threads or wires exactly as on the abacus. This doubtlessly increased the learning capacity of medieval pupils, who readily dealt in terms of physical objects because they were working with their hands at the same time as they studied.

THE ROMAN COUNTING BOARD WITH LOOSE COUNTERS

After this long detour through the ancient world, in the course of which we retraced the paths taken by the idea of the Roman hand abacus, let us complete the circle by going back to Rome itself.

For more extensive and complicated calculations, such as those involved in Roman land surveys, there was, in addition to the hand abacus, a true reckoning board with unattached counters or pebbles. The Etruscan cameo and the Greek predecessors, such as the Salamis Tablet and the Darius Vase, give us a good idea of what it must have been like, although no actual specimens of the true Roman counting board are known to be extant. But language, the most reliable and conservative guardian of a past culture, has come to our rescue once more. Above all, it has preserved the fact of the *unattached* counters so faithfully that we can discern this more clearly than if we possessed an actual counting board.

What the Greeks called *psêphoi*, the Romans called *calculi*. The Latin word *calx* means "pebble" or "gravel stone"; *calculi* are thus little stones (used as counters). The connection with the word "calculate" is immediately obvious. But the Romans had no such word as *calculare*; it appeared for the first time around A.D. 400 in

Spain. Latin expressed the action of computing as *calculos ponere* or *subducere*, "placing" or "drawing pebbles"; when he meant to "settle accounts with someone," the Roman would say, very expressively, *vocare aliquem ad calculos*, "to call someone to the counting stones." When a person measured his obligations exactly against his benefits, and thus set one off against the other, he "called friendship to the calculating stones": *vocavit amicitiam ad calculos*, or he *parem calculum ponit*, "put down an equal counter." Teachers of arithmetic computation who were slaves were called by the Romans *calculones*, whereas those of higher birth were *calculatores*; the *arenarii* or "sand reckoners," on the other hand, were geometers who, like Archimedes, drew their figures on boards or tablets covered with sand. According to the Emperor Diocletian's schedule of wages and prices in A.D. 301, a mathematics teacher of this sort received 200 denarii for instructing a pupil, but an abacus teacher got only 75 denarii.

The name *calculi* was retained throughout the Middle Ages in the West. But the other medieval technical term, to "purify" the computation, *purgatio rationis* or *purgare rationem* also goes back to the Romans:

Decimo quoque die numerum puniendorum ex custodia subscribens rationem se purgare dicebat,

says Suetonius about the Emperor Caligula,

As he signed the order condemning to death the prisoners who were to be punished on the tenth day, he said that he was settling his account.

The pebbles that first served as counters were later replaced by small discs of ivory, metal, or glass (Fig. 156).

For all our lack of direct evidence, we shall surely not go too far wrong if we describe the larger Roman counting board and its operations as follows: Instead of grooves, it had parallel columns incised on it like the Salamis Tablet. To be sure, it had the quinary grouping, because a counter placed above the row of numbers meant 5 units. This simplified not only the representation of numbers (for the number 7, for instance, only 3 *calculi* were needed instead of 7), but also computations, for multiplication then required only "half" the multiplication table, 1×1 up to 5×5. In multiplying 37×26 ($= 962$), the largest computation called for was only $3 \times 2 = 6$, for the problem was set up as follows:

	X	V	I
(37 =)	3	1	2
(26 =)	2	1	1.

Fig. 156 Roman ivory *calculi* discovered in Wels, the old Roman city of *Colonia Aurelia Ovilava*. Diameter 15 to 20 mm. Städtisches Museum, Wels.

Das Multiplicirn zu oberst anheben ["Raising multiplication to the highest level"], as Adam Riese put it, meaning that the operation of multiplication is begun with the highest rank, is one remarkable peculiarity of calculations on counting boards generally. Thus in our example above, the first step was to multiply the tens, 2×3; but in which column is the result, 6, to be placed? The rules of position in multiplication on counting boards have been puzzled over since time immemorial, for in contrast to the small multiplication table that is required, they constitute the only "difficulty." I shall cite here only the most important rule for the products of multiplication, which essentially goes back to Archimedes and which refers to Greek written computations (see p. 273), but which originated in computations on the reckoning board.

The number of ranks p of a product is equal to one less than the sum of the ranks $a + b$ of the factors; to state this in terms of an equation: $p = a + b - 1$.

Examples: $20 \times 30 = 600$; $a = b = 2$; $p = 2 + 2 - 1 = 3$; thus, the product has three digits. Yet according to this rule $20 \times 70 = 1400$ also has three digits — that is, the "core number" 14 and only this is to be placed in the third column (the hundreds) on the counting board. Thus if we replace the word "rank" or "digit" by "column number" or simply by "column" (3) on the reckoning table, the rule now reads simply: The column (number of columns) of a product is equal to one less than the sum of the columns of the factors.

With the aid of this rule, the reader should now be able to follow through the computation 2703×45 on the Roman abacus (see Fig. 157). If he does not allow himself to be put off by the slight trouble of carrying through the computation, his reward will be a deeper understanding of the numerical concepts underlying computations on the counting board and a recognition of some of the basic characteristics of the abacus, which we shall, in conclusion, attempt to bring out once more as clearly as possible.

If we recall the early written numerals of the Greeks and Romans, we shall see at once that the vast difference between these on the one hand and the counting board on the other hand is exactly equivalent to the same advance that was made in spoken numbers! This alone led to a mature, perfected system of numerals (see p. 53). Thus in comparison to their cumbersome, primitive written numerals, ancient peoples possessed in the counting board the most highly perfected symbols, so to speak, for computation. Although arithmetic computations had still to undergo much refinement, they were at least based on essentially the same laws and rules as our own modern computations with Indian numerals. The procedures and the effectiveness of ancient computations were essentially no different from those of our written computations of today.

Numbers were represented according to the laws of place value. There was no symbol for an omitted rank, no "zero"; the appropriate column was simply left empty. The counters were the units which indicated the "number" of the rank. But they were still ordered in a series and grouped. In this archaic feature lay the peculiar advantage of computations performed on the counting board: the two fundamental operations of all computation, addition and subtraction, were simplified on the basis of the two original and most primitive rules of numbering, ordering and grouping. This wonderful mixture of mature and primitive qualities, of rank gradation with ordering and grouping, imparts to the reckoning board the double fascination of

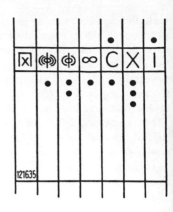

Fig. 157 Calculation on a Roman counting board. 2703 × 45.
a. Recording the two factors (left): 45 and 2703 are formed at the bottom.
b. Computation (with the rule of position):
    1) 2 × 4 = 8, in column 4 + 2 − 1 = 5 (●)
        2 × 5 = 10, in column 4 + 1 − 1 = 4 (○)
The counters for 2000 are removed from the formation of the number, so
that only those for 703 remain.
    2) 7 × 4 = 28, in column 3 + 2 − 1 = 4 (×)
        7 × 5 = 35, in column 3 + 1 − 1 = 3 (☉)
The 700 is then eliminated.
    3) 0 × 45 = 0
    4) 3 × 4 = 12, in column 1 + 2 − 1 = 2 (+)
        3 × 5 = 15, in column 1 + 1 − 1 = 1 (♦).
The 3 is then eliminated.
c. Combining all the above then gives the correct answer: 121,635 (right).

intellectual interest and of closeness to primitive human culture which
always captivates its students.

    We shall feel this even more intensely as we now trace the develop-
ment of the counting board through the history of western culture.

The Counting Board
in the Early Middle Ages

THE WEST

At this point we shall do well to glance at the general trends of history which combined the Old and the New to create the phenomenon known as "the West." For things and events will ring truer and fuller in the reverberations of history than if they are merely set, however skillfully, against the bare backdrop of the development of science.

During the last century B.C. the Roman Empire pushed its frontiers far into the west and north. Julius Caesar conquered Gaul, its Celtic inhabitants succumbed to the permanent influence of Roman culture, and France became a "Romance" land. The Rhine and the Danube became the Empire's northern boundary. At about the same time that the Roman legions of Quintilius Varus were forced to retreat behind the old frontier by the still unconquered tribes who lived beyond the Rhine, far to the east Christianity was beginning to spread from its embryonic origin. Thus both of the major historical forces with which the old Rome was to wage her final struggle emerged simultaneously at opposite ends of the Empire.

The Empire reinforced its Germanic borders with soldiers and settlers, and it persecuted the Christians. Yet as time passed, Rome felt her own inability to maintain its farthest boundaries, and divided her Empire into East and West. At about the same time the same paralysis overcame the last great persecution of the Christians (in A.D. 303, under Emperor Diocletian). But the last suffering of Christianity was also its first victory: about a dozen years later, under Constantine I, it became the official religion of the Empire.

Now, too, the Germanic tribes broke through the boundaries confining them and stormed violently into the dying Empire. Rome once more drove the Visigoths back into Gaul and Spain, but ever since then Gaul and Spain were no longer Roman. The Western Roman Empire consisted only of what is now Italy. And into this last central remnant of the former world-wide Empire thrust the Ostrogoths: with only nominal allegiance to Rome, their leader Theodoric established a Germanic state in Northern Italy, with its capital in Ravenna, which lasted for over half a century (see p. 108). When this was overthrown about A.D. 550 by the armies of Byzantium, when the Gallic areas occupied by the Visigoths fell to the Frankish kings, and when 200 years later their Spanish possessions were also lost to the Arabs, the first features of a new era began to emerge dimly out of the debris. Old Rome was no more.

Now an equilibrium had to be found between these three new western forces: the Church, the Frankish realm, and the Arabs. Slowly and with fierce struggles the kingdom of the Franks strengthened itself, and its inhabitants became settled and adhered to its rules. In 732 Charles Martel defeated the Arabs and put an end to their thrust into Europe. Now only the Church and the Empire were left to seek a *modus vivendi* with each other. In their overlapping unity each saw its complement in the other: The Church placed its inner development under the Empire's wordly protection, the Empire was strengthened by the intellectual and spiritual fertilization provided by the Church. When on Christmas Day in the year 800 the Pope crowned Charlemagne Emperor in Rome, the knot was finally tied, and the union was then still blessed because the Church was not yet aware of the troublesome heritage of the Caesars which it acquired at that time.

Now the foundations of the West were laid. Out of the dying spasms of a declining world-wide empire, out of the still vigorous but untamed strength of younger peoples, out of the intellectual stirrings of a new faith which opposed and finally overcame this strength, arose a new empire, within which the forces which made for learning and art could again live and evolve.

"And [the seed] fell on good ground, and did yield fruit that sprang up and increased; and brought forth, some thirty, and some sixty, and some an hundred" (Mark 4:8). The seeds of Christendom, which were scattered all over the whole world of ancient culture, nowhere brought forth so plentifully as on the young and fertile soil of the peoples of northern Europe. Itself a new plant which required the most painstaking care for it to germinate, it was at the same time also the bearer of the old cultural inheritance which the millennia had gathered together in the Mediterranean basin.

Let us look back over the rise and decline of this Mediterranean culture. Just as the main features of the wind's strength and direction, the temperature and precipitation can be read from the high- and low-pressure areas on a weather map, the historical table of Fig. 158 should help to make clear the extension of political power ($\twoheadrightarrow$) and the movements of culture ($\rightarrow$) in the horizontal segments representing seven successive periods in the history of the West:

1000 B.C.: Babylonia at the height of her power; Egypt and the Minoan-Mycenaean cultures on the rise. Greece, Rome, and the West still dormant in the darkness of their prehistory.

500 B.C.: Greece in her flower; Cretan culture, itself remaining without lasting historical significance, overflowed its boundaries and left to the young Greek peoples a gift which they soon brought to a magnificently flourishing state. Through their settlements in southern Italy the Greeks had already exerted an influence on old Rome. Babylonia had recently fallen to the invading Persians; Egypt was losing her importance. The West and the Arabs still remained in darkness.

300 B.C.: Rome grew politically powerful and began to expand; she had already subjugated all of Italy and was soon to conquer the eastern Mediterranean. But her culture was still greatly dependent on Greece. The brief drama of Alexander the Great's conquests was the precursor of Rome's.

Year 0: The Roman Empire embarked on her period of world rule, while Eastern culture flowed toward Rome in a powerful countercurrent, with which Christianity merged for a short time. Only northern Europe resisted Rome's might, although Caesar's legions had conquered Gaul. But shortly the Roman legions were to be expelled from the area of the Germanic people once and for all.

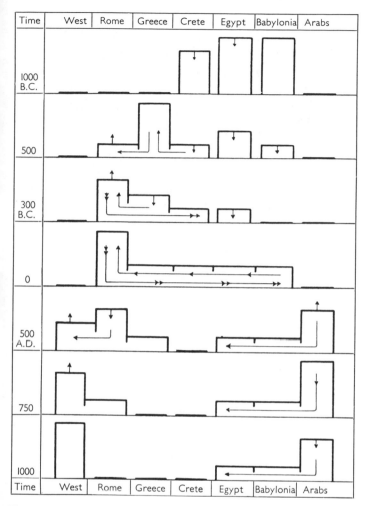

Fig. 158 The movements of Mediterranean cultures from 1000 B.C. to A.D. 1000. The flow of political power is represented by a double arrow ⇢ and that of culture by a single arrow →. The height of each rectangle symbolizes the level of the culture it represents at a given time; the arrows pointing upward indicate a rising level of culture, those pointing downward a declining level.

A.D. 500: Rome's political power at an end; with the Germanic tribes crossing her Empire's boundaries at will, Rome for some time had not been master in her own house. The inheritance which she had received now moved slowly over into the West. In the East the Arabs now were soon to grasp the reins of power as Babylonia and Egypt came under their rule and culture.

A.D. 750: Movement of political power and culture in full flow, in the same direction. The rise of western power was becoming increasingly clear (under Charlemagne); the Arabs stood at the pinnacle of their power, although the battles at Tours and Poitiers (in A.D. 732) had put limits to their incursions into Europe.

A.D. 1000: The apex of Mediterranean culture moved from the East and was now firmly fixed in the West. While the Arabs were on the wane, the West was breaking out into the full light of history.

The veins through which the old culture and knowledge flowed into the young body-of the West during the first millennium A.D. were thin and brittle. The heart of western culture was Rome, which was now Christian. But around A.D. 1000 the West began to stir and to lead its own life. Its rise was extraordinary. The West, which hitherto had required many long centuries to tame and harness its crude strength, now took flight, and, with its rich heritage of knowledge and ability passed rapidly over long stretches of slow development through which other cultures had to struggle laboriously. Now, at the beginning of the second millennium, the West was about to demolish the poor hovel of its earlier culture and erect a shining palace in its place. This was the sign of its youthful vigor, which did not succumb to the soporific poison that accompanied the rich and alien gifts from the East; it was also testimony of the greatness of western faith and the beauty of western thought.

THE MONASTIC ABACUS

It is hardly surprising that our information about the abacus declines sharply during the centuries of the decline of Rome and western Mediterranean culture. Some knowledge of mathematics was at least preserved and handed on by the late Roman scholars Boëthius (died 524) and Cassiodorus (died 570) and by the early medieval monks such as the Venerable Bede (died 735) and Alcuin Flaccus (died 804), who was a teacher in Charlemagne's palace school and the founder of a famous monastery school at Tours. But until the end of the first millennium, we do not have a single clear word anywhere about computations on the abacus. Although it may have been used or practiced in a monastery here and there, the monk Gerbert was the first who through his writings restored computations on the reckoning board to any extensive, if purely monastic and learned, use.

Gerbert

To gain some understanding of the origin and development of knowledge in the early Middle Ages, it will be useful to look briefly at the life and career of Pope Sylvester II, whose real name was Gerbert, to explain the great influence he exerted on later times.

Gerbert was born around 940, the son of poor parents in the Auvergne, and obtained his education in a monastery. About the year 967 he left the monastery in order to accompany a Count Borel of Barcelona on his journey home. His stay in the Spanish border country, where he was able to continue his studies as a result of his friendship with a Bishop Hatto, greatly enriched Gerbert's mathematical knowledge. It was probably here that he first became acquainted with the Indian numerals, for the Arabs had been in Spain since 713 (see p. 407).

Three years later Gerbert traveled to Rome in company with the Bishop and the Count, and there he was presented by the Pope to

Emperor Otto I. He declined a position at Otto's court which was offered to him; he had sufficient knowledge of mathematics, but required further training in dialectics. With the Emperor's blessing he then traveled to Rheims, where he studied this art and in return taught his own mathematics. From the evidence of his pupils we know that he took great pains with his teaching and that as a preparation for mathematics he taught computations on the abacus.

Gerbert remained at the cathedral school in Rheims for ten years; he then went to Ravenna and there gained the favor and admiration of Emperor Otto II. For a while he was an abbot in Ravenna, but after the Emperor's death (in 983), due to the unpleasantness to which he was subjected because he was a foreigner, he returned to Rheims, where he was elevated to Archbishop in 991. Theophano, mother of the young Emperor Otto III, was acquainted with Gerbert; she induced her son, who had been tutored by the well-known Bishop Bernward of Hildesheim, to write to Gerbert inviting him to come to Ravenna to instruct him in the Book of Arithmetic (perhaps that of Boëthius). Gerbert agreed. Later he traveled across the Alps with the Imperial court; at that time the Pope died, and after the Emperor's departure Gerbert remained in Rome as adviser to the new Pope. Thus he was appointed Bishop of Ravenna and a year later, in 999, himself became Pope Sylvester II:

Scandit ab R Gerbertus in R, post Papa viget R!

"From R Gerbert ascended to R, and then reached the summit as Pope at R" — Rheims — Ravenna — Rome!

In May of the year 1003 Gerbert died at the age of 63. The career of this learned man, his summons to the Imperial court, and his elevation to the Papacy inevitably promoted the spread of his teachings.

From one of Gerbert's students we know exactly what *Gerbert's*, and hence, the *early medieval counting board* looked like. Gerbert himself produced only "Rules for Computation with Numbers on the Abacus." His monastic abacus had parallel columns, 27 of them in the case of Gerbert (3 for fractions), which were sometimes closed off at the top by an "arch." This was called *arcus Pythagorei*, the "arch of Pythagoras" because in the Middle Ages this Greek was erroneously believed to be the inventor of the abacus (see Fig. 182).

"And every arch contains its name" — *et unus quisque arcus nomen suum contineat* — is written above a picture of a 15-column abacus from a monastic manuscript of the 12th century (see Fig. 159). The

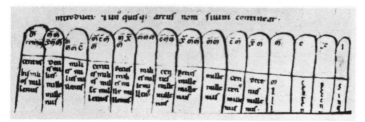

Fig. 159 Early medieval monastic abacus, from a 12th-century manuscript. Bayrische Staatsbibliothek, Munich.

"name" is the designation of the rank of the units in Roman numerals. Here the thousands were no longer ordered in series, but counted and topped by a stroke, $\overline{CM} = 100,000$. Beneath these Roman numeral designations the copyist again wrote the respective number-words, to aid in learning "numeration," the writing and reading of numbers. Thus $\overline{X}\,\overline{M}\,\overline{M}$ is *decies mille millenus*, 10 millions; M is the highest rank symbol, which occurs four times in the two highest columns (the designation is wrong in columns 11 and 13).

Although the abacus in our illustration is crudely drawn and has only 15 columns, those in other manuscripts have 27 columns and designate the whole units at the top and sometimes the half-units just below, corresponding to a quinary grouping. At the top of the sixth column (10^5) for example, is the sign $\bar{c}$ ($= 100,000$) and beneath it $\overline{L}$ ($= 50,000$). This probably indicates that computations were made here not with *apices*, but with *calculi*, for 2 *calculi* below were equivalent to 1 *calculus* above.

The Roman 5-grouping has disappeared. In its place Gerbert for the first time used, instead of the seven pebbles that once had to be lined up in a row, a single counter marked with the appropriate number symbol (for 7) (see Fig. 160). But for this purpose he did not use the Roman numerals, but

novem numero notas omnem
numerum significantes disposuit,

or, as a pupil of Gerbert's said, "he used nine symbols, with which he was able to express every number." These were the highly peculiar and alien *characteres*, which no one then knew. But we can see in them "our" Indian numerals, which thus made their first appearance in the Christian West here on Gerbert's counting board. Their forms, to be sure, were somewhat strange, at least as they appear in medieval manuscripts (see Fig. 161):

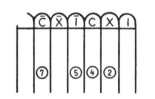

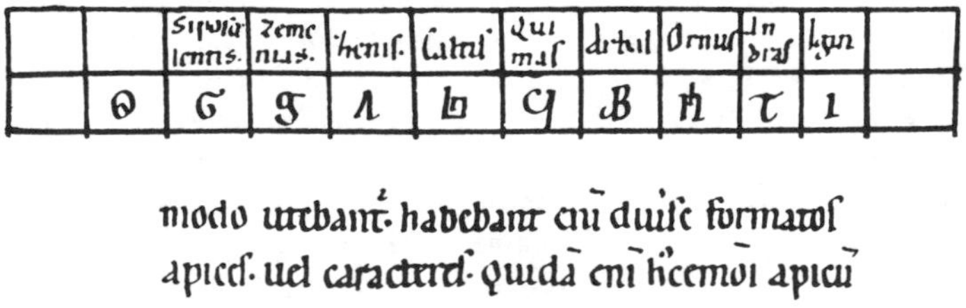

Fig. 160 Gerbert's reckoning board, with numbered *calculi*, called *apices*. The number 705,420 is shown here (in our numerals!).

		Sipoia lcnns.	Zcmc nus.	hcnif.	Catral	Qui maſ	dɪʜɪl	Ormuf	In ɔɪaſ	lɣꞁ	
	Θ	G	ꙅ	Λ	ɓ	Ꝙ	ꞵ	ʜ	Ƭ	1	

modo utrbant͗ habcbanr cñ duiſc formatoſ
apiccſ͗ ud caracttrd͗ Quida cñ hcemoi apicñ

Fig. 161 The *apices* were the first Indian numerals in the Christian West. Beneath this table we read: "... *modo utebantur.*) *habebant enim diverse formatos apices vel caracteres (quidam huiusmodi apic...*,," — "... were used in this manner.) For they had variously formed *apices* or signs...". At the top are the curious names of the individual symbols. From a 12th-century monastic manuscript. Bayrische Staatsbibliothek, Munich.

The Apices

Apices was thus the name for these counters marked with numerical symbols, and later came to be applied to the symbols themselves (from Latin *apex*, "tip of a cone," perhaps after the conical form of one of the counters). Gerbert had some 1000 such counters carved out of horn. This was his innovation: the replacement of seven *calculi* by a single counter or *apex* marked with the symbol sign for 7.

And where did he obtain these signs? Since it is known that Gerbert spent some time in Spain, where the western Arabs used the so-called *gubar* figures for their computations, it is more than likely that he became acquainted with them there. There was no zero sign — those columns on the abacus that had no digits were simply left empty (see Fig. 160). The zero began to play its important role only in written computations; since Gerbert and his disciples had no knowledge of written computations, they did not grasp the essence of the "new" system of numerals. They merely adopted the symbols, whose mysterious origin and significance were beclouded by their obscure and partly Arabic names (see Fig. 161, names above the symbols):

1 *igin*	6 *caltis*
2 *andras*	7 *zenis*
3 *ormis*	8 *temenias*
4 *arbas*	9 *celentis*
5 *quimas*	0 *sipos.*

If we compare these with the Arabic number words from 1 through 9 (see p. 115), we shall perhaps find some of their ancestors in *arba* 4, *pamanin* 8 and possibly also *tis* 9. The name of the zero-like sign (the Arabic zero) which was used as a marker in multiplication was corrupted from the Greek word *pséphos*, "pebble, counter," to *sipos* (in Fig. 161 it was mistakenly placed in the wrong column, above the sign for 9).

We have now arrived at a significant point in the cultural history of numbers. The mature and perfected system of written numerals which the Arabs already possessed, and which was in the future to be of such inestimable importance to the West, was adopted here at one point in the early Middle Ages, but blindly and without grasping its essence. The new acquisition thus remained dead, because its spirit was not understood. It was merely wasted on numbered counters, which could just as well have been marked with other signs, with Roman numerals, or perhaps the first nine letters of the alphabet, as indeed they sometimes were. For Gerbert's innovation was actually a step in the wrong direction, from the standpoint of true abacus computations. Even if the counters were properly and neatly arranged before being used, it required the substitution of *apices* 1 and 4 for *apices* 9 and 5 instead of simply collecting 14 counters in a single column.

Yet it would be wrong to see in the *apices* nothing more than a trivial innovation introduced by Gerbert. The truth is that he did adumbrate the use of the new numerals; he had heard marvelous things about the new computations which they made possible but which he, and perhaps also his informants, did not essentially understand.

The misunderstanding of the new numerals by the literate persons of the Middle Ages can be seen in a fascinating multiplication table from a manuscript of the 11th century (Fig. 162).

Fig. 162 A medieval multiplication table (beginning) with *apices* and Roman numerals, drawn up by a Pater Othlo at the St. Emeran Monastery at Ratisbon in the 11th century.
Bayrische Staatsbibliothek, Munich.

We see that the *apices* have been placed in the first column to identify the individual lines of the multiplication table, at left, namely: 1 ... 9, then 2, 3 (... 9, then 3 ... 8); at right: (3 ... 8) 9, then 4 ... 9, 5 ... 8 (9 and so on, as far as 8 ... 9 and 9).

The peculiar thing about this table is that each multiplication series, for example the 4 × ..., does not begin at the bottom with 4 × 1, then 4 × 2 and so on, but with 4 × 4; correspondingly the 5-series begins with 5 × 5, etc. The product of 4 × 1 is not found in the 4 × series (on the right side in the Figure), but in the 1 × series (on the left side) under the identification symbol 4 — in general, thus, in the series corresponding to the smaller factor:

(*Apex*) *4 Semel quatuor · quatuor s(unt) v(el) quatuor digiti s(unt)*,
"One times four makes four or four fingers (units)."

The table proceeds in this manner. One more example: 4 × 6 is found in the 4-series under the *apex* 6 (line 4 on the right side in the Fig. above):

quat(uor) seni fac(iunt) XXII duo artic(uli) et II dig(iti),
"four times six make 22, 2 tens and 2 units;"

(the two missing units are placed over the line, above *digiti* and above *articuli*; see p. 205).

If we compare these decoratively written *apices* with those in Fig. 161, we see the surprising similarity of their forms (only 3 and 4 differ, while 0 is missing from the latter). We soon discover that the sign for 4 has merely been rotated by ninety degrees, so that the long slanting stroke is now horizontal. Once we have become aware of this rotation, we can see in these digits our own 2 and 9, which as *apices* have merely been stood on their heads. And because the *apex* 9 is in this (upside-down) position, *apex* 6 is drawn with square corners to differentiate it. Now the reader is urged to compare the *apices* with the West Arabic (*gubar*) numerals in the genealogical table of the Indian numerals (Fig. 239); he will see that our numerals actually imitate their Arabic ancestors in a heavier, "more decorative" style.

The rotation, or different orientations, of the individual number symbols and *apices* may be due to the fact that the counters were customarily placed on the counting board in a particular manner in one monastery and differently in another. We shall have more to say later about the rotation of the digits (see p. 418).

The *apices*, as the misunderstood offspring of a very sensible and practical system of numerals, led a dreary life in the monastic cells of early medieval Europe, and disappeared the moment the true Indian numerals — this time, properly understood — once more found their way into the West. Indeed the "unadorned" West Arabic forms of the numerals make a significant contribution to our understanding of this portion of the history of written numbers.

But the counting board on which the Indian numerals, disguised as *apices*, played their role as strangers in our culture, did not disappear. On the contrary — as production and trade intensified and strengthened life outside the cloister, it was to flourish once again in a new form. But before we go on to speak of this, let us first examine two more methods of division on the early medieval counting board with *apices*, whose procedure was described by Gerbert and which will increase our understanding of the later "operations with erasure." At the same time we shall round out our knowledge of computations on the abacus (see pp. 302 and 317).

Iron Division and Golden Division
(*divisio ferrea and divisio aurea*)

Both of these procedures were extremely involved as performed on the counting board, and it is these which prompted the medieval writer's famous reference to "sweating abacists": *regulae quae a sudantibus abacistis vix intelleguntur*, "rules which the sweating abacists scarcely understand." It is not hard to divide 7825 by 43 in Indian numerals, but the problem becomes very difficult as the result of the everlasting combinations and changes of apices that are necessary on the counting board; thus the small computational steps are constantly interrupted.

Let us begin with "iron division." It was so named because this division, one of the procedures using supplementary numbers of which the Middle Ages was so fond, was "so extraordinarily difficult that its hardness surpasses that of iron," as one medieval manuscript puts it.

Example: 7825 divided by 43: first the divisor 43 is increased by the supplement $e = 7$ to the next complete decade, the level $s = 50$; $43 + 7 = 50$.

Procedure: divide through the rank (50) and again equate the error with the balance (7×100) (see line 1″).

$$7825 \div 43(50 - 7) = 100$$

$1' - 50 \times 100$	5000	(1)
	2825	
$1'' + 7 \times 100$	700	
	$3525 \div 50$	$= 70$
$2' - 50 \times 70$	3500	(2)
	25	
$2'' + 7 \times 70$	490	
	$515 \div 50$	$= 10$
$3' - 50 \times 10$	500	(3)
	15	
$3'' + 7 \times 10$	70	
	$85 \div 50$	$= 1$
$4' - 50 \times 1$	50	(4)
	35	
$4'' + 7 \times 1$	7	
	42	181

The quotient is thus 181, with a
remainder of 42.

Fig. 163 Example of an "iron" division. In Indian numerals.

		Th	H	T	U	
Divisor				4	3	b
Supplement					7	e
Rank				5		s
Dividend		7	8	2	5	p
	1'	2	8	2	5	
	1''		7			
		3	5	2	5	p'
	2'			2	5	
	2''		4	9		
			5	1	5	p''
	3'			1	5	
	3''			7		
				8	5	p'''
	4'			3	5	
	4''				7	
				4	2	R
Result			1	7	1	a
			(1)	(2)	(3)	
				1		
				(3)		
			1	8	1	

$$7825 \div 43 = 181, \text{ remainder } 42$$

Fig. 164 Same as example of Fig. 163. On the counting board.

As always in computations on the counting board, the rules of place value are extremely important. 600 divided by 20 gives 30 — three counters, but in which column do they belong? Since 30 (a) × 20 (b) = 600 (p), we recall the rule for the multiplication products $p = a + b - 1$ (see p. 317), in which a, b, and p are the numbers of the columns of the previous numbers. From this we can derive the place-value rule for division $a = p - b + 1$, in which $p = 3$ would be the column number of the dividend 600 and $b = 2$ that of the divisor 20. The quotient 3 goes into column $a = 3 - 2 + 1 = 2$, and thus into the tens column.

Now, in our problem 7825 $(p) \div 43(b)$. As previously we shall use the abbreviations U for units, T for tens, H for hundreds and Th for thousands; also b = divisor 43, e = supplement 7, s = rank 50 = 5T, and p = dividend 7825.

The reader would do well to perform the steps of this procedure himself on a reckoning board, using coins or buttons as counters. In Fig. 163 the same operation is carried out in our own numerals.

Step 1. We have $p = 7825$.
Division 7Th ÷ 5T = 1H; thus the quotient 1 is placed below, according to (a), in column $4 - 2 + 1 = 3$;
(1′) Remainder 2; take away 7, put down 2.
(1″) Remainder supplement $e \times a = 7U \times 1H = 7H$ is combined into the new number p′ = 3525.
The reader should practice this step until he understands it thoroughly, since he will be repeating it until the very end.

Step 2. Divide 35T ÷ T5 = 7T, place value according to (a).
(2′) Remainder 0, take away 35, 25 remains.
(2″) Remainder supplement $e \times a = 7U \times 7T = 49T$ or 4H and 9T, combined into the new number $p'' = 515$.

Step 3. Divide 5H ÷ 5T = 1T, place value according to (a).
(3′) Remainder 0, take away 5, 15 remains.
(3″) Remainder supplement $e \times a = 7U \times 1T = 7T$, combined into the new number $p'' = 85$.

Step 4. Divide 8T ÷ 5T = 1U, place value according to (a).
(4′) Remainder: take away 8, replace with 3, 35 remains.
(4″) Remainder supplement $e \times a = 7U \times 1U = 7U$, which thus gives the final remainder = 42.
With this the division ends, and the final quotient is 181 with a remainder of 42. (For more experienced reckoners the author has presented this division in his book *Rechenkniffe*, Stuttgart 1953, as a "division by levels," which has surprising advantages.)

The "golden division" uses the divisor directly, as we do; it is, in fact, our customary method of division. The divisor always appears on the same level as the highest digit of the dividend, and the rank of the quotient is ascertained by the aforementioned rule. Again, to divide 7825 by 43:

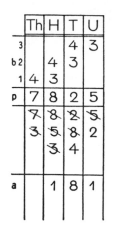

Th	H	T	U
3		4	3
b2	4	3	
1	4	3	
p	7	8	2	5

Fig. 165 Golden division:
7825 ÷ 43 = 181, remainder 42

1st Division:
> 7 ÷ 4 = 1H (*a*);
> 1 × 4 = 4Th, remainder 3; take away 7, leave 3.
> 1 × 3 = 3H, remainder 5; take away 8, leave 5.

2nd Division: the divisor moves 1 place to the right (*b*2).
> 35 ÷ 4 = 8T (*a*);
> 8 × 4 = 32H, remainder 3; take away 35, leave 3,
> 8 × 3 = 24T, remainder 8; take away 32, leave 8.

3rd Division: the divisor 43 moves into the units position (*b*3).
> 8 ÷ 4 = 1U (*a*);
> 1 × 4 = 4T, remainder 4; take away 8, leave 4,
> 1 × 3 = 3U, remainder 2; take away 5, leave 2.

Final result: quotient 181, with a remainder of 42.

This method of division is thus like our own with Indian numerals except that the divisor keeps moving to the right one place with each step, so as to make the rank determinations easier.

To the operator on the counting board the composition of a number was much clearer than it is to us in our written computations. Numbers were broken down distinctly into their component hundreds, tens, and units. The number 3000, for instance, had only one digit, namely 3, which was placed in the 1000-column. This of course called for quite a different way of dealing with numbers on the counting board from the later methods to which we are accustomed, with the Indian place-value notation in computations on paper.

One more advantage of division on the abacus: If one should choose too small a quotient, for example 48 ÷ 9 = 4, the discrepancy immediately reveals itself; and the quotient is off by only 1. In written long division the mistake could easily become 41 (instead of 4 + 1).

On the hand abacus, the smaller divisions were performed by successive subtractions, or repeatedly taking away the divisor from the dividend until the latter was used up, just as a Russian peasant would do if he wished, for example, to divide 300 rubles among 35 persons.

Strike-out Operations and Erasure Operations
Instead of taking away the finished numbers again and again, as happens on the counting board, the writer has here crossed them out in the illustration below as an introduction to the "strike-out operations" which were used in Europe when the new Indian numerals first came in and for quite a long time thereafter. Thus our computation

begins thus: and finally looks like this:

The individual divisions were performed exactly as in the previously described methods, except that here the divisor was placed beneath and the remainder was sometimes written above the dividend and in its column, where the space was free (compare the final remainder 42, where the 4 is all the way at the top and the 2 below it; the same is true of the digits 3 and 4 of the divisor 43 moving to the right). The digits were struck out as they were finished with. Because a computation of this kind looked like a "boat" or a "galley" with its sail spread, the Italians called it *divisione per batello* or *per galea* (cf. Figs. 166 and 221).

These "strike-out" or "cross-out" operations were derived from the Indian "erasure operations" on a sand-covered table, on which the finished numbers digits could be lightly rubbed out and thus disappeared completely, so that after the division was completed, only the result [42/181] was left. The reader should also try this method of computation on a slate or blackboard, to see how each partial step is carried out.

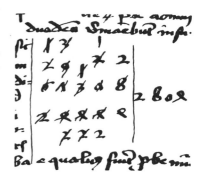

Fig. 166. A "strike-out" division, from a 15th-century manuscript, which the reader can figure out for himself (cf. Figs. 182 and 221). Bayrische Staatsbibliothek, Munich.

Now let us look back again. The old calculations on the abacus and the "new" operations with digits, crossing them out or erasing them as they are finished with — we are surprised to realize that they are essentially the same. The operations with digits merely translate the steps on the counting board into written calculations on paper or a sand tablet. They do away with the abacus and *calculi*, but only by introducing the zero into the writing of numbers. One might say, in a nutshell, that the zero overcame the abacus. But its victory, which started in the early Middle Ages, took a long time.

Thus beginning with ancient times, we have learned how to perform the four basic operations of arithmetic on the abacus or counting board: addition and subtraction on the Greek reckoning board (p. 302), multiplication on the Roman (p. 316), and division on the medieval abacus (p. 327).

In the early Middle Ages we find the counting board used only in monasteries. Its rules were learned and practiced more for the sake of scholarship than for practical usefulness. But the monks, and especially Gerbert, deserve the credit for tying the threads together again and continuing to weave the fabric which had been torn apart with the fall of Rome. In the 11th and 12th centuries the word *gerbertista* was equivalent to *abacista*, computer on the abacus.

Now, in the later Middle Ages, from the 13th century on, we shall again see the counting board used constantly in ordinary life.

The Counting Board
in the Later Middle Ages

THE EVIDENCE FOR ITS USE

La vient le duc en la chambre des Finances, bien souvent et ne se cloent nuls comptes sans luy ou sans son sceu. Luy mesmes siet au bureau à ung bout, jecte et calcule comme les autres, et n'y a différance en eulx en iciluy exercice sinon que le duc jecte en jectes d'or et les autres de jectes d'argent.

The duke himself often came into the treasury, and no computations were ever concluded without his presence or his official seal. He himself would sit at one end of the table and would move counters and calculate like the others; and there was no difference between their reckoning and his, except that the duke worked with golden counters and the others with silver ones.

We owe this interesting bit of information about the court of Charles the Bold of Burgundy to one Olivier de la Marche, who wrote his memoirs in the year 1474. Thus it seems that the administration of the Burgundian court very often was visited by the Duke, who kept track of his financial affairs with almost painful exactness and frequently joined in and calculated with his own hands.

What was this *bureau* (table), at whose "end" he would sit and *jecte* (literally, "throw")? The room in which the officials of the court worked? No, the *bureau* was merely a reckoning table, a table whose surface was marked off appropriately to serve as a counting board! And the word *jeter* becomes intelligible as meaning "to throw" the counters, which however are no longer mere pebbles, but coins, *jetons* in French. Thus the French verb *jeter*, just like *psēphízein* in Greek and *calculare* in Latin, came to mean "to compute, to reckon" (see p. 316).

Thus the abacus abandoned its monastic cell and went back out into the wide world. When did this happen? The earliest evidence for the secular use of the abacus is indirect: it consists of French counters dating from the 13th century (see Fig. 210).

The first of these counters are thought to have come from the household of Queen Blanche of Castile, the mother of Louis IX of France (St. Louis, died 1251). Shortly thereafter there was a whole series of such counters, indicating that about this time the royal treasury of France used them in its computation. Then the counting board was adopted by the nobility for the management and calculation of their own possessions, and from them it gradually came into wide use by people in general. The habit of using counters then traveled from France into all the surrounding countries.

That the first secular use of counters was in France may perhaps have been due to Gerbert himself and his disciples, who in Rheims and elsewhere raised the monastic schools to a high level of excellence in this respect. Moreover the inhabitants of Lorraine had always been reputed to be skillful computers, as an old 10th-century document testifies.

Where there are *jetons*, there are naturally also counting boards. But whereas we have counters from this period by the thousands, the few reckoning boards which are still preserved can be literally counted on the fingers of two hands. The reason is unfortunately clear. The counters, being intrinsically useless and made of base metal, outlived the period in which they were used and at the same time escaped the melting pot. But the tables were of wood. Once they were worn out as computation instruments, they could still serve as ordinary tables, until they finally ended up in the fireplace. But we do have evidence of their continuous use until the end of the 15th century, the time of the oldest preserved medieval reckoning table; from then on we have numerous illustrations and accounts of reckoning tables in printed books.

Counting boards are very frequently mentioned in household inventories and in wills. There was, for example, among the rooms in an English monastery built in 1330, a

camera quaedam cum mensa quadrata ad calculandam,
a chamber with a four-cornered table for reckoning.

In a monastic inventory of 1491 it is stated that:

In camera domini Prioris j cownter cum j covering de rede say,
In the Lord Prior's room [is] a counting table ("counter") with a covering of red silk.

In the year 1493 a dyer bequeathed to his wife his reckoning table, with the stipulation that she buy another worth 8 shillings for his daughter Anne. This sounds as if the counting table were to be a part of the daughter's dowry:

Also I bequethe to kateryn my wyff my countour standing in my parlour, with this condition, that she bye another for my doughter Anneys, to the value of viij s.

Some 250 years later, in fact, a Frenchman stated that some skill in using the counting board was one qualification of a marriageable daughter. This opinion is further supported by the will of an English alderman:

I gif to Christofer Nelson, my sone and heir, the grete cownter in the haull, also I gif to William Nelson my sone, the cownter in the parlour ... to Richard the cownter in my bedchamber.

The alderman had thus owned a whole series of counting boards which had perhaps come into his possession through inheritance, just as he was now passing them on to his own heirs. At any rate, from that time, in the 15th century, the ability to reckon on the counting board was an important part of the background of an educated man of substance, as shown by the fact that it was deemed worthy to be handed down to one's heirs. The same can be seen in the will of a Low-German testator drawn up in 1433:

Item gheve ik Hans Brunes myn kuntore, dat steyt in Lambert Eykeyes hus,
Item: I give to Hans Brunes my counting table, which stands in Lambert Eykeyes' house.

With the counting board went the counters. In an inventory of 1556 the owner says: "In my own chamber is a counting board of chestnut wood, worth 5 shillings; *item a lyttle purse with a cast of counters in hit ij d.*" The necessary equpiment for computation also included a set (properly, a "throw") of coins or counters, just as a set of various weights went with a balance. (Purses for holding counters are shown in Figs. 176 and 200.)

A remarkable custom may be seen in the following reference from the year 1556:

Among Newe Yere's Guiftes gevon to the Quenis Maiestie [Mary Tudor] were by Mr. Surton a peire of tables, thre silver boxes for compters, and fourtie compters (cf. Fig. 211).

Thus counting tables and counters were among the New Year's presents to the Queen. These particular counters were certainly made of precious metal, and the Queen would scarcely have used them for reckoning. But the gift of counting boards show that this by then common custom originated in actual computations. This usage recalls the opposite French custom, according to which the King at New Year's would present a purse full of counters to high officials of the royal household (see p. 378).

With the invention of printing by movable type (around 1450), as a result of which all scholarship and science, which had hitherto been concealed in expensive and inaccessible hand written manuscripts, now came out into the open in a broad stream, direct evidence for the existence and use of the counting board is also multiplied.

Arithmetic Books
Such text books show not only that computations were made, but also by what methods and how the counting board was set up. These

Fig. 167 First page of the oldest printed arithmetic textbook, published in Treviso in 1478: *Incommincia una practica molto bona et utiles a ciaschaduno chi vuole uxare larte dela merchadantia, chiamata vulgarmente larte de labbacho.* ("Here begins a very good and useful book of instruction for everyone who wishes to learn the mercantile art, which is popularly known as the art of the abacus.")

books of computations, which were among the first popular printed works, appeared in large numbers in all countries in the 16th century. I shall describe only one or two of the best known.

The oldest is an Italian textbook of arithmetic which appeared in Treviso in 1478 (Fig. 167). Unfortunately we now possess only one page of the oldest German arithmetic book, which was written by the Nuremberg teacher Ulrich Wagner and printed in 1482 in Bamberg. But another such book of computations, written by the same author in 1483 (although here he is not named) and likewise printed in Bamberg, is still preserved in three (and apparently only three) copies (one each in Augsburg, in Zwickau and, according to information kindly supplied by J. J. Burckhardt, in Zürich), so that we may properly speak of this as the oldest preserved German textbook of arithmetic (Figs. 168 and 169). But there is also the Bamberg

Fig. 168 The Bamberg arithmetic textbook of 1483, by the Nuremberg master Ulrich Wagner. This is the oldest fully preserved German printed book of computations. It teaches computations with numerals in 77 pages; see Figs. 152 and 223 below. The closing sentence, reproduced here, reads: *Im Jahre Christi* 1483, *am* 17. *Tag vor den Kalenden des Meyen* ... ("In Anno Domini 1483, on the 17th day before the Calends of May," i.e., April 15). Size about 11 × 11 cm.

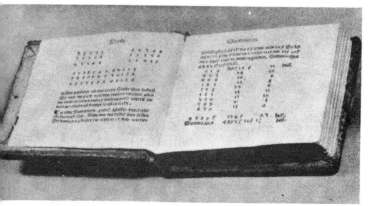

Fig. 169 The Augsburg copy of the Bamberg arithmetic book. It begins with p. 4, the first three pages being lost.

block-print book of 1470, which is thus still older and is a priceless treasure (Fig. 170).

Instead of being printed with individual movable or interchangeable letters, each page of the Bamberg block-printed book was cut entire from a block of wood. (This block book was very kindly brought to the author's attention by its finder and interpreter, K. Vogel of Munich.)

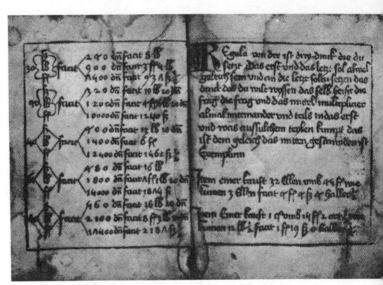

Fig. 170 The Bamberg block-printed book, dating from the second half of the 15th century, was not printed with movable type, but every page was cut whole (and very beautifully) from a single tablet of wood. 14 leaves, size: 9 × 10 cm.
Staatsbibliothek, Bamberg.

Left page: a table for the conversion of shillings *s*, pounds *lb* and gulden *f* into pennies *dn* (*facit* = "makes"):

"Shilling	240 pennies, that is 8 pounds
"30 pounds make	900 pennies, that is 3 gulden 5 pounds
gulden	7500 pennies, that is $937\frac{1}{2}$ shilling," etc.

Right page: the beginning of the Rule of Three (*Regeldetri*):

Regula von de dre ist drey dink die du setzt . . .
Item einer kauft 32 *Ellen umb* 45 *Gulden wie kummen* 3 *Ellen — facit* 4 *Gulden* 4 *Schilling* $4\frac{1}{2}$ *Haller . . .*

The rule of three is three things (magnitudes) which you place . . .
Item: a man who buys 32 Ells for 45 Gulden; how much will 3 Ells cost? They come to 4 Gulden 4 Shillings $4\frac{1}{2}$ Heller . . .

It is most indicative that these oldest textbooks of arithmetic are devoted exclusively to the new art of written computations with numerals, which, as we know from the old Augsburg municipal account books, was introduced into German offices around 1470 (see p. 288). Not until later, around 1500 it seems, did popular books appear, which only or also taught the old and familiar counting-board methods of computation "on the line," usually as an

introduction to the new written calculations "with ciphers," like
the much-read textbooks by the Oppenheim city clerk, Jacob
Köbel (1470–1533; see Figs. 171 and 172). The contents of his books
can be seen from their titles. One feature typical of the times was the
rhymed self-recommendation addressed by the author to his
potential buyers and readers.

Fig. 171 Jacob Köbel's arithmetic textbook: title page. The woodcut by
H. S. Behaim shown here was also used in other books on arithmetic. Edition
of 1544.

Among the most generally used books of arithmetic, however, were
those by Adam Riese (see p. 436). There are three of these altogether, printed in 1518, 1522, and 1550, respectively; our illustrations
show the title page of the second (in a later edition) (Fig. 175) and of
the last (Fig. 257).

One of the "seven liberal arts" discussed by the Carthusian Prior
Gregor Reisch in 1503 in his *Margarita Philosophica* ("Pearl of
Philosophy") was arithmetic (Fig. 173).

In Leipzig in 1490 there was printed an *Algorithmus linealis*, a
"Book of Computations on the Line (board)."

A very popular and widely read book in England was *The Ground
of Arts* of 1541, in which Robert Recorde dealt with arithmetic
computations among other topics. Another well known English
book was the St. Albans Book of Computations of 1537: *Introduction
for to Lerne to Recken with the Pen or with the Counters.*

Fig. 172 Title page from another of Köbel's textbooks on arithmetic, with one of the then customary author's recommendations to the (potential) reader.

As early as the end of the 15th century a French book of arithmetic appeared in Lyons, by an unknown author: this was the *Liure des Getz* (that is, *Livre des jets* — the "Book of Counters"). From this book we have taken our illustration (see Fig. 200) of the bag of counters, about which we shall have more to say later. The *Ars supputandi tam per calculos quam per notas arithmeticas* by the French Doctor Clichtovaeus (died 1543) shows in its title, like its German cousins, that it offers instruction in the two known methods of calculation: "The Art of Computation both with Counters and with Numerals."

An "Arithmetica" was also written in Latin in 1513 by Joannes Martinus Silicius ("Gravel-stone"), a Spaniard who was Philip II's tutor and later became Archbishop of Toledo and ultimately a cardinal. The *Tratado de Mathematicas* by one Perez de Moya, which appeared in Alcala in 1573, was a Spanish "Adam Riese."

Many such textbooks of computations could be found in other countries as well, such as in the Netherlands (by Gemma Frisius), in Denmark (by Niels Mickelson), and elsewhere, all of them written with the express purpose of making a knowledge of the counting board available to everyone (see Fig. 193).

Among the unequivocal preserved documents for the use of the counting board are the many medieval counters that show reckoning tables, some even with an image of a calculator at work (Fig. 174).

Fig. 173 Title page of the *Margarita Philosophica* by Gregor Reisch, 1503. Surrounding the three-headed figure in the circle are the seven liberal arts; the personification of arithmetic is seated in the middle, with a counting board.

Fig. 174 Counters showing reckoning tables. *Top row:* Nuremberg "school counters," the one at right showing a "counting cloth." *Bottom row:* left, the winged lion of St. Mark (Venice), which instead of the Bible perhaps holds a counting board in its claws. It is not quite clear whether this is a counter or some other token. *Center:* the reverse of the middle counter at the top, with the alphabet. *At right:* a German counter of 1691; the lines on the counting board are marked with Roman numerals, showing the number 1051.
Diameter of counter in center of both rows: 27 mm.
Collection A. König, Frankfurt/Main.

These are the medieval western descendants of the Greek treasurer on the Darius Vase and the calculator on the Etruscan cameo (see Figs. 132–134).

THE NEW RECKONING BOARD

Columns and Lines Perpendicular to Each Other
Some of the books we have just mentioned very definitely speak of *calculi* and *jetons*, so that their relationship to the reckoning board is clear without further consideration. But the German arithmeticians Adam Riese and Jacob Köbel also speak of *die Rechnung auff der Linihen vn Federn* ("reckoning on lines and [with] quills") (Fig. 175). What did this mean? "Reckoning with the quill of course referred to written computations using the Indian numerals, which were now becoming generally familiar. But what about "reckoning on the lines"? Our illustration offers a clue. In the drawing on the title page, a man is seated in the background before a pattern of horizontal and vertical lines with counters placed on and between them. The contrast between the two methods of calculation is even clearer in an illustration (see Fig. 182) from Reisch's *Margarita Philosophica*. But the reader is mistaken if he thinks this is the old early medieval monastery abacus with its parallel vertical columns; a review of the last half dozen figures will show quite clearly that since the time of Gerbert and his disciples the old counting board with vertical columns had been rotated through a quarter of a turn into one with horizontal lines or strokes (see Fig. 176). Just when this happened, we do not know; perhaps in the 13th century, if we are correct in assuming that the change took place at about the time when metal counters came into use. The board with horizontal lines had definitely been long in use by the 16th century, for all the books and pictures of that date show the columns horizontal and contain not a word about the "vertical" monastic abacus.

Fig. 175 Title page of Adam Riese's second book of computations, edition of 1529, alluding to the competition between the old and the new, between reckoning "on the lines" and "with the quill."

Fig. 176 Counting table with specified coin rows (called a coin-board or number-table). This woodcut probably from Strasbourg.

Why the change in orientation? Probably because it was more comfortable to "read" the counters horizontally. Long horizontal rows are easier to "grasp" than long vertical columns. Because of the restoration and great expansion of trade about the 12th century (as a result of the Crusades and the Hanseatic League, among other things), there was an enormously increased demand for money changing. This meant that ordinary, everyday computation, which had been reduced to a small trickle at the time of the monastic abacus, which itself was more a subject for scholarly study than an object of daily use, now began to flow again in a mighty stream. Merchants, shopkeepers, and officials may well have adopted the monastic abacus or been guided by it; at any rate, they improved it and made it handier, among other things by turning it to the horizontal position.

Apart from its greater convenience, two other possible reasons for the new horizontal arrangement of the counting board may be mentioned. In the 11th century Guido of Arezzo invented the horizontal lines which are still used for the writing of musical notes and which are fully analogous with the parallel lines of the counting board. Guido's lines also made possible an easy method for the "gradation" of tones: the writing of musical notes is, in fact, a form of place-value notation. To be sure, because of their varying lengths, musical notes are not used exactly like *calculi* on a counting board but rather are "grouped" in a certain manner, like the *apices*.

The horizontal lines of the new reckoning board may even have been suggested by the monastic abacus itself. The latter used *apices*: A number such as 705,420 would thus have been laid out in a horizontal line across the vertical columns, and the next number again in a horizontal row beneath it, and so on (as in Fig. 160). In this manner the abacus was "divided up" into horizontal bars. Such lines of "rows" actually do occur here and there in old manuscripts.

But merchants and money changers again took up the old, un-differentiated *calculi*, and thus did not adopt Gerbert's innovation of *apices* with different values. Thus they would represent the number 4 no longer with one counter bearing the numeral 4 but again by putting down 4 counters of equal value. In addition to counters made from the usual materials such as bone, wood, and metal, we also find, from the 13th century on, stamped or embossed counters which, however, had no monetary value. This new custom shows more strikingly than anything else how important the counting board had become in daily life.

The horizontal reckoning table existed in two forms: the first had lines of unspecified value, upon and between which the counters were placed (the "line-board"); but the other had specified "coin rows" each for a specific denomination, within which the counters lay as in the columns of the old monastic abacus (the coin-board or number-table; see Fig. 176).

The latter form of counting board was used only for calculations involving sums of money, primarily the addition (and subtraction) of receipts and outlays (taxes, expenditures) and the conversion of one denomination into others (such as heller or farthings into pennies, shillings, pounds, or gulden). This was a true number-table. The line-board, on the other hand, was intended for true computations with abstract, unnamed numbers, and could also be used for multiplication and division. Money calculations on the line board could be made only in terms of a single denomination — only in gulden, for example. The reason for this will be given presently.

Are some of the old reckoning tables still in existence? The few that are still preserved, which the author was able to locate, are "official"

Fig. 177 The Basel reckoning tables of the "Three Masters."
Each of the tables has three different counting areas, with coin rows, which are divided into two areas on the later table, shown below. Lines and coinage symbols are inlaid in different-colored wood (*d* for pence, *s* for shillings, *lb* and *lib* for pounds and X, C, and M pounds). The counters were kept in drawers. The lower reckoning table has a raised edge to keep the counters from accidentally falling off. The tables are 130 × 98 cm and 209 × 85 cm in size, respectively, with the marked areas 75 × 45 cm on the upper and 62 × 43 on the lower table.
Historical Museum, Basel.

coin-boards from the municipal offices of Dinkelsbühl and Basel and from the Cathedral of Strasbourg.

Each of the two reckoning tables from Basel has three counting areas on its surface, with lines and letters inlaid in darker or lighter wood. The more ornate one is also the more recent; it has two counting tables side by side and only one at right angles to it. It also has raised edges, to prevent the counters from falling off. Both Basel tables have drawers in which to keep the counters. Although the older of the two has only three simple plain counting tables, the tables on the newer one are divided by a vertical line down the middle into two fields.

At such tables worked the *Dreierherren* ("Triple Masters") of Basel, who until the uprising of 1798 were charged with the administration of the city treasury and supervised the city's taxes and coinage. Their deputies were placed in charge of the various branches of production and trade; for example, the *Weinherren* ("Wine Masters") saw to the proper collection of the taxes from the wine trade. Their offices went back to the 14th century. Our Basel reckoning tables (Fig. 177) date from the 16th and 17th centuries.

The letters standing for the various denominations of coins are, from bottom to top, *d* for pennies (*denarius*), *s* with a flourish for shilling (*solidus*), and *lb* or *lib* for pound (*libra*) followed by X, C, and M for 10, 100, and 1000 pounds, respectively. The three Masters all carried out the same computations simultaneously, to guard against both errors and fraud.

The three Dinkelsbühl reckoning tables from the 16th century, the only German ones known to the author, were also coin tables. One of these is illustrated for the first time here (Fig. 178). Each of the three has two counting areas with two different coinage areas deeply incised into its surface. Unlike the Basel tables, these have two different scales of coinage, one at left for calculations with pounds: *hl* heller, *d'* pennies, *X* 10 pennies, *lb* pounds, and then 10, 100, 1000, and 10,000 pounds; and another at the right for calculations involving gulden, which at the end of the Middle Ages were carried on side by side with pound-reckoning: *d* for Halbort (= $\frac{1}{8}$ gulden, from the Latin *d-imidium*, "half"), *O* for Ort, *d* for half-gulden, *f* for gulden, and lines for 10, 100, 1000, and 10,000 gulden. We have already seen amounts stated simultaneously in both gulden and pounds in the Augsburg city account books; the Dinkelsbühl reckoning tables are quite consistent with these.

Currency Standards

In Franconia during the 14th century people began to use the "small" pound of 30 pennies, which also appeared in the Dinkelsbühl accounts and computations. According to the Carolingian currency scale, 1 pound was equal to 240 pennies. Theoretically a gulden should also have had the same value, but it was worth 252 pennies in Nuremberg and 168 in Würzburg and Augsburg. One Ort was equal to $\frac{1}{4}$ gulden (see p. 148); 1 heller = $\frac{1}{2}$ penny.

Pounds and shillings were computational units, not actual monetary values; in other words, there were no "pound" or "shilling"

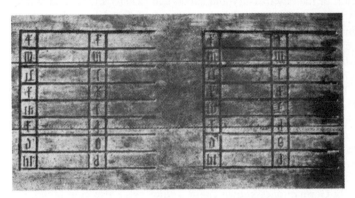

Fig. 178 One of the three known German reckoning tables in existence, from Dinkelsbühl (now in the Museum there). Both counting areas are carved into the surface of the table; each has two areas for computations, with pounds (*lb*) at left and for gulden (*f*) at the right (see the fourth horizontal row from the bottom).
Photographed by P. Hammerich.

coins, but both were numerical units into which pennies were grouped (whence the numerical pounds and shillings; see p. 289).

Another very handsome reckoning table dating from the end of the 16th century, on which the comptroller of the Foundation of the Virgin Mary in Strasbourg calculated the rent income from payments for the institution's houses and from the sale of wood from the forests owned by the chapter, and also the payments made to the cathedral builders' guild, is still *in situ* in its old location, the present Musée de l'Oeuvre Notre Dame (Fig. 179). On the walnut table top with a raised edge there are two reckoning boards inlaid in ivory; in contrast to the Basel and Dinkelsbühl boards these do not have horizontal bands but are "line-boards," as the symbols on the left side show. Each of the two line-boards has four vertical columns, for pounds, shillings, pennies and heller (four *banckirs*, according to Fig. 150).

The same division into pound and gulden units and their fractions as on the Dinkelsbühl tables, but with three separate areas for computation (for pound and gulden) may be seen in two Bavarian reckoning cloths (Fig. 180). These could be easily carried about by

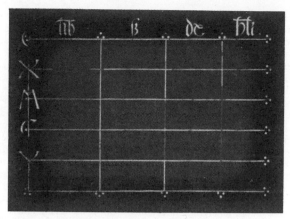

Fig. 179 A "line-board" with four columns for different coins, on the Strasbourg reckoning table, which has two such boards. The manager of the cathedral chapter worked out his accounts on this. Date around 1600. Table size 153 × 98 cm, line-board 65 × 47 cm.
Musée de l'Oeuvre Notre Dame, Strasbourg.

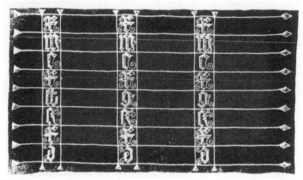

Fig. 180 Bavarian reckoning cloth with three areas for computations with pound and gulden, of the type carried about by inspectors who checked on municipal computations throughout the province. Pale yellow lines and yellow symbols are embroidered on a green cloth. Size 71 × 41 cm.
National Museum, Munich.

the officials whose duty it was to check the calculations made by the mayors of the towns and other administrative centers throughout the province. These reckoning cloths were probably used around 1700. On these the horizontal spaces were, from bottom to top, for 1 and 10 pennies (*d*), for shillings (*s* with a flourish) of 30 pennies each, for pounds (*lb*) of 8 shillings or gulden (*g*) of 7 shillings each, and then for 10, 100, 1000, and 10,000 of these last units.

Along with these counting cloths, we have a contemporary document which very clearly reveals their exact use:

Notice: This cloth was formerly used by the provincial inspectors of finances in their annual inspection of computations in which everything was calculated in black counters with gulden, pounds, shillings, and pennies, as follows: the Lord Mayor would read the amounts and call out the numbers, for example, 10 gulden, 5 shillings and 2 pennies. The first or the second Cavalier had the cloth and a dish of silver counters before him. As the mayor called out the amounts, the Cavalier would place a silver counter in the space marked 10 gulden, 5 counters in the space for the shillings and 2 counters in the space for the pennies. Then they would proceed in this manner. But after there were 10 *tantes* [counters; see p. 385] in the 10-gulden space, he would take these away again and put down one in the space where the hundreds were marked. But if there were 7 *tantes* in the shilling space, these were also removed and converted with one counter into 1 gulden, which could always be multiplied by the number ten, hundred, or thousand. The same was done with the pennies, so that as often as there were 30 counters together in the pennies space a shilling was made, and from the latter likewise a gulden. The second Cavalier had the certificates or proofs at hand for his examination. The prelate, however, read the second item or, if it was a simple matter as from the Lowland, the finished computation. As soon as the space in which a sum was placed was complete, the Cavalier would so state, according to the information on the reckoning cloth and the counters lying on its spaces. But it was the mayor's duty to see that this was consistent with the written accounts.

This notice describing the way the computations were actually made with counters is of immeasurable value. It provides indisputable evidence against the view, which some readers may still hold, that computations in the old days were made with Roman numerals.

Among the few items of this kind still preserved are also four reckoning tables in Switzerland (in Zürich, in Thun and in the castles of Chillon and d'Oex, according to information very kindly provided by Colin Martin). Since the necessary lines were for the most part merely drawn temporarily on the table surface (Adam Riese, for example, says: "Make some lines for yourself"), the scarcity of old reckoning tables is easy to understand. Those which have been preserved, and probably also those mentioned in bequests (see p. 333), were of the kind with *horizontal spaces* marked with various coin denominations. The only exception is the Strasbourg reckoning table, which is a line-board with *columns* for the various coins. In our own ledgers and bank books, which are $ and ¢ in the United States and £, *s*, and *d* in England, we very likely have the descendants of the old horizontal spaces and vertical columns marked with various coins.

THE NAMES OF THE COUNTING BOARDS
Such names are extremely revealing of the development of such boards; to some extent these still live on, with changed meanings.

We have already seen that the *bureau* of the Dukes of Burgundy was a reckoning table (see p. 332). But this was not the original meaning of this word: the French *bure* is a "coarse wool cloth"; cf. the Italian *burato* < Latin *burra*, "tuft" or "flock" of wool. The plural form of this word, *burrae*, was used by the Romans to mean "buffoonery," whence the later diminutive form *burrula* and finally the Italian word *burlesco*, which mean the same thing. The coarse woolen cloth was then probably used as a covering for the reckoning table, like the "red silk" on the English equivalent (see p. 333), and as a table covering then came to be called *bureau* (which is actually a diminutive of *bure* by way of *burel*). A curious change of meanings: the name of the material from which the cloth was made was first transferred to the cloth cover on the table, then to the reckoning table itself, from that to the room in which it stood, then to the counting room or office and finally to the staff of clerks and officials themselves, the "bureau." A Middle-Eastern counterpart is the Turkish-Arabic word *divan*, whose reference was gradually expanded from "register" or "collection" (as in Goethe's West-East Divan) to the official's room, particularly the customs office (whence the Italian word *dogana* and the French *douane*), then to the furniture in it (our own "divan" or "sofa"), and finally to the entire Ottoman government administration (the *Divan*).

The word *Exchequer*, referring to the British royal treasury, has a similar origin. This name goes back to the time of Henry II (12th century), in whose reign half of France was subject to the English kings. Thus the English state treasury was called *échiquier* even in French. This was derived from the Latin word *scaccarium*, the "checker"-board, which was in turn a Latin translation of the Persian word *šah*, "king," because the "king" is the chief piece of the Eastern game of chess, which is played on a "chequered" board. From this we have the modern English words *chess* and *check* and the French *échec*.

But what does the checkerboard have to do with the royal treasury of England? The officials who were charged with the administration of the state's receipts and expenditures used to gather around a table which was covered with a woolen cloth marked by perpendicular lines into a checkered pattern. The *Dialogus de Scaccario* of 1186, which we have already met in connection with the tally sticks (p. 238), says:

Superponitur scaccario pannus non quilibet, sed niger virgis distinctus ...
In spatiis autem calculi fiunt iuxta ordines ...,

On the reckoning table was placed not just any cloth, but a black cloth marked with parallel lines at equal distances.... In the spaces between the lines the counters were placed according to the rules .... .

the order of the columns from right to left was pence, shillings, £1, £20 (a *score* of £), £100, and £1000, just as notches were carved in the tallies for these very same amounts. This checkered cloth, which is represented in an English drawing from the 14th century, was the origin of the name (*Ex*)*chequer* (see Fig. 181).

A long history of cultural development is embodied in these changes in the meaning of a single word: The *chequer* counting board was used in conjunction with the tally sticks on which the amounts of the Exchequer were recorded. Thus the act of comparing the board with the tally sticks was called *to check*, whence the *check* in the sense of a bank draft finally acquired its name (see p. 233 above).

On the English checkered counting cloth, spaces were provided for the four English coin denominations

1 pound = 20 shillings, 1 shilling = 12 pence, 1 penny = 4 farthings.

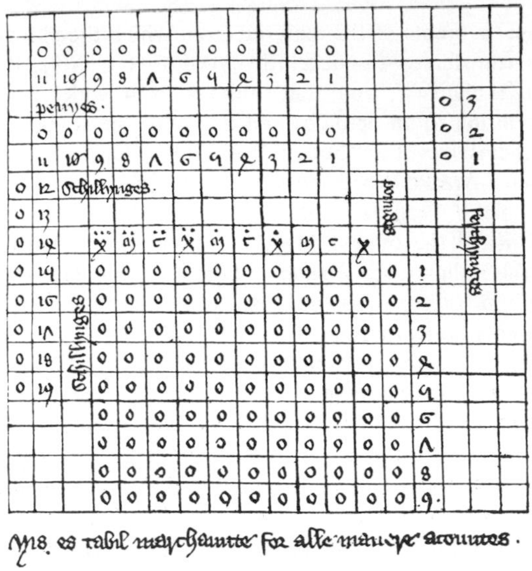

Fig. 181 A checkered counting board, from a drawing in a 14th-century English manuscript, with the lines drawn in red ink and the numbers and letters in black. The circles indicate counters placed on the board. The checker-like symbols where there are no counters refer to a checkered cloth. Beneath the drawing are written the words: *es tabil marchantte for alle manere accountes.*

In the drawing (Fig. 181) these spaces are marked by Indian numerals: at the top are 11 spaces for pence, below these are 19 spaces for shillings, and at the far right are 3 spaces for farthings. For pounds there are 9 rows, one below another, which are also arranged in columns marked (I), X, C, M, $\dot{X}$, ..., $\ddot{X}$, etc. To add 7 + 8 pence, the merchant using this board would proceed as follows:

He would place 7 counters in spaces 1 to 7 of the pennies row, followed by 8 more counters. But as soon as he had put down 4 of the 8, which with the 7 previous counters made 11, the row of pence was filled. Thus he would lay down the next, or 5th, counter in the first shilling space and take away all those in the penny spaces, for 12 pence make 1 shilling. Then he had only to put down the last 3 counters, for 3 pence. Thus 7 pence + 8 pence = 1 shilling 3 pence. The other coin denominations were handled correspondingly. The pounds were treated thus: up to 9 pounds counters were placed in the last or unit column at the extreme right, the 10th pound went into the first space in the X-column, and the 9 counters in the units column were then removed.

Since the English coinage system was built not upon the number 10, but upon the inconvenient numbers 4, 12, and 20, the money counter here would guard against making mistakes by marking the spaces for the individual denominations (in Indian numerals). And since small squares were the most suitable shapes for the individual spaces, the result was the *chequered cloth* after which the English Royal Exchequer was named.

A third example: instead of red silk, the cloth covering the reckoning tables used in the counting rooms of the English royal household was green (see p. 333). Perhaps it was even a green counting cloth, similar to the Bavarian cloth we have seen. Thus this office was known informally as the *Court of Green Cloth*. This recalls the German expression referring to the "green table," which probably likewise is connected with the reckoning table; the "Triple Masters" of Basel also determined the assessments levied on the inhabitants of the city on their "table" — that is, from their calculations.

Gettours pour servir ès comptouers des clercs ("Counters for the *comptouers* of the officials") were mentioned on one occasion in France. These *comptouers* were, of course, counting boards (cf. p. 333). If we replace this old word by the later form *comptoir*, we shall see in it its German derivative *Kontor*, "office." The *Kontor*, "counting house" or "counting-booth" (from the Latin *computatorium*) was thus a room or booth with a counting table, and originally the reckoning table itself. The English word *counter* retained this last meaning, but was also used for the small round *counters* placed on the board.

Bureau, Kontor, counter — who would have guessed that the medieval reckoning table lies behind these words?

The old German name for the counting board was *Rechenbank* (whence the word *Bankier*, "banker"); in Bavaria it was also called *Raitbrett*, since *raiten* is *rechnen* ("to reckon") in the Bavarian dialect (cf. p. 383 and *Raitholz* on p. 224; the word is related to *Rate*, "rate, installment," p. 130). The old word *abacus* continued to live on in computation books written in Latin. Quite often the reckoning board was not even mentioned in German, and the text would merely speak of reckoning "on the lines."

RECKONING "ON THE LINES"

The Line-board

Four parallel horizontal lines were drawn on a board or table, with a vertical line down the middle dividing them into two columns or *bankire* (see Figs. 182–184). The topmost of the four horizontal lines was marked with an X at the center. Unlike on the reckoning board with spaces for denominations of coins, the counters were here placed directly upon the lines themselves. These were unspecified, for the first time in the history of the counting board, for they imparted to the counters only an abstract decimal place value, independently of any specified system of coinage or weights: a counter on the bottom line had the value of 1, on the second line 10, on the third line 100, and on the top line, which was marked with an X, 1000. Now let us see what the arithmetician Johann Albrecht of Wittenberg had to say in his *Rechenbüchlein auff der linien* ("Little Book of Computations on the Lines") of 1534:

So that the lines may be known / they are to be marked in the following manner: the line that is called the first means one / the next line above it ten / the third hundred / and the fourth thousand / mark this last line with a small cross / and count off on the same line again [as on the first] one / on the second above ten / on the third hundred / and on the fourth thousand. But mark this one with a small cross. But starting with the first small cross / for each line you must say thousand. When there are one

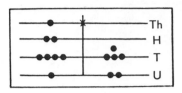

Fig. 183 Basic form of the line-board
(after Fig. 182).

thousand / ten thousand / a hundred thousand / a thousand times a
thousand / and as many small crosses as there are, so many thousands
must you always say [cf. p. 143]. You must also know / that any space
means five times as many as the line next beneath it [to which it belongs],
and the space under the first line / means a half. . . .

In the left column of the line-board in Fig. 183 the number 1241 is
displayed; the space between the lines, called the *spatium*, groups the
units of the line beneath it into groups of 5; thus the number placed

Fig. 182 Calculators using the counting board and written numerals. On the
left hand of the female figure personifying Arithmetic sits Pythagoras at a line-
board on which the numbers 1241 and 82 have been formed; at the right hand
of Arithmetic Boëthius faces his computations in Indian numerals. The Middle
Ages believed erroneously that these two men were the inventors of the respective forms of computation. (On the garment of Arithmetic are the two
geometric progressions 1 — 2 — 4 — 8 and 1 — 3 — 9 — 27). From the *Margarita Philosophica* of Gregor Reisch, 1503 (Fig. 173).

in the right side of the line-board in the same illustration is 82. A
counter placed beneath the bottom line means $\frac{1}{2}$. In a few instances
the X marking the thousand line is in the wrong place (as in Fig. 175).

In this very simple manner the more elegant representation of
numbers with the aid of the quinary 5-grouping was reintroduced
after its abandonment by the early medieval abacus with its numbered
apices.

Whereas the ancient Roman counting board could form several
numbers only one above another, the late medieval board, with its
columns, or *Bankire*, could form them next to each other. Now the
operator really had his numbers "at hand." This was the advantage
of the columns and also the basic reason for drawing the lines

horizontally instead of vertically, as before. But now the similarity of these numbers with the Indian numerals was again abandoned, and in the view of those who reproach medieval Europe for failing to devise a place-value notation, this was a step backward.

The reckoning boards shown in the preceding illustrations have generally had two columns side by side. But in Fig. 184, taken from the title page of a book of computations, the reckoning table has only one working area, whereas the counting board hanging on the wall has four columns. This suggests that boards were divided into numbers of vertical columns according to the particular problem to be solved, and that this could always be done easily by merely drawing a line or lines with chalk.

Fig. 184 A reckoning table, and on the wall a reckoning board with four columns. From a 16th-century book of computations.

Now let us try our hand at calculation on the new reckoning board. Many textbooks of the time list the following branches, or operations, of arithmetic: *numeration* (the placing of the numbers on the counting board), *addition, subtraction, duplation* and *mediation* (that is, the doubling and halving of numbers), *multiplication*, and *division*. Furthermore, along with each of these operations there was *elevation* (in French *déjeter*), or "purification," the grouping of units into smaller numbers of higher units; and with subtraction there was also *resolution*, the "de-grouping" of higher units into lower ones. Both procedures together constituted the operation of *reduction.*

Addition. What is the sum of 3507 and 7249?

1. Numeration: both of these numbers are placed with counters on the board (see Fig. 185).

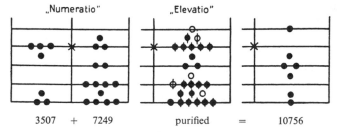

„Numeratio"   „Elevatio"

3507 + 7249   purified   =   10756

Fig. 185 The sum 3507 + 7249 = 10756 as worked out "on the lines." The counters with a vertical stroke through them were elevated to a ○.

2. Elevation:

Schau wo zwen pfennig im spatio
Ligen hebs auff und leg ayn do
Auff d'ncegsten lyni vbersich |
Desgleich so fünf liegen halt dich
Auff eyner lyni merk vnd guck |
Ins spatium drüber ayn ruck.

See where two counters lie in the space;
Pick them up and lay one down
On the next line above /
Likewise if five counters lie
On one line, take note and look /
And put one in the space above.

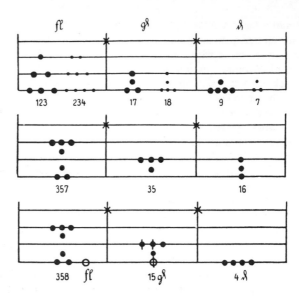

Fig. 186 A line-board with columns for various coins:
to 123 gulden 17 groschen 9 pennies are added
to 234 gulden 18 groschen 7 pennies, each gulden amounting to 21 groschen
and each groschen to 12 pennies. After Adam Riese (Fig. 187).

The reader is urged to work out this problem for himself on a line-board; he will be astonished at the ease with which the operation can be grasped visually, without actually performing any true computations at all.

If several numbers are to be added together, one begins by combining the first two, then adds the third to the sum of the first two, and so on, at each step adding one more number to the running subtotal as on an adding machine. Then if a merchant had to add various amounts of money on a line-table, for example 123 gulden 17 groschen 9 pennies to 234 gulden 18 groschen 7 pfennige (see Fig. 186), Adam Riese described the procedure to be followed:

Addieren heisst zusammen thun | leret wie man viel vnd mancherley zahlen von gülden groschen pfennigen vnd hellern in eine Summa bringen sol. Thue jm also: mache für dich linien | die theil in soviel feld | als Müntz vorhanden | lege die f besonder | g allein | vnd d auch jeglich allein | hl vnd d mach zu g | was kömpt leg zu den g. Als denn mach die g zu f | leg es zu den andern f | nach art jeglichen Landes.

Addition, which means putting together, shows how one should bring together many different numbers of gulden, groschen, pennies, and heller into a single sum. Do it thus: Make lines to divide the board into as many areas as there are different kinds of coins. Put down the gulden separately, the groschen alone, and the pennies likewise alone, and convert the heller and pennies into groschen and add the result to the groschen. Then convert all the groschen into gulden and add these to the other gulden, as is done in every land.

Adam Riese computed an example with several numbers to be added together and then represented the grand total of 1344 gulden 19 groschen 3 pennies on the line-board (see Fig. 187).

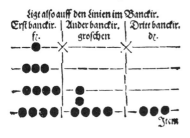

Fig. 187 A line-board for money computations, as illustrated by Adam Riese.

Thus the counting bank was divided into designated columns, each coin denomination was placed in its column and added together. Then the pennies were converted into groschen and the groschen grouped into gulden. This procedure was followed because the coinage scale was not based on gradations of ten. In Adam Riese's example one gulden was worth 21 groschen and each groschen had 12 pennies. In this clever manner calculations involving various denominations of money could be made as easily on the abstract line-board as on a counting board with named spaces for the different coins. The Strasbourg table is an example of this kind.

The coinage symbols *f*, *g*, and *d* and those on the reckoning tables, *s* and *lb*, call for a brief discussion of their nomenclature.

Names of Coins and Their Abbreviations and Symbols

This will reveal the reasons for some of the peculiarities of usage which still live on today.

In ancient Rome, when a person acquired something through legal purchase, the expression *per aes et libra*, "through bronze and balance," was placed next to the amount of the transaction. Early money systems had no minted and struck coins with a face value stamped on them which was generally greater than the value of the metal they contained; instead they merely weighed out the appropriate amounts of the pure metal (copper, silver, bronze, etc.).

The old Roman bronze coinage system was built upon the *as* as the basic unit, which contained 288 scruples (see p. 158). This was originally called the *as liberalis*, the "weighed one," and also the *solidum*, "the whole" (< *solidus*, "pure, unmixed, solid"), or the *libra*, which was properly the "balance" and then came to mean "pound." The meaning of "pound" came from the Roman expression *libra pondo*, *pondo* meaning "according to weight"; thus *libra pondo* was literally a "scale according to weight" and thus meant "properly weighed."

Then the Germanic peoples some time in the first centuries A.D. borrowed from the Romans not the word *libra* but the participle *pondo*, "pound," although as a symbol for it they used the abbreviation of *libra*, which was *lib* or *lb*, as they appear on reckoning tables and in the Augsburg account books (see Figs. 120–124 and 178–180). Then through Charlemagne's currency scale it finally became fixed in monetary terminology, where, for example, it appears today in the English *Pound* (sterling) and its abbreviation £. From *Libra*, on the other hand, comes the French word *livre* and the Italian *lira*; its abbreviation lives on in the German abbreviation for the pound (avoirdupois) *℔*, which is a cursive *lb* crossed by a flourish like the English £. From the old custom of weighing out money (1 pound silver or sterling) and its consequent division (into 240 pennies), the pound acquired a third meaning. It may be a pound by weight (as a pound of flour), a numerical pound (1 pound of pennies = 240 single pennies), or a unit of money (1 English pound sterling, as either a banknote or a coin, then generally a *sovereign*). In Germany the *Pfund*, like the *pound* in America, is now only a unit of weight.

Money has always ruled the world, from the very beginning and certainly from the time when it was weighed out in a scale as raw metal, like any other commodity. There are still many current phrases and turns of speech that go back to this custom of weighing out metal for use as money (Latin *pendere*, "to hang" on the steelyard, and thus "to weigh"). Hence the German word *Pensum* ["task, quota"] is actually the "weighed" work that has been contracted for; in the French and English word *pension* it refers to a "weighed out" or fixed money payment. The Roman soldier called his pay a *stipendium* (< Latin *stips*, a "sum of money"), which in modern German and English has come to mean an amount of money given as a form of scholarship. A *Pensionat* was originally a boarding house in which the food was weighed out and divided among the boarders.

A surprising class of words has descended from the Latin *ex-pendere*, "to weigh out, give out, pay out." In Medieval Latin the prefix was contracted, and from the resulting *spendere* came, among others, the Modern German words *spenden* (to spend, distribute) and *spendieren* (to treat someone to something; to squander, waste), both originally having the meaning of "giving out money." And there is still another etymology here. In Holland a *spind*, which in German is a soldier's footlocker, is specifically a cupboard for food, from the Medieval Latin *spenda*, "pantry" or "storage room for eating utensils and food." But why just this kind of storage container? Because monasteries once used to "weigh out" or "pay out" (*spendere*) their alms to the poor, but the donations they "weighed out" (*spesa*) were not sums of money but what people spend money to buy — food. In this way *spesa* became the German *Speise* (food, provisions) and also *Spesen* (charges, costs, expenses), whose original meaning was "expenses for eating." The word itself came into the German language, along with other terms connected with trade and banking, around 1500 (see p. 428).

The money which was weighed out also lives on in the *peso*, the old Spanish coin which became the present-day Mexican and South American basic monetary unit, while the mother country is now content with its diminutive, the *peseta*. Although the "weight" can still be discerned here, it has become well hidden in the Latin *pensare*, French *penser*, "to think"; but it will come out in the open again if we think of the synonym "to weigh, to consider."

From the Biblical parable in the Gospel according to St. Luke (19:13ff.) about the servant who in the absence of his master multiplied the "pound" (= "money") with which he had been entrusted into 10 pounds, and the other servant who buried his pound securely but profitlessly in a napkin, comes the German expression *mit seinem Pfund wuchern* (to lend out one's pound at usurious interest) and *sein Pfund begraben* (to bury one's pound (= money)).

The same idea is also extended to the intellectual and spiritual realm: man is supposed to use and develop his God-given abilities. Exactly the same expansion of meaning may be seen in the ancient Greek unit of weight and money, the *talent*, which has become our word "talent" in the sense of a person's mental gifts or faculties.

The word *Pfund* was Martin Luther's felicitous translation of the word *mnâ* in the original text of this passage; this was the Greek weight *mina:* ... *eis dedit decem mnas* ..., "... *und gab ihm zehn Pfund* ..." ("... and delivered them ten pounds..."); in the same parable as related in Matthew 25:15. Luther translated the Greek word *tálanton* as *Zentner* (["centner," "hundredweight"]: ... *et uni dedit quinque talenta* ..., ... *und einem gab er fünf Zentner* ... ("... and unto one he gave five talents ...") (see p. 268 and p. 162).

Now let us return to the pound, the pound of weight whose symbol in German, ℔, looks like a cursive *u* crossed by an elongated flourish. One should have no difficulty in seeing in this the old *lb* of *libra* crossed through by the final stroke of the pen, in the same curious fashion as in the English abbreviation £ for the pound sterling (compare the flourishes in the *lb* in Figs. 121, 122, and 124).

This cross stroke, the Roman *perscriptio*, did not originate, as one might well suppose, in a mere decorative flourish made by the writer or copyist; it goes back to the Roman coin symbols. Since initially the only Roman coin was the *as* of 12 *unciae* (ounces), the number XXV, for example, quoted in the laws as the penalty for some infraction, required no other identification: the 25 could be understood as nothing but 25 *as*. X was a denarius of 10 *as*. But later on (in 268 B.C.) the Romans adopted a silver coinage system in which the *denarius* now became a coin worth 10 (later 16) *as*. The numeral I now meant not 1 *as* but 1 *denarius*. To avoid misunderstanding, the monetary symbol X (= 10) was prefixed, and because it meant 10 *as*, not in the old bronze but in the new silver coinage system, a horizontal stroke was drawn through it:

Roman coin symbols:

denarius   5 denarii   5 sesterces.

Thus the original 10-*as* sign X was now used no longer as a numeral but as a symbol for a denomination of money (like our $ sign), which always was placed in front of the amount (like the 5 denarii or 5 sesterces in the examples above). Then in the still later coinage system based on the sesterce, the basic unit was one sesterce worth $2\frac{1}{2}$ *as*, abbreviated IIS; as a monetary symbol it was crossed through (see Fig. 188) and placed in front of the numeral for the amount (for the word itself, see p. 78).

This Roman cross stroke was retained throughout the Middle Ages in almost all symbols for coins or units of money, but only as a flourish whose meaning was no longer understood. The reckoning tables we have seen provide excellent examples. In the bottom rows or spaces we see a *d* with the flourish (see Figs. 177 and 179), the *denarius*, used as an abbreviation for pence, which is still written as a *d* with a flourish. The English also use *d* as their abbreviation for penny. The horizontal stroke is clearly to be seen in the abbreviations for pound, *lb* and *lib*. The long Gothic ſ with a flourish is, of course, the initial letter of *solidus*, which became the standard abbreviation

Fig. 188 A Roman sesterce (36 B.C.).

for shilling (see Fig. 179). The *g* for gulden also has its ornate form
on the counting cloth (see Fig. 180). Adam Riese often wrote the *g*
for groschen with a flourish like the *d* for penny, and then abbreviated gulden, as was the custom, as an *f* with a flourish (not an *l*; see
Fig. 187). But what connection does *f* have with "gulden"?

In 1252 Florence minted a gold coin with the city's heraldic arms,
the lily; this coin was called the *fiorino d'oro* (from the Italian *fiore*,
"flower"), which became the Medieval Latin *florenus* and the French
and English *florin*. Such florins with the figure of the lily were, in the
14th century, also struck by many mints north of the Alps, including
that of the four Rhenish principalities which had joined together in a
monetary union, Mainz, Cologne, Trier, and the Palatinate. Their
coins, however, because they no longer bore the symbol of the lily,
were called Rhenish gold gulden, or gulden for short, or also *güldene
schilling* (golden shillings). But as an abbreviation the long *f* with the
flourish was retained; for the Rhenish gulden the abbreviation was *rf*,
and in Holland to this day it is still *fl* (see Figs. 114 and 125). As with
the pound, shilling, and penny, the German or English name was
indicated by a foreign symbol.

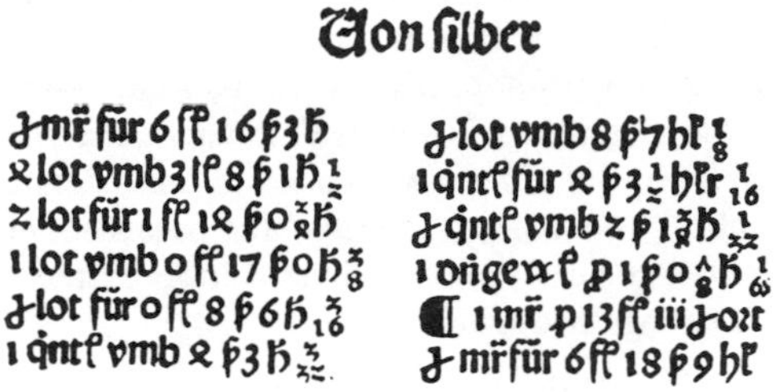

Von silber

Fig. 189 Table of values for various weights of silver from the Bamberger
Rechenbuch of 1483 (see Figs. 168 and 169). The weights rise in half-steps from
1 mark to 1 pennyweight for each definite base price of silver: The table reads:

$\frac{1}{2}$ Mark (silver) for 6 gulden 16 shillings 3 heller
4 Lot ("weights") vmb 3 gulden 8 shillings 1$\frac{1}{2}$ heller, etc.

A different base price begins in the second line from the bottom on the right:

1 Mark per 13 gulden 3$\frac{1}{2}$ ort,
$\frac{1}{2}$ Mark for 6 gulden 18 shillings 9 heller, etc.

The weights are: *mr* = Mark, Lot (ounce), *qnt* = quent, *dnge* = pennyweight;
the denominations of money are *f* = gulden, *s* = shilling, *h* = heller and Ort
(see p. 148). Note that *j* stands for $\frac{1}{2}$ and that the fractions are always placed
after the monetary symbol, unabbreviated and without the stroke denoting
a fraction.

At the end of his Bamberg book of computations, the arithmetician
Ulrich Wagner appended a table of values for the various standard
weights of silver (Fig. 189), which contains a whole series of interesting old symbols for weights and coins. The confusion and uncertainty
caused by the superabundance of various coins minted by different
authorities and belonging to different systems frequently required
"de-coining" — melting down the different coins into bars of metal
and stating the value in terms of the amount of pure silver they
contained. Hence for merchants such a table of silver weights and
values was an absolute necessity. In Wagner's table the weights
increase by half-steps from the mark (*mr*) to the pennyweight (*dn*),

calculated according to the scale of 1 mark ($= \frac{1}{2}$ pound) containing 16 lots, 1 lot containing 4 quents, 1 quent containing 4 pennies; $j = \frac{1}{2}$ was the "halved" I (see Fig. 116). The weight and value in each succeeding line are always half of those in the preceding line.

Line 1, left: $\frac{1}{2}$ mark for 6 gulden 16 shillings 3 heller;
line 2 (half): 4 lot are worth 3 gulden 8 shillings $1\frac{1}{2}$ heller.

As symbols for money, we find the gulden (f) worth 20 shillings (s) and the shilling worth 12 heller (h'), and at one point the ort ($= \frac{1}{4}$ gulden, fifth line down on the right). The reader is urged to check a few of these calculations for himself, taking note of the unabbreviated fractions without the fraction-stroke. This will provide a very instructive example of medieval monetary calculations.

"Shilling" is a Germanic word, for the *skilling* or *skild-ling*, "shield-ling," was the name the Germanic tribesmen gave to the East Roman golden *solidus*, which they first carried about on their persons as ornaments (compare the French *écu*, from the Latin *scutum*, "shield"). The original meaning of "penny" is not clear, for the word already occurs in Gothic (*penniggs*), so that its commonly accepted derivation from *Pfännchen* ["little round pan"] cannot be right. The pan-shaped hollow coins known as *Brakteates* did not appear until the 12th century. The *Mark* was just that: a mark or symbol, an official sign stamped on bars of silver, which finally became the specified weight of the bar ($\frac{1}{2}$ pound). As a result of the general debasement of money and coinage standards, silver was again weighed out for quite a long time during the Middle Ages, as it was in very ancient times. Smaller amounts then had to be cut off from the weighed portions (the so-called *Hacksilber*, "chopped silver"). The Russian ruble (Russian *rubitj*, "to chop, cut") took its name from this practice (see p. 225), and people in Russian banking and financial institutions still speak of "cutting" money. A tiny chopped-off piece thus came to be called a *Duet* in German (from Old Norse *þveita*, "to chop off," whence also Dutch *duit* and English *doit*).

The groschen or "thick penny," a Bohemian expression derived from the Italian *grosso* ("fat, thick"), and the ducat, are also foreigners in origin. The gold coins known as ducats were minted for the first time in 1284 in Venice and obtained their name from the words stamped on them:

Sit tibi Christe datus, quem tu regis iste ducatus,

"May this duchy which thou rulest, Christ, be consecrated to thee."

The ducat was also sometimes called *zecchino*, from the mint known as La Zecca in Venice (which in turn derived its name from the Arabic word *sekkah*, "coin").

The heller obtained its name from the town of Schwäbisch-Hall, where it was minted for the first time in the Imperial Mint during the reign of Frederick Barbarossa (1152–1190). The first Joachimstal guldengroschen was struck in 1519 from the silver mined at Joachimstal in the Erzgebirge mountain range, which forms the border between Czechoslovakia and the present East Germany; the name

"Joachimstaler," shortened to "taler" or "thaler," migrated from there into many countries of Europe and today is still known throughout the world in the form of its descendant, the American "dollar."

The history of the remarkable abbreviation for the dollar, the $, is now known as a result of an article published by Cajori in 1912. The Spanish conquistadores brought the *peso* (the Spanish thaler) to the New World in the 16th century. This became the chief coin used in the Spanish colonial possessions, including those in North America. To indicate a sum of money, the word *pesos* was first written out, or else abbreviated p^s, with the initial p and the raised s of the plural ending. The elevation of the last letter in abbreviations occurs throughout the Middle Ages (for instance, compare c^o, Fig. 115). The abbreviation p^s was first written as two distinct letters, but then as a single letter plus the flourish into which the final s had degenerated. When the English-Americans around 1780 came into contact and had dealings with the Spanish-Americans (in Mexico), they based their own *dollar* on the model of the Spanish thaler, and along with it adopted the Spanish abbreviation p^s, which they made by first writing the terminal s and then the p merely as a double vertical line through it, thus producing the dollar sign $. The $ appeared in a printed document for the first time in 1717. Thus the foreign "peso" sign for the dollar embodies a fascinating segment of commercial history (see Fig. 190).

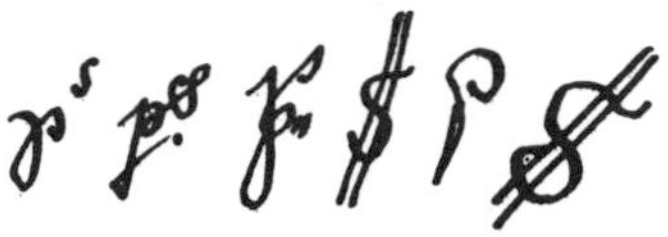

The coin known as the *Kreuzer*, which was first struck in the Southern Tyrol during the 13th century, took its name from the cross that was stamped on it. Finally, the last of this series of old coins, which could very easily be continued much further, is the *Scherflein*, which comes from Martin Luther's translation of the Gospel according to St. Mark (12:42):

Und es kam eine arme Witwe und legte zwei Scherflein ein; die machen einen Heller,

And there came a certain poor widow, and she threw in two mites, which make a farthing.

The original Latin text reads *duo minuta quod est quadrans*, "two small bits, which make a [Roman] *quadrans*" (see Fig. 32). Since 1480 the smallest coin in Erfurt (where the Wartburg is located) had been the *Scherf*. Its name comes from the Old High German *scarbon*, or Dutch *scharven*, carve, and this Germanic descriptive term probably goes still further back, to the Roman silver coins with serrated or toothed edges which, according to Tacitus, were preferred above all others by the Germanic tribes (see Fig. 67). Particularly in times of war the Romans would often make payments in debased "sheathed" coins which had a core of copper and an outer coating of silver. The Germanic tribesmen naturally did not want these, but they were very eager to obtain the coin known as *serratus*, because in such serrated ("saw-toothed") coins the base inner core and silver coating were easy to detect.

After this excursion into the history of coins, we must now return to computations on the line-board.

Fig. 190 The dollar sign developed from the abbreviation p^s for pesos, which were the Spanish-American "thalers." The symbols shown are, left to right, from manuscripts dated 1672, 1768, 1778, 1778, 1793, and 1796.

Subtraction

We should no longer have any difficulty in performing the operation 425 − 279 on the calculating board (Fig. 191): (*a*) *Numeratio*: place

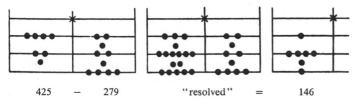

425   −   279   "resolved"   =   146

Fig. 191 Subtraction: 425 − 279 = 146.

the numbers in the two fields; (*b*) *Resolutio*: the "solution" or "de-grouping" of the number 425; (*c*) *Subtractio*: the "drawing away" of as many counters from the number in the left field as there are in the number on the right. In the second *spatium* or space between the lines there is one counter (= 50) on both the left and the right side; these are now removed and the space remains empty. From this procedure comes the German expression in computation: 5 − 5 *hebt sich auf*, "is picked up" (for canceled). Perhaps before resolving his numbers, the computer would first "pick up" all the counters that he could, in order to clear the board, and then "de-group" only when he could no longer go on subtracting without doing so. Our illustration shows that in this problem, after first picking up all possible counters, only the partial problem 200 − 54 remains to be solved.

Doubling and Halving

These operations were first introduced as separate, independent procedures by the then Master General of the Dominican Order, Jordanus Nemorarius (died ca. 1236); after his time they reappeared frequently in the writings of arithmeticians. It was not until the 16th century that they were generally recognized as special cases of multiplication and division, respectively.

The counting board directly promotes doubling and halving, because these procedures can be carried out without any computation at all, simply by adding or removing counters according to a fixed rule. To learn how halving was done, let us again hear what our master arithmetician Adam Riese has to say (Fig. 192):

Always for two counters place one over into the other field / and like-wise on the line on which you have your finger [the 3rd line; the index finger was always kept on the line on which the computation was being made, see Fig. 193]. And for one counter alone you shall place one in the next *spatium* beneath the line on which you have your finger [4th and 1st line] / but where there is no counter / none is transferred / and that is all there is to halving.

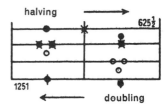

Fig. 192 Doubling and halving

He failed only to say that one counter in the space between becomes two on the line below, and one counter in the space next below that. The procedure in doubling, of course, was just the reverse.

Doubling and halving are primitive operations, early forms of multiplication and division. They occur even in ancient Egyptian computations (see p. 219) and are thought to be still in use among Russian peasants. Unlettered people would naturally substitute, for the difficult multiplication of 56 × 83, the easier operation of addition and then successively double and halve both numbers to find the answer. This results in a very curious computation:

56 ×   83

half 28 × 166 double

14 × 332

7 × 664-

3 × 1328-

1 × 2656-

4648

In halving an odd number, the remainder of 1 is disregarded; thus $\frac{7}{2} = 3$. After the halving has been carried all the way down to 1, all the doubled numbers with odd multipliers (marked by a hyphen) are added together: 4648.

The procedure immediately becomes understandable if the only factor is some power of 2, such as $64 = 2^6$. Then, apart from the last 1, there are no odd numbers — that is, the last doubling gives the product directly. For example, 64 × 83:

Half: 64   32   16   8   4   2   1
Double: 83   166 332 664 1328 2656 *5312*

The procedure in the case of the first example of multiplication, 56 × 83, may be explained as follows: each time a remainder of 1 is discarded, the number next to it is added on: $14 \times 332 = 7 \times 664$ $= 3 \times (2 \times 664) + 1 \times 664 = 1 \times (2 \times 1328) + 1 \times 1328 + 1$ $\times 664$. These remainders are lost in the progressive halving and must therefore again be gathered together. Every progressive series of halving ends at 1; thus the last double, which is here 1 × 2656, must be added in.

Multiplication

This was taught by Adam Riese in three steps, which the reader is urged to perform for himself on the line-board:

(a) the multiplier has one digit, for instance 28 × 6: "Know that you have two numbers to be multiplied. One / which is to be multiplied / lies always on the line. [When placed on the board, the number 28 = (3 + 5) + 20 looks like 3 − (1) − 2; that is, on the first line at the bottom lie 3 counters, on the second line are 2 counters, and in the space between them is 1 counter]. The other number with which you wish to multiply / write down [or form it on a second board]. / Now if you are to multiply with one digit [here 6] / then take the line above [the second] / where one or more counters lie / and place your written number as many times / as there are counters lying upon the same line [6 × 2 = 12, which placed on the board beginning with the second line is 2 − 1]. But where a counter lies within a space / take the next line above the same space / and put down only half of your number which you have written down." [Instead of 6 × (1) only 3 × (1) = 3 is formed on the second line. Now the problem to be solved here is 6 × 8 = 48 = 4 − (1) − 3, so that as a final answer we obtain 1 − (1) −1 − (1) − 3 = 168.]

(*b*) the multiplier or second factor has two digits, for example 28 × 34:
"But if you wish to multiply a number by one with two digits / take the other line above the counters [the next higher one] / and place the other digit of the number you have written down as many times [3] / as there are counters on the line / below / then take the line upon which the counters lie / and form the first digit of the written number as many times [4] / as there are counters lying upon the line."

The expression "take the other line" means that the reckoner is to place the index finger of his free hand on the next higher line (see the operators at work in Figs. 184 and 193); for the task here is to set down out of 28 × 34 the product 20 × 30 = 600. The 2 lies on the second line; thus the 6 goes on the third. Thus the rules of place value are observed in this simple manner. In general: a number consisting only of units goes on the same, the "first" line; if it is in the tens it goes on the second; thousands, on the third line above, and so on. This is just what Riese says in his next rule:

(*c*) the second factor has more than two digits, for example 28 × 354:
"So proceed also with three / four / five and more digits / in such a way that you always place the fifth digit of your written number on the fifth line / from the line / on which the counters lie / and begin counting by placing your finger on this line. Place / the fourth number / on the fourth line / and so on / to the bottom line. But with the spaces proceed / as shown in doubling.

Then follows the important admonition: "Above all, you must learn the multiplication table so that as soon as possible you shall know it by heart:

Fig. 193 Reckoners at their table. The man at the left marks with his finger the line on which he is working at the moment. From the arithmetic book of Grammateus, 1518.

Lern wol mit fleis das Ein mal ein /
so wird dir alle rechnung gemein!

Take care to learn the multiplication table well,
And you shall master all computations.

If the second factor of a multiplication was also formed on a reckoning table, because of the quinary 5-grouping one had to know the multiplication table only through 4 × 4.

The place value or order of magnitude of the final answer is very cleverly specified by the index finger, which "holds on to" the proper line. This line always becomes the unit line. Now let it be said once more that a number like 28 is broken down into 3 − (1) -- 2 counters and is thus no longer felt to be a single number. In such stepped-down numbers the individual steps of a computation are much more easily lost sight of than in our present-day operations. This must always be kept in mind when dealing with medieval computations. Then one can understand the difficulties which counting-board reckoners had with the Indian numerals and the completely new procedures they demanded.

Division

If we simultaneously perform the division 42 + 2 on the line-board, we shall much more readily understand the words of our arithmetic master:

Place the finger of your left hand on the top line above / and mark whether you can take the number by which you wish to divide / if you cannot take it / then place the finger on the other line / and do this as long as you can take / the numbers by which you wish to divide / then take away [that is, remove the counters] the same number / as often as you can and each time place a counter at the finger [which is pointing to the proper line] / do this until you can no longer take the number [by which you are dividing] / and what then remains lying next to your finger / is the part of the number you have taken away."

To make the solution fully clear: place the number $42 = 2 - 4$ on the line-board; on the second line the divisor 2 can be taken twice from 4; thus 2 goes on the same line, and correspondingly 1 on the line below. Result: $1 - 2 (= 21)$. The reader should now easily solve such problems as $36 + 2$ or $462 (= 2 - 1 - (1) - 4) + 3$.

Thus we have obtained at least a glimpse of ordinary computations "on the lines" as they were used by townsmen and petty tradesmen for their household accounts. Köbel and other arithmeticians also taught more complicated and difficult computations on the reckoning board, such as currency conversions and even the extraction of cube roots.

But it was its visual quality and simplicity which earned the counting board its great popularity among ordinary people. Now let us see just what role it played in their lives.

THE COUNTING BOARD IN EVERYDAY LIFE

Inventories and wills, about which we shall have more to say a little later (pp. 375 ff.), show that reckoning boards were used in a great variety of places: in monasteries, in royal treasuries, in the offices of town officials and the counting rooms of merchants. The French "Book of Counters," the *Livre de Getz*, says that a knowledge of its contents is urgently necessary *purce que il ja plusieurs marchans qui ne sceuent lire ne escripre, et leur est necessayre de bien scauoir comter* — "for there are many merchants who can neither read nor write, but who must know well how to reckon".

Yet the situation was not always quite as bad as that. Contemporary illustrations show pen and ink sitting on the table of the merchant who is turning over the pages of his account book (see Figs. 194 and 195). At the right end of the table in Fig. 194 there is a counting board with the spaces marked for the various coin denominations, on which the counters from the last calculation are still lying around. The merchant works at his table, writing and reckoning in his warehouse, surrounded by his goods, which his employees outside are tying up into bales and loading on a wagon.

The other illustration (Fig. 195) shows beautifully just how the merchant double checked the numbers written in his account book on his counting board, and thus how the writing of numbers was separated from computation in the Middle Ages.

The evil counterpart of the merchant and money-changer was the usurer, who is identified in our illustration both by his sack of coins and by his reckoning board (see Fig. 196):

Ich bitt eüch jud leicht mir zu hand
was eüch gebürt gebt mir verstand
Bargelt auff bürgen oder pfand,

I beg thee, Jew, to let me have,
Giving me to know what is your charge [interest],
Money against security or pledge,

Fig. 194 A merchant at work, with his counting board and his account book.

Fig. 195 A clerk with his reckoning table and his book.
Figures 194 and 195 clearly show how merchants, the Western counterparts of
the Greek treasurer of Darius and of the Etruscan reckoner (Figs. 133 and 134),
computed their written numbers on the counting table. Both woodcuts are by
Hans Weiditz, Augsburg, 1539 and 1531.

stands above this woodcut from a *Buch vom Schimpf und Ernst*
(Book of Shame and Truth) which appeared in Augsburg in 1531.
The moneylender is identified as a Jew by the Hebrew characters on
the hem of his robe. Quarrels and disputes over money matters were
so frequent in those days as to cause Luther to include the following
passage in his *Deudsch Catechismus 1530* (Chapter IV): "For we live
in the flesh and carry about with us the old Adam, who tempts us
and rouses us daily to deceive and cheat our neighbor." In the
accompanying illustration (Fig. 197) we see a poor wretch who has

Fig. 196 A Jewish moneylender; woodcut by Jörg Breu (died 1537).

Fig. 197 Illustration from Martin Luther's German Catechism of 1530.

Fig. 198 The parable of the unjust steward (Luke, Chapter 16). Woodcut by Urs Graf in a Basel Book of Sermons of 1515. Actual size.

been cheated out of his money complaining to a rich moneylender, whose helper is figuring the "correct" amount of the debt on the reckoning board as the creditor reads the numbers from the credit book; outside the building the devil attempts to prevent a man from praying.

Urs Graf (died 1528), who is well known for his portraits of German soldiery, illustrated the parable of the unjust steward (Chapter 16 of the Gospel according to St. Luke) very interestingly in a book of sermons printed in 1515 (see Fig. 198). Christ told his disciples the story of a rich man who had been cheated by his steward and asked him, "How is it that I hear this of thee? give an account of thy stewardship" — *redde rationem villicationis tuae* (fifth line from the bottom). The picture shows the steward figuring his accounts on a reckoning table (which is marked with medieval coin denominations).

Let us follow this example with the fascinating French representation of Arithmetic personified on a very large pictorial tapestry (Fig. 199). Arithmetic, as one of the seven liberal arts, is seated at a counting table (a *bureau*), where she shows her distinguished pupils how to

Fig. 199 Arithmetic personified, instructing noble pupils in the methods of computation with counters. In the open book is the number 1520, which is probably the date of the tapestry. The inscription embroidered at the bottom is in praise of the art of numbers. French tapestry about 3 × 3 m. Musée de Cluny, Paris.

use counters (*jetons*) to form numbers taken from a book. Twice in this illustration we read the number 1520, which is probably the date. The striking thing is that the cloth on the table has no lines; but such blank reckoning cloths existed, as we shall see shortly (Fig. 200).

The popularity of counting boards is often expressed in the literature as well. Here, however, it is primarily the counters that are referred to, either actually or symbolically. They are frequently mentioned by Shakespeare. There is the fine passage in *The Winter's Tale* (Act 4, Scene 2) where the young shepherd has to work out a sum (see p. 40):

Let me see: every 'leven wether tods; every tod yields pound and odd shilling; fifteen hundred shorn, what comes the wool to?

and then cries in despair:

I cannot do't without counters.

Here we see counters in their proper "role" — as aids to computation. With the same purpose in mind, Luther mentions them in one of his colorful sayings: "Thus the Jews placed the counters on the lines and reckoned how many Canaanites there were and what a small number of Israelites there were."

Will you with counters sum
the past proportion of his infinite?

asks Shakespeare in *Troilus and Cressida* (Act 2, Scene 1), and Iago speaks contemptuously of Cassio as a "counter caster" (*Othello*, Act 1, Scene 1). Counters often appear in Shakespeare as symbols of something worth very little, as in *Julius Caesar* (Act 4, Scene 3):

When Marcus Brutus grows so covetous
to lock such rascal counters from his friends
be ready, gods ...

Polybius' old simile about value depending entirely on position (see p. 300) now appeared again in various forms in the West. In Luther's words: "To the counting master all counters are equal, and their worth depends on where he places them. Just so are men equal before God, but they are unequal according to the station in which God has placed them."

Another writer of the 16th century puts it somewhat differently:

Life in this ephemeral world and all men in it are like a counter, which is worth so much (and no more) according to the line on which it lies and designates a sum. Now it lies on the top line. Why waste words? Before it can look around, the reckoner takes the counter away again, and again it is no more than just another counter, a mere piece of brass.

Then we suddenly come upon the old metaphor again, in French costume of the mid-18th century:

Les courtisans sont des jetons,
leur valeur dépend de leur place:
Dans la faveur, des millions,
et des zéros dans la disgrâce!

"Courtiers are but counters;
Their value depends on their place:
In favor, they're worth millions,
And nothing in disgrace!"

This verse has been attributed to various persons, including even Frederick the Great.

Martin Luther found a new expression in the counting board; on one occasion he thundered: "They have a strange way of reckoning — they follow no order, but throw the hundred into the thousand." And on another occasion he said: "The Devil is in a rage and he throws the hundred into the thousand, and thereby creates so much confusion that no one knows what to think."

What is the meaning of this expression, *das Hundert ins Tausend werfen*, which is still used even today in Germany? Whoever places the counter not on the 100-line where it belongs, but on the 1000-line, follows no order and creates confusion. The line-board clearly provides the explanation. Luther, who often drew on the counting board for figures of speech in his vivid sermons, is thus an unimpeachable witness for the common people's daily familiarity with the reckoning board.

As long as we are calling on great men for evidence, we should not omit Goethe. Here is a passage from a letter by Bettina von Arnim: "No toy fascinated him more than his father's counting board, on which he would reproduce the constellations of the stars with counters." This is already well into the 18th century, and yet Goethe's father, the municipal councillor, was still using it for serious computations. But it was his son, the great writer, who brought the counting board into the realm of poetry. In *Faust* (Act 2, Scene 2) Pluto, the god of wealth, is dividing up his treasures among the masses. As the mob reaches greedily for the loot, only to find that it is all fool's gold, the Herald declares:

Glaubt ihr, man geb euch Gold und Wert?
Sind doch für euch in diesem Spiel
selbst Rechenpfennige zuviel.

Did you think they'd give you real money and goods?
In this game even worthless counters
Are far too good for you.

And finally, it will happen to all of us, as a Swabian saying puts it, that "we shall have to make an account of ourselves on the red-hot counting board in God's chancery."

THE COUNTING BOARD AND WRITTEN NUMERALS

Using Counters to Write Numbers

Now we shall return to the development of the counting board itself and examine two noteworthy forms, one of which provided the stimulus for the elaboration of a system of written numbers.

The "number tree" or *arbre de numération*, as it was still called in a French textbook of arithmetic published in 1753, occurs as early as the old *Liure de Getz* (see p. 338) of the 15th century (Fig. 200).

We see that the reckoner has laid down the counters on the left side of the table (as he faces it), one above another in a vertical column, a "tree"; the man standing in front of the table points to it. This is the simplest conceivable preparation for computation, for the individual coins or counters stand for the lines and take the place of the specific counting table (see Fig. 201). We must also visualize such a "number tree" in the French picture of Arithmetic (Fig. 199), although the artist has merely suggested its presence by the lineless tablecloth.

There was also a special way of placing and indicating the value of counters on the board in England, where during the 16th century it came into common use in government offices (along with the line-board). This used a board with horizontal bars which in turn were divided into individual vertical columns or fields, for £20, £, *s*, and *d* respectively (see Fig. 202).

The coefficient for conversion from £20 to £1 and from £ to *s* is 20, but from *s* to *d* it is 12; thus there could be up to 19 counters in the first or top three fields and as many as 11 counters in the last. This form of counting board itself has no provision for decimal or quinary

Fig. 200 The number tree (*arbre de numération*) replaced the lines on the counting board. It consisted of a vertical row of counters which were set down and not moved again; these marked the positions where the lines would have been. The *arbre de numération* appeared primarily in French arithmetic textbooks. From the *Liure de Getz*, 15th century.

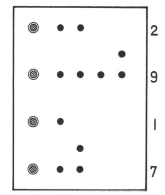

Fig. 201 The number 2917 formed on a counting board on which the lines are replaced by place counters or coins.

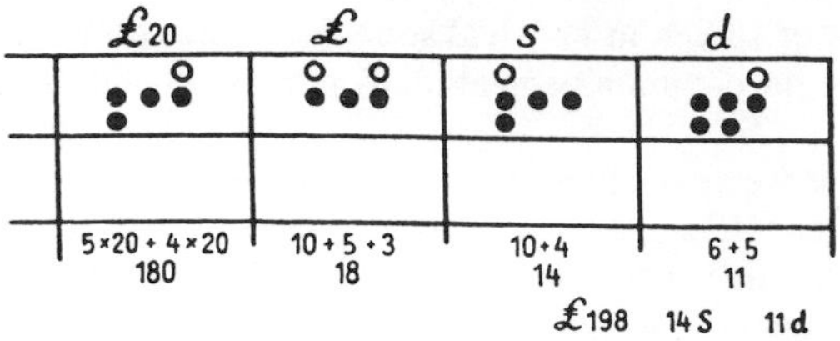

£20	£	s	d
5×20 + 4×20 180	10 + 5 + 3 18	10+4 14	6+5 11

£198   14 S   11 d

Fig. 202 A method of placing counters on the board that was common in England.

groupings; these were very cleverly introduced by special arrangement of the counters themselves. If an individual a single counter (○) was placed above another at the left end of the field, it meant 10; likewise a counter placed above at the right end meant 5, except in the *d*-field, where it stood for 6. Thus the grouping entered only when the number of coins or counters in a space was greater than 5 (respectively 6), and thus 5 counters were laid out.

Through this place-value arrangement, the counters actually formed true number symbols. These "counter numerals" were actually taken directly from the counting board and were written down on paper until well into the 17th century. How this was done is shown by an example of an English computation (Fig. 203).

This system of written numerals derived from the counting board is, in our sense, an "early" or primitive system, for it makes use only of the laws of ordering and grouping; on the other hand, it achieves its grouping through a place-value arrangement. This feature makes it extremely interesting as an intermediate stage in the development of the mature place-value system of numerals. Even in England, however, these written "counter-numerals" never came into universal use; although they had a somewhat wider and even partially official recognition, they belong essentially to the class of "peasant numerals".

The Chinese Stick Numbers

The history of written numbers shows one instance in which a system of numerals derived from the counting board achieved universal validity and recognition. Long before the appearance of the modern *suan pan*, the Chinese used little bamboo or wooden sticks as calculating pieces (*chou*) on a reckoning board which was likewise called a *suan pan*. Around A.D. 600 the Japanese adopted this form of counting board (which they called the *sangi* or *san-ju*) and used it until very recently not only for ordinary computations but even for solving algebraic equations. The oldest documents testifying to the use of these small sticks for computation date from the era before the birth of Christ, and the old Chinese character for "compute" is a representation of these calculating sticks (Fig. 204), as we shall see presently.

A Japanese textbook from the late 18th century shows how these were used (Fig. 205). We see, at work in front of a gentleman of high rank, a reckoner kneeling on the floor and arranging sticks on a checkerboard-like *sangi* table. The sticks themselves, which were made of cherrywood, were some 4 cm long and had square or triangular

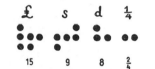

£   s   d   ¼

15   9   8   2/4

Fig. 203 A sum of money (£15, 9*s*, 8½*d*) represented in English "counter-numerals."

Fig. 204 *Suan*, "to compute": the Chinese character representing calculating sticks.

Fig. 205 A reckoner working with sticks on a counting board: from a Japanese book of the year 1795.

cross sections about 0.6 cm wide; thus they resembled our old-fashioned kitchen matches. In earlier times they were much longer. A set of such calculating sticks consisted of 200 sticks, which if they were to be used for "higher computations" were divided into 100 red sticks for positive and 100 black sticks for negative numbers.

The writer has been unable to obtain even a picture of some authentic examples of these calculating sticks. A friend in Japan who went to much trouble in attempting to obtain such a specimen or illustration reports that today even the memory of them seems to have vanished.

The counting boards on which these sticks were used had ten marked or unmarked vertical columns (see Fig. 206). Here we see, from left to right, the columns for 1000, 100, 10, and 1, to which columns for the decimal fractions $\frac{1}{10}$, $\frac{1}{100}$, etc., have been added. Thus up to 9 sticks could be placed in each square. But here again the

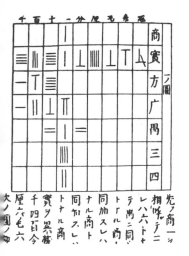

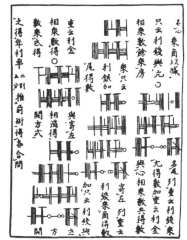

Fig. 206 (*Left*): *Sangi* table with marked longitudinal columns, in which the numbers were represented by sticks. Along with a zero symbol (for an empty column), these developed into (*Right*): true "stick numerals" with which any number which could be "written" in a place-value notation. — An invaluable example of the intrinsic connection between the counting board and written numerals. From an 18th-century Japanese book.

operator introduced a quinary grouping for greater ease in reading the "digits," by the way in which the sticks were arranged. They could be put down in two orientations: vertical and horizontal, or longitudinal and transverse (Fig. 207). The number 5 was still

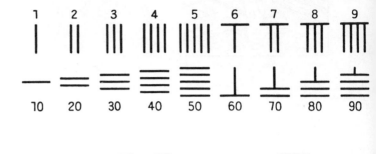

Fig. 207 *Sangi* numerals made up of sticks (or strokes) in two directions, vertical and horizontal. Within a given number the orientations alternate; the units are always arranged parallel to each other. Note the 5-groups.

represented by five sticks arranged vertically, but higher numbers were formed by attaching vertical sticks to one or more horizontal sticks representing the quinary group or groups.

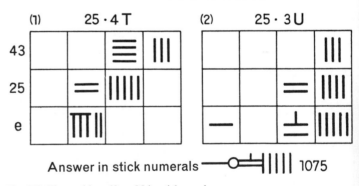

Fig. 208 The problem 43 × 25 in stick numbers.
(1a) Formation of the numbers 43 and 25, the units digit 5 being shown under the higher rank 4 to make the positional principle fully clear.
(1b) Multiplication: 4 × 2 = 8 and 4 × 5 = 20; the number 4 is thus disposed of and removed; 8 + 2 at the bottom are combined into 10 (row 2e).
(2b) Multiplication: 3 × 2 = 6 and 3 × 5 = 15. Result e = 1075.

In forming a number, the arrangement of the digits was changed in successive columns: the units, hundreds and ten thousands were placed vertically and the tens and thousands horizontally (see Figs. 207 and 208). Thus even a large number could be easily read. From top to bottom on the board illustrated at left in Fig. 206 we read the following numbers with surprising ease: 1; 4351.65666; 1650; 1267; 21.2; 4; and 2.

On *sangi* boards with marked columns these alternate changes in direction were obviously unnecessary; thus they are thought to be evidence of an earlier stage in which computations were made on a table without lines, similar to the medieval European number tree.

As an example of the way computations were made on this form of counting board, we shall multiply 43 × 25 in the Japanese manner (see Fig. 208); the reader should follow this himself with matches or toothpicks.

The entire manner of representing numbers directly promotes their expression in written form, for the *sangi* sticks, even though they are ordered and grouped instead of being arranged by gradations, are a form of digit. As soon as the Indian symbol for the lack of a digit, the 0, became known and was adopted for an empty space, the *sangi* numbers could immediately be removed from the reckoning board and represented in pictorial, or written, form. The device of changing the direction of the sticks or strokes making up the alternate individual digits is particularly felicitous; thus, for example, on the page from the Japanese book shown on the right side in Fig. 169 we can read the following numbers from top to bottom without much trouble: 46,431; 5,399,856; 614,585,664; 295,949,808; etc.

After the Buddhist teachings penetrated in a powerful stream into China and Japan and were translated into the indigenous languages in the 5th and 6th centuries A.D., there was constant communication between Indian and the Far East. The zero appeared in China around the middle of the 13th century.

These Chinese stroke or stick numbers represent the only mixture of primitive and mature written numbers that is known in cultural history: the digits themselves were based exclusively on the principles of ordering and grouping, but were then combined in a place-value notation to form numbers!

Their arrangement was derived from the counting board. Had the zero symbol in this system of written numerals been an invention of the Chinese themselves, this would have been the only case in the whole history of human culture in which the clear gradational arrangement of the counting board, based on ranks of 10, could be seen embodied in a mature, fully developed place-value notation. This is what in fact did happen — not in China, but in India. But here the whole development remains hidden in obscurity, although the existence of the *sangi* numerals makes it in the highest degree likely. From this we see also that the development of a place-value notation rested on the invention of a symbol for the absence of a digit, of a zero sign. Without it, the place-value principle, so clearly and easily visualized on the counting board, becomes lost in the confusion of a system of abstract written numerals.

The Counting Board and Computations

Let us now look back briefly over our course. In the Far Eastern counting board with its stick-digits we have seen the last of the many forms which the reckoning board took in its development and wanderings since classical antiquity (see Fig. 209). In spite of the great apparent differences between them, they all show a surprising

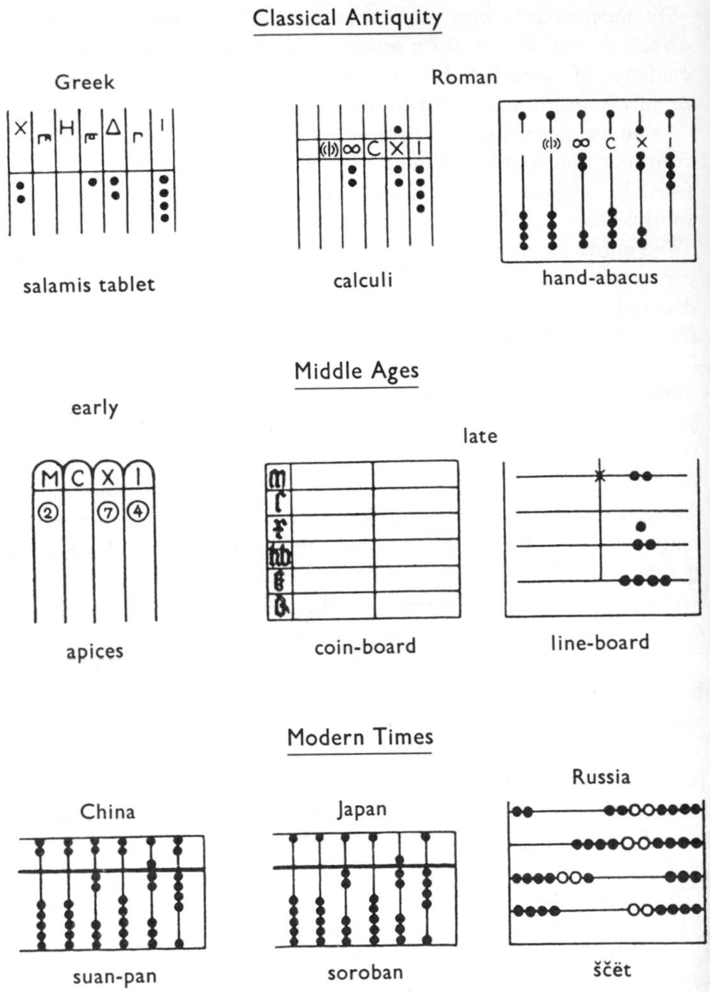

Classical Antiquity

Greek

Roman

salamis tablet

calculi

hand-abacus

Middle Ages

early

late

apices

coin-board

line-board

Modern Times

China

Japan

Russia

suan-pan

soroban

ščёt

Fig. 209 Various forms taken by the counting board throughout history; the number 2074 is represented on all of them.

degree of fundamental similarity in their completely clear place-value notation, in their gradations based on the number 10, to which the only exceptions were the counting boards specially adapted to nondecimal monetary systems like the English, and finally, and less important, in the quinary grouping derived from the monastic abacus which is possessed by all counting boards except the Russian *ščёt*.

In the late Middle Ages western Europeans developed a unique form of the reckoning board, with horizontal instead of vertical lines or columns. The boards developed by all other peoples and in all other times have had vertical columns.

The decimal place-value notation follows the decimal gradation which is present in the spoken numbers of all the peoples who developed or used counting boards. Yet neither ancient nor medieval Europe evolved a system of written numerals based on a similar principle. This failure, more clearly than anything else, reveals the profound difference between computation and the writing or representation of numbers — a difference which we, in possession of the Indian numerals, have completely forgotten. But it is equally clear, now that we have learned something of the history of the abacus, that ancient cultures as well as those of the Middle Ages, despite their cumbersome and undeveloped written numerals, possessed through their counting boards the same fully developed arithmetic procedures which we do today. If we glance rapidly over all the illustrations from Fig. 128 through Fig. 208 we shall very clearly understand the importance which the counting board has had, and to some extent still has, all over the world.

The various languages reveal the same thing. If we trace the root meanings of the word "to compute", we shall meet the very oldest computation instruments, the tally stick and the fingers:

Language	Word	Original Meaning	Derived from
Greek	*pempazein*	"to make fives"	Finger counting or, possibly, a counting board with a quinary grouping
	pséphizein	"to move pebbles"	counting board
Latin	*computare* *supputare*	"to cut"	notched tally stick
	calculos ponere	"to place stones"	counting board
Medieval Latin	*calculare*	"to move pebbles"	counting board
French	*jeter*	"to throw"	counting board
	compter *calculer*	< Lat. *computare*	
English	*to cast*	"to throw"	counting board
	to count	< Lat. *computare*	

And what about our word *reckon*? Let us look at its various relatives: Modern German *rechnen*, Middle High German *rechenen*, Old High German *rehhanon*, Gothic *raþjan*, Anglo-Saxon *ge-recenian*, whence English *reckon*. These all have the meaning of "to compute, calculate" and the further meaning of "to think, to believe." Then we have the Friesian *rekon*, "in order," from the Latin *rex, regis*, "king," and the Greek *arégo*, "to help" — thus the Indo-European root *reg* is thought to mean "to order, to put straight." It can be seen clearly in the German word *aus-recken*, "to stretch out, extend (oneself)." It is probably also related to the Germanic root *rek*, from which comes *Rechen*, "rake, rack, screen" — to create order out of chaos by raking together and collecting the pebbles and sticks on the board, which is what the very first primitive "reckoning" amounted to. For the mother of all computations is addition, putting together, which we have already seen as the basis of many ancient and medieval arithmetical procedures (see pp. 129, 218 and 360).

Thus computations using sticks and stones, with which mankind first began to count, evolved on the counting board into a highly developed and efficient arrangement that satisfactorily met the needs of cultures and people for many hundreds of years. Even the old *calculi* flourished to an unprecedented degree in western Europe; in the next section we shall see how this happened.

The Counters

'OI. VOI. TES'
"Hear, see, and be silent":
Words on an old French counter

The French Origin of Counters

The Roman *calculi*, which were made of stone, glass, or metal, bore
no symbols or writing by which they could be recognized as counters
(see Fig. 156). Gerbert, on the other hand, marked his small horn
counters with the symbols known as *apices*. Then in the 13th century
counters first appeared as small coin-like metal discs with designs
stamped upon them, and thus, like their more noble sisters, began to
reflect the cultural history not only of the counting board, but quite
often of their time in general.

The custom of stamping counters like coins first began in France.
These counting-board "pennies," which from that time on were made
only of metal, were of course not actually money, but they could
easily be mistaken for it and were even sometimes passed off as
money. Thus many of them bore a warning; often in the form of a
cryptic statement: *Je suis de laiton, je ne suis pas d'argent* ["I am
made of brass, not of silver"].

And we can again call on Martin Luther referring to a common
deception: *Als were ich ein Kind oder narr, der sich mit zalpfenningen
für gulden effen liesse* ["... as if I were a child or a fool, who could
be fobbed off with counters instead of gulden."].

The oldest known medieval counters, dated about the middle of the
13th century, come from the fiscal offices of the royal government of
France (Fig. 210).

These bear no inscription, but the French fleur de lys clearly indicates
their origin. The scale or balance on one of the counters is the symbol
of the royal mint, while the key stands for the royal treasury of
France. Other examples similarly show that officials of the French
court, such as the master of the royal stables, the master of the
hunt, the chief gardener, and the head chef, all had their "royal"
counters. From them the feudal lords and the king's relatives, the
noble houses of Valois, Anjou, Artois, and others, introduced the
custom of using stamped counters into the administration of their
own lands and possessions, whence they passed in turn into the
offices of town clerks and the counting houses of merchants. Madame
de Sévigné wrote to her daughter in 1671:

*Nous avons trouvé, avec ces jetons qui sont si bons, que j'aurai eu cinq cent
trente mille livres de bien, en comptant toutes mes petites successions,*

With these counters which are so very useful, we have found that my
property amounts to 530,000 livres, counting all my small inheritances.

There is a counter bearing the heraldic arms of the City of Paris, a
sailing ship at anchor with its sails furled (see Fig. 219). Its inscription,
Enrico tenente salva 1599, may be translated as "May [Paris]
prosper under the rule of Henry [IV]."

From France, where the coinage system decreed by Louis IX in 1266
became the model for many countries of Europe, the use of stamped
metal counters soon spread to the Low Countries (see Fig. 215), to
England, and to Germany (Figs. 217 and 218). The only important
exception was Italy, although here too, the coins bearing the Venetian

[375]

Fig. 210 The oldest counters come from France and were made in the 13th century.
Top row: counter adorned with fleur de lys and a squirrel, the heraldic symbol of Blanche of Castile;
counter with the scales used in the royal mint;
counter with the key of the royal treasury.
Bottom row: reverse of the above counters, showing a castle between four flowers and four rings;
the three fleurs de lys, the coat of arms of the kings of France;
a cross of the type that often appears on French counters.
Diameter of counter at left, 20 mm.
From the collection of A. König, Frankfurt/Main.

image of the Lion of St. Mark may very well be a counter (see Fig. 174); yet this may also be some other kind of sign (*tessera*) which has nothing at all to do with the counting board. At any rate, in Italy counters do not occur with such certainty and in such quantities as in the other countries named; we shall discuss the reason for this later (p. 424). There are a few counters with the coats of arms of the Florentine families which controlled the great banking houses of Europe at that time; however, these "Italian" counters were probably used not in Italy but in the foreign branches of the Italian banks. Thus there are, for example, counters from Nuremberg with the Venetian emblem of the Lion of St. Mark that were most likely made for the use of German merchants resident in Italy (see p. 385).

France, which originated the minted metal counters, also remained a leader in their development. For a long time only the royal mints were allowed to produce such counters. Princes and noblemen had to obtain the permission of the *Cour de Monnaie* if they wished to have counters of their own. In 1531 the latter gave its approval to the Duchess of Ferrara to have some counters made *pour servir aux gens de ses comptes*, "for the use of the officers of her counting chamber." In 1457 Margaret of Anjou, the consort of Charles VII, paid 5 *sous tournois* for 100 *gectouers* (= *jetons*, "counters") for use by her kitchen staff.

Like the French mint, the Flemish mint at Ghent also struck counters to order. In 1334 it sold 120 silver counters to a tax collector of Flanders for the sum of 10 shillings and 8 groschen:

A Nickolay Guidüche recheveur de Flandres bailliet VI^{xx} getoirs d'argent; costerent x sols viij deniers de gros,

a highly significant transaction from our own standpoint because it documents the use of both the 20-group vigesimal grouping and the "great hundred." The former also occurs in the following very revealing account made in 1413 by the *Hôtel des Monnaies des Pays Bas*, which were then under Burgundian rule, for the Duke of Burgundy:

A Jehan Gobelet, maistre particulier de la monnaie de Gand: pour ij^c iiij^{xx} xix jettoirs d'argent par lui délivrés en la Chambre des comptes à Lille pour jetter et besogner illec (?) pour les affaires de Monseigner, pesans ensemble iiij marcs vj onces et xij esterlin, qui valent iij^c lxij sol v d.

To Jehan Gobelet, (presumably) master of the mint at Ghent: for 299 silver counters delivered by him to the counting chamber in Lille, for the purpose of "casting" and keeping account of the Duke's business; total weight 4 marks 6 ounces and 12 sterling, to the value of 362 shillings 5 pence.

This entry beautifully complements our previous information about the computations made personally by the Duke of Burgundy (see p. 332). A very old item from England, dated 1290, states:

Item pro xj de counturs et i hannapero emtis pro compoto executorum Reginae,

Likewise for 11 shillings counters and 1 bowl bought for the counting chamber of the Queen's officers.

It is quite clear from all these documents that the princes did obtain the necessary *jetons* or counters for the use of their court officials. Over the course of time, moreover, the counters wore out and had to be replenished, like pens and ink:

10 pfennig 3 heller umb ein hundert rechenpfennige und ein dindenhorn und kalamaren,

10 pence 3 heller paid for a hundred counters and one inkhorn and writing gear,

reads an entry in the books of the city of Frankfurt for the year 1399.
  In Bohemia an ordinance of 1608 stated

dass nachdem bei den Prägerischen und anderen Münzwerken in Behaimb die Verordnung gethan, dass jederzeit mit Eingang des neuen Jahres eine Anzahl von 2–3000 Kupferrechenpfennigen angefertigt und auf die Kammer und Buchhalterei überschickt werden sollen, diesem Befehl unverzüglich nachgekommen werden solle,

that since it has been decreed for the Prague and other coin works in Bohemia that at the beginning of each new year they shall prepare an amount of from two to three thousand copper counters and deliver them to the counting chambers and financial offices, this command is to be implicitly obeyed.

Gold and Silver Counters

From these regulations, which were necessary for the individual accounting offices, it soon came to be the custom in France for the king at New Year's to present a sack or box of *jetons* to the chiefs of the mint and all other branches of the royal government, who assuredly did not spend much of their time sitting at a counting board. And of course these presentation counters were no longer made of copper, but of silver or gold. These were naturally not used for making computations, but for the most part were melted down, since they had no face value as money. These *jetons d'étrennes* (New Year's gifts) reached their high point in the reign of Louis XIV, from whose time the following list of jetons gives some idea of the extent and the cost of this royal custom:

Pour			For	
le trésor royal 800 *j. d'or*	26000 *d'argent*		the royal treasury . . .	
la maison de la reine	6100		the Queen's household . . .	
les revenus casuel 100	3500		miscellaneous revenues (?) . . .	
l'amirauté	4500		the admiralty . . .	
les galères	2800		the ship's officers . . .	
les bâtiments	1600		the construction offices . . .	
900 *d'or*	44500 *d'argent*		900 gold 44500 silver	

Nine hundred gold and 44,500 silver counters! These New Year's presents, which originated in the Roman custom of the *strenae* (> *étrennes*), cost the royal treasury of France a pretty penny.

In return, however, the king also received gold and silver *jetons* from his high officials at New Year's. Perhaps this was originally an oblique reference to the officials' neatness and efficiency, the implication being that "you can rely on our accounts, everything is in order." Later on, however, this was certainly no longer the case — the reverse, if anything, was true. Sully, the minister in charge of the French finances in the early 17th century, says in his memoirs that every year, on New Year's Day, he brought to his master, the King, a present of golden counters, and that one year he was received by the two Majesties in bed. The custom was already established in the 15th century, for we have information that King Louis XII upon his entrance into Tours was presented with 60 *gettoirs d'or* by the city. Louis XV received so many gold counters each year that he was able to have six golden platters made from them after they were melted down, and a contemporary of his reported that he already had 42 such plates.

We have seen earlier that England took over this custom from France (see p. 334). When Mary Stuart was held prisoner in 1585 at Chartley, a list of her possessions included, among other things,

bourses de velours vert, garnyes de jetons d'argent aux armes de sa Majesté,

purses of green velvet, filled with silver counters bearing Her Majesty's coat of arms.

In Holland, too, as early as the end of the 15th century the custom began of presenting silver and copper counters in an ornamented silver can to officials of the government, at public expense. Our illustration (Fig. 211) shows three such cylindrical containers, each containing a *cast* or *set* (French = *jet*) of counters.

Fig. 211 Containers for ornamental counters, the two at left of copper and the one at right of silver. Height 12 cm, diameter 3 cm.

Counters as Mirrors of their Times

These elaborate presentation counters exerted a great influence on the development of coins: their minting became more meticulous and the motifs became more varied. The *jetons d'étrennes*, in addition

Fig. 212 A *jeton d'étrennes* (New Year's counter). *Carolus VI. Rom(anorum) Imp(erator) Dux Brabantiae C(omes) Fland(riae)* — "Charles VI [1711–1740], Holy Roman Emperor, Duke of Brabant and Count of Flanders."
Restaurata et Protecta — "[The mercantile fleet] rebuilt and protected." In the bottom segments: — *Strena Kalend. Januar* 1722 — "New Year's present, 1722."

to the image of the reigning prince, began to bear representations of the events and achievements of the previous year or the recent past, with the purpose of glorifying the ruler or his house. Thus they generally developed into commemorative medallions whose value, apart from this, was still reckoned in terms of the metal they contained and their worth as objects of art. The *jetons*, the name now universally applied to these display medals, were completely divorced from the reckoning board and in France were still given at New Year's even in the mid 18th century, when the government bureaus certainly no longer used any form of the counting board for their accounts.

The presentation counters also caused their humbler working cousins, which remained on the counting board, to appear with a greater variety of representations, so that these too began to reflect the events of contemporary history, especially in the Netherlands. When the plague was rampant in Brussels in 1488, for example, counters appeared with a representation on one side of Death carrying a coffin under one arm:

Heus quid gestis? En hic te manet exitus,

Alas, why dost thou rejoice? For behold, thine end is at hand!

On the reverse side, however, there is a completely different verse·

Tempora leto tristia risu tempera,

Through laughter thou shalt moderate the sad time of death.

In the Netherlands, under Spanish occupation, counters became equivalent to political pamphlets. Upon the abdication of the Emperor Charles V, the Netherlands came into the possession of Charles's son, Philip II of Spain. This land, which from time immemorial had enjoyed important rights and privileges, rose up in arms against the persecution of "heretics" by the Inquisition and against the Spanish occupation generally. To suppress the rebellion, the Duke of Alba marched into the Netherlands with 20,000 Spanish troops and began his reign of terror. In 1568 the Counts of Egmont and of Hoorn were executed. Alba was finally recalled in 1573, but too late, for afterwards his successor Alexander Farnese, the Duke of Parma, was only gradually able to restore quiet to the southern, Catholic provinces (which make up the modern Belgium). The northern, or Protestant, provinces joined together in 1579 in the Union of Utrecht and declared their independence from Spain. England aided them in their rebellion against her greatest enemy. To punish England for her interference, Philip II in 1588 sent the great Spanish Armada northward, where English ships aided by a violent storm brought about its defeat and destruction. Philip II died ten years later, the Spanish Empire went into decline under his successors, and the full independence of the Netherlands was at last officially recognized by the Treaty of Westphalia in 1648.

During the epoch of the Spanish occupation and the struggles of the Dutch against the Spanish occupation it was necessary to keep the people aroused for long periods against their oppressors and to encourage them in their struggle. Counters with special motifs circulating among the people served as impressive and lasting reminders of their situation.

The Duke of Alba's arrival in Holland was greeted by the issuance of the commemorative counter illustrated in Fig. 213, left, on which the image of Death, encircled by snakes, is cutting down flowers in bloom with his scythe:

Justitia is geslagen dood, veritas die lyd in nood,
falacit(as) is geboren, de fides heeft di Strit verloren,

Justice is struck dead, the truth lies in distress,
Untruth is born, faith has lost its struggle.

But the Spanish authorities availed themselves of the same political weapon: a counter *Leggelt van s. C(onings) rekencamer (in) Gel(der land)*, "issued by His Majesty's counting office in Gelderland,"

Fig. 213 Political commemorative counters marking the arrival of the Duke of Alba in the Netherlands in 1568;
top: counter issued by the party of the Netherlands;
bottom: one put out by the Spanish party.

bearing the head and bust of Philip II (Fig. 213, bottom) carries a warning on the reverse side:

Vae genti insurgenti contra genus meum,

Woe unto the people which rises against my house.

The arrival of the Duke of Alba, was followed in the same year by the execution of the Counts of Egmont and Hoorn. The Dutch never allowed themselves to forget this atrocity; eleven years later a commemorative counter appeared which treated them as if they were still alive (right-hand counter on Fig. 214).

Fig. 214 Political commemorative counters from the Spanish Netherlands; struck on the occasion (*left*) of the religious persecution of the Dutch by the Spaniards in 1580, (*center*) of the destruction of the Spanish Armada in 1588, and (*right*) of the execution of the Counts of Egmont and Hoorn in 1579. Diameter 3 cm.
From the collection of A. König, Frankfurt/Main.

At the top right in Fig. 214 we see a representation of the Dutch fighting against the Spaniards, on horseback above and on foot below; on the reverse are the decapitated corpses of the two murdered leaders, whose heads have been hung up on poles for public display. The inscription reads:

Praestat pugnare pro patria | quam simulata pace decipi 1579,

Better to fight for one's country / than to be deceived by a false peace.

The first of the counters (of Fig. 214) was aimed at the Inquisition: on the obverse, standing on a pedestal is the Inquisition itself; the Belgian lion is chained to the column by a ring around its neck, on which is engraved the word *inqui(sitio)*. A small mouse gnaws through the ring to free the lion, in a reference to the fable of the lion and the mouse. The mouse symbolizes the Prince of Orange, the first ruler of the Netherlands after the uniting of the seven northern provinces in the Union of Utrech of 1579. The reverse shows Pope Gregory XIII and King Philip II of Spain, with the Lion of Belgium in the foreground (bottom left of Fig. 214). With his right hand the king holds out the palm of peace, but in his left hand he conceals the neck ring of the Inquisition behind his back. This is a reference to the peace negotiations then being held at Cologne. The inscription reads:

Rosis leonem loris mus liberat | liber revinciri leo pernegat,

Freed by the mouse from his rose-decked chains, / the lion refuses to be bound again.

This medallion was struck in Doordrecht in 1580. Then, eight years later, when the Armada, which had been sent to punish the ally of the Netherlands, was totally destroyed without accomplishing its objective, the Dutch saw the helping hand of God on their side. Our counter (center) shows a large ship being broken up in a storm, with the sailors' arms raised in vain supplication to Heaven:

Hispani fugiunt et pereunt, neminem sequete,

The Spaniards flee and perish, and there is none to save them.

On the reverse side a Dutch family — father, mother, son, and daughter — kneels and gives joyful thanks to God:

Homo proponit, deus disponit,

Man [Philip II] proposes, but God disposes

— a representation of terrifying contrasts!

The Names of the Counters

In no country except the Netherlands did the counters emerge from their humble seclusion to take on the role of political commentaries, in addition to their "official" duties. When they did not, as on the French *jetons*, merely refer to the images represented, the inscriptions frequently tended to be pious expressions such as *Ave Maria, Gratia Plena* (Hail Mary, full of Grace) or mottos like *Oi voi tes si tu veus vivre en pes* (Listen, look, and keep still, if to live in peace is your will). In other cases they referred to computations: *Der hat selten guoten Muot, der verloren Schuld raiten tut* (He who finds himself in debt / is rarely in good spirits) are the words on a counter from the Tyrol (Fig. 217); or *se.lui.qui.ne.scet.bien.son.compte | en.vient. acheter.come.il.si.monte* (He who does not know how to reckon well must pay the penalty as high as the error runs); or *Leg recht, greif recht, sprich recht, so kumt das facit recht* (Put down correctly, take away correctly, speak the numbers correctly, and the answer will come out correctly), which has a French counterpart, *Que bien jetera son compte trouvera* (Whoever casts well shall also find the right answers), an expression whose meaning is reflected in a modern proverb, as you make your bed, so shall you lie in it. Sometimes, again, counters were merely labeled or identified, like a counter of Philip II issued in the Spanish Netherlands (Fig. 215): *Jectz du bureau des Finances du Roy* (I)I)LXXVII (Counter belonging to the King's financial office 1572).

Fig. 215 Counter of Philip II in the Spanish Netherlands.

Similarly, but in very different style, another counter from the Tyrol (see Fig. 221) identifies itself in the following jingle:

Recen·Pfenning·bin ich·gena(nnt),
Zaig·an·gros ehr·vnd·schan(d),

I am called a counter,
I reveal both honor and disgrace.

In this example we finally see the German name *Rechenpfennig* stamped on an actual counter. Other German names for counters were *Raitpfenni(n)g* and *Zahlpfennig*, as well as *Raitgroschen* in Bohemia. In Dutch or Flemish we find counters called *telpenning* or *reckenghelde*, but more often, from the word *werfen* (to cast), *worpgelt*, *leggeld* (see Fig. 213) or *legpenning*. In England they were called "counters," as we have already seen in quotations from Shakespeare (pp. 365 ff.). The English expression, *to cast accounts*, meaning "to reckon," still preserves the old conception of throwing counters on the counting board. The Spanish language used both names, *contos* (or *contador*) and *giton*, the hispanicized form of the French word *jeton*. Latin books of the period incorrectly call counters either *jactator* (literally "the thrower") and properly *projectile* (from "thrown" or "cast"), in addition to the already well known names *calculus* (Fig. 216), *denarius*, and on one occasion *abaculus*.

German Counters

We know, from the accounts of the City of Frankfurt in 1399 (see p. 377), that counters were in use in Germany during the 14th century. But not until the reign of Emperor Maximilian I (1493–1519) do we find minted or stamped metal counters produced for the official provincial and municipal financial offices in Austria. Emperor Maximilian almost certainly became acquainted with this custom in the lands of his consort Mary of Burgundy, and from there introduced it into his own administration. The two Tyrolean counters we have already mentioned were made in Maximilian's time. On the one shown in Fig. 217 we see a crowned M with the symbol of the Order of the Golden Fleece beneath it, and on the reverse the Tyrolean eagle.

In southern Germany minted counters can be identified with certainty from about 1450 on. From a master of the mint at Würzburg

Fig. 216 French counter of the year 1659. *Ne calculus erret,* "May the counter never err." Here a hand "casts" counters on a table (or counting cloth); cf. Fig. 199. Overseeing both book and reckoning table is the eye of God.

Fig. 217 Tyrolean counter from the time of the Emperor Maximilian I, ca. 1500. Diameter 22 mm.

Fig. 218 The oldest German counter, with the date ano (14)58. Actual size. From the collection of A. König, Frankfurt/Main.

we have what is probably the oldest dated German counter now in existence (Fig. 218). On one side this bears the mint mark, and on the other side a three-cornered star within a trefoil with the legend

Rechen · sere · navwe
vnd · beczal · war · ano (14)58,

Reckon · very · accurately
and · pay · in cash · year (14)58.

This date is, moreover, one of the earliest European numbers written in Indian numerals to appear on a German coin (see p. 439). We also know from written documents that metal counters were minted in Nuremberg at the beginning of the 15th century.

The German *Rechenpfennig*, however, never attained the noble status of the French *jeton*. Although the custom of presenting counters as New Year's gifts may occasionally have been observed here and there, it was evidently not deemed worth recording. Germany was then relatively poor and was still recovering from the wounds of the Thirty Year's War at the time when *Le Roi Soleil* in France could scarcely think of enough costly projects to enhance his own and his country's glitter and glory. The lack of a uniform authority, recognized throughout a country split into hundreds of large, small, and minute independent sovereignties, together with Germany's poverty, prevented the country from achieving any significant political success, in contrast to the Netherlands. New counters in Germany thus tended to be merely stamped out from the old dies. An entry of the year 1569 in the *Oberkammeramtsraitung* (Chief Accounting Office) in Vienna states:

Item den 29. *November Niclasen Einigl sigilschneider allhie von stock und eisen zu gemeiner statt raitpfenningen so in der münz umbgestanden wiederumben zum andern und dritenmal davon zu schneiden und zuezurichten inhalt quitung bezalt* 4 *gulden,*

Further: on November 29 to Nicholas Einigl, seal engraver, here, for cutting and adjusting counters for common use as were on hand at the mint, for the second and third time, paid for attached receipt four gulden.

At any rate, the semiofficial counters minted by specially licensed mints for the personal use of an individual customer or an office (such as the Board of Mines) were far more varied. And yet Germany once led in the manufacture of counters — not official counters, but those intended for commercial use.

When *jetons* were introduced in France, the government very soon issued a decree that only the royal mints were authorized to produce them (see p. 376). From 1672 on their legitimate production was further limited to the mint at the Louvre. The reason for this decree is perfectly clear: Anyone who had the tools and materials for making authorized counters could just as easily use them to produce counterfeit coins.

This monopoly in the production of counters naturally increased their price, so that counters made in the Flemish towns, and above all in Nuremberg, came to be commonly bought. The Nuremberg counters flooded Europe from the 16th to the 19th centuries. This happened first in the Low Countries, where the merchants of Nuremberg enjoyed special trading privileges in individual towns such as Bruges and Ghent. Soon afterwards, however, enormous numbers of Nuremberg counters were brought into France, where there was a great demand for counters and where people were, due to habit, unwilling to calculate with blank counters, to which they had to resort because of the high cost of the French *jetons*. The government of course prohibited the importation of golden, silver, and copper *jetons*. But the Nuremberg merchants struck counters out of brass; at any rate, the prohibition against importation was seldom enforced. Later the French market for counters declined somewhat, but England soon more than replaced it. In England the stamped counters were so highly favored that they were even used as money, and in some places old pieces dating from the 17th and 18th centuries continued in use even into the 19th century. The whole of Germany, of course, was also a market for the Nuremberg counters, but Russia and Poland too had their Nuremberg "dantes" (we have seen this name in the reckoning cloths of Munich, see p. 346; the word itself derives from the Latin *tantum*, "so much," which made its way into medieval commercial jargon, as in the Spanish noun *tanto*, "price" and the Middle High German expression *uf den tant*, "on credit," which later degenerated to "play money" and finally came to mean only a "worthless object"). Later, when the use of counters had already ended completely in western Europe, Nuremberg counters still found their way into Turkey, where, however, they sank to the level of cheap discs used for costume jewelry because they looked at first glance like gold coins (since they were brass). Just as the *jeton*, the presentation and the commemorative counters were the noble relatives of the ordinary counter used for computations; the false "coin" used as play money was its common and degenerate descendant. Nuremberg, continuing the old tradition of manufacturing decorative stamped counters, is still the chief producer of such counters today.

Why precisely Nuremberg? *Nürnberger Tand geht durch alle Land* — "Nuremberg wares are seen in every land." The powerful, increasingly flourishing Imperial City had active commercial relations with all parts of the world and harbored numerous busy craftsmen. Thus traveling merchants soon realized that there was good business to be done with *jetons*, especially in the west. The *Spenglers* ("tinsmiths" — producers of *Spangen*, "clasps and buckles") and workers in brass were the first to produce such counters for the export trade.

But one can sell one's goods abroad only if they are cheaper than everyone else's. This the Nurembergers achieved in three ways. First, they took baser metals, like copper and brass, and they used less metal and made their counters thinner. They also did not stamp them with the same meticulous care as did the French and the Netherlanders. A comparison of the pieces shown in Fig. 219 — a sumptuously ornamented French counter made in Paris and a Nuremberg imitation after the French model (bottom center) — reveals this strikingly. The far more careless surface finish of the Nuremberg specimens can be seen at once, even in the photographs.

Fig. 219 French (*top row*) and Nuremberg (*bottom row*) counters:
upper left: French original, and below it
lower left: Nuremberg imitation. "H. Krawin:[ckel]" is stamped below the bust.
upper center and right: French *jeton* with the ship of the coat of arms of the city of Paris and with the coats of arms of France and Navarre.
lower center: Nuremberg imitation, also with a ship, copied from another French model.
lower right: Another Nuremberg counter, with a representation of the market bridge. Diameter of middle counter in bottom row: 33 mm.
From the collection of A. König, Frankfurt/Main.

Secondly, the Nuremberg traders, with a few exceptions, naturally (lower right), made no designs of their own, but for the most part imitated foreign originals. Thus they killed two birds with one stone: The designs cost them nothing, and the buyer was offered no strange foreign goods — factors that could not fail to increase sales. Of the many such known specimens, our illustration shows an original *jeton* with the bust of Henri IV (upper left), and below it the Nuremberg imitation thereof. Even the earliest known counter made in Nuremberg goes back to a Dutch original, as do the so-called *Rechenmeisterpfennige* (reckoner's counters) or *Schulpfennige* (school counters) (see Fig. 174). Perhaps these were really used in schools for the computation to which the picture alludes; then the alphabet stamped on the reverse side could actually have served as a model to be memorized (lower center in Fig. 174). Likewise there are many exact Nuremberg copies of Dutch counters (see Fig. 214). In the third place, the cheapness of the Nuremberg counters was due in considerable measure to the low wages paid in that city. The Nuremberg craftsmen were very diligent and were kept busy from early morning to late evening.

As business really began to flourish, a regular guild of counter-makers was established in Nuremberg; it had strict by-laws, like all other guilds, and was subject to supervision by the "Honorable Council of the City of Nuremberg." The Town Council kept an especially sharp eye on this particular guild because of the constant danger that its members could counterfeit coins and counters if they were so inclined. Its membership was "closed" and no others were permitted to make counters, and the authorized makers of counters moreover were not allowed to leave the city. Thus the Council, through a number of ordinances, constantly forbade people to "possess stamps that resemble coins and to strike counters" (see Fig. 220).

In 1616 the Nuremberg Town Council ordered:

Es soll auch kein Meister einigen Rechenpfennig, er sey gleich unter ein französisch oder andern gepreg geschlagen, in seiner Werkstatt daraus kommen lassen, es stehe denn sein ganzer Tauff- und Zuname, samt dem Wort Rechenpfennig mit ganzen Worten und ziemlich erhobenen Buchstaben darauf, bey Straff zehen Gulden,

Further: no master is to issue any counters from his shop, whether struck with French or other engravings, unless his full baptismal and family name, together with the word "Counter," unabbreviated, appear thereon in fully raised letters, under penalty of ten guilders.

In fact, below the royal bust of the French ruler on the Nuremberg copy of the first of the counters (Fig. 219) we see one of the ironies of cultural history — the name of the honest German coiner "H. Krauwinckel." The "calculator's counter" also bears his full name, along with that of the city of Nuremberg (see Fig. 174). To Egidius Krauwinckel, the brother of Hanns, the aforementioned Town Council in 1583 sent the following directive:

Egidi Krauwinckel, Rechenpfenningschlager soll man beschicken und seiner geschlagenen neuen Rechenpfennig halber mit der Cron Frankreich / Wappen und der königlichen Pildnuss zu red halten und ihm auf legen, keine weiter von handen kommen zu lassen. [Denn] das Prägen der Bildtnuss hoher potentiaten und was sonst der Müntz ähnlich sei, ist bey eines erbaren rats straff abzuschaffen und zu verpieten,

To Egidius Krauwinckel, counter engraver, word is to be sent, and he is to be reprimanded because of his newly struck counter with the French coat of arms and the Royal portrait and to be enjoined not to let any further ones out of his hands, [for] the engraving of likenesses of high potentates or anything else that resembles the coins is to be done away with at a penalty for the honorable Council and it is to be prohibited.

But the admonition had little effect, and the images continued to be stamped; Nuremberg-made counters with the likeness of Louis XIV of France were sold in large numbers. Still, it speaks volumes not only for the strict and effective supervision by the city's authorities, but also for the honesty of the workmen in this peculiarly enticing industry that, in spite of strong temptation, only one man was ever hanged, in the year 1692, for counterfeiting during the more than four centuries that the makers of counters plied their trade.

The business must have been very profitable, for we find whole families in which the counter-making trade was handed down from one member to another — Lauffer, Krauwinckel, Schultess, and others. Among the best and most prosperous was Hanns

Der Müntzmeister.

Jn meiner Müntz schlag ich gericht/
Gute Müntz an kern vnd gewicht/
Gülden/Cron/Taler vnd Batzen/
Mit gutem preg / künstlich zu schatzen/
Halb Batzen/Creutzer vnd Weißpfennig/
Vnd gut alt Thurnis / aller mennig
Zu gut/in recht guter Landswerung/
Dardurch niemanb geschicht gferung.
J iij   *Der*

Fig. 220 How coins and counters were struck: Woodcut by Jost Amman (died 1591), verse by Hans Sachs.
The Coin Master
In my shop I strike truly
Coin sound as to content and weight,
Guldens, crowns, talers, and pennies
With fine engraving, worthy as art,
Ha'pennies, pence, and farthings.
Also good old turnings; for all men
Of Benefit, in true current value
Through which no one will be cheated.

Krauwinckel, whom we have already met, for one of his counters bears the proud boast:

Hanns Krauwinckel bin ich bekont
in Frankreich und auch in Niederlont,

I, Hanns Krauwinckel, am known
in France and also in the Netherlands.

The counter makers did not always market their wares themselves. Many firms offered their counters for sale through agents, while others, such as Lauffers, produced them in their workshops and also sold them themselves. The new "samples" were offered for sale in the markets of Frankfurt and Leipzig. Nuremberg counters are constantly being found in the sands along the banks of the Main River near Frankfurt. They came into the hands of merchants in Frankfurt through the market there, and later, when they were worn out or outdated, were given as toys to the children, who finally lost them in the sand.

The last of the German counters comes from the Boards of Mines of the Harz Mountains, which used counting boards for their business until the beginning of the 18th century.

In 1748, a hundredweight of ordinary counters cost 78 guilders.

The counter-makers' guild continued to exist until the middle of the 19th century, although by then no counters had been used for a long time. For more than two hundred years the Nuremberg counters had played a dominating role throughout Europe. But as the counting board disappeared, they too came to an end. There is an unconscious historical irony in the fact that a Tyrolean counter (Fig. 221) shows a counting board with counters on one of its sides, while on the reverse appears its mortal enemy — the Indian numerals and a problem in the new style of computations: $178 \div 2 = 89$! In France, it was the Revolution that gave the *coup de grâce* to the *jetons*, which were quite correctly regarded as a relic of the *ancien régime*; under Napoleon, however, the presentation and commemorative counters again attained a highly flourishing state, for the last time.

With the death of the *calculi*, we finally come to the end of our discussion of the counting board and the abacus. This has taken us on a long journey along paths that were obscure and unexpected, but rich in wonderful and rare insights into the history of our culture. Along with "its own" Roman numerals, the counting board for a long time tenaciously resisted the incursions of the Indian numerals.

But to understand the struggles and the final victory of the Indian numerals, we must first go back in time once more.

Fig. 221 Tyrolean counter, showing a counting board (321; 3260?) and Indian numerals (in the division $178 \div 2 = 89$). Diameter 22 mm. From the collection of A. König, Frankfurt/Main.

Our Own Numerals

Place-Value Notation

*Sed hoc totum quasi errorem
computavi respectu modi Indorum,*

"But all this I have regarded as a
transgression against the operations
of the Indians."

Leonardo of Pisa (Fibonacci), 1202.

1 2 3 4 5 6 7 8 9 and 0 — these ten symbols which today all peoples use to record numbers, symbolize the world-wide victory of an idea. There are few things on earth that are universal, and the universal customs which man has successfully established are fewer still. But this is one boast he can make: The new Indian numerals are indeed universal.

We have just finished discussing the Roman numerals and the counting board of the Middle Ages; now we suddenly find ourselves face to face with the new written numerals. What has happened?

The ordering and grouping that governed the old systems of written numerals gave way to gradation: the nine unit steps ran up each successive level of the first rank, from 21 through 29, then from 31 through 39, and so on (see Fig. 14); the levels of the first rank, together with the unit-steps, then mounted up through the second rank from bottom to top, and so forth. Now the structure of the written numerals finally corresponded to that of the spoken numbers, in which there are two kinds of "numbers" — units and ranks. It is the units which impart number to the ranks.

What was the structure of an early system of written numbers, such as the Greek (see Fig. 97), the Egyptian, or the Roman (see Fig. 8)? Here there was only one kind of symbol, I, X, or C, which was used only for the ranks or gradations. These numerals are essentially symbols for groupings, not numbered by units but themselves counted by lining them up in order: CCCXXIIII = 324.

China, on the other hand, had separate symbols for the units and for the ranks or gradations. The Chinese — here we use Indian and Roman numerals as substitutes for ideograms — would write not CCCXXIIII, but 3C 2X 4. Our own system, using the Indian numerals, by contrast puts down only the units: 324. The ranks are not actually written down, but are represented by their "positions."

Thus the two systems of written numerals, the Indian and the Chinese, are in essence no different: both arrange the numbers by gradations, and both clearly differentiate the units from the numerical ranks. Both are forms of gradational or "place-value" notation, if we consider a numeral's place as just another way of indicating its rank. The Chinese is a "named," the Indian an abstract place-value notation; the former specifies, or "names," the ranks (= gradations), the latter does not. These two varieties of developed number systems, as we have said, differ fundamentally from the primitive forms of numerals, all of which recognized only ranks or gradations (= groupings) and indicated the number of these not by placing a numeral before or behind them, but merely lined them up in order to any desired number (or to the next higher grouping).

Today in China and Japan numbers are also written in abstract place-value notation, either horizontally or vertically (see Fig. 275). But, surprisingly enough, up to 100 this also occurs on the very ancient Chinese fork coins (Fig. 222).

Only the abstract place-value notation with no symbols for the different ranks needs to introduce a zero, a symbol for a missing

Fig. 222 Chinese fork coin, with the number 34 written in abstract place-value notation. Ca. A.D. 20.

[391]

rank. Since none of the ranks are written down, it must somehow indicate that one of the ranks which are never written is missing. Let us recall the very instructive example

$$I \cdot V^c \cdot V,$$

which shows clearly that the medieval European scribe had already heard about the "new" place-value system and now tried to find it in the Roman numerals (see p. 285). Since the meaning of the zero was still not fully clear to him, I V O V = 1505, at the critical point he yielded and retreated into the "named" place-value notation.

Medieval Europe never achieved a logical "named" or expressed place-value notation, although this was not only at hand in visual form on the counting board but was heard constantly, day after day, in the spoken number sequence. From our modern vantage point, this looks like a peculiar form of mental block. This curious failure of imagination has already been mentioned (see p. 298). The fact remains, of course, as an anomaly in the history of the development of numbers.

Spoken numbers, being older, also served as the conceptual model for the counting board — where else could the abacus have derived its basic pattern? Most cultures were able in one way or another to reproduce spoken numbers in "palpable" form on a counting board. But only the Chinese succeeded in representing them in "written" form in a named place-value notation (see p. 53), and only the Indians in an abstract place-value notation.

Now let us examine the evolution of this Indian system of written numerals and its migrations through cultural history.

The Antecedents of Our Numerals

This history of India for the most part lies beyond our ken. Nevertheless we should be familiar at least with its highlights, so that we may fully realize how late — as compared to Egypt and Babylonia, and also Greece — it was when India finally made her great contribution to the development of modern written numerals.

The Indus Valley culture flourished around the middle of the third millennium B.C. In the ruins of Mohenjo-Daro, which were excavated for the first time only fairly recently, a pre-Hindu form of writing has been found which has yet to be deciphered. Nothing of significance is known about any numerals.

Some thousand years later, around 1500 B.C., the Indo-European Aryans streamed into the Ganges River region from the northwest, drove back the peoples already living there, and forced them into the lowest, servile castes. The only rulers were the members of the warrior castes and the caste of priests, the Brahmins. The latter made themselves the guardians of all knowledge and saw to it that not a drop of education filtered down to the common people. Sanskrit, the language of the Aryans, which was once predominant in India, soon died out as a common vernacular and lived on — as it does to this day — only as an arcane language of learning. The exclusiveness of the Brahmins was the cause of their opposition to writing; the Vedic religious hymns were handed down for the most part orally. To make them easier to memorize and to guard against change, they were composed in verses, into which, as we have already seen (see p. 275) numbers also had to be fitted. If we add to this circumstance the Indians' complete lack of any sense of their own history, which, until recently, has been the chief cause of their lack of political independence, we can understand why we have so little knowledge of the development of Hindu culture in India itself. What a contrast to the Egyptians, who did not build a single temple or monument or burial vault without recording on it events of history and of everyday life in the fullest detail!

The 6th century B.C. saw the beginnings of Buddhism, one of the folk religions aimed against the exclusiveness of the Brahmins. Now, finally, a rich literature began to appear. Beginning with the life of Buddha (560–483 B.C.), reliable records of historical events were kept in India.

Buddhism reached its zenith as a flourishing official religion around 250 B.C., in the reign of the great king Asoka, who established an empire extending over almost the whole of India. Monuments bearing his proclamations are to be found everywhere.

Of no small importance for the history of numbers was the fact that the northwest part of India (Gandhara) was a part of the Persian Empire from the 6th centry B.C. on. For when Alexander the Great between 327 and 325 B.C. overran this land in the course of his conquest of Persia, not only Persian but also Greek culture (including mathematics and astronomy) took refuge in India. The resulting mixture of East and West can best be seen in the so-called **Gandhara** art which continued to flourish well into the 5th century A.D. By this route, as well as others, India came into contact with Egyptian, Babylonian, and Assyrian ideas.

Indian culture reached its highest level between the 4th and 9th centuries A.D. Kalidasa, the poet of the "Sakuntala," lived in the 5th century.

Two developments decisively affected the intellectual life of India during the period following. From the 8th century on, Buddhism began to give way to Hinduism and to a resurgent Brahmanism, and ultimately to Islam, which the Arab invaders introduced into India in A.D. 712. By around 1200 the religion of Buddha had almost completely disappeared from the land of its birth.

THE KHAROSTHI NUMERALS

The Hindu written numerals made their first appearance in the period from the 6th to the 8th century. How much is known about their origin? In general, the following: There were two forms of writing in India, the *Kharosthi*, which originated in the northwest and was in use only from the 5th century B.C. to the 3rd century A.D., and the much more important *Brahmi* writing, which was the mother of all (200 different) Indian alphabets, including the (Deva-)Nagari alphabet which is most widely used today (see Fig. 243). After the 11th century Sanskrit poetry was in general written in the Devanagari alphabet.

The Kharosthi writings are read from right to left, the Brahmi from left to right. The former, which was expressly a scribes' and merchants' alphabet, developed by way of Persian from Aramaic. The origin of the Brahmi script is disputed. In the Indian view it was indigenous, but it is more likely that it developed from the northern Semitic (perhaps Phoenician) group of alphabets (see Fig. 94). The majority of Asoka's decrees (including the Gvalior inscription shown in Fig. 226) were written in Brahmi characters.

Numbers appear in both forms of writing, although quite different in nature. Thus there have been essentially three kinds of Indian written numbers — the Kharosthi, the Brahmi, and, in the third place, the familiar place-value notation with the zero sign which we use today, which made use of the Brahmi numerals and developed directly out of them.

Now let us look at these Indian written numbers in detail. We mentioned the Kharosthi numbers previously when we discussed the formation of the number sequence, because they show a peculiar 4-grouping, and then a decimal and a vigesimal grouping (see Fig. 12). The symbol for 20, which looks like our 3, is made up of two 10 signs. The hundreds were counted, not ordered (2C, not CC)— evidence that there was an old counting limit at 100.

THE BRAHMI NUMERALS

The Kharosthi numbers, however, are not among the ancestors of our present numerals, while the Brahmi numerals are. From the time of Asoka on, many Brahmi inscriptions have been found all over India, on copper plates, on temple walls, and on rock faces, in which individual numerals occur in isolation. Over this entire vast territory and throughout the long span of almost a thousand years, the numerals have retained essentially the same forms. A still more important aspect is the constant structure of these written numbers. Their underlying principle is no longer ordering and grouping; now each of the units has its own individual symbol, a "digit," just as every number has its own number word in the spoken language (Fig. 223).

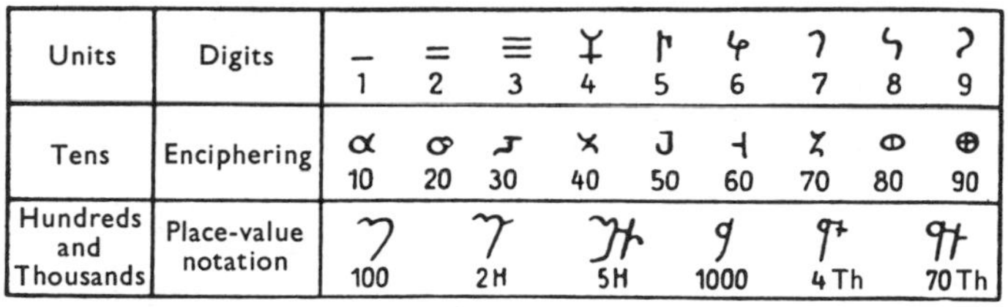
Units
Digits
1 2 3 4 5 6 7 8 9
Tens
Enciphering
10 20 30 40 50 60 70 80 90
Hundreds and Thousands
Place-value notation
100 2H 5H 1000 4Th 70Th

Fig. 223 Indian Brahmi numerals.

The long step that has been taken toward a mature system of written numerals is perfectly clear. But in very early times something happened which partially negated this great advance: the tens, 20, 30, 40, 50, 60, 70, 80, and 90, were also "enciphered"—each of them also acquired its own separate symbol (see second line in Fig. 223). This may have been the fault of the language itself, which did not express the connection between the units and the tens as distinctly as that between the units and the hundreds (see p. 81). And in the case of the hundreds and the thousands, the Brahmi script fashioned a true "named" place-value notation out of the digits for the tens and units. A really outstanding example of the slow step-by-step evolution of a system of written numerals! A form like the last one shown (70,000) embodies the clear limit of counting at 1000 (see bottom line in Fig. 223).

The history of written numbers provides an example of a still more striking "encipherment" of numbers. This is the Egyptian "hieratic" script which developed from the pictographic hieroglyphs carved on monuments (Fig. 224); it will be readily seen that many of the

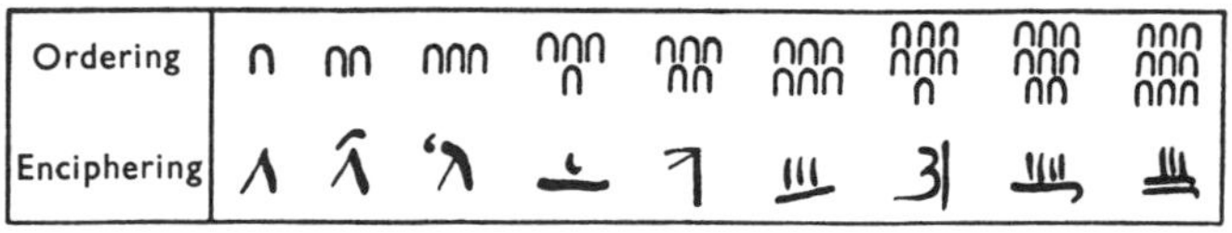
Ordering
Enciphering

Fig. 224 Egyptian ordered pictographic numerals (top) and "enciphered" letter numerals (bottom).

numerals, such as 20, 60, and 90, are merely abbreviations of the ordered pictographic numerals used earlier. It is a very instructive example from the standpoint of the development of numerals, since here, because of the Egyptians' limited feeling for mathematics, they went directly from the "too little" of the ordering and grouping principles to the "too much" of "encipherment."

This Egyptian hieratic script itself used digits only through the number 9000; then from 10,000 through 40,000 it suddenly reverted to ordering, and from 50,000 went on to a "named" place-value notation: a unique case of advance and retreat in the development of written numerals, which we have previously seen in spoken numbers as well (Fig. 225).

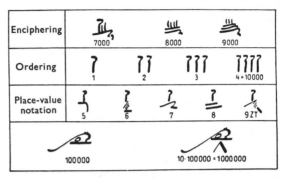

Fig. 225 Egyptian written numerals of the so-called "hieratic" script of the papyrus rolls (as distinguished from the hieroglyphs used on stone monuments, as in Figs. 9 and 10).

The same stage of development is also represented by the letter numerals (such as the Greek) and the Siyaq script, since these too assign individual digits to all the units, tens, hundreds, and thousands (see table on p. 270 and Fig. 109).

PLACE-VALUE NOTATION

The springboard for the positional principle was not ordering and grouping but "encipherment," i.e., the assigning of an individual digit to each of the first nine numbers. The Indian and the Egyptian numerals both overshot this target. But whereas the Egyptian system came to a premature halt because the hieroglyphs which preceded it already contained a set of numerals that served as a model, the Indian system was merely an early stage, a first growth capable of developing further. One hundred was already a "high" number, and here the governing principle of the number system changed to a place-value notation; at 100, too, the numerical rank clearly made its appearance. As soon as this concept reverted back only one level, to the rank represented by 10, a "named" place-value notation was born.

And this is just what happened. Around A.D. 600 a system of numerals appeared which used only the first nine digits of the Brahmi numerals — that is, only the Brahmi digits for the units. In this new system "nine hundred thirty-three" was no longer written 900'30'3 as in the Brahmi manner (using the symbols illustrated in Fig. 223), but now only with the units in a place-value notation, 933, as in the inscription from Gvalior (see Fig. 226). With this step the transition to an abstract place-value notation was complete, and we now have

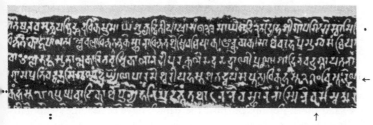

Fig. 226 The Gvalior inscription, with the earliest known written zero. In the fourth line from the top (marked by the arrow) there is the number 270 in place-value notation; in the top line (marked by a dot) there is the date 933 (equivalent to A.D. 870); and in the fifth line (marked by two dots) there appears the number 187.

in the first nine Brahmi numerals the oldest ancestors of our own digits (top row of Fig. 223). From now on the only differences were to be the inevitable changes in form which resulted from the fact that these numerals passed through many hands — Indian, Arabic, and Western — before they finally took on the appearance which they have today (as we shall show on p. 418).

How did this change to a place-value notation come about? The fact that it took place in India is not disputed. But Indian researchers believe that the place-value notation arose sometime around 200 B.C. in India without further stimulus from outside, while non-Indian scholars have substantial grounds for seeing an external stimulus behind this development. There is no conclusive evidence for either view, and there probably never will be. We can merely put down the various points in favor of the "Indian" and the "foreign" hypotheses and weigh them against each other.

The Indian number-towers which go back in time to the 5th century B.C. and lay one "story" of numerical rank upon another, each with its own name, clearly build up a number after the manner of a tall building with many levels. Thus they "write down" numbers in a form of "named" place-value notation: 26432 = 2 *ayuta* 6 *sahastra* 4 *śata* 3 *dasa* 2 (see p. 137).

A further stimulus was provided by the counting board, which in its columns precisely mirrored this verbal pattern. This comes close to representing a number in terms of a place-value notation — to "grasping" the number in this form. The Romans, who built no number-towers, also had the abacus, to be sure, but a people as fascinated with numbers as the Indians doubtlessly sensed the close relationship between the abacus and place-value notation.

The Indians made computations on tablets covered with a thin layer of sand. This we know not only from documents, but also from the fact that the Indian expression *dhuli-karma*, literally "sand-work," is used to mean "higher computations." It is shown even more strikingly by the "erasure operations" from which, as we have already seen, the medieval European "strike-out operations" developed (see p. 330). But on a counting board on which the columns were drawn in sand, one could hardly push counters back and forth without soon erasing the lines. Thus the Indian computers probably wrote down two numbers to be added, such as 1803 and 271, on the reckoning board in Brahmi numerals. Their combination into the sum 2074 could then easily be accomplished by erasing and rewriting individual digits.

And how was the result, 2074, then recorded? It was written down either verbally or symbolically or — what is just as likely — in the same Brahmi numerals with which the computations were made on the sand-table. For this purpose there must have been some symbol to indicate a missing place, such as a dot: 2·74.

Thus it was not the celebrated zero which made possible the invention of a place-value notation using digits, but the digits themselves. These constituted the most important aspect of the new invention. The zero, to be sure, first liberated the digits from the counting board and enabled them to stand alone.

Finally, we must go back to the peculiar method of representing numbers by symbolic objects, which appeared around A.D. 600. In this symbolic system 1 and 2 were expressed by "moon" and "wings," respectively, so that things or concepts were transformed into digits. These symbol-digits were then put together in an abstract place-value notation, in increasing order of magnitude from left to right, to form numbers (as the example on p. 120 shows). Not only the first nine, but some fifty numbers in all were represented by their own individual symbols (32, for example, was expressed as "teeth"). This came close to a thoroughgoing "encipherment" like that of the early Brahmi numerals (example: 321 was written as "moon-teeth").

Thus a place-value notation was already present, but with the order of magnitudes reversed and with more than just the units represented by their own digits. Although this method of representing numbers prepared the ground for an abstract, "unnamed" place-value notation, it also prevented its further development.

Now for the hypothesis of an external stimulus, for which there are also some weighty arguments. There is documentary evidence that Babylonian astronomical writings had a significant influence on Indian astronomy in the century following the campaigns of Alexander the Great, during which Hellenistic culture spread farthest to the east and deep into India. In these late Babylonian texts numbers were no longer written on a sliding scale, but with a fixed order of magnitudes, since a zero symbol was now used to indicate an empty space not only within a number but also at the end (for instance, "thirty thousand four hundred" would be written 30400 and not 3 4 as in the old Babylonian style: see p. 167). This Late Babylonian method of writing numbers is the exact equivalent of a mature abstract place-value notation and thus of the Indian numerals, the only difference being that the base number was 10 in the Indian and 60 in the Babylonian system. But this Babylonian method of writing numbers was used only in astronomical and not in mathematical writings (perhaps to avoid confusion in reading the numbers), and only after about A.D. 200, the very time when, according to the results of Indian researches, the Brahmi place-value numerals made their first appearance in India. Since Indian scholars borrowed concepts and expressions from these Babylonian-Greek texts, why could they not also have developed a place-value notation using Brahmi digits based on this model?

As far as the zero is concerned, the Greek astronomer Ptolemy was familiar with the symbol o — an abbreviation of the Greek word *oudén*, "nothing" — as a sign indicating a missing place, and used it in writing Babylonian sexagesimal fractions, for example $\overline{\mu\alpha}\ o\ \overline{\iota\eta}$ 41°00'18" and $o\ \overline{\lambda\gamma}\ \overline{\delta}$ 0°33'04". Thus the symbol o indicated not only the absence of a fractional group, but even the absence of an integer (a degree). This means that the Greco-Babylonian model already possessed a zero symbol which could have been the stimulus for the creation of a zero in the Indian numeral system — perhaps it even influenced the form of the Indian zero, for later on the zero in India was written as a small circle instead of a dot.

Such are the sets of facts to be weighed against one another. No culture ever takes over a foreign concept for which it is not already intrinsically prepared. Cultural history shows this over and over again, as in the abortive career of the *apices* on the monastic abacus in the early Middle Ages. Indian mathematics was ripe for a place-value notation. The use of the counting board must have brought this mathematically adept people directly to the threshold of a place-value notation as soon as individual digits had been developed and assigned to the first nine numbers, the units. A conspicuous argument against this hypothesis is the reverse order of magnitudes in the system of verbal symbolic numerals; this indicates that the last step from the abacus to a place-value notation had not yet been taken. In writing numbers, the old custom prevented the new concept from breaking through. Hence it is quite conceivable that the final impulse toward the development of the Indian place-value notation came from the Greco-Babylonian model. For the Indians this step cannot have been a very big one.

This is how it could have happened — and with this "could have" we shall have to be satisfied. One can scarcely expect to draw a more distinct picture from the mists of the reconstructed past. But we shall soon be on firmer ground again. With the adoption of the zero, India finally attained the abstract place-value notation which was about to begin its journey of conquest through the world as the most mature and highly developed form of numerals.

The Westward Migration of the Indian Numerals

THE ZERO

What do the Indian numerals have to do with migrations? Almost immediately we run into a unique connection. The zero in a number, such as 1505, which as an empty column on the counting board had been familiar to the Greeks, the Romans and the medieval Western monks from earliest times, remained an obstacle which none of the three was able to overcome in a columnless system of numerals. The conceptual difficulty may have been this: The zero is something that must be there in order to say that nothing is there. And just as the early calculator on the counting board put down twenty as two tens, XX, because ten is present twice, he was accustomed to representing "nothing" by an empty space or column, because there is nothing in it. Then the place-value principle got in the way. For safety's sake, it had to be made a "named" system, as in the highly revealing example $I \cdot V^c \cdot V = 1505$. For the primitive reckoner number is always *a number*, a quantity, and only a number can have a symbol. Thus it is easy to see how the zero came to be the great stumbling block for the medieval arithmeticians in the West. They found it very hard to give up the old principle of ordering and grouping, in which every unit has its own symbol, which appears when that unit is present and is lacking when the unit is absent.

The spoken language very neatly avoided this conceptual difficulty presented by the zero: the whole system of Indian numerals maintains its names — after the zero. And not only that: In the changes that these names for the units underwent as they passed from one culture to another we can trace their entire migration from east to west, milestone by milestone.

This long journey begins with the Indian inscription which contains the earliest true zero known thus far (Fig. 226). This famous text, inscribed on the wall of a small temple in the vicinity of Gvalior (near Lashkar in Central India) first gives the date 933 (A.D. 870 in our reckoning) in words and in Brahmi numerals. Then it goes on to list four gifts to a temple, including a tract of land "270 royal *hastas* long and 187 wide, for a flower-graden." Here, in the number 270, the zero first appears as a small circle (fourth line in the Figure); in the twentieth line of the inscription it appears once more in the expression "50 wreaths of flowers" which the gardeners promise to give in perpetuity to honor the Divinity.

Now let us follow the wanderings of the zero through space and time. There are few things to which the following quotation from *Faust* applies better than to this: *Bei euch, ihr Herren, kann man das Wesen gewöhnlich aus dem Namen lesen*, ["In your case, my lords, the nature can generally be seen in the name"]. The zero, of course, is no Devil, but during the Middle Ages it was often regarded as the creation of the Devil. Here it is not the nature of the zero which can be deduced from its name, but the countries through which it traveled as it migrated from east to west:

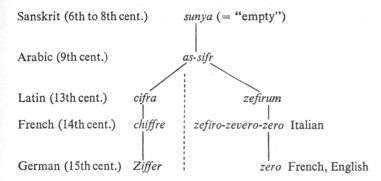

Sanskrit (6th to 8th cent.)   *sunya* (= "empty")

Arabic (9th cent.)   *as-sifr*

Latin (13th cent.)   *cifra*   *zefirum*

French (14th cent.)   *chiffre*   *zefiro-zevero-zero* Italian

German (15th cent.) *Ziffer*   *zero* French, English

In Sanskrit the zero was called *sunya*, "empty" (also *sunya-bindu*, "empty-dot") after its physical meaning: the position (originally on the counting board) is empty. The modern custom of indicating a missing word or line of verse by a row of dots goes back to this Indian practice.

The intermediaries between India and the West were the Arabs. When they became acquainted with the zero in the 9th century, they translated the Indian name *sunya* literally into the Arabic *as-sifr*, "the empty." (A more precise transcription of the Arabic word would be *aṣ-ṣifr*, indicating that in pronouncing this sound ṣ the tongue touches not the top row of teeth, but the front of the palate; the simple consonant *s* indicates a different meaning in Arabic.)

The West, when it became aware of the new digit, no longer translated its name but took over the symbol as well as its name, from the Arabic, and transformed the latter into the learned Latin form *cifra* and *cephirum* (as Leonardo of Pisa wrote it; see p. 425) and even the Greek word τζίφρα (after Maximus Planudes, whence the abbreviation τ for zero in the example on p. 287).

The two Latin words then gradually worked their way into the vernacular: in Italian the second form was changed to *zefiro*, *zefro* or *zevero*, which was shortened in the dialect of Venice to *zero* (like *libra* > *livra* > *lira*).

The French formed the word *le chiffre*, but also took over the Italian form *zero* and thus had two different names for "zero." The first of these was soon extended to the other nine symbols and came to mean "digit" (cf. "cipher"). A French arithmetic textbook for merchants dated 1485 says:

Et en chiffres ne sont que dix figures, des quelles les neuf ont valeur et la dixieme ne vaut rien mais elle fait valloir les autres figures et se nomme zero ou chiffre,

The digits are no more than ten different figures, of which nine have value and the tenth is worth nothing [in itself] but gives [a higher] value to the others, and is called "zero" or "cipher."

Two names for the same thing (the zero), and then one name (*chiffre*) for two different things (for zero and for the units which have numerical as well as place value) and both still in use today — there could be no better symptom of the confusion and insecurity which the zero produced in the minds of people in the Middle Ages. It is likewise manifested in the Italian *cifra* and *zero* and in the English *cipher* and *zero*. From "cipher" derives the English schoolboy term "ciphering" in the sense of "computation."

In German, however, the numerals were then generally called *Figuren*, so that the word *cifra* was confined to the zero, and continued to be used thus in scholarly Latin, even as late as the German mathematician Gauss (1777–1855). The German arithmetician Köbel in 1514 spoke of the nine *bedeutlichen Figuren* ("significant figures") and the one *Zeiffer* ("cipher").

This expresses the distinction which people commonly drew in the 13th century. Around 1240 the scholar Johannes de Sacrobosco, one of the most celebrated teachers at the University of Paris which had been founded in 1206, composed an introduction to the new Indian numerals and their use in arithmetic computations, a so-called *Algorismus*, which came to be used all over Western Europe and continued to be copied in manuscripts until the 17th century. In this *Algorismus* Sacrobosco presented the new numerals as follows (Fig. 227):

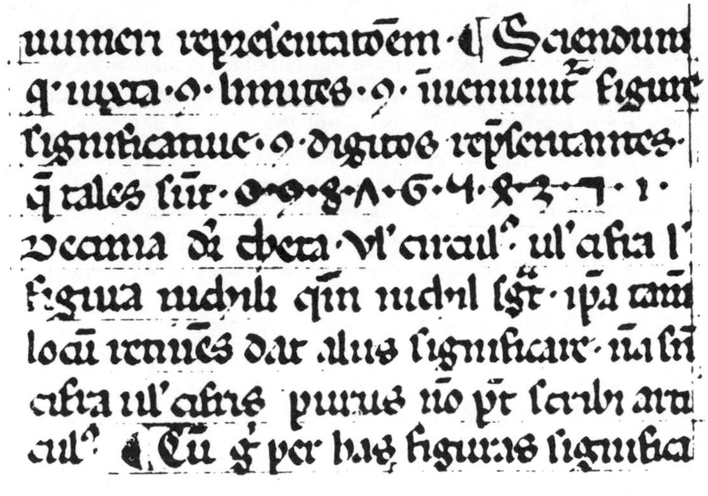

Fig. 227 The Indian numerals and the zero as presented in the *Algorismus* of Sacrobosco (died 1256); in the fifth and sixth lines from the top are the medieval names for the zero. The numerals themselves are written in red ink in this beautiful 15th-century manuscript.
Hessische Landesbibliothek, Darmstadt.

Sciendum quod iuxta 9 *limites* 9 *inveniuntur figure significative* 9 *digitos representantes qui tales sunt*

0 9 8 7 6 5 4 3 2 1

decima dicitur theca vel circulus vel cifra vel figura nihili quoniam nihil significat. ipsa tamen locum ten/n/ens (?) *dat aliis significare. nam sine cifra vel cifris purus non potest scribi articulus,*

Know that corresponding to the 9 units there are 9 number symbols, as follows:

0 9 8 7 6 5 4 3 2 1

The tenth is called *theca* ["hollow cup" in Greek] or *circulus* or *cifra* or *figura nihili*, because it stands for "nothing." Yet when placed in the proper position, it gives [higher] value to the others. For a pure *articulus* number [one that is divisible by 10] cannot be written without one or more zeros.

The passage here quoted is significant in many respects for the history of numbers. Quite apart from the fact that it is from one of those few but important writings which introduced the numerals into the West for the second time, after the decline of the *apices*, it is documentary evidence that the nine Indian numerals were called *figurae* in the 13th century. This name was retained in English and French, but not in German. Moreover we can see how the zero acquired its name of "null": It is the "figure of nothing" and thus no numeral, no figure at all, *nulla figura*, in Latin. The word *nulla* as a noun probably appeared for the first time in an Italian arithmetic textbook of 1484.

The other names given for the zero refer to its form: *theca* was the circular round brand that in the Middle Ages was burned into the cheek or forehead of convicted criminals; in reference to the Latin name *circulus*, "little circle," the master arithmetician Köbel called the zero the "*Ringlein 0, die Ziffer genannt, die nichts bedeut,*" "the little ring 0, that signifieth nothing."

Such were the names for the zero: they reflect almost a thousand years' movement and change in cultural history which we shall now attempt to trace in greater detail.

As a sign for zero the Indians used a dot, later a circle and quite often also a cross. They had customarily used dots as a pledge to carry out unfulfilled tasks — this may perhaps have influenced the shape of the zero sign. The small circle may have been suggested by the Brahmi numeral for 10 (see Fig. 223) or perhaps by the abbreviation of the Greek word *oudén*, "nothing."

Poets and philosophers have both been fascinated by the zero, by its curious changes in numerical value and by the magic which it can create, although in itself it is nothing. So it is called in a Hindu saying:

Ten men live in such a way that they allow one to take precedence. Without this one they have as little meaning as zeros [unless they are preceded by a "one"].

King Lear's abandonment is made clear to him by his Fool in the following words (*King Lear*, Act 2, Scene 4):

Now thou art an O without a figure. I am better than thou art now. I am a fool, thou art nothing.

The Zero in Other Cultures

There was a symbol for a missing rank used within numbers in the later Babylonian numeral system from the 2nd century B.C. on (for the numerals themselves, see pp. 163 ff.). This consisted of two small slanted wedges, which however could indicate a missing numerical rank only within a number, not the order of magnitude itself. Thus the Babylonians were able to write 304 (in their numerals), but not 340 or 3400. Hence in the sexagesimal system $46{,}821 = 13 \times 60^2 + 0 \times 60 + 21 = 13'0'21$ (see Fig. 228).

The symbol for a missing digit in the Persian *siyaq* script resembles the Babylonian (see Fig. 109); there are probably solid historical

Fig. 228 The Babylonian vacancy symbol.

grounds for this, since prior to the advent of the Arabs the Persians wrote in cuneiform characters. The old symbol for a missing digit used by scribes and merchants may have been preserved from this era.

Then there is the *o*, the initial letter of the ancient Greek word *oudén*, "nothing," which may well have influenced the form of the Indian zero (see p. 399).

Now we come to a missing-digit symbol used by a culture far removed from the civilizations of the Old World, that of the Mayans. We have already discussed their peculiar number sequence based on gradations of twenty (pp. 59 ff.). The calculations we find today on their monuments and in their documents are concerned exclusively with the calendar. The Mayan year was divided into 18 months of 20 days each, and then 5 more days were added on to fill the interval between this and the solar year: $18 \times 20 + 5 = 365$. For their calendar computations the Mayan priests used two forms of numerals, a very extraordinary kind of pictorial numeral carved on monuments, and an earlier set of numerals based on ordering and grouping, which was unique in being an abstract place-value notation based almost entirely on gradations of twenty, along with a zero.

Corresponding to the subdivisions of the year, the Mayan vigesimal ranks were as follows: 1, 20, 18×20 (= 360), 360×20, $360 \times 20 \times 20$; the artificial device patterned to suit the calendar derives from the deviation in the second level of rank, which the number sequence does not follow. The ranks of the numbers carved on monuments and buildings were represented by grotesque, headshaped pictographs, but the counting was done by means of the 19 numerals from 1 through 19, the "units" (Fig. 229). These are quite simple: 1, 2, 3, and 4 are represented by the appropriate numbers of dots, grouped into 5 by a horizontal stroke; thus, for example, $\overset{\bullet\bullet}{\rule{1em}{0.4pt}} = 7$, $\overset{\bullet\bullet\bullet}{\rule{1em}{0.4pt}} = 13$.

These early numerals can be seen to the left of the bizarre head-numbers. The pictographs are read as follows (R = rank):

$9 \times$ 4th R $(20^3 \times 18) = 1,296,000$;	$14 \times$ 3rd R $(20^2 \times 18) = 100,800$;
$12 \times$ 2nd R (360) $= 4,320$;	$4 \times$ 1st R (20) $= 80$;
17×0 R (1) $= 17$;	

the two numbers 12 and 5 at bottom right are combined with designations of the year and the month.

There is no doubt that the original numerals, which are combinations of dots and short lines, are not synthetic, "learned" constructions but arose from the people. Their 5-grouping up to 20 is remarkably consistent with the spoken number sequence of the ancient Aztec language, so that they may well have originated somewhere in the neighboring region (see p. 62).

These early numerals in the Mayan manuscripts generally lose their connection with the secret priestly numerals and proceed from a "named" place-value notation — the one illustrated in Fig. 229 — to an abstract place-value notation with a zero sign (Fig. 230). But the vigesimal gradation retains its deviation at the second rank (18×20 instead of 20^2). It too was patterned on the calendar, which was understood and calculated only by the priests. This method of writing numbers was never used by ordinary people. Thus we have an enigma of cultural history: an abstract place-value notation with a zero,

Fig. 229 The Mayan "named" place-value notation. The heads are rank levels which are numbered by units from 1 to 19. The vertical beams are 5-groups; curiously enough, there was no decimal grouping.

Fig. 230 Mayan abstract place-value notation containing the oldest zero in the New World. The zero sign looks like a snail's shell. The units are the early numerals with a quinary grouping, and there are no numerical ranks. The three Mayan numbers illustrated, read from top to bottom, are as follows:

left: Maya $\quad$ 820 = 8 × (20 × 18) + 2 × 20 + 0 = Indian 2920;
center: Maya $\quad$ (16)40 = 16 × (20 × 18) + 4 × 20 + 0 = Indian 5840;
right: Maya 9(10)502 = 9 × (20^3 × 18) + 10 × (20^2 × 18) + 5 × (20 × 18) $\quad$ + 0.20 + 2 = Indian 1,369,802.

based on the number 20, occurring in apparent isolation, far away in the New World. A native invention, or an indirect borrowing from India? Or was it even borrowed directly? Since the extinct Mayan culture was at its height during the period from the 6th to the 11th centuries A.D., China may be eliminated as the intermediary, for the zero sign was first brought from India to China in the middle of the 13th century.

To sum up: The Mayan numerals were not developed anonymously for general use, but were rather a special system created artificially for purposes of calendar computations, an arcane and "holy" system of numerals belonging to the priests alone. It was the counterpart of the "sacred" number-towers based on levels of 20 in the Mayan language (see p. 62). Similarly, the simple "unit" numerals made up of dots and lines and based on a quinary grouping correspond to the everyday structure of the lower numbers.

Now let us turn back to the Old World and follow the Indian numerals from their home in India through their migrations to the West.

ALEXANDRIA

A glance at the map of the lands bordering the Mediterranean Sea (see Fig. 231) will show that at its Eastern end the cultural ground was fertilized by the ancient empires of the Egyptians, the Babylonians, and the Greeks. When toward the end of the 4th century B.C. Alexander the Great left his tracks in all these countries, as he made his way past Mesopotamia through Iran and all the way into India, the scanty contacts previously maintained by trade now swelled into a lively exchange of intellectual and cultural goods as well. In Egypt the city of Alexandria, founded at the mouth of the Nile, became the center of Hellenistic learning and civilization. Even the earliest rulers of the city established the famous library and the *Museion*, where scholars, foreigners as well as Greeks, were supported by the state in their dedication to the increase of knowledge. Here, in the 3rd century B.C., we find the mathematicians Euclid, Eratosthenes, Apollonius, and Aristarchus. Around A.D. 150 the famous astronomer Ptolemy (Claudius Ptolemaeus), whose great work, *Hé Megálē Sýntaxis* ("The Great Compendium"), of mathematical and astronomical knowledge, best known by its Arabic title of *The*

Almagest, laid the foundations for the medieval pre-Copernican cosmology (Arabic *(al)-magest* < Greek *hè megístē*, "the greatest compilation," as distinguished from the "small" compilation of works by Alexandrian scientists which served as an introduction to his "great" synthesis).

It is quite probable that because of the active commercial relations with India, the first Indian numerals became known in Alexandria sometime in the 5th century, shortly after their invention. At any rate, they did not arrive in Egypt from India as a scientific treasure, but rather like the written numerals of alien peoples that become known in the harbors and ports everywhere where foreign wares were imported and traded. They do not appear at all in scientific writings, unlike the Babylonian sexagesimal gradations which Ptolemy had adopted for writing fractions. But the merchants of Alexandria may very well have known the 9 Indian numerals, which of course without a zero would not have been considered at all remarkable; from Alexandria they would have penetrated farther westward in the same way that they arrived from the Orient.

Yet this path west by way of Alexandria remains completely obscure. There is not a single shred of documentary evidence for the use of Indian numerals in Egypt at that time. It can only be inferred as highly likely because of the great economic and cultural importance of Alexandria during these centuries, and also because the "Arabic numerals" which were used in 10th-century Spain had forms somewhat different from those which were later written by the Arabs in the East.

THE INDIAN NUMERALS IN ARAB HANDS

A brief survey of the history of the Arabs will bring out their importance for the culture of the West. An unusual series of events raised them from their former historical obscurity to masters of a universal empire within the short span of a single century.

Vast desert wastelands separated the few fertile areas of the Arabian peninsula and thus divided its inhabitants into isolated tribes. A few of these lived settled lives as tillers of the soil, but most wandered from one pasture to another. Their intense competition for grazing land and for water made the small Arab tribes mutual enemies and hindered their uniting as a nation. The only acknowledged allegiance was to the clan and the family, but even within these the Arab would admit obedience to no one. Yemen, in the southwestern corner of the Arabian peninsula, was the oldest area of Arab culture and had become rich and flourishing as a result of trade.

In the 5th century A.D. the closest neighbors of the Arabs were the Byzantine (East Roman) and the Persian Empires. Asia Minor, Syria, and Egypt belonged to the former, Byzantium, while the latter, including Mesopotamia, bordered the flanks of Arabia, so that the two great powers were separated only by the wedge of the Syrian desert. The outcome of their struggle over which was to rule Arabia seemed to be only a matter of time.

Then in Mecca, the chief city of Arabia, a Prophet arose who proclaimed the doctrine of the unity of God and taught that He would support all believers against other men. Although there was nothing new in the first of these doctrines, with which the Arabs were familiar from the Jews and Christians who lived among them, the second — the idea of a war between believers and infidels — revolutionized the Arabs' outlook and character: it was no longer just the family that claimed the individual's loyalty and obedience, but the community of all believers, which far transcended tribe or clan.

With these religious doctrines Mohammed laid the first and the firmest foundation stones of the Arabs' political dominion. His judicious mixture of religious beliefs and politics not only enabled his teachings to gain a foothold among the previously divided Arabs, but also brought them, within the unbelievably short time of a single century, to such a high pitch of energy that they made themselves the rulers of all the peoples from Turkestan to Spain.

In A.D. 622 Mohammed fled from the proud city of Mecca, where his projects were received with hostility, to Medina. The rise of Islam dates from the year of his flight, the hegira. Mohammed transformed all wars into Holy Wars. Soon after his troops, who had been taught to welcome death in their desire for paradise, enjoyed their first triumphs over Mecca, almost all the Arabs were under the green flag of Islam. Then began a series of conquests without parallel in history (see Fig. 231): the Prophet died in A.D. 632 after uniting the Arab

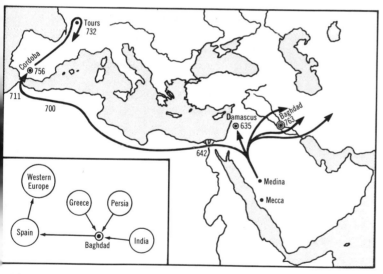

Fig. 231 The Arab conquests. The inset indicates the direction of movement of cultural influences.

tribes under the new faith and arousing them militarily. A Byzantine army was defeated in 635, although the attempt to conquer Byzantium itself failed; two years later, in 637, the might of the Persians was broken and the Empire of the Sassanids was at an end. In 642 Egypt, which had hitherto belonged to the Eastern Roman Empire, was under Arab rule. Toward the end of the century the conquest of North Africa was completed, and in 711 the Arab commander

Tariq crossed the Straits of Gibraltar (whose name, *Djibel-al-Tariq*, means "the mountain of Tariq"). In the same year the numerically far superior army of the Visigoth ruler Roderick was defeated by the Muslims at Jerez de la Frontera in Spain, the Visigothic kingdom was overthrown, and it was not until the Arabs had penetrated beyond the Pyrenees, deep into the Western world, that the Frankish leader Charles Martel put a halt to their advance. This happened in 732, exactly one hundred years after Mohammed's death.

Just as unique as the rapid Arab conquest itself was its ultimate significance for the history of Western culture. The old and by then dying cultures of the ancient world could not longer withstand the onslaught of this young and vigorous people. The conquerors came not as destroyers, however, but as preservers of culture, who rejoiced in the brilliance of the cities they had conquered and rekindled the intellectual life that was dying out in them. In 635 Mohammed's successor, the Caliph Omar, transferred his capital from the remote city of Mecca to the middle of the East-West cultural stream, Damascus, and it became the brilliant center of Islamic life. Here the ruling dynasty of the Ommiads (661–750) transformed the Arab rule into a world empire. Jews and Christians, who like the Muslims believed in a single God, enjoyed religious freedom; the pagans, however, were compelled to adopt Islam unconditionally. Since a Muslim was forbidden to read the Koran in any language other than Arabic, they also thereby became "Arabs." This prohibition, which had the effect of using religion to do away with the differences between peoples and races, harbored within itself the political disintegration of the Arab "state," because the latter was not made up of and supported by a single unified people; but from the standpoint of the flow and exchange of culture it was of inestimable significance because it eliminated all linguistic barriers. Thus it happened that the Ommiad Caliphs, and afterward even more so the Abbasid Caliphs who in 763 caused the fabled city of Baghdad to rise on the soil of the ancient Babylonian culture, encouraged and supported learning and science in every possible way. This was one of the most important factors behind the role of "cultural middlemen" that the Arabs played in world history.

The Arabs gained enormous wealth as a result of their conquests. Under its Abbasid rulers Baghdad competed with Byzantium for the status of principal city of the civilized world. Because of its location on the banks of the Tigris River, it was also a port with access to the Persian Gulf. Through it passed the goods of China and India, and also of Russia and Europe. From the Chinese the Arabs learned to manufacture paper — an accomplishment whose importance for the spread of knowledge, including computations with Indian numerals, cannot be overestimated. Their intellectual demands expanded along with their world-wide trade. Going far beyond their studies of the Koran, the Arabs desired to master the intellectual accomplishments of Greece, Persia, and India. Under the early Abbasid Caliphs, especially al-Mamun (813–833), the son of Harun al-Rashid, numerous scholarly translations began to appear in Baghdad: Aristotle, who suited the intellectually curious temperament of the Arabs better than Plato, was their greatest teacher, as he was to be of the medieval Europeans. The mathematical writings of Euclid and the astronomical works of Ptolemy met with an eager reception. The words and names of Arabic origin still used in astronomy testify to the Arabs' important contributions to this field: zenith, nadir, azimuth, Betelgeuze, Algol.

Thus it would have been extraordinary if the rich gifts of learning which India then had to bestow had not also undergone some transformation by the Arabs, especially since they had been established on India's soil since 712. But before we follow the fortunes of the Indian numerals in Islamic lands, let us first take a look at Moorish Spain.

Family feuds, the old and dire heritage of the desert tribes, flared up again soon after the conquest, even under the Ommiads. In 750 this dynasty was overthrown and almost totally destroyed, except for one small branch of the family which fled to the West and there, far removed from the new rulers, founded its own Caliphate in Cordoba in 756. As under the Abbasids in Baghdad, the Arab lands in the West also experienced an unexpected intellectual and scientific flowering. The empire of the Arabs flourished for 300 years. Baghdad and Cordoba, the Eastern and Western Arab Caliphates, which were each others' political enemies, were nevertheless like the two terminal points of a gigantic intercontinental system between which the intellectual current which had been won from the old tired cultures and brought to life flowed from east to west through the superconductive cable of a single Arabic language understood everywhere. The flow was from east to west, because — to carry on the metaphor — in general the Orient was the transmitter and the West the receiver.

But for the remainder of Europe, Arabic Spain was the provider of culture for a very long time. Where were the rich libraries, the paper manufactories in France and Germany in the 8th, the 9th, and the 10th centuries — a period during which the West was slowly beginning to form? It is ironic that France owed her political significance and her recognition by the Pope in large measure to her victory in 732 over the archenemy of Christendom, the Moslem, to whom she was at the same time deeply indebted intellectually. For it was in the schools in Cordoba, and, after their overthrow around 1000, in Seville and Toledo, that western monks obtained the spiritual riches of Greece and the Orient. About the end of the first millennium a refreshing breeze blew from Moorish Spain through the monasteries of Europe, which had gradually sunk into torpid stagnation by merely collecting and copying the same things over and over again. Western Europe did not acquire Euclid through any succession from Alexandria to Rome to the West; the work of the Greek mathematician was translated into Arabic, perhaps in Baghdad, and then traveled the long westward road to Spain, where it was again translated into Latin by a monk and so introduced into Europe.

In the same manner Greek, Persian, and Indian cultural treasures clothed in Arabic dress made their way to the West and here, changed into Latin garb, spread throughout Europe (see inset in Fig.

231). If we go back to our abbreviated survey of cultural movements through history (see Fig. 158), we can understand why the Arab "block" on the three bottom diagrams is so prominent during the first millennium A.D.

In 1258 Baghdad fell to the Mongols, and in 1492 Granada, the last Moorish enclave in Spain, was conquered by Ferdinand and Isabella. Arab political rule was at an end, but Islam, with the exception of Spain, replaced the old political empire, and along with it came the Arabic alphabet. This is one of the most outstanding examples of a remarkable fact of cultural history, that writing is the token of the confession of the pacemakers' faith. But the reader will be left to find other instances of this for himself, for we shall now turn back to the development of written numerals.

Indigenous Arab Numerals

Did the Arabs have any numerals of their own? No, at least none with which they could administer the world empire they had so rapidly created. Thus the native administrations which already existed in the conquered lands were allowed to go on functioning. Greeks and Persians continued to keep their books in Greek, and wrote numbers in their Greek alphabetical numerals (see p. 270).

In the conquerors' edicts to their subject peoples, which at first had to be issued in two languages, Arabic and Greek, the numbers in the Arabic portion were written out in words, and likewise in the Greek text, but in these they were also repeated in Greek alphabetical numerals.

We already know what importance the Arabs attached to the spread of their language. Thus the Caliph Walid I in the year 706 forbade the use of Greek in his financial offices' administration in favor of Arabic, but decreed that the Greek alphabet must be used for writing out numbers. This decree is highly significant because it also shows that at this time the Indian numerals had not yet penetrated as far as Damascus.

The Greek model probably induced the Arabs also to assign numerical values to the letters of their alphabet (which was based directly on a Semitic predecessor). Thus they arrived at the alphabetical numerals which we have already discussed earlier (see Fig. 75). Along with these, however, the Arabic custom of writing out numbers in words, even in books on mathematics, continued for centuries, until a new method, the Indian numerals, suddenly presented itself.

Al-Khwarizmi and Algorithmus. In 773 there appeared at the court of the Caliph al-Mansur in Baghdad a man from India who brought with him the writings on astronomy (the *Siddhanta*) of his compatriot Brahmagupta (fl. ca. A.D. 600). Al-Mansur had this book translated from Sanskrit into Arabic (in which it became known as the *Sindhind*). It was promptly disseminated and induced Arab scholars to pursue their own investigations of astronomy.

One of these was Abu Jafar Muhammad ibn Musa al-Khwarizmi, "Mohammed the father of Jafar and the son of Musa, the Khwarizmian" (from the Persian province of Khoresm, south of the Aral Sea, which the Greeks had called Khorasmia). This man, who was

probably the greatest mathematician of his time, wrote among other
things a small textbook on arithmetic in which he explained the use of
the new Indian numerals, as he had probably learned them himself
from the Indian writings. This was around A.D. 820.

Muhammad ibn Musa al-Khwarizmi was also the author of a book
showing how to solve equations and problems derived from ordinary
life, entitled *Hisab aljabr w'almuqabala*, "The Book of Restoration
and Equalization" (that is, expressions arrived at in the course of
solving equations). Its translation into Latin, *Algebra et Almucabala*,
in the 12th century, then ultimately gave its name to the discipline of
algebra.

The orignal of al-Khwarizmi's book on arithmetic is lost, but it still
occupies an important place in the history of our numerals, for it
made its way to Spain by the route already sketched out and it was
there, at the beginning of the 12th century, that it was translated into
Latin by the Englishman Robert of Chester, who "read mathematics"
in Spain. Another Latin version was produced by the Spanish Jew,
John of Seville. Robert's translation is the earliest known introduction of the Indian numerals into the West. The manuscript, discovered
in the 19th century, begins with the words:

Dixit Algoritmi: laudes deo rectori nostro atque defensori dicamus dignas

Algoritmi has spoken: praise be to God, our Lord and our Defender.

At about the same time (around 1143) an epitome of this book was
written which is now in the Royal Library in Vienna; the illustration
of Fig. 249 is taken from this. The Codex of the Salem monastery is
also one of the oldest witnesses to the presence of al-Khwarizmi's
book in the Germanic part of Europe; its origin has been placed
around the year 1200. This Salem Codex begins (in very abbreviated
Latin; see Fig. 232):

*Incipit liber algorithmi: omnis sapientia sive scienta a domine Deo; sicut
scriptum est: Hoc quod continent omnia scientiam habet, et iterum: Omnia
in mensura et pondere et numero constituisti, . . ."*

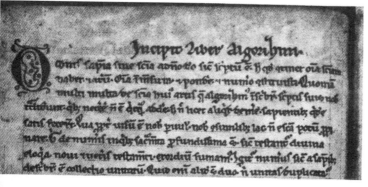

Fig. 232 The Salem manuscript of the 12th century is one of the oldest in the
West in which computations are described with Indian numerals. 15 pages.
University Library, Heidelberg.

Here begins the book of Algorithmus. All wisdom and all knowledge comes from God our Lord; as it is written [referring to Ecclesiastes 1:7]: that which embraces all things is full of wisdom, and further: thou hast established all things by measure and weight and number...

But al-Khwarizmi, or Algorismus as he was known in Latin, had not only spoken but also given his own name to the new art of computation, as we learn from the *Carmen de Algorismo*, "Song of the Algorismus" (Fig. 233):

Hinc incipit algorismus.
Haec algorismus ars praesens dicitur in qua
talibus indorum fruimur bis quinque figuris
         0 9 8 7 6 5 4 3 2 1,

Here begins the algorismus.
This new art is called the algorismus, in which
out of these twice five figures
         0 9 8 7 6 5 4 3 2 1
of the Indians we derive such benefit...

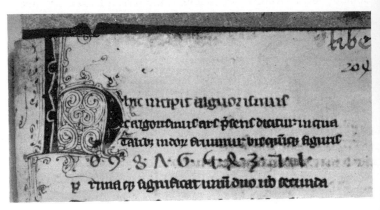

Fig. 233 The *Carmen de Algorismo* by the French monk Alexander de Villa Dei presented the new computations with numerals in verse form. The manuscript dates from the 13th century.
Hessische Landesbibliothek, Darmstadt.

The French Minorite friar Alexander de Villa Dei, who taught in Paris around 1240, taught the methods of computation with the new numerals in two hundred forty-four widely read (but not always very good) verses of dactylic hexameter. In his version an Indian king named Algor figures as the inventor of the new "art," which itself is called the *algorismus*. In this manner the word "algorithm" was tortuously derived from Muhammad's surname al-Khwarizmi, and has remained in use to this day in the sense of an arithmetic operation.

The West and East Arabic "Indian" Numerals
Al-Khwarizmi described these numerals in detail. Concerning the zero, he taught (according to the Latin version):

Si nihil remanserit pones circulum ut non sit differentia vacua: sed sit in ea circulus qui occupet eam, ne forte cum vacua fuerit minuantur differentiae, et putetur secunda esse prima,

When nothing remains [in subtraction], put down a small circle so that the place [differentia] be not empty, but the circle must occupy it, so that the number of places will not be diminished when the place is empty and the second be mistaken for the first.

As regards the forms of the number symbols al-Khwarizmi stated that the new numerals, particularly 5, 6, 7, and 8, were written differently by different peoples but that this circumstance was no obstacle to their use as a place-value notation.

Al-Khwarizmi may have had in mind only minor individual variations, but he may also have intended to distinguish between the two different Arabic versions of the Indian numerals, which later became differentiated geographically into West and East Arabic numerals. Both are still in use today (Fig. 234).

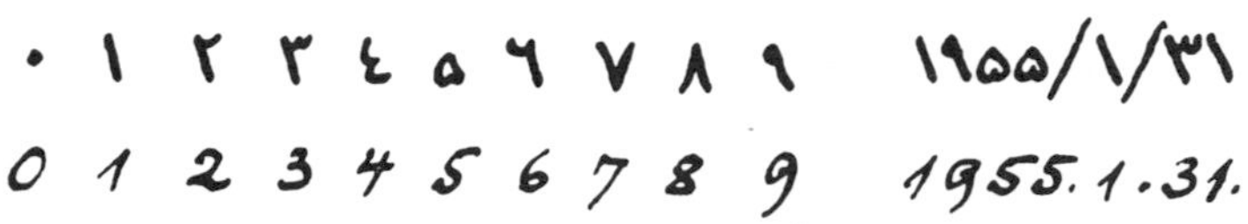

Fig. 234 The East Arabic and our own numerals, which derived from the West Arabic version.

The East Arabic numerals are used by all Arabic-writing peoples of the Orient (Egyptians, Syrians, Turks, Persians), and are known there as "Indian" (*huruf hindayyah*); the West Arabic numerals were the direct ancestors of our own Western numerals, which are popularly called "Arabic." The "Western Arabs" in Morocco today still use these numerals rather than the East Arabic.

In our table of numerals (see Fig. 239), we see the nine digits with the zero derived from the first nine "enciphered" Brahmi numerals which appeared very early in India, on the Gvalior inscription (see Fig. 226). These may well have become known to the Arab merchants quite soon after their development. Although the West Arabic style of writing script may have gone directly to the West by the sea route, the East Arabic perhaps underwent its transformations in the course of its travels overland, by way of Kabul and Persia. The numerals 5, 6, 7, and 8 actually do diverge more sharply.

The East Arabic 1, 2, and 3 consist mainly of vertical strokes, and as compared to the West Arabic equivalents they are rotated through a quarter of a circle, so that 2 and 3 "lie on their backs." The digit 4 is contracted by present-day Arabs, as may be seen from three modern Egyptian postage stamps (see Fig. 235 and also Fig. 238). Since the East Arabic 5 looks like a small circle, the zero has again become a dot. In the numeral 6 one can see the similarity between the two forms: In the East Arabic 6 the loose tail has become a stroke pointing downward, so that today it looks almost like our 7. The derivation of both from the Brahmi numerals is easy to see. The numeral 7 is the same in both forms, except that the East Arabic 7 is, from our point of view, upside down. Only the difference in shape between the West and the East Arabic 8 cannot be explained. The transformations which these numerals have undergone will be discussed more comprehensively a little later (see p. 418).

Fig. 235 Egyptian postage stamps with both West and East Arabic numerals:
the East Arabic zero is a dot, and our "zero" is the East Arabic 5.

Fig. 236 Turkish coins with East
Arabic numerals. The bottom line
generally gives the date of the then
reigning Sultan's accession, 1203 on
the bottom coin and 1223 on the top
(date after the hegira; i.e., 1788 and
1808 in our reckoning). The numbers
10 and 15 in the first lines are the
years of the Sultan's reigns. The East
Arab writes five as "zero," and zero
as "five."
Badisches Münzkabinett, Karlsruhe.

The oldest Arabic number expressed in Indian numerals, so far as
we know, is the year ٢٦٠ = 260 in an Egyptian papyrus written in
the year 873 — a number in which, as al-Khwarizmi said of the
numerals in his own time, the West (2) and the East (6 and 0) Arabic
forms still appear together.

In the two examples to follow, the reader may transcribe into our
own numerals the East Arabic digits that appear on two old Turkish
coins (Fig. 236) and in a multiplication table from a modern Egyptian
textbook on arithmetic (Fig. 237).

On the other hand, the East Arabic problem 435 − 389 = 46 from
the same arithmetic textbook (Fig. 238), shown once in a horizontal

$$\frac{١٢ \times ٧}{٨٤} \quad \frac{١١ \times ٧}{٧٧} \quad \frac{١٠ \times ٧}{٧٠} \quad \frac{٩ \times ٧}{٦٣} \quad \frac{٨ \times ٧}{٥٦} \quad \frac{٧ \times ٧}{٤٩}$$

$$\frac{١٢ \times ٨}{٩٦} \quad \frac{١١ \times ٨}{٨٨} \quad \frac{١٠ \times ٨}{٨٠} \quad \frac{٩ \times ٨}{٧٢} \quad \frac{٨ \times ٨}{٦٤}$$

$$\frac{١٢ \times ٩}{١٠٨} \quad \frac{١١ \times ٩}{٩٩} \quad \frac{١٠ \times ٩}{٩٠} \quad \frac{٩ \times ٩}{٨١}$$

$$\frac{١٢ \times ١٠}{١٢٠} \quad \frac{١١ \times ١٠}{١١٠} \quad \frac{١٠ \times ١٠}{١٠٠}$$

$$\frac{١٢ \times ١١}{١٣٢} \quad \frac{١١ \times ١١}{١٢١}$$

$$\frac{١٢ \times ١٢}{١٤٤}$$

Fig. 237 Multiplication table in a present-day arithmetic textbook from Cairo.
This goes from 1×1 to 12×12, each row beginning with the product of
equal factors (thus the first line begins, reading right to left, with 7×7, not
1×7):

$$\frac{12 \times 7}{84} \quad \ldots\ldots\ldots\ldots\ldots\ldots\ldots \quad \frac{8 \times 7}{56} \quad \frac{7 \times 7}{49}$$

$$\ldots\ldots\ldots\ldots\ldots\ldots\ldots \quad \frac{12 \times 12}{144}$$

A striking anomaly to the Western eye is the fact that the lines are read from
right to left, but the numbers themselves from left to right.

And here the operation is performed as follows:	ولذا بجري العمل هكذا : ـ
"The remainder	الباقي ٤٦ = ٣٨٩ ـ ٤٣٥
That from which the subtraction is made	المطروح منه ٤٣٥
"The subtrahend	المطروح ٣٨٩
The remainder after the subtraction	باق الطرح ٠٤٦

Fig. 238 Subtraction: 435 − 389 = 46, from the Arabic textbook of Fig. 237.

line and again vertically, the figures one above the other, should alert us to a peculiarity of cultural history. This involves the Indian, Arabic, and Western methods of arranging the orders of magnitude in writing numbers. Arabic is read from right to left, and thus one might well read the digits in the number also from right to left: 534 − 983 = 64! The absurdity of this result, of course, at once shows that the numbers themselves, with their decreasing orders of magnitude from left to right must have been incorporated into the Arabic script as foreign borrowings. In the Indian writing, which is read from left to right, the orders of magnitude of the digits also consistently decrease from left to right, but in a right-to-left form of writing like the Arabic alphabet they should also decrease from right to left, as do the ranks of the alphabetical numbers written with the Arabic alphabet (see Fig. 112).

Then when we left-to-right-reading Westerners adopted the "Arabic" numerals, the succession of ranks was once more consistent with the Indian and with the direction of writing and reading. Actually, we should have taken over the opposite order of ranks, because of the reversed order of script writing, but we did not do so any more than did the Arabs in their own time. Many a neophyte has tried to reverse the digits this way when he first attempted to understand these curious numerals.

The peculiar reversal of the order of the numerals, 9 ... 1, as they are presented in the early medieval algorisms (like that of Sacrobosco, Fig. 227, of Alexander de Villa Dei, Fig. 233, and the Salem Codex, Fig. 232) goes back to the Arabic archetype. For the Arabs certainly counted from low to high as we do now, 1 2 3 4 5 6 7 8 9, but wrote the numerals from right to left according to their alphabet. Yet the early Western translators evidently missed this point.

Up to now we have dealt primarily with the East Arabic digits; now we must turn to the West Arabic numerals, which the Arabs themselves called *huruf al gubar*.

The Gubar Numerals

These "dust numerals," from the Arabic word *gubar*, "dust," are both formally and historically the direct ancestors of the numerals which we use today.

But how did they acquire their peculiar name? This refers to the sand that was strewn over the counting board used in India which in the 14th century in Europe was considered to be Indian (according to Maximus Planudes; see p. 301). Of course in the columns drawn in the sand or dust no counters were put down, but rather numerals were written (see p. 397). Thus no zero was needed. And, in fact, the *gubar* numerals had no zero (at least not at first; see Fig. 239), and whenever they appear in manuscripts they are written in the following simple and sensible method:

$$\overset{..}{4}\overset{...}{5}6 \,(=456),\ \overset{...}{4}\overset{..}{5}\,(=450),\ \overset{..}{4}6(=406).$$

Dots were placed over the digits to indicate the rank or order of magnitude: one dot for the tens, two for the hundreds, three for the thousands and so on. This, in our terms, amounts to a "named" place-value notation, which symbolically states 406 as 4C6 and in which the ranks are indicated by the number of dots over the digit. In the Arabic alphabet dots are commonly added to individual letters, and this custom was extended to the numerals as well.

Thus the *gubar* numerals are Indian in their external forms, and yet again not Indian because they neglect the essential advantage of the place-value system and instead use another, non-Indian, principle of organization. They lack the characteristic feature of the "Indian" numerals — the zero.

Historically, moreover, they constitute a puzzle in that they occur nowhere but in Moorish Spain. There is no evidence testifying to their migration there from India. In this regard, however, it is highly significant that they came into widespread use quite early, not through learned, scholarly books but by practical merchants and men of affairs. In the course of trade and barter and also by direct contact (perhaps especially in Alexandria) these numerals traveled westward from hand to hand. The new principle they embodied, the place-value notation with a zero sign, was not understood, but everyone appreciated their other characteristic advantage, in that they enabled numbers to be written down rapidly and clearly. Moreover, the Arabs had no numerals of their own. By adding the dots they adopted the new Indian symbols and made them their own — that is, they transformed them into an easily readable "named" place-value notation. Later, when the true, abstract place-value notation finally arrived from the Eastern Arab world, all that remained to be done was to add the zero to the nine digits already in use and do away with the superscribed dots. Yet the dots over the numerals persisted for a long time afterwards, and appeared even in Byzantine manuscripts of the 15th century.

In general, Western culture lagged behind that of the East. Around Baghdad, which itself was located on fertile cultural ground, the sources of learning flowed in directly from its immediate neighbors; here in Cordoba, on the other hand, culture and people were still immature and the land itself was cut off and isolated; the ties to the Arab motherland were weak and tenuous. Here in the West the first school of higher learning was founded in 976, almost two centuries after the peak of Baghdad's most flourishing period, and paper first came to be manufactured some 350 years after it was made in the East. Thus it should be easy to understand that an old set of numerals, to which the zero was not added until much later, should have survived for so long a time.

It was these *gubar* numerals, with which Gerbert first became acquainted during his visit in Spain, that he carried back home with him and there, in the intricate form of the *apices*, inscribed on the counters of the monastic abacus (see p. 324 and Fig. 162). In this form the Indian numerals thus made their first tentative excursion into the West, around the year 1000. But Europe was not ready for them; neither their nature nor their advantages were appreciated, and they soon retreated into the cells of learned monks as they failed to survive in ordinary use.

Two hundred years later the Indian numerals (now including the zero) were carried northward from Spain once more, in the translations from Arabic of the "Algorisms" and in treatises of university scholars like Sacrobosco and Villa Dei (see pp. 402 and 412). But even these did not bring the new numerals into common use; they remained esoteric for the time being, even though since the 13th century a path had been prepared for them "from above." The third path, "from below," by which they came to be the numerals now in universal use — although not without having to overcome much resistance — was through their gradual acceptance by merchants and tradesmen and their popularization in the textbooks written by calculators and arithmeticians.

This third path, the "low road," will be the subject of the next section of this book, after one more look at the development and spread of our numerals from their first beginnings in India.

THE FAMILY TREE OF INDIAN NUMERALS

From the Brahmi numerals of India there ultimately developed the East and West Arabic numerals, of which the latter, after the brief episode of the *apices*, in turn gave rise to our own modern numerals. But the Brahmi figures also were the ancestors of the many native forms of numerals in India, such as those used in Tibet, Bengal and elsewhere, of which the Devanagari are the most important; they are used by most Hindus today. Just as the Arabic numerals represent the "western" branch of the family, the Devanagari and the many other sets of numerals in India are the "Eastern" descendants of the Brahmi figures.

Concerning the significance of the numerals there is very little concrete evidence apart from the first three, which require no explanation. It is futile to try to derive them from Chinese, Phoenician, Babylonian, or Egyptian predecessors. It is far more likely that they began as the initial letters of the corresponding number-words, like the early Greek row numerals (see Fig. 97) or as contractions, like the Persian *siyaq* numerals (see Fig. 109). Or were they perhaps devised purely as abstract numerals in the first place? If so, this would be an unparalleled instance in cultural history, because then they would have originated after the rise of the alphabetical writings.

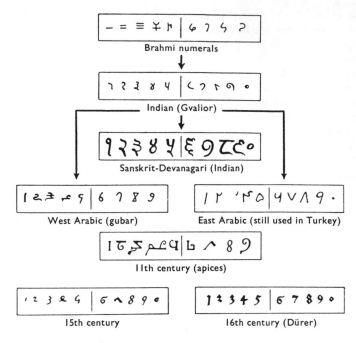

Fig. 239 The Family Tree of the Indian Numerals

One way in which they definitely did *not* originate is shown by an 18th-century "explanation" (Fig. 240). This illustration could just as well have been taken from our own time, for the desire to hold in one's hand the key to a mystery which experts have sought for in vain remains alive in all periods of history.

Fig. 240 Jacob Leupold's fanciful explanation of the origin of Indian figures.

The Form Consistency of the Indian Numerals

During their long migration from one culture to another through the centuries the Indian numerals remained astonishingly constant in their form. How new and mysterious they must ever have seemed to those who became acquainted with them for the first time, for they have continued to resemble their original forms! The small changes that have taken place from the Brahmi (B) to our modern (M) numerals, as well as to the Devangari (D) and to the East Arabic (E) forms, brought together in Fig. 241 for ready comparison,

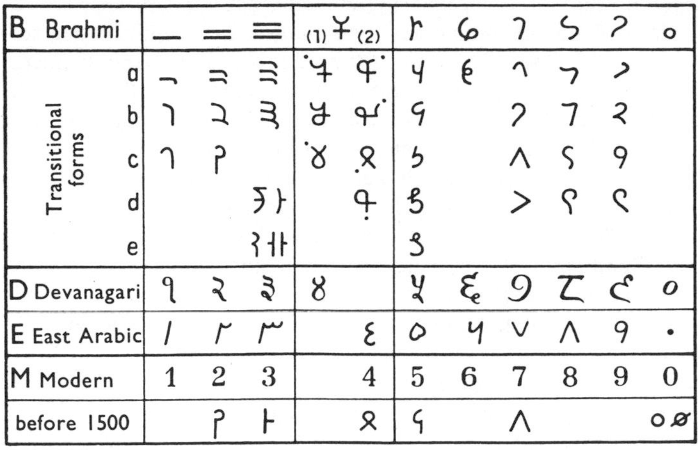

B Brahmi	—	=	≡	(1) ⊁ (2)	↾	৬	੭	৸	?	o
Transitional forms a	⌐	=	≡	⅄ ⅁	५	६	∧	⁊	⟩	
b	٦	੨	३	⅄ ঀ	੧		?	੭	੨	
c	੧	?		४	७		∧	ৎ	੧	
d			⅁⊦	੧	ঃ		⟩	੬	ৎ	
e			੨⊦		३					
D Devanagari	९	੨	३	४	੫	६	੧	८	੬	0
E East Arabic	/	↾	↾	ε	০	੫	٧	∧	੧	·
M Modern	1	2	3	4	5	6	7	8	9	0
before 1500	?	⊦		੨	੫	∧				o ⌀

Fig. 241 Transformations of the Indian numerals from the Brahmi (B) to the Devanagari (D), the East Arabic (E), and our modern (M) forms.

amount to no more than shortening or blurring of the main stroke (as with 2 and 3), rotation about the vertical or horizontal axis (in the case of 1, 2, 3, and 7), and being written from right to left or left to right (as with 4, for which the side from which the pen stroke begins is indicated by a dot). The reader is urged to trace the alterations of one form or another himself with pencil and paper; thus he will be able to follow the transformations from the Brahmi to our own modern numerals without much further explanation:

Numeral 1 — horizontal in B, vertical in E. D and M arose from the comma-like form into which it developed when written rapidly.

Numeral 2 — the curving and connection of the two lines led to D and M, and the vertical stokes when connected and written right to left resulted in E. In Germany the numeral 2 often took the form Z in inscriptions carved in stone (see Fig. 118 and also Figs. 189 and 260 from the Bamberger Rechenbuch). For the old 2-form, see 2c as shown in Fig. 247.

Numeral 3 — connecting the curved ends of the three horizontal lines resulted in D and M, and the same written right to left led to E. In old manuscripts of the 12th century there appeared a peculiar 3-form which probably arose from the old numeral 2 plus an additional stroke and a further rotation (3c developed from 2c, see also Figs. 247 and 248; this evidently had already happened by the time of the *apex* 3d, as in Fig. 162).

Numeral 4 — the inward curving elision of B resulted in two forms (1) and (2) in Fig. 241, depending on whether the top curved stroke was bent around "underneath" (1) or "above" (2); (1) written left to right and with the top hook eliminated led to 4c and to D, and (2) written right to left in the Arabic style resulted in the older East and West Arabic form 2b with a horizontal main stroke, which today is abbreviated to E. In western Europe this form was pulled around to be written from left to right, so that it began from behind with the loop and lost its "Arabic" head. This led to the medieval form 2c, which around the year 1500, when the digit 4 no

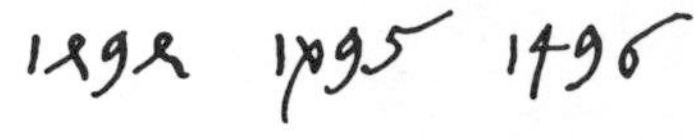

Fig. 242 Albrecht Dürer's year dates. In writing the dates of the years around 1495, Dürer illustrated the development of the 4 into its present form. From three of his drawings dated in successive years.

Fig. 243 Devanagari numerals in instructions for Yoga exercises from Kashmir, 18th century. Below the Indian letters one can read the numerals from 8 through 17 (bottom to top). Ethnographic Museum, Munich.

longer appeared in every single year's date that was written, became angular in form like the modern digit and was rotated ninety degrees to produce 2d. One milestone on the road to this form is the Bamberg block-printed book (see Fig. 170), while the Bamberger Rechenbuch used both variants side by side (see Fig. 260). The numeral 4 also appears in flux in Albrecht Dürer's work (see Fig. 242).

Numeral 5 — the symbol 5a is merely B upside down, whence comes 5b. Turning the latter upside down again and wearing it down led to D and M. The old form 5b predominated before 1500, as for example in the Bamberger block-printed book and the Bamberger Rechenbuch. The circular shape of the numeral 5 in E is probably derived from the Arabic alphabetical numerals, in which the fifth letter with its more or less circular shape had the value of 5 (see Fig. 112).

Numeral 6 — here D extends the two curves of the B-form, while E suppressed the loop and let the second bend die out.

Numeral 7 — rotation makes the difference between M and E; D is formed from the head of 7b. The old form 7c lasted in Germany until about 1500 (see Fig. 260).

Numeral 8 — to form M, the B-form was extended to an 8, whereas it was shortened in the case of D and E.

Numeral 9 — the E and M forms extend the head of the B-form; D turned 9b and 9c around and arrived at a curious 3-form for the numeral 9.

The zero 0 — in E this is still a dot; it then became a small circle which in the case of M at first often had a slanting line drawn through it.

In the Bamberger Rechenbuch the old and new forms of the digits 4, 5, and 7 occur interchangeably, often in the same problem (see Fig. 260).

Now let us end our historical survey of the form changes of the Indian numerals symbolically with the following two pieces of evidence: From the East we have a set of instructions for Yoga exercises from Kashmir, written in the 18th century (Fig. 243) with digits which strongly resemble the Devanagari numerals, the personal seal of a Chinese tax collector who received revenues from Siam (Figs. 244 and 245), and a small road marker (the 80th kilometer) from Syria (Fig. 246). From the West, we have the so-called Regensburg Annals of Hugo von Lerchenfeld with what is probably the oldest example of Indian numerals used in a medieval European manuscript. In addition to the historically significant entries for the years 726 through 744 (Fig. 247), those for the years 1002 and 1056 are also shown (Fig. 248), revealing the surprising fact that this copyist of the 12th century correctly understood the new numerals with their zero. These should be compared with the multiplication table from what is probably the oldest Algorism manuscript in Europe, dated 1143 (Fig. 249).

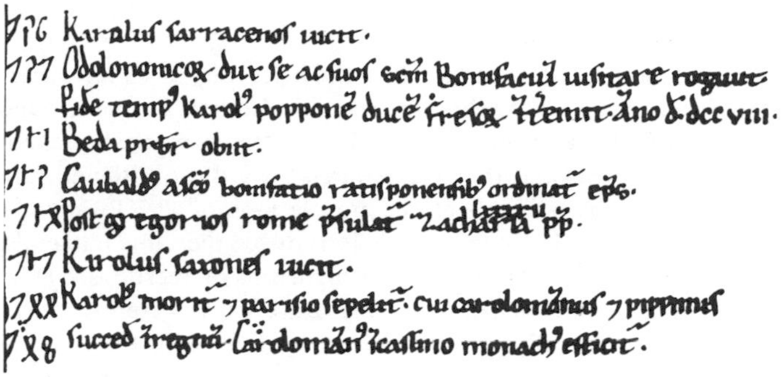

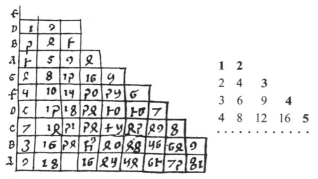

Fig. 247 Entries in the Regensburg Annals.
726 Charles (the Great) made war on the Saracens (Arabs) ... 727 ...
731 Bede the Priest died [the same Venerable Bede whose finger gestures we
have already seen much earlier, see p. 201] ... 732 ... 734 ...
737 Charles made war on the Saxons
744 Charles died and was buried in Paris

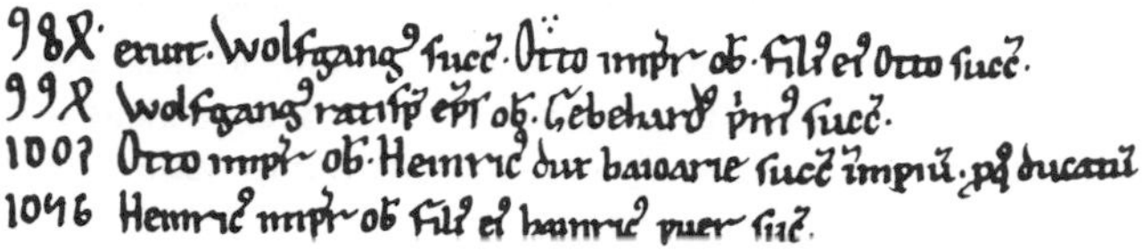

Fig. 248 Later entries in the Regensburg Annals.
 984 ... the Emperor Otto [II] died and was succeeded by his son Otto.
 994 Bishop Wolfgang of Regensburg died ...
1002 the Emperor Otto [III] died [the one who nominated Gerbert to the
papacy, see p. 323] ...
1056 the Emperor Henry [III] died and was succeeded by his younger son
Henry ...
Bayrische Staatsbibliothek, Munich.

Fig. 244 The ivory seal of a Chinese
official who collected revenues from
Siam. Carved in magnificent style and
colored red is a horse, with the
owner's name in Chinese characters
above and below; enclosed in a long
rectangle, the date 1218 (1856 in our
reckoning). This number is recorded in
Indian place-value notation, but with
peculiar Siamese digits — the whole
being an outstanding example of the
way in which an interrelationship of
different cultures can be manifested
in numbers. Diameter 4.3 cm.
Ethnographic Museum, Munich.

Fig. 245 The same seal as Fig. 244.
9.5 cm high.

1	2			
2	4	3		
3	6	9	4	
4	8	12	16	5

Fig. 249 A multiplication table from 1 × 1 through 9 × 9, from one of the
oldest German algorism manuscripts, from the 12th century.

Fig. 246 Road marker indicating the
80th kilometer along the Aleppo-
Latakia road.
Photograph by U. Dubs.

The Indian Numerals
in Western Europe

The Two Early Encounters

The Indian numerals played their first and brief role, as we have seen, in the form of *apices* (without a zero) in Gerbert's time around the year 1000. They appeared for a short while in monastic manuscripts (see Fig. 162) and then disappeared toward the 12th and 13th centuries when, as a result of the translations of al-Khwarizmi's book on arithmetic, Latin texts again introduced the same numerals, such as in the Salem Algorism and the algorisms of Sacrobosco (Fig. 227) and of Alexander de Villa Dei (Fig. 223). These were "learned" treatises which basically recorded the new along with the old and thus never filtered down to ordinary people, who went on making their practical everyday computations as they always had. For this reason the "abacists" and the "algorithmicists" formed two opposing camps for a long time.

Today we can no longer understand the stubborn resistance to the new numerals during the early Middle Ages; to us they seem so much easier to work with than the cumbersome Roman numerals. But from our previous discussion it is clear that the counting board served medieval Europe as a perhaps slow but essentially equivalent and above all highly visual means of computation. Computations with the new numerals, in contrast, were certainly not as easy to visualize. But most important of all they embodied an intellectual obstacle that was scarcely overcome during the first few centuries of their presence in the West: the zero! We have already touched on this subject (see p. 401).

The Zero Again

What kind of crazy symbol is this, which means nothing at all? Is it a digit, or isn't it? 1, 2, 3, 4, 5, 6, 7, 8, and 9 all stand for numbers one can understand and grasp — but 0? If it is nothing, then it should be nothing. But sometimes it is nothing, and then at other times it is something: $3 + 0 = 3$ and $3 - 0 = 3$, so here the zero is nothing, it is not expressed, and when it is placed in front of a number it does not change it: $03 = 3$, so the zero is still nothing, *nulla figura*! But write the zero *after* a number, and it suddenly multiplies the number by ten: $30 = 3 \times 10$. So now it is something — something incomprehensible but powerful, if a few "nothings" can raise a small number to an immeasurably vast magnitude. Who could understand such a thing? And the old and simple one-place number 3000 (on the counting board) has now become a four-place number with its long tail of "nothings" — in short, the zero is nothing but "a sign which creates confusion and difficulties," as a French writer of the 15th century put it — *une chiffre donnant umbre et encombre*.

Thus the resistance to the Indian numerals by those who used the counting board for calculations took two forms: some regarded them as the creation of the Devil, while others made fun and ridiculed them:

Just as the rag doll wanted to be an eagle, the donkey a lion, and the monkey a queen, the *cifra* put on airs and pretended to be a digit, wrote an educated man in France as late as the 15th century. According to another French source, an "algorism-cipher" is a term of abuse of the same class as blockhead. Astrologers, however, gladly adopted the new numerals; like every form of secret writing, they helped to raise their status. The Algorism of the Salem Monastery correctly interpreted the new numerals and used them for computations, but they still created such confusion in the mind of their author that he appended the following mystical interpretation:

Every number arises from One, and this in turn from the Zero. In this lies a great and sacred mystery — *in hoc magnum latet sacramentum* —: HE is symbolized by that which has neither beginning nor end; and just as the zero neither increases nor diminishes / another number to which it is added or from which it is subtracted / so does HE neither wax nor wane. And as the zero multiplies by ten / the number behind which it is placed / so does HE increase not tenfold, but a thousandfold — nay, to speak more correctly, HE creates all out of nothing, preserves and rules it — *omnia ex nichillo creat, conservat atque gubernat.*

In this way the zero acquired its profound "significance" and began to represent something.

But the learned men too were not sure whether the zero was a symbol, a numeral, or not. According to the name *Null* which they gave to it, it was not; and so medieval writers would frequently present the "9 digits," to which they would add one more, which was called a *cifra*:

Welcher lernen will anfänklich rechnen durch dye zyffer yst not das er wysse dye figurenn der Ziffer / darnach lerne dye crafft und bedeutnus der stätt daran dye ziffer gesetzt werden / Vnd seyn der bedeutlychen figuren newen / vnd ain Figur ausserhalb dero wirt genant nulla / 0 / die nichts für sich selbs bedeut / aber dye andern bey ir mer bedeuten macht,

He who wishes to learn to reckon with digits must begin by knowing the figures of the digits / and then learn the force and meaning of the place-values according to which the digits are set. And there are nine figures that have value meaning / and one more figure outside of them which is called null, 0, which has no value in itself / but increases the value of the others.

Another manifestation of the same confusion and insecurity was the many names given to the zero (see p. 401). What was the point of forsaking the old reliable counting board for something so full of contradictions that only a few learned men could understand it, and even they just barely? Even today the expression *faire par algorisme*, "to do it with the algorism," is still used in France in the sense of "to miscalculate."

This popular disinclination to use the new numerals was also behind the attempt to make these strange new concepts, the zero and the place-value principle, comprehensible by presenting them in verse form; thus Alexander de Villa Dei said of the zero (see p. 412) that

cifra nil significát, dat significáre sequénti

the zero has no value, but gives value to the next [digit of higher rank];

and he explained place value in the following lines (of which only the beginning and end are quoted here):

unum dat prima, secunda decem, dat tertia centenum
quarta dabit mille, milia quinta decem ...
chifra nil condit, sed dat signare sequentem,

The first [place] makes [the digits there worth] units, the second tens, the third hundreds, the fourth thousands, the fifth ten thousands ... the zero itself makes nothing, but it makes the following digits have [greater] value.

From all this we infer that the new numerals were adopted in the early Middle Ages not because of any conception of the advantages of place-value notation but merely as a new and exotic means of writing numbers. The Indian numerals were seen as nothing but abbreviations for numbers established on the counting board. People were so strongly tied to the Roman numerals with their ordering and grouping that very few grasped the significance of the place-value principle by which the ranks no longer appeared visually but were expressed by the positions of the digits and thus by something which was not intrinsic to them. The place-value notation gradually came to be understood, naturally, but this would have happened much more slowly than it did were it not for a strong external stimulus. This stimulus was provided not by the scholars and scientists, but by the merchants.

ITALY

After the consolidation of the West by the end of the first millennium, its energies were now free to foster its intellectual growth. The translations from the Arabic liberated the machinery of the intellect from dead center; the spirit of the crusades fired the European imagination. A new world, never dreamed of, suddenly blossomed forth in the dark and narrow regions of the North. The journeys to Rome made by the German emperors pushed open the gates to the South, through which the northerners gazed in wonder at the worldly and sophisticated life in the older cultural region of the Mediterranean. The Phoenicians and the Greeks had established settlements in Italy and for many centuries had been weaving their network of trade between the Orient and the West. The Romans strengthened this network and made their capital city its center. The Arabs who had trading outposts in Sicily and Amalfi on the mainland maintained the movement of commerce well into the 9th century. Italy, thus endowed by nature and geography, remained awake and culturally alive, and grew rich in both money and mind.

Now, in the high Middle Ages, Italy became the foremost maritime power. Italian cities provided the ships that carried the crusaders to the Holy Land, Italian bankers lent the money, and all the trade and commerce, resulting from the crusades, which surged back and forth from West to East and from East to West for three hundred years, passed through the hands of the Italians. Venice, Genoa, and Pisa all became rich and powerful. The furrows which the "holy" ships plowed through the seas were used by less holy ones for commerce. They founded trading colonies along the whole Mediterranean coast and took control of the strategic points along their routes.

LEONARDO OF PISA

The governor of the Pisan trading station at Bugia in Algeria was the father of the greatest and most prolific mathematician of the Middle Ages, Leonardo of Pisa, who lived from 1180 to about 1250 and was thus a contemporary of the Hohenstaufen Emperor Frederick II. The name of Leonardo's father is unknown; we have only his nickname *Bonaccio*, "man of good spirit," because his son (*figlio*) Leonardo was also known as Fibonacci. The father allowed Leonardo, as he himself says at the beginning of his book, to travel to Bugia in order to take up the *studium abaci* — to learn the art of computation. He goes on:

Ubi ex mirabili magisterio in arte per novem figuris Indorum introductus,

There I was introduced by a magnificent teacher [perhaps an Arab instructor] to the art of reckoning with the nine Indian numerals.

This art of computation, he continues, appealed to him far better than any of the others he had learned during his travels to Egypt, Syria, Greece, and Provence:

Sed hoc totum et algorismum atque arcus pictagore quasi errorem computavi respectu modi Indorum,

But all this, the agorithm and the arch of Pythagoras, I regarded as an error as compared to the methods of the Indians.

A decisive and fruitful judgment! Leonardo did not approach the new methods of computation superficially, as "just another procedure," but tested them thoroughly and came to regard them as a vast improvement. It was in this spirit that he wrote his great Book of Computations, the *Liber Abaci* of 1202, which prepared the ground for the widespread adoption of the Indian numerals and the new operations in the West.

He introduced the new numerals in the following words:

Novem figure Indorum he sunt 9 8 7 6 5 4 3 2 1. Cum his itaque nouem figuris, et cum hoc signo 0, quod arabic cephirum appellatur, scribitur quilibet numerus,

The nine numerals of the Indians are these: 9 8 7 6 5 4 3 2 1. With them and with this sign 0, which in Arabic is called *cephirum* [cipher], any desired number can be written.

The *Liber Abaci* was a compendium of the art of computation which filled 459 pages; only a second edition of 1228 is preserved in manuscript. It came to the attention of the Emperor Frederick II, whose book on falconry was mentioned in connection with finger counting, and it was also read by the Emperor's court astrologer Michael Scotus, a man of great learning who knew Arabic and to whom Leonardo dedicated the second edition of his own book.

The title of Leonardo's book is confusing: He wrote an introduction to the methods of computation with the new numerals, which in the West were called *algorithm*, but he avoided this word. He also rejected the arch of Pythagoras, the monastic abacus (see p. 323), and yet he called his work the "Book of the Abacus." Thus there is some obfuscation of terms. Primarily in Italy the term *ars abaci* referred to computation in general, regardless of method. But what did "algorithm" mean? Probably the learned treatises concerning the new numerals, such as those of Sacrobosco and Villa Dei; whereas Leonardo was the first to bring the Indian computations into ordinary, commercial use.

After introducing the digits themselves, Leonardo explained counting on the fingers (beginning with the left hand; see pp. 202 ff.), the four arithmetic operations, and operations with fractions, in which he followed the "Arabic" style in placing the fraction in front of, not after the integer: $\frac{1}{2}2$ instead of $2\frac{1}{2}$. This was followed by his important section on commercial arithmetic — *practica*, as it was known in the Middle Ages — the rule of three and the chain rule, with all their "practical" applications (cf. Adam Riese's book, Fig. 251). Thus the German problem in exchange had its exact counterpart here (see p. 429). Then came combinatorial and mixture problems (alloys, and their contents of various precious metals) After this there was a long series of problems in higher mathematics (series, indeterminate and quadratic equations, square and cube roots, etc.) — a compendium of mathematics such as had never existed before. It embodied virtually all the numerical knowledge of Leonardo's time, including much Arabic science, and gave original interpretations of all this material, opening the doors to levels of knowledge hitherto undreamed of. Thus the *Liber Abaci* remained for centuries both a model and a source. In its breadth of knowledge, its superior exposition and its practicality, it differed essentially from the algorisms written during the same time in Northern Europe (the Salem Algorism, Sacrobosco, and others).

The New Numerals Used in Bookkeeping

In the account books kept by the great trading and banking houses generated and stimulated by the crusaders, however, the entries were still made in Roman numerals. To keep the amounts from being overlooked and to see that they were properly accounted for, some kind of orderly parallel system of entries was needed (see Fig. 195); it was not until the 15th century that for the first time the hundreds and thousands began occasionally to be written in vertical columns, as we have already seen in the account books of the City of Augsburg (Figs. 120–125). In contrast to this, it was much simpler just to enter the various amounts in the new numerals.

Now, however, an enemy suddenly appeared from an unexpected direction. As numbers began to be written in the new Indian numerals by some Italian trading houses, the City Council of Florence in 1299 issued an ordinance governing financial procedures. This *Statuto dell'Arte di Cambio* made it illegal, at a penalty of 20 solidi, to enter the amounts of money in the account books in numerals and to separate them from the text of the entry (*modo abaci*), and required them to be written out as before "in letters" (*per literam*) and to be placed within the entry itself. Why? To prevent fraud! An old Venetian work on bookeeping explains:

... lequal figure antique solamente si fanno, perche le non si possono cosi facilmente diffraudare come quelle dell'abaco moderno, lequal con facilita

*di una sene [segno] potria fare un'altra, come quella del nulla, dalla qual
sene potria far un 6 uno 9 e molte altre si potriano mutare,*

... the old figures alone are used because they cannot be falsified as
easily as those of the new art of computation, of which one can with ease
make one out of another, such as turning the zero into a 6 or a 9, and
similarly many others can also be falsified.

For this reason too in using the old numerals a good bookkeeper
should

*ben formarle e ben ligarle l'una con l'altra accio siano incatenate insieme,
con prestezza senza levar la penna de la carta,*

form them carefully and join them together like the links of a chain,
all the while writing rapidly and without lifting the point of the pen from
the paper.

We have seen the last point confirmed in our examples of Roman
numerals written in Germany (see Fig. 114). To be sure, the amounts
were here separated from the text of the entries, but the line was
terminated with a stroke for the same purpose of preventing falsification of the numbers. Likewise the last unit (i) in Roman numerals
was transformed into a j, again for security against tampering (see
Fig. 122).

People were still too insecure about the new numerals; it was not
only their forms that were unfamiliar but also the method of writing
them: *item sollen die rechenmeister sich hiefür mit zyffern zu rechen
maassen,* "moreover the master calculators are to abstain from
calculating with digits," ordered the Frankfurt Bürgermeisterbuch
("Mayor's book") of 1494, when a writer occasionally used the new
numerals. A hundred years later a book on arithmetic and the new
numerals was submitted to a deacon of the cathedral at Antwerp for
his *imprimatur*; his decision was:

*regulae haec ac rationes computandi ac summas conficidendi utiles quidem
sunt pro mercatoribus, in quorum gratiam imprimi possint, sed caveant sibi
ab usurariis ac aliis illicitis contractibus et cambiis,*

these rules and procedures for computation and for finding the answers
to problems are admittedly useful for merchants, and for their sake
permission is granted for them to be printed; but they [the merchants]
must see to it that they avoid usury and other illicit transactions and
exchanges —

that is, the new numerals are not to be used for dealings that are not
approved of. For a long time, not only here but also in Italy, documents written with Roman numerals carried greater weight in court.

Thus it happened that in the course of the 13th century computation
by the new methods and with the new numerals became familiar in
the commercial and trading establishments. They were even used
occasionally to transfer entries from invoices to account books, but
bookkeeping was still carried on in the old manner.

This, of course, was a serious obstacle to the spread of the Indian
numerals. If we add to this the scarcity of scrap paper cheap enough
to be thrown away after a computation was finished, and also the
undeveloped state of the operations of arithmetic, such as long
division in which the numerals constantly had to be crossed out as the

division proceeded (see Figs. 165 and 182) whereas they could be easily erased on a wax- or sand-covered counting board or simply removed by picking up the *calculi* on the old form of reckoning board, and then remember above all the conceptual difficulties which the zero created for most people, we shall understand why it took such a long time for the abstract place-value notation to come into general use even in Italy.

Italy as a School for Merchants
Despite all these difficulties, Italy was still far ahead of Northern Europe. For the South-German merchants of Augsburg, Nuremberg, Ulm, and Ravensburg Venice was the educational proving ground from the 14th century on. The *Arte dela Mercadantia*, the "mercantile art," as the oldest printed Italian arithmetic textbook called it (see Fig. 167), including bookkeeping, computation, and the *Welsche Practica*, "foreign practice," a merchant's son could learn nowhere better than in Venice. A citizen of Nuremberg urged his son in Venice to do three things above all: to rise early, to go to church regularly, and to pay attention to his arithmetic teacher. Then when the young merchant returned home, he brought back with him all the latest procedures and their Italian names: *agio*, "advantage, profit" for a payment on account, *disagio* for a discount; *conto* (*–corrent*) for a running account; *disconto* and its short form *sconto* for an extract from the reckoning; *giro*, "circle" for the cash circulation of money; *saldo* (from Latin *solidus*, "full, complete") for the closing of an account, *Bilanz* ("balance") for the "weighed" statement of credits and debits (see p. 175); *debit(o)* and *credit(o)* for "owe" and "have"; *ultimo* for the "last" day of the month; *spese* for "expenditure" (see p. 354); *valuta* for "value," especially for the value of foreign currency; *franco* for "free of cost"; posting constituting an entry in an account book from the Italian *posto*, "position, place"; *Muster* for "sample," from the Italian *mostra* and the Latin *monstrare*, "to show, display, allow to see"; *gross*, "coarse, rough" for goods with their packing, and *net*, "pure, clean" for the goods alone; *net cash* for an amount of money without deductions. *Cassa* was originally the name for a "money chest" (from the Latin *capsa*, "cask, chest"), so that a payment *per cassa* was a cash payment "for the money chest"; *incasso*, "in the chest," meant to collect money. A *Lombard*-transaction, a short-term borrowing on pledge of securities, was so named for the custom of the North Italian (Lombard) bankers who, for example, took over the money-lending business in England after the expulsion of the Jews; from this comes the English word *lumber-room*, "store-room," which was originally the room in which the "Lombard" accumulated the valuables pawned with him.

Then there was bankruptcy. This literally meant *banca rotta* (Latin *rupta*, from *rumpere*, "to break"), a "broken board," because dishonest moneychangers, who sat behind their "banks" or counting tables in the market place, had their boards smashed so that they "fled from the market," as in the Roman expression *foro fugere* ("to flee from the forum") meaning to go into bankruptcy. In addition,

Fig. 250 A money-changer at his "bank." Wood cut by Hans Weiditz, Augsburg, 16th century.

this "rescuing flight" (from debt) was also the underlying meaning of *pleite*, "business failure, bankruptcy" (from the Hebrew word *peleta*). Another term which came from the commercial world of Italy was "per cent" with its symbol %, derived from the abbreviation *p c°* for *per cento* by contraction of the *c°* into % (cf. Fig. 115); also the debit and credit in double-entry bookkeeping after the Venetian custom on two opposite facing pages of an account book, with the respective headings

dover dare — dover avere, "to be paid — to be received."

Most important of all, however, the young merchant apprentice from Northern Europe brought back home with him the Indian place-value notation and its living, practical use in solving his everyday problems: the four basic operations of addition, subtraction, multiplication, and division, the "golden" rule of three and the chain rule, an Oriental and probably Arabic gift to the Venetians, and the *welsche Praktik*, the "foreign procedures" of business arithmetic and money exchanges (Fig. 250):

Eyner geet zu wyen yn eyn wechsselpanck vnd hat 30 ℈ Nurmberger. alsso sprechen zy dem wechsseler liber wechssel mir die 30 ℈ vn gieb mir wiener darfür als vil sy dan wert seyn. also weyss der wechssler nit wie viel er ym wiener ssol geben. vnd begert der muncz vnderrichtung. also unterweyst jener den wechssler vnd spricht 7 wyner gelten 9 linczer vnd 8 linczer geltn 11 passawer vnd 12 passawer gelten 13 vilsshofer vnd 15 vilsshofer gelten 10 regensperger vnd 8 regensperger geltn 18 neumerker vn 5 neumerker geltn 4 nurmberger wie vil kummen wiener umb 30 nurmberger,

A man goes to a money-changer in Vienna with 30 pennies in Nuremberg currency. So he says to the money-changer, "Please change my 30 pennies and give me Vienna pounds for them as much as they are worth." And the money-changer does not know how much he should give the man in Viennese currency. Thus he goes to the money office, and they there advise the money-changer and say to him, "7 Vienna are worth 9 Linz, and 8 Linz are worth 11 Passau, and 12 Passau are worth 13 Vilshofen and 15 Vilshofen are worth 10 Regensburg and 8 Regensburg are worth 18 Neumarkt and 5 Neumarkt are worth 4 Nuremberg pennies." How many Viennese pennies do 30 Nuremberg pennies come to?

To solve this complicated and contorted problem from Johannes Widmann's arithmetic textbook of 1489 one really had to know how to calculate, in order to find out that 30 Nuremberg pennies are worth 13 Viennese (actually $13\frac{23}{429}$).

The chain rule is so named because the individual exchange values were tied together like the links of a chain, as in the example cited. Let us solve the shorter problem given by Adam Riese (Fig. 251) by

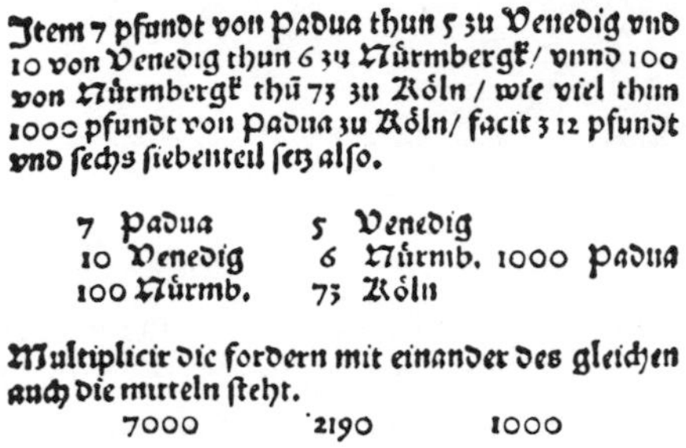

Item 7 pfundt von padua thun 5 zu Venedig vnd
10 von Venedig thun 6 zu Nürmbergk / vnnd 100
von Nürmbergk thū 73 zu Köln / wie viel thun
1000 pfundt von padua zu Köln / facit 3 12 pfundt
vnd sechs siebenteil setz also.

7 padua	5 Venedig
10 Venedig	6 Nürmb. 1000 padua
100 Nürmb.	73 Köln

Multiplicir die fordern mit einander des gleichen
auch die mitteln steht.

| 7000 | 2190 | 1000 |

Fig. 251 A chain problem from Adam Riese's second arithmetic textbook (1532 edition, see Fig. 257).

the procedure which was described in essence by Leonardo of Pisa and which amounts to arranging the equivalents in the proper order:

(1) Write down the desired number of Cologne marks, x, on the left and their given value on the right side, thus:

"x Cologne   1000 Padua,"

as the top line of Adam Riese's chain;

(2) link the individual values together (like dominoes) so that the amount at the left is always the same as the previous amount at the right (as Riese does);

(3) the last amount on the right is the equivalent of the first amount on the left (the x Cologne), and the chain is complete. Then x turns out to be a fraction whose numerator is the product of the amounts on the right and whose denominator is the product of the amounts on the left; thus

$$x = (1000 \cdot 5 \cdot 6 \cdot 73)/(7 \cdot 10 \cdot 100) = 312\tfrac{6}{7},$$

as Adam Riese also indicates. From his solution we see also that a person could, with some difficulty, have solved such a problem by himself without instructions. We are never told just why such an apparently unnecessarily complicated procedure was followed, but this is always the method given in arithmetic textbooks, without any indication of the reason. The reader will find it easy to follow if he proceeds upward step by step from the last line of the chain problem.

German cities also began to grow wealthy and rise in importance as a result of trade in the 13th century; here too they became focal points of power and culture, like the Hanseatic League in the North and Augsburg and Nuremberg in the South. During the 13th century, the time of the Scholastics, the first universities were founded at Paris, Oxford, Padua, and Naples, and although their teachers and students kept alive the knowledge of the new Indian numerals, which migrated about widely in algorithmic treatises during the following epoch, only in the rarest cases did merchants in Northern Europe use them for their accounts. Melanchthon's lecture to the students at the University of Wittenberg (see p. 434) is significant evidence for this. It was not from the universities but from Italy that the new numerals were taken into common use toward the end of the 15th century in the mercantile houses and the offices of the great German towns.

THE GERMAN ARITHMETICIANS
Place-value Notation versus Counting Board
The invention of printing in the 15th century enabled all the people for the first time to share in the knowledge of their time. It is thanks to this "black art" that the contest between the new place-value notation and the old counting board becomes visible in all its impact around the turn of the century (ca. 1500), when the mists of the Middle Ages began to lift. For the *Rechenbücher* "reckoning books," the algoristic textbooks of arithmetic, were among the first popular didactic works to be printed. We have already had to become familiar with them because they are important sources of information about computations on the various forms of the counting board (see p. 334).

The old and the new are symbolically represented in the book by Gregor Reisch (see Fig. 182): Next to Pythagoras with his sorrowful face working at his reckoning table sits a cheerful and serene Boëthius contemplating his computations in the new numerals. Arithmetic, personified, hovers with her books between them, looking at the computer with digits and indicates her approval of him by the two geometric series in Indian numerals on her garment.

The title page of a book by Adam Riese (Fig. 175) has a picture with a view of an actual medieval calculator's booth. The three men there are busily contemplating both methods of computation and appear willing to judge thoughtfully the advantages and disadvantages of both, corresponding to Riese's view:

Ich habe befunden in Unterweisung der Jugend, dass alleweg die, so auf den Linien anheben, des Rechnens fertiger und lauftiger werden, denn so mit den Ziffern, die Feder genannt, anfahen. In den Linien werden sie fertig des·zelens und für alle exemple der kauffhendel und Hausrechnung schöpfen sie einen bessern grund. Mügen alsdann mit geringer Mühe auff den Ziffern ihre Rechnung vollbringen,

I have found, in teaching young people, that it was always those who begin on the lines who are more adept and quicker than those who work with ciphers and a pen. On the lines they get finished counting, and in all examples of commerce and domestic trade they stand on firmer ground. They may therefore with but little trouble complete their calculation with ciphers.

There is a more lively encounter of this kind, in which each of the two reckoners seems to be asserting the merits of his own procedures

Fig. 252 Winged Arithmetic, shown here as the fourth of the seven liberal arts, in a woodcut by the Nuremberg artist Hans Sebald Beham (died 1550); she turns her back on the counting board (cloth) and points emphatically to the tablet with the new Indian numerals.

(Fig. 253) A third man not involved in the dispute, computes quietly in the background. In addition, a fourth man takes part in this controversy, for in such pictures there is often a merchant, a city clerk, or someone else who is concerned with numbers present in the calculator's office as the latter explains the new Indian numerals and argues with someone else over the superiority of one method of computation over the other.

Fig. 253 "Abacist" and "algorithmicist" dispute the advantages of computations on the counting board and with place-value numerals. The man at the rear works on paper with ink and pen, the one at front right with chalk. The letters on the wall stand for *verbum domini manet in eternum,* "God's Word remains forever." From an English work on the seven liberal arts by Robert Recorde, personal physician to the King.

What we see here are the echoes of the long struggle between the "abacists" and the "algorithmicists" which continued beyond the 16th century, ending in the final victory of the latter. The contest began in the time of the Scholastics, as evidenced by an interesting passage from the *Jüngere Titurel,* a much-loved knightly romance of the 13th century; here the characters Algorismus and Abakuc (for Abacus) appear as experts in computation:

Nun ist auch hi gesundert
Lot vurste von Norwege
Ichn weyz, mit we vil hundert,
Ob Algorismus noch lebens plege
Unde Abakuc de geometrien kunde,
De heten vil tzo scaffen
Solten se ir aller tzal da haben funden,

And here too comes out
Lot, Prince of Norway,
With I know not how many hundreds;
If Algorismus were still alive
And Abakuc, learned in geometry,
They would have much to do
To find the number of them all
[that is, to count the great number of knights accompanying Lot].

But we know also how much life there was still in the counting board and the Roman numerals, for even in the 18th century the ability to reckon with *jetons* was considered one of the qualifications of a marriageable daughter (see p. 333). When Molière had his *Malade Imaginaire* check his apothecary's bill *ayant une table devant lui comptant avec des jetons,* "counting with counters at a table placed before him," this was not intended to appear comical in mid 17th-century France.

Fig. 254 A view of a 16th-century German counting house, in which book-keepers and reckoners are at work. Woodcut by Jost Ammann.

But the *Rechenmeister* ("master calculators") with their Rechen-bücher, arithmetic textbooks, gave the *coup de grâce* which finally ended the counting board, even though it continued to be used here and there for a long time afterward. Who were these master cal-culators? Up to the time of Martin Luther and even later, German schools in which children learned to read and write the vernacular were few in number and poor in quality. An education in the "hu-manities" was provided only by the ecclesiastical schools, which taught in Latin; but with the rise of the cities and towns they were increasingly forced to meet the requirements and demands of the times. The mercantile classes needed teachers who would instruct their children in the kind of reading, writing, and arithmetic that were required for everyday use. Here and there an occasional "German" school was established to meet the first two needs, but arithmetic was not taught here either. Who, after all, knew how to calculate, and in the "new" arithmetic at that? The students and graduates of the universities? Let us see what Philip Melanchthon had to say in a lecture delivered in the year 1517 to the students at the University of Wittenberg (translated from the Latin):

Now that I have discussed for you the usefulness of the art of computation, of which there cannot be the slightest doubt, I believe that I should also make some brief remarks about the ease with which it can be done. I believe that students allow themselves to be frightened away from this art because of their preconceived notion that it is too difficult. As far as the elements of computation are concerned, which are already generally taught in schools and used in daily life, those who think that they are too difficult are greatly in error. This knowledge springs directly from the human mind and appears with full clarity. Therefore its elements cannot be obscure and difficult; on the contrary, they are so clear and evident that even children can grasp them, because everything proceeds so naturally from one point to the next. The rules for multiplication and division, to be sure, require more diligence for their mastery, but their meaning will still be understood very quickly by those who give to them their full attention. These skills, of course, like all others, must be sharpened by practice and experience.

Even the youngest German children who are taught arithmetic are not treated as kindly as Melanchthon here treated the adult students of the seven Liberal Arts.

Thus the would-be merchant either had to obtain his practical training in Italy or else learn his trade in the counting house of a German commercial establishment, until the cities' demands for a general education that included computation became louder. Toward the end of the 15th century we have the first evidence about a German teacher of computation, Ulrich Wagner in Nuremberg, the author of the first printed German arithmetic textbook, which we have already met several times (see Figs. 168, 189, and 260). In the 16th century we find teachers of computation in quite a number of cities. The schools of Nuremberg and Ulm were especially famous, the latter so much so that there was a saying in the 17th century in the Empire that *Ulmenses sunt mathematici,* "men of Ulm are skillful calculators."

The document with which the Hanseatic city of Rostock appointed a *Rechenmeister* in 1627 reveals something about his activities and obligations:

We, the Mayor and Council of the City of Rostock, hereby testify that we have appointed the honorable and learned Jeremias Bernstertz master of writing and calculating for our city; and by force of this letter he is directed to spend one hour every week on Mondays, Tuesdays, Thursdays, and Fridays in the Latin school for the instruction of the young without distinction and without charge, and others outside the school every day of the week, whether boys, girls, or any others who may apply to him, he is authorized for a fair and reasonable weekly or monthly fee to instruct in writing Latin and German, in computations, in bookkeeping and other useful skills and good habits, and he is authorized to perform everything else besides which an honest and industrious master of writing and computation may properly do according to his understanding and his best abilities.

I cannot resist the temptation to show, along with a "Western" computation master who is undertaking to instruct a young pupil, a picture of his Far Eastern counterpart (Figs. 255 and 256).

Fig. 255 A father apprenticing his son to a calculator, who appears to be fully in the camp of the abacists. Hans Weiditz, who made this woodcut, used it in *Petrarcas Trostspiegel* ("Petrarch's Mirror of Consolation"), 1535, to illustrate an unrighteous guardian: *Als bosslich solcher vormund thuut | Der stilt von seines pflegkinds gut*, "When such a guardian does wrong, he steals from his ward as well."

Fig. 256 A Japanese student learning the *soroban* from a master calculator. Woodcut from a Japanese textbook.

In 1613 there were in Nuremberg 48 such computation "schools" — that is, 48 authorized teachers. Small wonder then that they were not too well disposed toward one another and competed vigorously for their pupils' pennies, until the Nuremberg city council finally ordained that "no teacher of computation shall establish himself too close to another, but shall hang his shingle at least two streets away." In the larger cities they, like masters of other crafts, joined together in associations and guilds, some of which lasted into the 19th century. In the smaller towns, especially in the 16th century, a teacher of computations would at the same time occupy the position of town clerk, with all its computational duties such as the supervision of weights and measures and testing the contents of barrels and other

containers, and so forth. Not everyone could do what master calculators could, and they were valued and honored, as we may read
on the title pages of many of their books (see Figs. 171 and 172).

Adam Riese

One of the *Rechenmeisters* who have become immortal was Adam
Riese, who is still called on in Germany to testify for the correctness
of a computation: "2 × 6 = 12 according to Adam Riese." He was
born in 1492 in Staffelstein near Bamberg; at the age of 30 he was
Rechenmeister in Erfurt and in 1525 *Rezessschreiber*, "settlement
writer" in the Bureau of Mines in the Saxon town of Annaberg —
that is, he kept the records of the receipts and expenditures of
the mines. In addition, he maintained his "very large and famous
school (for computation)" as the chronicle puts it. He died in
Annaberg in 1559.

Why did Riese in particular become so famous? Not just because
he was an expert calculator, but above all because his books on
computation were widely used and imitated for their clear instruction
and their "charming" problems:

*Item ann eynem tantz seyndt 546 personn darunder seyndt eynn ditteyll
Jungegeselln | ¼ burger | ⅙ edelleut ⅛ pauern vnd ¾ Junckfrawen | Nun
seyndt der Junckfrawen nit so vil | darmit sie alle tzu gleychenn tantzenn
mügen | dann so offt 6 Junckfrawen tantzen | so manches mal muss 1
person under gemelten geschlechtern feyern | Frage ich | wieviel eynes
ytzlichenn geschlecht in sonderheyt sey,*

And again: at a dance there are 546 persons; one third are bachelors,
one fourth are burghers, one sixth are nobility, one eighth are peasants
and three fourths are unmarried young maidens. Now there are not so
many maidens that all (the people) can dance at once; as often as 6
maidens dance, 1 person of the groups mentioned must be idle. So I ask,
how many are there in each separate group?

Riese calculates 112 (16) bachelors, 84 (12) burghers, 56 (8) nobles,
42 (6) peasants, and 252 maidens who are dancing. Most readers will
probably misunderstand Riese's problem. It is not the proportional
fractions of 542 that are to be found, but rather: out of some unknown number x of people, one third are bachelors, etc., amounting
to 546 altogether. (a) How great is the quantity x? The sum of the
fractions when added together comes to $\frac{39}{24}x = 546$, so that $x = 336$.
(b) How many persons of each group take part in the festivities? One
third of the bachelors $\frac{1}{3} \times 336 = 112$, etc. according to Riese's numbers above. (c) Those who are actually dancing are to be calculated
in proportion to the excess of men (the proportional number $= \frac{1}{7}$);
thus this means $112 \cdot \frac{1}{7} = 16$ bachelors, etc. (the numbers in
parentheses above).

Adam Riese had never studied formally. Thus he was one of the
unlearned writers of computation textbooks, which of course were
also composed by scholars in the universities. Perhaps for this very
reason Riese better understood the nature and requirements of the
everyday calculations that merchants had to make. His books are
based essentially on these and on the *Arte dela Mercandantia* ("The
Mercantile Art"), so that they were in effect patterned after Italian

models. It is not merely by chance that the oldest known book of computations was printed in Italy in 1478 and that the first such German book was written in 1482 by a master calculator of Nuremberg. The oldest German computation books also include the *Behende und hubsche Rechnung auff allen kauffmannschafft*, "Speedy and Elegant Computations for all Commercial Purposes," by Johann Widmann of Eger, which appeared in Leipzig in 1489 and was later reprinted many times. A surprising source of problems in "Practica" composed in the manner of the Italian computation books and used by German *Rechenmeister* (such as Widmann, the Bamberg Rechenbuch, etc.) was a Regenburg Algorismus of the mid 15th century (cf. also the *Regeldedri* — "rule of three" — in the Bamberger Blockbuch, Fig. 170).

Adam Riese (also spelled Ryse, Ries, Ris) was the author of the following books:

(1) the computation book of 1518, which taught only the use of the counting board: "Computation on the lines, by Adam Riese of Staffelstein, / in the measure that it is usually taught in all schools of computation, beginning with the basic elements, 1518." There were later editions in 1525 and 1527.

(2) the computation book of 1522, which was devoted especially to computations with numerals: "Computations on the lines and with the quill: numbers, measures, and weights for all purposes, performed and brought together by Adam Riese of Staffelstein, master computer at Erfurt, in the year 1522." This was printed in 76 octavo sheets, with many reprintings and editions, of which those of 1527, 1529 (our Fig. 175), 1530, 1532, 1535, 1544, 1556, 1574, and 1579 are known.

(3) his largest and most popular computation book of 1550, which, on 196 quarto sheets, was concerned especially with "Practica" (see Fig. 257): "Computation on the lines and with the quill / set forth at length / including the advantages and speed of the *Proportiones Practica* / with thorough instructions in mental arithmetic. By Adam Riese, in the year 1550." This was considered the best computation book of its time; a new edition was prepared in 1611 by Riese's grandson, and a new printing came out in 1656.

(4) the so-called *Annaberger Brotordnung* ("Annaberg Bread tables") of 1553: "A booklet / of computations concerning the bushel / the pail, and the pound weight / for the wise and gracious council of St. Annaberg, by Adam Riese, 1533." From these computational tables, which were prepared at the bidding of the town council, one could figure the proper weight of a penny loaf as the price of grain rose from 20 to 80 groschen, and also how many penny loaves could be baked from one bushel of grain at such and such a price. Hence this was not a textbook of instructions in computation, but a "precomputed" booklet with numerical tables.

The surprisingly large number of books on computation which appeared after the beginning of the 16th century, some of which went through numerous reprintings and new editions, is testimony to the general need for skill in computation; it also shows that this was the time when the new Indian numerals first entered Germany on a large

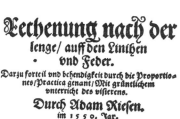

Fig. 257 Adam Riese aged LVIII, in the year 1550. From a woodcut in his third computation textbook.

scale. Even if here and there the author of a computation book expressed his preference for the counting board and for the "good old German numbers," as the Roman numerals were called, he still at least had to explain the new place-value notation, "the cipher numbers," if he wanted to compete on equal terms with his many rivals. This competition was quite fierce, as we see from the self-advertisements on many a title page: "... the same had never before been printed either in German or in any foreign language ..." or "... the same so artfully and beautifully composed and never seen in print ...." Some writers even took the trouble to compose their instruction in verse (like the *Carmen de Algorismo* some three hundred years earlier):

So du magst von der obern nit
Ein ziffer subtrahirn mit sitt
Von zehen solt sie ziehen ab
Der nechst vnder addir eins knab,

So if you cannot properly subtract
A digit from the one above it,
You must take it away from ten
And add one to the next digit in the line below.

In this manner the guild of master calculators was the last harbinger of the modern numerals we use today.

THE "NEW" NUMERALS

But there were resistance and obstacles, apart from those mentioned earlier, still to be overcome, even in the mere writing of the numerals:

Unum dat zungel (1), *kruck* (2) *duo significabit,*
suswanczque (3) *tria, wuerstfuel* (*) *dat tibi fiere*
reffstab (*) *dat funfe, widder* (6) *dat tibi sechse*
Süben gesperre (*) *ethwe kette* (8), *nün colb* (9) *significabit.*
Ringel (0) *cum zingel* (10) *tibi decem significabit.*
Si zingel (1) *desit, ringel* (0) *nihil significabit,*

The little tongue gives "one," the hook means "two,"
The pig's tail "three," the sausage (*) "four,"
The crook (?) (*) "five," the ram's horn means "six,"
"Seven" the lock (bolt ?) (*),
The chain means "eight," the club "nine,"
And the little ring with the tongue means "ten."
If the tongue is missing, the ring means nothing.

(*) refers to the old forms of the numerals 4, 5 and 7 (see bottom of Fig. 241).

I shall leave it to the reader himself to check the forms of the digits in the family tree of the Indian numerals (see Fig. 239) and see how much they actually resemble their mnemonic words. However effective it may have been, this passage from a 15th-century manuscript from Strasbourg is striking evidence of the difficulty which the beginner had merely in writing the forms of the new numerals properly, let alone using them in computations. And then there was always the peculiar shape of the zero sign, which always required a special explanation:

Hab achtung neun sein der figur /	Take care when writing figure nine
On all beschwer auszusprechen pur.	In any case to express it clean.
Bei solchen ferner merk auch mich	And further note in any case,
Ein nulla steht vnaussprechlich	The zero stands without a sound;
Rund vnd formit recht wie ein o /	Round and formed just like an o
Wirt dann dasselb versteht also	It will then be so understood
Ayner deutlichen fürgemalt	When it is writ before another
Bringts zehenmal so vil als bald.	It brings ten times as much to it.
Mit den kanstu recht numeriren	Thus canst thou truly numerate
All zal aussprechen vnd volfüren,	And mark and fill all cyphers well.

All this, along with the concept of place-value notation, was quite a lot to swallow!

Nevertheless the numerals soon began to appear everywhere, apart from account books. The year especially was often inscribed in Indian numerals in books, on buildings, and on pieces of furniture (Fig. 258), just as it was in bookkeeping.

This can be seen in the financial accounts of the city of Augsburg for 1430; the entries in the accounts of the Rüsselsheim office, which are written entirely in Roman numerals, are preceded by the year's date in the Indian place-value notation (see Fig. 123). A number which in any event recurred all year long, was both more easily and more prominently written in the new style. Because of their brevity and also their special position, the page numbers in books soon came to be printed in the new numerals as well.

Fig. 258 The date 1472. Inscribed on Eberhard von Württemberg's prayer stool in the cloister chapel at Urach.

The Indian numerals, as we know, made their earliest appearance in Europe as *apices* in monastic manuscripts of the 10th and 11th century (see Fig. 161). Among the oldest manuscripts with these numerals are our two examples from the 12th century (see Figs. 247, 248, and 249). The oldest year date, 1138 (= 533 after the Hegira), to appear in Europe in the new place-value notation (East

Fig. 259 Coins with the oldest dates in Indian numerals (East Arabic):
(1) The oldest date of all in Europe is on a Sicilian copper coin, which bears the head of Christ on the obverse, and on the reverse in Arabic script: "At the command of the great King Roger, mighty with the help of God, 533" (after the hegira, in East Arabic numerals; diameter 15 mm).
(2) The oldest German date, 1424, occurs on a medallion from St. Gall, at the upper left next to the figure of the saint, who is shown standing on a bear and begging; diameter 23 mm. Collection Münzen und Medaillen A. G., Basel.

Arabic numerals) occurs on a Sicilian coin of the Norman King Roger II, in whose reign the Norman state in the Mediterranean reached the height of its power. This coin (Fig. 259) is also an excellent example of the close ties between the Norman kingdom and the Arab world.

The oldest German coin which gives the date of the year in the new numerals is a *Plappart* ("blaffert,") a base silver coin struck by the town of St. Gall in 1424. This Swiss German name for the $\frac{1}{26}$-gulden coin goes back to the Old High German *bleih-faro*, "pale color," from which was derived, through the French *blafard*, the Medieval Latin term *blaffardus*, "white penny."

The new methods of computation helped to promote the ultimate victory of the new numerals. The old medieval cleavage between writing numbers in numerals (Roman) and doing computations with counters on the counting board disappeared; with the new numerals one could write numbers and make computations at the same time. We have already seen in the case of division how the cross-out and erasure operations developed from the counting board after the introduction of the Indian numerals (see p. 331). Hence these represent an intriguing transitional form of operation between the counting board and pure written operations. Now let us follow this through again, in multiplying 765 × 321 = 245,565, so that we may better understand the nature and impact of the new procedures. The reader is urged to follow these steps himself.

The essence of the old-fashioned method lay in the fact that the factor 321 had to be moved (1 ... 3) so as to establish the order of magnitude of the partial product each successive time, and that the carry-overs were immediately combined with the partial products.

	(A)					
	2	4	5	5	6	5
3'			1	5		
			1	0	5	
2'		1	8			
			1	2	6	
1'	2	1				
		1	4	7		
				7	6	5 (b
1		3	2	1		(a
2			3	2	1	
3				3	2	1

	(B)					
	2	4	5	5	6	5
			1	5		
		1	8	1		
		1	1	2	0	
	2	1	4	7	6	5
				7	6	5
	3	2	1	1	1	
	3	2	2			
		3				

(A) First let us take each of the steps one by one:

1) The initial numbers are marked a and b. The units 1 of 321 are placed beneath the highest order 7 of the factor b. The 7 is then multiplied successively by 321: 1 × 7 = 7, then 2 × 7 = 14, then 3 × 7 = 21. The partial products are placed above the successive factors from a (Line 1').

2) The factor 321 then moves one place to the right, under the 6 of b, and we obtain the successive partial products 1 × 6 = 6, then 12 and 18 (Line 2').

3) The factor 321 moves one more place to the right, and the partial products are in Line 3'.

The final product after adding everything together is 245,565. But the medieval calculator performed the operation differently:

(B) He began by writing the numbers not separately in different lines but by placing each succeeding digit in the next open space below. He ended by pushing the numbers in the Lines 1' to 3' down to the bottom. Correspondingly, he moved the successive digits of the "wandering" factor 321 upward.

```
        (C)                          (D)
         5                2  2  4  7   6   5
         4  5                3  2  1
      4  3  0
      2  5  9              2  4  3  2  6  5
   2  1  4  7  6  5           3  2  1
   ────────────────
            7  6  5
   ────────────────
      3  2  1  1  1        2  4  5  5  4  5
         3  2  2              3  2  1
            3
```

(C) In the cross-out operation procedure he immediately combined the carry-overs and struck out the finished digits, so that the final product 245,565 was read off from the "peak numbers" that were not crossed out.

(D) In the erasure procedure the digits crossed out (in C) and also the successively used up digits in the factor 765 were erased, so that the computation after three steps gave the final result as the only number remaining.

Fig. 260 "*Multiplicirn ins creucz*," "multiplying by the cross," from the Bamberger Rechenbuch of 1483. The old and new forms of the digits 4 and 5 appear simultaneously in this same problem (see Figs. 168 and 169).

In Italy the very first computation textbooks taught other procedures that had no traces of the counting board and were much more suitable for written numerals and true computation "with the quill." For multiplication especially there was a whole series of procedures that had long been familiar in India.

1. Multiplication by the cross — *multiplicare per crocetta* — which was the method normally used in Venice:

```
2   3   The positions of the same rank were combined immediately:
| × |   units: 3 × 4 = 12; tens over the cross: 4 × 2 + 1 × 3 + 1 =
1   4   12; hundreds: 1 × 2 + 1 = 3; final result: 322.
─────
3 2 2
```

This procedure was taught as far as two four-place factors; it was also shown in the *Bamberger Rechenbuch*. Adam Riese remarked in regard to it: *Sie nimmt viel kopffs*, "It takes much skill."

2. Multiplication by factors — *multiplicare per repiego*. A divisible (nonprime) number in Italian is *un numero di ripiego* (from the Latin *plicare*, "to fold"). Problem: $23 \times 14 = 322$; the multiplication is done not by the number (14), but by its factors $(2 \cdot 7)$.

$$
\begin{array}{r}
23 \cdot 14 \\
\cdot 2 \quad \overline{46} \\
\cdot 7 \quad 322
\end{array}
$$

3. The method commonly used today, which was called *multiplicare per colonna*, "multiplication in columns" or *per scacchiero*, "on the checkerboard."

4. Multiplication by the diagonal lattice — *multiplicare per gelosia*, literally "in the manner of a jalousie." Again, $765 \times 321 = 245,565$.

On a square subdivided and with diagonals drawn in one direction, the factors 765 (marked by dots above) and 321 (prime superscripts) are written down along two sides next to the spaces, as in the upper illustration of Fig. 261, with the second factor in reverse order. Each digit is multiplied by every other, and the results are written in the square, the two digits of each partial product separated by the diagonals as shown $(3 \times 5 = \frac{1}{5}$, then $3 \times 6 = \frac{1}{8}$, $3 \times 7 = \frac{2}{1}$, and so on). In this very ingenious manner the carry-overs are eliminated. The operator can simply take each of the partial products from a multiplication table and write them in the spaces provided; thus the lattice methods requires *wenig kopffs* — "little thinking." The temporary sum of the digits in the same diagonal bar forms one place in the final product: from bottom right, we have 5, then $6 + 0 + 0 = 6$, etc. The Arabic writer had to place his second factor on the left side of the large square so as to be able to write the three last digits of the product, 565, in the diagonal bars at right. The reader will find these East Arabic numerals easy to read with the aid of Fig. 234.

All these various procedures for the one operation of multiplication reveal the great effectiveness of the new numerals: the duality of computation on the counting board and writing the numbers in Roman numerals were thus finally overcome by the single-step procedure of written computation using the new Indian numerals.

This was but one effect of the new digits, one which was both obvious and important. But there was also another effect, of great future significance, which is seldom seen or thought of. This was the much greater skill in computation that was made possible by the new numerals.

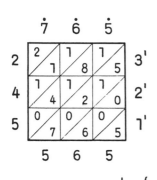

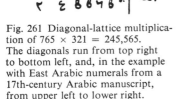

Fig. 261 Diagonal-lattice multiplication of $765 \times 321 = 245,565$. The diagonals run from top right to bottom left, and, in the example with East Arabic numerals from a 17th-century Arabic manuscript, from upper left to lower right. Bayrische Staatsbibliothek, Munich.

Skill in Computation

The Indian digits and place-value notation also resulted in another "mathematical" development: computational dexterity increased, so that calculations became easier and could be run off on a calculating "machine."

Glance back over the history of the new numerals: Among all the various peoples to which they ultimately became familiar, they found the most fertile soil for their establishment and further development in the "Faust-like" Western culture. Just as soon as their strength was unleashed, number and its controversial offspring, technology, set out to transform nature and life on this planet to an unprecedented and fateful extent. There can be no doubt that the new numerals have contributed to determining the spirit and the shape of life in our century.

Now let us return to the beginnings of this development. People who dealt with numbers had scarcely mastered the four basic operations of arithmetic in the new style, and merchants and tradesmen had scarcely adopted the new "Practica," when such celebrated men of science as Kepler, Bürgi, and Napier developed their almost miraculous computations with exponents which Napier as their first inventor in 1617 presented to the world under the name of logarithms. Who, in the time when Roman numerals were used exclusively, even approached such a concept? But now, since these numerals made it possible to leap rapidly and with agility from an idea to its consequences and corollaries, logarithms were invented barely a hundred years after Adam Riese's first book of computations on the counting board. Logarithms were a magic tool: how, for instance, could anyone previously manage to extract the seventh root of 5? Now, with exponents, this computation is no more difficult than ordinary long division.

The concept of the logarithmic multiplying rod or primitive slide rule was invented in 1624, only five years after Napier's table of logarithms, by the Englishman Edmund Gunter. In 1657 it appeared with a movable "tongue" — in other words, in the modern form in which it is now used so extensively in mathematics, the physical sciences, and engineering that one can no longer imagine them without it.

This is, of course, beyond the scope of this book. But what is pertinent to our discussion is the early development of the calculating machine, the so-called Napier's bones (or rods). Apart from the multiplication table, some form of which probably existed in every period of history, these were the first "machines" that could calculate independently once the numbers were "given" to them.

It is profoundly significant that they were thought of by the same Napier who also (although not alone) invented logarithms, and thus perceived these two possibilities inherent in the Indian numerals: their potential ability to increase man's skill and capacity for computation, and their capacity for making numerical calculations much easier.

The "calculating rods" were conceptually and physically simple — so simple, in fact, that they found enthusiastic users and proponents throughout Europe, and elsewhere. Perhaps the friars in the monastery at Andechs used them for their calculations as late as the 19th century (Fig. 262).

The surprising thing is that these calculating rods were obviously invented in conjunction with the Italian procedure of lattice multiplication (see p. 442). On each rod, which was subdivided by successive diagonals into a slanting lattice pattern, were inscribed the first nine multiples of a unit: 4, 8, 12, . . ., 36 for 4. Then if one wanted to calculate 479 × 83, for example, he would compose the factor 479 out of the rods for 4, 7, and 9, place the guide rod with the Roman

Fig. 262 "Calculating rods." This, the oldest calculating "machine" (invented by the Scotsman Napier in 1617), for any number (in this case, 479) immediately gives the partial product when the number is multiplied by some particular factor (here 2395 for 5) inscribed (in Roman numerals) on the left side of the guiding rod. The digits in each successive slanting row are added together to find the product. Each rod is 8 cm long. From the Andechs monastery in Bavaria.
Deutsches Museum, Munich.

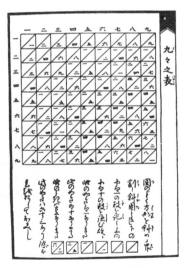

Fig. 263 "Calculating rods" in the Far East: The first column on the extreme left is the guiding rod with the numerals 1 ... 9; then follow the multiples of the units from 2 through 9. In this order, as shown, they form the multiplication table; example: 5 × 7 = 35 where the fifth horizontal line intersects the seventh vertical column. These Chinese numerals can be read with the aid of the table on p. 457. From a Japanese work on mathematics of the early 19th century.

numerals from I through IX at the left, and then read off the partial products without effort: 38320 + 1437 = 39,757. These rods are also a great help in division (for instance, 1,837,625:479), because the multiples of the divisor can be seen immediately. Thus on each of the four sides of a rod there is a multiplication series (see Fig. 262); the unit of which they are formed is indicated in the topmost square 3.1.7.9. Thus on the last extracted 9-rod, the left side contains the multiples of 3, the right side the multiples of 7 (shown), and the bottom side the multiples of 1; R indicates the guiding rod with Roman numerals.

Napier also extended the use of these calculating rods to other mathematical operations, for which, of course, the rods then had different numbers.

Thus we have seen that the introduction of the Indian numerals meant far more than merely the change from the old Roman numerals to a new and improved means of writing numbers. Place-value numerals — lattice multiplication — calculating rods: this succession of words symbolizes the effects and the consequences of the new numerals. Their influence was not only germinating but also fateful. For as this seed grew, as this effect became increasingly powerful, it ended by changing the essential features of Western culture.

RETROSPECT

We have finally come to the end of our wanderings through the history of written numerals. If we take another look at the "family tree" (Fig. 239) showing the descent and development of the Indian numerals, we shall be able to trace the path we have followed for so long as if on a map:

They sprang up in India. In the 8th century they reached the Islamic lands, where they are still written in East Arabic form. From the West Arabic *gubar* numerals descended the *apices*, which in the 10th and 11th centuries first carried the news of the new numerals to Europe. But the *apices* mistook the essence of the Indian place-value notation and thus followed the Roman abacus into oblivion.

The translations of al-Khwarizmi's textbook on computation (in the 12th century), on the other hand, reintroduced this place-value notation into Europe rationally and with a full understanding of its real nature. Nevertheless, it remained throughout the Middle Ages more or less tucked away in learned treatises on the Algorism, and met with no acceptance on the part of the general public.

In Italy, however, through Leonardo of Pisa's *Liber Abaci*, the new place-value notation worked its way into the counting houses and bookkeeping offices of the great mercantile houses, whence merchants and traders carried the new methods of computation with them northward across the Alps. The *Rechenmeister* helped them to take root by teaching and by writing computation textbooks, so that from the 16th century on, after a thousand years of migration from their initial home, their ultimate victory was assured.

This was the victory of an alien culture, to be sure, but also a victory for the mind of man, who finally, in the long history of written numerals, arrived at a mature, abstract place-value notation: the Indian numerals, which have become our own, were now fully developed.

Spoken Numbers and Number Symbols in China and Japan

Far Eastern Number Systems

Now that we have completed our journey through the history of spoken and written numbers, let us turn our attention to the culture of China and Japan, which — apart from our discussion of the abacus and the counting board — we have thus far touched on only very lightly.

China, a land of mystery! Yet Chinese culture was already in existence when the Pharaohs of Egypt were building their pyramids, when Greece and Rome flourished and declined, when western Europe arose from their ruins — and it continues to live on today. The barriers formed by the Chinese language and Chinese writings have protected China against foreign invaders far more effectively than the Great Wall of stone behind which she attempted to isolate herself.

Thus Chinese spoken numbers and written numerals are unique. We shall meet traits and trends which will surprise us not only by their very early appearance, but also by their maturity and completeness. In Chinese culture the old stands unquestioned and even unnoticed next to the new which the Orient absorbed as a result of its contact with the West.

It is intriguing that Japan and Korea, although both are located geographically within China's sphere, each have a language that is totally unrelated to the Chinese, and that each developed its own number sequence, the Japanese more sophisticated than the Korean, but then abandoned it and adopted the Chinese number sequence, which they assimilated into their own languages.

We shall now take up the Chinese, Japanese, and Korean spoken number sequences and then the Chinese written numerals, which until the introduction of the Indian numerals were the only ones used widely in the Far East. The peculiar features of the spoken and written numbers of these peoples will be more easily seen and understood against the background of the Western development we have just finished studying.

Spoken Numbers

THE CHINESE NUMBER WORDS
The four most important numerical ranks, the tens T, the hundreds H, the thousands Th and the ten thousands TTh are:

T *shih* 10, H *pai* 100, Th *ch'ien* 1000,   TTh *wan* 10,000.

The "last" rank, *wan*, which is clearly an old limit of counting, forms the following higher ranks (like our own "thousand"):

HTh hundred thousand — *shi-wan* (T TTh),
M million — *pai-wan* (H TTh)
TM ten million — *wan-wan* (TTh TTh),
five hundred thousand — *wu shih wan* (5 T TTh),
5 M five million — *wu pai wan* (5 H TTh),
50 M fifty million — *wu wan wan* (5 TTh TTh).

The Chinese, Japanese, and Korean Number Sequences

	1	2	3	4	5
		Japanese		Korean	
	Chinese	pure	Sino-Japanese	pure	Sino-Korean
1	*i*[1]	*hito-tsu*	*ichi*	*hana*	*il*
2	*erh*	*futa-*	*ni*	*tul*	*i*
3	*san*	*mi-*	*san*	*sed*	*sam*
4	*szu*	*yo-*	*shi*	*ned*	*sa*
5	*wu*	*itsu-, i-*	*go*	*tassöd*	*o*
6	*liu*	*mu-*	*roku*	*yösöd*	*ryuk*
7	*ch'i*	*nana-*	*shichi*	*nilkop*	*tchil*
8	*pa*	*ya-*	*hachi*	*yöltöp*	*phal*
9	*chiu*	*kokono-*	*ku*	*ahop*	*ku*
10	*shih*	*to (-so-)*	*ju*	*yöl*	*sip*
11	*shih-i*		*ju-ichi*	*yör-hana*	*sip-il*
12	*-erh*		*-ni*	*-tul*	*-i*
20	*erh-shih*	[*hata-chi*	*ni-ju*	*sümul*	*i-sip*
30	*san-*	*mi-so-ji*	*san-*	*sörhün*	*sam-*
40	*szu-*	*yo-so-*	*shi-*	*mahün*	*sa-*
50	*wu-*	*i-so-*	*go-*	*sün*	*o-*
60	*liu-*	*mu-so*	*roku-*	*yesün*	*ryuk-*
70	*ch'i-*	*nana-so-*	*shichi-*	*nirhün*	*tchil-*
80	*pa-*	*ya-so-*	*hachi-*	*yötun*	*phal-*
90	*kiu-*	*kokono-so-*	*ku-*	*ahün*	*ku-*
100	*pai*	*momo,(ho)*	*hyaku*	*päk*	*päk*
1000	*ch'ien*	*chi*	*sen*		*tchön*
2000	*erh-ch'ien*		*ni-sen*		*i-tchön*
10,000	*wan*	*yorozu*]	*man, ban*		*man*
10⁵	*shi-wan*		*ju-wan*		*sip-man*

Pronunciation:

ch as in "*ch*ild"
sh as in "*sh*ould"
sz: sharp *s* as in "*s*ee"
h as in German a*ch* (guttural, formed at rear of mouth)
hs as in German i*ch* (formed at front of mouth)
j as in French *j*our
w as in "*w*ould"
y as in "*y*ear"
' indicates aspiration: *ch'ien* pronounced *ch-hien*.

In addition to these above, there are also the "new" ranks:

H Th hundred thousand i^4 — but this is also the word for 10^{11}, an indication of how seldom it is used. It is differentiated from the word i^1 by the "tone," which is indicated by the superscribed number (see below). There are also

M million – *chao*, TM ten million – *king* and
HM hundred million – *kai*
    hundred million – *oku* (Japanese)
    hundred million – *ök* (Korean).

Buddhism, which came to China around the beginning of the Christian era, had a predilection for very large numbers, as manifested in the number-towers of India (see Fig. 30). In China, too, it extended the number sequence with native numerical ranks beyond *wan*, going up to 10^{14}, but apart from *chao* and the Japanese *oku* these never took root.

The word for a composite number, such as 24,789, for example, is formed by decimal gradations and without exception follows the rule of succession of magnitudes: the units *a* come in front of the ranks *R* and count them, and the rank groups *aR* appear in descending order of magnitude:

Symbolically	2-TTh	4-Th	7-H	8-T	9(-U)
Chinese	*erh-wan*	*szu-ch'ien*	*ch'i-pai*	*pa-shih*	*chiu*
Japanese	*ni-man*	*shi-sen*	*shichi-hyaku*	*hachi-ju*	*kyu.*

The number 980,000 in Chinese is *ku-shih pa wan*, in Japanese *ku-ju ya man* — 9'10'8 TTh.

The individual words are not subject to any grammatical alteration, so that the Chinese words are themselves true models of an ideal number-word formation. Moreover, Chinese does not distinguish between nouns, verbs, adjectives, and adverbs. A single word, like *shih*, for instance, can thus serve as a number word "ten," as a noun "the ten," and as a verb "to multiply by ten, increase tenfold"; *pai chih*, "to do something a hundred times," means literally "to 'hundred' something." The particular meaning of the word in each case is determined by its position and its context.

Like every member of the Indochinese family of languages (to which Japanese does *not* belong!), Chinese is a tone language consisting entirely of monosyllabic words. Every word is pronounced in one of the four (in some dialects six) "tones," each of which gives it a totally different meaning. The first tone is high and even, the second is high and rising at the end, the third is low rising to high, and the fourth tone is half high and rises sharply. The particular tone required is indicated, in transcription, by a superscribed number from 1 to 4. To give a phonetic (but not semantic) analogy in English, the word *so* is pronounced with different tones in the expressions: "You must do it *so*" (1st tone, so^1) and "Is that so?" (3rd tone, so^3).

The Chinese number-word *shih* is pronounced with the second tone; in the first tone it means "to lose," with the third tone "history," and with the fourth tone "city." The spoken number sequence differentiates i^1, "one," from i^4, "a hundred million," only by the tone; thus the tone is indicated only for these two words, since for the others it is irrelevant to our purposes.

The fact that Chinese words are monosyllables limits the number of basic words to some four hundred and twenty. Although a language generally expresses some fifty thousand different ideas and concepts, Chinese with its four tones can express no more than 4×420 — about 1700 different concepts in all. Since the remaining 48,000-odd ideas must then be divided among the two thousand or less phonetically different words, almost every Chinese word has multiple meanings. The word *i*, for example, has forty unrelated different meanings, including its two numerical values. The problem of distinguishing between the individual meanings of words with the same pronunciation gave rise to the use of paraphrases, augmentative combinations, and the like, and it is one of the great difficulties that the Chinese language poses for foreigners. In Chinese writing, on the other hand, each concept or meaning has its own unequivocal character; here, in contrast, the difficulty is that there are some 45,000 different characters! The Chinese used by the public at large for general purposes (as in the newspapers) needs only about 2000 different characters. Too few words, too many characters: a curious paradox that sums up the strangeness of Chinese as compared to our familiar European languages.

China, moreover, has no single unified language spoken and understood all over the vast and populous country. The dialects and variations of Chinese are so many and so different, that a man from Peking cannot communicate with a person from Canton without some intermediary. This is where written Chinese comes in. A Chinese character has one universal and unequivocal meaning, just as the numeral 4 is understood by every European in the same sense, whether he pronounces it *vier*, *four*, *quatro*, *tesseres*, or *četyre*. Recently (in 1957) written Chinese has been "transcribed" into a phonetic alphabet, which, because of the many similar-sounding words that differ only in tone, is itself a difficult undertaking. Since 1955 newspapers are no longer printed in vertical columns but in horizontal lines read from left to right.

JAPANESE NUMBER WORDS

The Japanese number words are clear evidence of China's cultural influence on Japan: the native Japanese number sequence today is used only through *to*, 10 (see Column 2, Table on p. 450); from there on the whole sequence is made up of number-words taken over from Chinese (Column 3). The words have been altered phonetically, of course (for instance, *ch'ien > sen*); *hyaku*, 100, is a Japanese word.

The old units are still used in counting, and the Japanese readily replace *shi* by the native word *yo*, "four," because *shi*, although written differently from the form in which it means "four," is also the word for "death." We have seen such avoidance of "delicate" or "tricky" numbers in other cultures as well (p. 266). Of the old number words above 10 a few are still in use in various expressions: *hatachi*, 20 (years old); *yorozu*, 10,000 in *yorozu-ya*, "10,000-house," meaning warehouse; *chi*, 1000 in *Chi-shima*, "Thousand Islands" — Japanese name of the Kurile Islands and in the Japanese national anthem for "forever and ever":

chi –   yo   ni ya – chi –   yo   ni,
1000 – generations (of) 8 – 1000 – generations (of).

Here we also see "eight" used as an indefinite number meaning "many": this is a Japanese peculiarity. Thus *ya-o-ya* is literally an "8-10-house" and means a "greengrocer" because he has "many" things in his storehouse. "All the gods" in Japanese is

ya-o-yorozu no kami, "8-100-10,000 gods,"

a remarkable expression because here the number concepts that once became consolidated from the meaning "many" to the specific fixed numbers 8, 100, and 10,000 have again been eroded semantically to the vague "many" (cf. *yorozu*, p. 134). The lumping together of the numbers 8–100–10,000 to express a very large quantity or measure is like an early stage of the "sacred" number towers erected by the Hindus and the Maya (see pp. 62, also 136).

In Old Japanese the sequence of units was

hi fu mi yo i mu na ya kono to,

without the ending *tsu*, which transforms the number word in the case of the units, like *-chi* or *-ji* in the case of the tens, into a noun (see Column 2 of table). This ending is also the old word for "things." Example: the number word stands alone as a noun in

"3   and 4 make (together)   7,"
mitsu   to yotsu wo   yoseru nanatsu,

but before a numbered object or verb it appears without the *tsu* as an adjective or adverb, as in the proverb:

nana-korobi ya-oki, "fall down 7 (times), get up 8 (times)."

The syllable *-to* or *-ta*, as in *hi-to*, "one," and *fu-ta*, "two," means literally "place in which a man is"; it is the number word for "person." Thus *hito* means actually "one person," and *futa* "two persons." But this word formation is limited to these two alone; from "three" on it is no longer used: this recalls the very first forward step in the history of the number sequence, from "one" to "two" (see p. 12).

Fig. 264 The Chinese number 41957 = 4TTh 1Th 9H 5T 7, in basic (left) and in official (right) numerals.

In Old Japanese the units were raised to tens by the word *amari*, "above, beyond": thus 11 was *to-amari-hi*. In compounds 10 was *-so* and 100 was *-(h)o*, as in 50 *i-so* and 500 *i-ho*. Today, however, as we have stated, the native Japanese number words are used only through *to*, 10; from there on they are replaced by the Sino-Japanese words with which all compound number words are also formed, after the Chinese pattern, in descending order of magnitudes. We shall disregard phonetic elisions like *ippyaku* for *ichi-hyaku*, "one hundred."

The old counting limit at 10 is also manifested in the following peculiarity of the Japanese multiplication table:

ni-nin ga shi, 2 ($\times$) 2 ($=$) 4, but *shi-shi jo-roku*, 4 ($\times$) 4 ($=$) 16.

If the product is higher than 10, the particle *ga* that precedes the noun is omitted. Thus the numbers are treated as nouns through 10, but as number-words thereafter.

There is a noteworthy counterpart in ancient Egyptian. The problem 4 $\times$ 5 was expressed in Egyptian literally as "bend the head (count, calculate!) by fours until five times." The plural form "times" was used for factors less than 10, but for factors larger than 10 only the (elided) singular form "time." In the first instance what is being expressed is the concept of addition, which of course underlies multiplication.

Buddhism was introduced into Japan in the 7th century; about a hundred years later the Japanese adopted the Chinese ideograms in writing, although to some extent they used them differently since their language is unrelated to Chinese (it belongs rather to the Altaic family, like Turkish and Mongolian).

Unlike Chinese, Japanese has words of more than one syllable, it does not distinguish words by tone and it possesses a large number of prefixes and suffixes that may be attached to a word. These attached syllables consist of one of the vowels *a, e, i, o*, or *u*, or of a combination of one of the fifteen consonants with a vowel, such as *ka, ke, ki, ko, ku, sa, te, yo*, and also *-n*. These combinations amount to a possible 5 $\times$ 15 + 6 = 81 such syllables, but actually only about 50 of them are used.

For these fifty attached syllables the Japanese have created 50 syllabic written symbols, which amount to a syllabary (*kana*). For these they use symbols derived from Chinese characters that have the same or related sounds. With this syllabary they can approximate a foreign word phonetically: the English word "cigarette" becomes *shi-ga-ret'-to*, the name MacArthur becomes *Ma-kē'-sa*, and "baseball" becomes *bē'-so-bo-ru* (there is no l). In addition to these syllables, the Japanese have also adopted many Chinese conceptual characters such as that for "person" (Chinese *jen*) which they pronounce in Japanese as *hito*. This Japanese phonetic syllabary is essentially different from the Chinese (cf. p. 458).

KOREAN NUMBER WORDS

The Korean number words, like the Japanese, reveal an indigenous number sequence (Column 4 in the table), which is now used only for the numbers through 10, and a borrowed Sino-Korean sequence for all succeeding numbers (Column 5). The Korean formation of the tens is essentially different from the Chinese one. The Korean language is not related to Chinese at all, and to Japanese only remotely. Since the 15th century written Korean has employed an alphabet.

The Korean tens through 90 are all formed with a suffix -*hün*, the only exception being 20 — a familiar phenomenon that somewhat resembles the Latin *viginti* (see p. 14). But the fact that through 50 in Korean they are not formed with the units is a still more ancient primitive feature ("encipherment"; cf. p. 82). From 60 on (thus there is a break after 50!) the tens are formed regularly with the corresponding units, perhaps in imitation of the Chinese model; *päk*, 100, then makes the second numerical rank (reverse influence of a rank? — see p. 169).

SUMMARY

Now let us sum up. We have seen that each of the three peoples created a number sequence in its own language, with ancient features, which in the case of Korean and Japanese was later displaced by the Chinese sequence from 10 on. It should be kept in mind that the Chinese language is totally unrelated to Japanese and Korean. Thus — to invent a hypothetical European example — it is not as if the English had adopted all or most of the German or French number sequence in place of their own, but as if they had replaced their own by the Hungarian, after first adapting it to the phonetic qualities of English. But the borrowing of a whole number sequence is unknown in the Western world: we know only of isolated foreign borrowed number words, such as the Finnish *sata*, 100, or the Lithuanian *tukstantis*, 1000 (see p. 190).

If this wholesale borrowing could take place in the Far Eastern languages, it should not seem strange that the Chinese written numerals alone should be used in this part of the world. But before we go on to these, let us first discuss a few peculiarities of these spoken numbers.

Number Classes

We have met these previously; they represent an early stage in the development of the number sequence (see p. 31). The Chinese, for example, does not say directly "four tables," but "4 *chang* tables" — something like "4 leaves tables." *Chang* here is the number class in which all things characterized by an extent of surface are counted: tables, spades, sheets of paper, etc. The number word is placed before this, not directly in front of the object. Thus it cannot count or number any object or concept at all but is limited to a particular class of objects: the number sequence is not yet completely abstract. Chinese has about 100 and Japanese about 50 such number classes; Korean uses them as well.

As with most languages, the *meanings of the individual number words* are now impossible to explain with certainty. For number words, at least the first few in the sequence, are among the oldest verbal expressions of mankind. But we may mention two peculiarities of Japanese.

The suffix *tsu* for the units is associated with *tu,* or *te,* "hand" (see Column 2 of the table). This would then seem to go back to an early method or custom of counting on the two hands. The number 5, (*i*)*tu* also means "hand," the initial vowel *i* having no meaning by itself.

Another rare and very ancient Japanese way of forming number words, which also is found occasionally among primitive cultures, creates the words for "two," "six," "eight," and "ten" by doubling those for "one," "three," "four" and "five" respectively:

1 *hito* – 3 *mi* – 4 *yo* –   5 (*i*)tu
2 *huta* – 6 *mu* – 8 *ya* – 10 *to.*

Doubling can also take the form of keeping the same vowel but changing the consonant: *futa* replaces *huta* in some dialects of Southern Japan.

Concealed Number Words

These are words such as "drill," whose meanings are derived essentially from a number word (in this case, "three" – *drei*). These often provide fascinating insights into the history of a number sequence (see p. 171). Often, as in the German word *Samt* ("velvet"), the number root *sechs* ("six") can no longer be recognized: it has become concealed.

The words of this kind in Chinese (and Japanese) can be listed — words in which a number word as the basic semantic core of meaning has not been verbally camouflaged but remains clearly visible. Because of the innumerable like-sounding words with different meanings in Chinese, the added number often serves to clarify the meaning. Thus we have (as in Japanese also) a great quantity of such number compounds, most of them using the number 100, 1000, or 10,000 in the sense of "many." Some of these are given below as illustrations; in Chinese, number words appear in the singular:

10 mouths	= old (what is in 10 = "many" mouths)
10 10'10 years	= a human lifetime
10 finished	= complete
100 things	= all, everything (*pai-shih*; Japanese *hyaku-ji*)
100 workmen	= the working class
100 (times) think	= Prime Minister
100 wares goods	= warehouse
100 years of life	= death (*pai-sui; hyaku-sai* – cf. "10,000 years")
100 miles capacity	= small horizon ("long circuit"), for 100 miles is no great distance in China
100 mouths	= wife, children, and all the relatives
1000 directions	= nimble, adroit, versatile
1000 old	= great antiquity (written "1000'10 mouth" — cf. "old" above)
1000 springs	= birthday ("many happy returns")

1000 mile glass = telescope
1000 mountains–10,000
                waters = remoteness, vast distance
1000 knives–10,000 cuts = to hack into small pieces
       10,000 nations = the world
    10,000 directions = everywhere, generally
      10,000 animals = zoo
    10,000 capacity = omnipotent
    10,000 mile wall = the Great Wall of China
    10,000 things rest = death
        10,000 years = Long Life! (*wan sui; ban-zai* — cf. 100-
                    sui, "death")
      8-100-house = greengrocer (Japanese)
8-100-10,000 gods = all the gods (see p. 453).

WRITTEN NUMERALS

The Chinese, and in imitation of them the Japanese, use five different kinds of numerals, of which four are indigenous to the Orient (Fig. 265): (1) the basic numerals, (2) the official numerals, (3) the commercial numerals, (4) the stick or stroke numerals and, in addition, (5) the Indian numerals, which have also penetrated slowly into the Far East.

The basic numbers are called *hsiao-hsieh*, the "small (common) writing." The Chinese form of writing is not a phonetic alphabet like our own, "t-r-e-e," but an ideographic form in which a single character represents the whole word "tree" (see Fig. 1). Thus, in

	(1)	(2)	(3)	(4a)	(4b)	
i	一	壹	Ⅰ	一	Ⅰ	1
erh	二	貳	Ⅱ	=	Ⅱ	2
san	三	參	Ⅲ	≡	Ⅲ	3
szu	四	肆	Ⅹ	≣	Ⅲ	4
wu	五	伍	𝄼	≣	Ⅲ ∂	5
liu	六	陸	丄	丅	丄	6
ch'i	七	柒	𠄎	𠔿	𠄎	7
pa	八	捌	𠔼	𠔽	𠔼	8
chiu	九	玖	夊	𠖊	𠖌	9
shih	十	拾	十	—○	⏋○	10
pai	百	百	𝟛𝟚	⏋○○	—○○	100
ch'ien	千	仟	千			1000
wan	萬 万	萬	万			1000
ling			○	○	○	0

Fig. 265 The four sets of Chinese numerals:
(1) the basic numerals, (2) the official numerals, (3) the commercial numerals, (4) the stick or stroke numerals (see Fig. 206).

the basic Chinese numerals each numeral is also the symbol for a whole number word and vice versa; but written number words and written numerals are one and the same, and are not distinct as they are generally throughout Western culture.

We write the number 41,957 in "words" with completely different symbols; that is, — with letters:

"forty-one thousand nine hundred fif-ty seven";

the Chinese, on the other hand, can write it only one way, as in Fig. 264, since their written characters are at the same time numerals and word-symbols.

In contrast, the Japanese can write not only the numeral for 2, but also the number word "two," *fu-ta-tsu*, with three syllables. This again brings to our attention the difference between written Chinese and Japanese.

The Chinese has 9 units *a* with which he numbers the ranks *R* (ten T, hundred H, thousand Th, ten thousand TTh, etc.), writing vertically from top to bottom. He begins each page at the right margin and places the second vertical column to the left of the first (Fig. 266). This custom of writing in columns instead of rows may well be a remnant of the ancient method of writing on tally sticks (see p. 246).

As a result of this coincidence between numeral and word-symbol, the Chinese very early attained the state of maturity and perfection

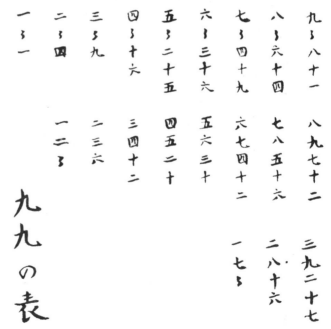

Fig. 266 Multiplication table in Chinese numerals. In Japanese this is named, not as in German, for the lowest product 1 × 1 (*Einmaleins-Tafel*, "one-times-one table"), but for the highest with which it begins: here *ku-ku-no-hyo*, "9 (times) 9-table" (bottom left). Excerpt from a document written by a Japanese lady with a brush in pure script and with a pen.
Through the kindness of Wolfram Müller, Tokyo.

in writing numbers which the West achieved only with the adoption of the Indian numerals, after a long and difficult development.

We in the West had a spoken number sequence: "one, two, three ... four hundred, etc." even before the beginning of recorded history but were not able to write these numbers until the letters of the Roman alphabet had been learned. These were able to represent the spoken number word, but not the numerical concept in a single numeral. Initially the Roman numerals such as CCCC = 400 were used to this end. But these were characterized by an essentially different law of formation (ordering: "hundred hundred hundred hundred") from that of the spoken number sequence, which proceeds by gradations ("four hundred"). Thus the Roman numerals failed to reflect the spoken numbers, and there remained a fundamental discrepancy between the two.

For this reason, around 1500 a new form of written numbers, the Indian numerals, displaced the awkward Roman numerals in Europe. Yet there was a long and difficult struggle, as we have just seen, before the new numerals were able to replace the old.

The structure of the Indian numerals corresponds to that of the spoken number sequence with its gradations of ranks: 41,957 = 4 TTh 1 Th 9 H 5 T 7 (U); in Europe, unlike China, the impetus toward this was provided by the counting board. The Indian digits number the ranks instead of ordering them: 4 H, not HHHH. To this extent the Indian and the Chinese numerals are alike. But they differ in that in the Chinese written numbers the ranks TTh Th H T E are expressly written down, whereas in the Indian numerals they are not expressed but indicated by the place value of the digits. Thus the Chinese is a "named" and the Hindu an "unnamed" or abstract place-value notation, if we equate position with rank; this has been discussed previously.

Although both these systems of numerals are essentially similar in reflecting the structure of the gradational spoken number sequence, the difference between them that was just mentioned becomes clear in the case of a number in which one or more places are left unfilled by digits:

Indian   4 0 8 9   Chinese   4 Th 8 T 9 U.

In other words, the Indian numerals need and have a zero sign, while the Chinese do not. Thus in China there was never a struggle in the popular mind against the concept of zero (and its cipher) such as there was in the West.

There is another difference between the Indian and the Chinese numerals: with the former it is possible to make written computations, but not with the latter. The fact that the ranks must be expressly written down makes computation too cumbersome. Thus the Chinese have always made their computations on the abacus, the *suan pan*, and the Japanese on the *soroban*, both of which we have discussed earlier (see Figs. 139–143).

A reckoner on the abacus also needs a multiplication table. To the medieval European (Figs. 113 and 162) and the Arabic (Fig. 237) multiplication tables which we have already met, we may add here another such table in Chinese numerals in two versions, one in the pure script drawn with a brush and the other written out by a Japanese lady in pen and ink (Figs. 266 and 267).

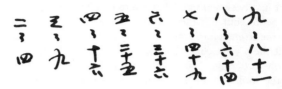

Fig. 267 Multiplication table: the top section of Fig. 266 written with pen instead of brush.

The arrangement of this Oriental multiplication table is unique; it runs in vertical columns from right to left, each column beginning with the product of equal factors, for example 7×7 to 1×7:

1×1	2×2	......	6×6	7×7	8×8	9×9	(a)
(1)	(4)		(36)	(49)	(64)	(81)	
	1×2		5×6	6×7	7×8	8×9	(b)
		......		5×7	6×8	7×9	
				....			
				1×7	2×8	3×9	(c)
9×9					1×8	2×9	
table						1×9	

(Only three excerpts, marked a, b, and c, are illustrated in Fig. 266.)

Thus any particular product (7×8) must be found in the column for the larger of its factors (8), so that only "half" of what we consider the normal multiplication table need be written out. As an interesting counterpart, compare the usual medieval European multiplication table, which began with the product of equal factors (7×7) and from there ran upward (8×7, 9×7; see Fig. 162).

In Figure 266 the symbol which resembles a 3 (second line from the top) is a sort of ditto sign indicating that the number above is repeated; thus 9_38T1 means $9 \times 9 = 81$ (marked (a) in top right corner of the table).

The numbers in this multiplication table are written in vertical columns, after the Chinese manner. But when we go on to read the numbers in the table of logarithms (Fig. 268), we are surprised to find that the numbers here run horizontally, and that the rank symbols are no longer written in the numbers so that the numerals have been transformed from a "named" to an abstract place-value notation with a zero sign (like the Indian numerals).

Now the clue to our problem is revealed: the pattern was the Indian system.

The contacts of Chinese culture with India, its books and its numerals, were brought about by Buddhism which took root in

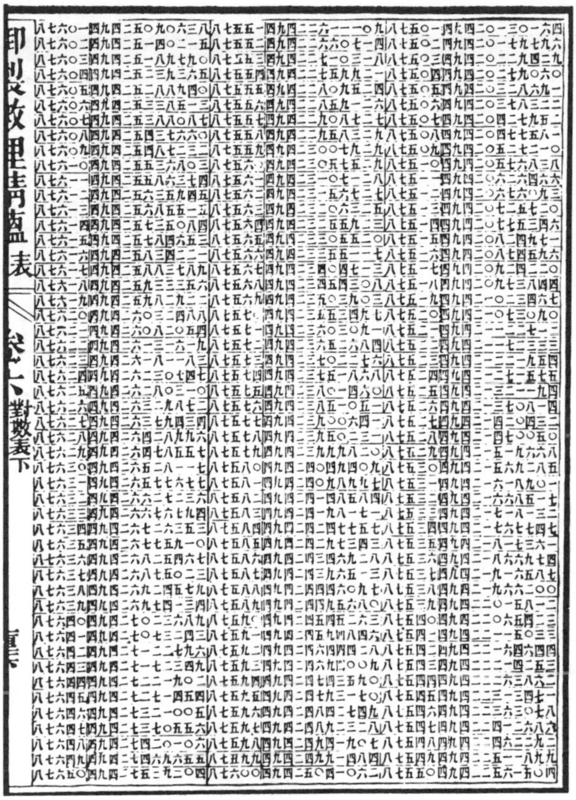

Fig. 268 A page from a ten-place table of logarithms of the Chinese Emperor K'anghsi, dated 1713, written in Chinese basic numerals horizontally in the Indian manner, without expressing the ranks and with a zero. This page begins at top right with log 87,501 = 4.9420130164 and runs in columns to log 87,650 = 4.9427529204 at bottom left (see detailed excerpt directly below). The whole page was cut entire from a wood plate (14 × 21 cm) and not set in movable type; thus this table of logarithms is a block-printed book like the Bamberger block book (Fig. 170).
Transmitted through the kindness of Wolfram Müller, Tokyo.

China in the 7th century and attained great importance there. In the 13th century the zero (Chinese *ling*, "gap, vacancy") first appeared in Chinese books and has continued to be used occasionally ever since. In the 16th century and later China became acquainted with Western mathematics through learned European missionaries, so that one can readily understand the 10-place logarithm table, which was probably patterned after that of the Dutchman Adrian Vlacq of 1633. The Emperor K'anghsi (1662–1722), who was a patron of learning and science, had it published in 1713 in a great "Collection of Mathematical Books Issued by Imperial Command." Our illustration shows a whole page, for the numbers 87,501 through 87,650; in the enlargement of the beginning of this page (Fig. 269), we see log 87,501 = 4.9420130164, without the symbol "log" or the sign of equality, but with the characteristic number 4., which is generally omitted in modern tables of logarithms.

8 7 5 0 1 ; 4 9 4 2 0 1 3 0 1 6 4

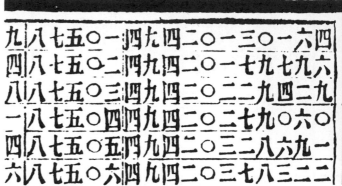

Fig. 269 Beginning of the page from the table of logarithms shown in Fig. 268: log 87,501 (=) 4(.)9420130164, etc.

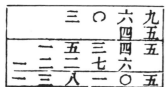

Fig. 270 3069 × 45 = 138,105, computed on paper in the "Indian" manner, with Chinese basic numerals and with the "check by casting out nines." From a Chinese work of 1355.

上院、民主17・共和5
下院、民主175・共和82

Fig. 271 Headline from a Japanese newspaper, announcing the results of an American election. To save space, Indian numerals are used in combination with Chinese characters: "Senate: 17 Democrats, 5 Republicans; House of Representatives: 175 Democrats, 82 Republicans."

We have seen the use of the Indian place-value notation, using not the Indian but indigenous numerals, once before in the Greco-Byzantine culture where the first nine letters of the Greek alphabet were substituted for the alien digits (see Fig. 68).

When the Chinese numerals are written in this manner, it now becomes possible to make written computations by the Indian method, as may be seen in the example 3069 × 45 = 138,105, from the "Computation of Ting Chü" of 1355 (Fig. 270).

Although today the Indian numerals have also penetrated the Far East and are being used increasingly in place of the Chinese numerals, here the ground has long been ready for the implantation of an abstract place-value notation, which was never so bewilderingly new and strange for the Chinese and Japanese as it was for the Europeans in the late Middle Ages; only the forms of the digits themselves were new. These digits stand wherever they appear, foreign but forbearing and, one might say, at home alongside the Chinese numerals (Fig. 271). They are even taught in Japanese schools, as we have seen from Japanese arithmetic textbooks written exclusively for the *soroban* — meaning for the old, non-Indian procedures on the abacus (see Fig. 141). Thus the Indian numerals created no disturbance: people in the Far East simply learned them, without giving up their own indigenous numerals and methods. (In China since 1955 the indigenous numerals have been officially replaced by the Indian ones).

This absence of discrimination in writing numbers and adapting to the Indian numerals, as we shall see again, is a peculiarity of the Far East. It is worth mentioning here that the basic Chinese numerals for 20 and 30 still use the old ordering principle TT and TTT, two and three crossed longitudinal strokes, in addition to the gradational arrangement 2T and 3T (see Fig. 240). This also occurred in the Han number-sticks, where we saw that they were pronounced in individual number words (see p. 247).

The Official Numerals.

Ta-hsieh ("great numerals"), also known as seal-writing, are highly

Fig. 272 Japanese 10-yen note. At left in the small circles the number 10 is written twice in Indian and twice in Chinese "great" numerals, and likewise once in each system in the square pattern at right. This paper money with its two sets of numerals is thus a counterpart of the Egyptian postage stamps (see Fig. 235).

ornamented forms (see Column 2 in Fig. 265 and Fig. 264 (right)). They are therefore used whenever a number needs to be protected under all circumstances against falsification: on banknotes, contracts, coins (Fig. 273), bills of exchange (see Fig. 277), and elsewhere. In such cases we write out the number in words; the Chinese can only write "numerals" since they are the same as "words." A few of the basic numerals are very easy to form and thus to alter: 2 to 3, 3 to 5, 1 to 7, or 10, and 10 to 1000. Thus remain only the basic characters for 100 and 10,000; those for 1, 2, 3, 5, 7, and 1000 and sometimes also 100 are "secured" by an additional sign.

Fig. 273 Chinese coin of 1674. Numeral 1 (fen) in official numerals on the right side. Collection of R. Schlösser, Hannover.

The Commercial Numerals.

Su-chou ma-tzu ("Weight numerals of Suchow (Kiuchuan)") are so named because they were used in the weighing of money (see Column 3, Fig. 265). They are employed by merchants and tradespeople to write numbers rapidly on less critical documents, such as price tags, minor notices and the like (Fig. 274). The numerals 1, 2, 3, 6, 7, and 8 are made up of combinations of simple strokes, 10, 1000, and 10,000 are the same as the corresponding basic numerals, and 5 and 100 are cursive, simplified basic numerals. The numeral 4 here appears to be an old group-form, a stroke crossed through (cf. the Indian Kharosthi form, Fig. 12), and 9 is a combination of 5 and 4.

The manner of writing a number is interesting; the numerals are written perpendicularly in an abstract place-value notation. Thus 1 or 2 or 3 when standing alone are vertical, but when there are several, the second numeral is written horizontally to differentiate it from the first, etc. Thus one can easily read numbers such as 31 and 231 (Fig. 274). The zero in 6080 is written, but the zero at the end is not; in its place is the symbol for the rank of the last digit: 608T. The rank is also often placed once again below the highest unit digit (here 1000). Thus

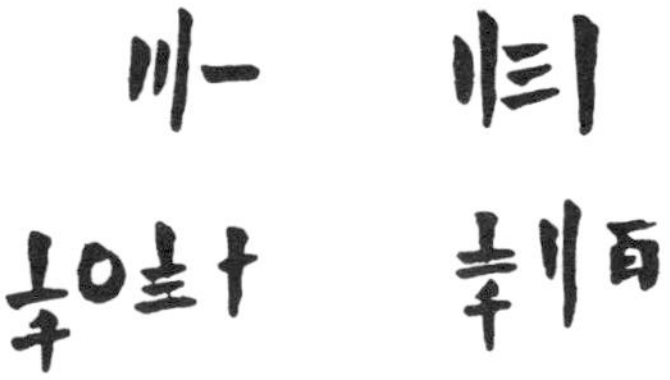

Fig. 274 Chinese commercial numerals:
top: 31 and 231
bottom: 6080 = 608T and 7200 = 72H, with the rank T written again under the highest digits 6 and 7.

Fig. 275 Amounts of money in Chinese commercial numerals:
72 (liang) 5 (chi'en) 3 (fen) 2 (li)
725 (chi'en) 3 (fen) 2(li).

Fig. 276 "1 box 320 yen": a pricetag (for Mandarin oranges) from a fruit store. In large stores the prices are generally indicated in Indian numerals. Size 22 × 8 cm. Supplied by W. Müller, Tokyo.

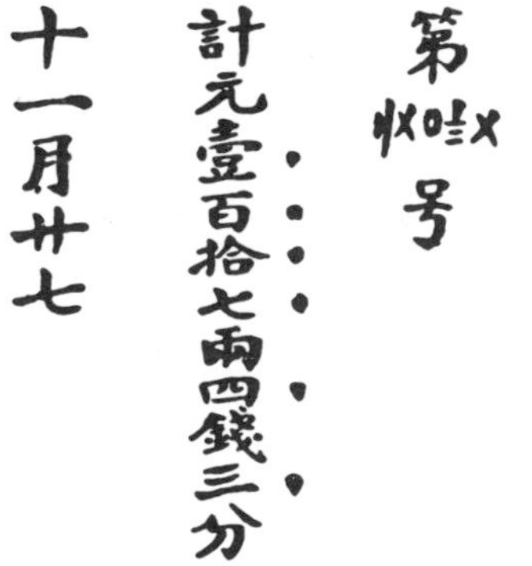

Fig. 277 A Chinese bank draft with three sets of numerals, basic, official, and commercial.
1st column (right): the check number 24,084 in commercial numerals; abstract place-value notation — 2nd column: the amount, 117.43 Taels (.), in official numerals; "named" place-value notation — 3rd column: in basic numerals, "the 27th day in the 11th month," with the 20 in the old form — "named" place-value notation.

these numerals represent a unique intermediate stage between a "named" and an abstract place-value notation.

In the case of sums of money, the Chinese write the denomination (such as Tael) beneath the appropriate digit and not behind, as we do (Fig. 275).

The Stick or Stroke Numerals

Stroke numbers are the Chinese equivalents of popular or peasant numbers: they are ordered and grouped by fives (see Column 4 on Fig. 265). It is known that they developed from the sticks used to form numbers on the counting board (see Figs. 206–208). Once removed from the board and "formed" by writing, they became an abstract place-value notation by adopting the zero sign and becoming digits. We have already seen how they are written alternately vertically and horizontally (see Fig. 208). The numerals 1, 2, 3, 6, 7, and 8 are the same as the Chinese commercial numerals.

The simultaneous use of the different Chinese numerals in the same document may be seen very nicely in the numbers written on a draft (Fig. 277). A surprising feature of this is the clarity and comprehensibility with which the different kinds of numerals are written and read. Each set of numerals is used for a specific purpose.

Much as they differ in form, they are all gradational. Thus they are all consistent in their internal structure. The Chinese were brought up with gradational numerals from the very beginning: they did not have to adopt them and learn to use them as something foreign and strange as we did after having been accustomed to numerals like the Roman, which are based on ordering. In comparison to this, whether the Chinese numerals were a "named" or abstract place-value notation is unimportant. In fact, on their ancient coins the Chinese themselves developed an abstract place-value notation for numbers under 100 (see Fig. 222).

There was probably another reason why the Chinese never had much trouble with their several sets of numerals used simultaneously: their art of calligraphy, which had been practiced from ancient times. Some thirty different forms of Chinese writing have been devised over the course of time: the "blade of grass," the "tadpole," and other scripts. As a result, the educated Chinese was accustomed to the different forms in which one and the same concept can be expressed in writing. An example is provided by the numerals 2, 9, 100, and 1000 in the "Writing of the Elevated Places"; the reader should try to "de-cipher" their basic forms for himself (using the table in Fig. 265).

The Meaning and the Changes in the Form of the Basic Numerals

The signs for the numbers 1, 2, and 3 are, as in many systems of written numerals, merely simple strokes — horizontal in this case, vertical in our own numerals (see Fig. 241). The symbol for 4 consists of four strokes that are joined at top and bottom, like the fingers of the human hand, as on ancient Chinese coins (Fig. 279). Today the frame enclosing these strokes looks like the principal feature of this numeral.

The Chinese numeral 5 was at first probably a form representing a grouping, a slanting crossed stroke that is carried around, somewhat like our numeral 8. This is how it looks on old coins (see Fig. 281). It is from this that the present-day form was derived.

The Chinese numerals 8 and 9 are probably images of finger gestures, as has already been suggested (see Fig. 47). The origin and meaning of the numerals 6 and 7 are unknown.

The symbol for 10 is clearly a group sign, a unit or one-stroke with another stroke through it; this is also true of the old forms of the numerals 20 and 30, which are familiar to us from the Han number-sticks (see Fig. 75). On very old coins the cross-stroke sometimes has the form of a hollow diamond or a small circle (Fig. 282); sometimes the whole numeral 10 is represented simply by a small circle. An unsuspecting student might be misled by the old numeral 50 into thinking erroneously that the Chinese had a zero hundreds of years before it was invented in India.

The numeral 100 may perhaps be a picture of a container of some specific capacity (see Fig. 280). The origin of the numeral 1000 is unknown; there is an old form of this on ancient Chinese coins and on the amulet shown in Fig. 281.

Fig. 278 The numerals 2, 9, 100, and 1000 in the "Writing of the Elevated Places."

Fig. 279 Chinese coin with the old symbol for 4 (bottom). Inscription: weight *liang 14 shu*. 6th century B.C. Natural size.
Collection of R. Schlösser, Hannover.

Fig. 280 Old Chinese numerals for 10, 50, and 100.

The symbol for 10,000 is a scorpion, a "number-beast," which we have already encountered earlier in its old and more naturalistic form (Fig. 26).

With the advent of Buddhism, the Indian swastika, the hooked cross which was the symbol for the "10,000" perfections of Buddha, came to China. From that time on it also appeared as the numeral for 10,000, primarily on amulets like the one illustrated here, which was supposed to bring to its bearer "10,000 × 1000 5-*shu*"-pieces — in other words, great wealth. This recalls the number "10,000" as a form of best wishes in China and Japan (*ban-zai*, see p. 457). The symbol shown for 10,000 is probably an altered form of the swastika (see Column 1, Fig. 265).

If we look back over the old forms of the Chinese numerals, we shall see here, too, as in the case of the Indian numerals, that the forms tend to persist over the course of time (see Fig. 241), so that numerals used even in very early times are not too difficult to read.

RETROSPECT: SPOKEN LANGUAGE — WRITING — WRITTEN NUMERALS

With this sketch we end our discussion of the spoken numbers and written numerals of the Far East. What is the most important lesson to be learned from our vantage point high above the distraction of

Fig. 281 Amulet or votive offering for a temple, with the Indian swastika as the sign for 10,000, and the old symbols for 5 (right) and 1000 (bottom). This piece is supposed to bring great wealth to the bearer or donor: "10,000 × 1000 5-*shu* pieces" — *wan ch'ien wu shu*. Diameter 62 mm. Rubbing of the original in the collection of R. Schlösser, Hannover,

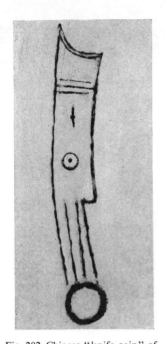

Fig. 282 Chinese "knife-coin" of
550 B.C., with an old 10-form; the
inscription reads: "3'10 (= 30) stars."
Length 18.5 cm.
Rubbing of the original in the collection
of R. Schlösser, Hannover.

small details? The wonderful interdependence of language, writing and numerals, of which China and Japan have made us aware.

Every culture has had a language, and thus at least the beginnings of a number sequence, "one," "two," and "three," from time immemorial. But not writing. Originally neither the Europeans nor the Japanese had a set of numerals that all members of the culture were obliged to use. Most people are not aware that this is an intellectual achievement that has nothing to do with the possession of a form of writing.

The various forms of writing words fall into two basic groups: an ideographic script in which the concept forms the word (as in China), and a phonetic form of writing in which the symbols represent the sound of the word. The latter is subdivided into a syllabary with the symbols standing for syllables (as in Japan, at least partly) and an alphabet consisting of letters, such as our own.

We have also seen many examples of the two fundamental classes of numerals:

(1) *ordered* numerals, in which the units (or ranks) are aligned in order like the Roman ones, and generally collected into groupings of 10 each, and

(2) *gradational* numerals, which number the ranks with nine digits; the latter class is subdivided into the subgroups of

(a) "named" gradational numerals (place-value notation) like the Chinese, and

(b) abstract gradational numerals or place-value notation (like the Indian numerals).

Let us look once more briefly at the two kinds of gradational numerals:

1 Th 9 H 6 T 8 (U)   1968

these are "named" and "unnamed" (or abstract) place-value notations, respectively, because the former write in the designations of the ranks, while the latter indicate them only by the place or the digits.

To mention a surprising paradox: the Babylonian numerals were a sliding abstract place-value notation that represented its digits in a form of ordering, and sliding because the fundamental unit on which the numerals were based was not indicated visually (see p. 167).

In addition, all written numerals based on the law of ordering, and also the "named" gradational numerals, require an abacus or counting board for computations: they are merely "number-representing" and not "computational" numerals. The only numerals with which it is possible to calculate freely — that is, without counting board and numerical tables — are the "unnamed" gradational numerals, the abstract place-value notation with 9 digits and a zero sign like those that arose in India. This is why these have now come into universal use all over the world.

Let us look back once more into history, so that we may recognize that these highly perfected numerals were not invented by Western man, although it was the West that made the greatest use of them and developed them to their highest state — they had their obscure beginnings in India and arrived in Europe only after long wanderings through other cultures. From the very beginning, however, the West did have the basic pattern of the Indian numerals in the *spoken* number sequence: "four *hundred* fif-*ty* three." The European spoken numbers are gradational, not ordered. We in the West thus had the spoken numbers but not the written numerals. The result is that we speak English (or German, or French), we write in the Roman fashion, and we compute by Indian methods.

This fact, which has been mentioned more than once, is now very clear. Sources far from each other in both time and space have come together in our own culture into a single stream of speech, writing, and calculating. The background of China reinforces our recognition: here language, writing, and numerals all arose out of the same cultural soil (as also in Egypt and Babylonia), and here, too, written number words are also numerals, as nowhere else in the world.

These two widely different worlds, ours and the Chinese, stand as symbols of the variety and the interaction of intellectual forces which we have seen manifested in our long discussion of

Number Words and Number Symbols.

Index

"Abacists": and the "algorithmicists," 422, 431–433 passim; "sweating," 327

Abacus, 301; Asian hand, 277, 306–315, 435; and the calculating match with modern machine, 309; division on, 330; and finger counting, 205–206, 216; monastic (Gerbert's), 322–331 passim, 341; Roman hand, 160, 305–306, 307, 310–312; teacher of, 316; in Western culture today, 315. *See also* Counters; Counting boards (tables)

Abstract number sequence, 7–8, 31, 37, 141

Addition: on new counting boards, 351–352; and ancient Egyptian multiplication, 454; on Roman hand abacus, 305–306; on Salamis Tablet, 302; word for, 302

Adjectives, numbers as, 18–32 passim, 81, 82, 111, 453. *See also* Attributes, numbers as

Age, counting of, 17, 24, 27, 33, 49, 72, 156, 210, 286. *See also* Time, measurement of

Ainu people and numerals, 69–70, 71, 75, 80

Akkadians, 134, 168

Albanian language and numerals, 68–69, 85, 91

Alexandria, 405–406; "logistics" in, 273

Algebra, 411

Algorismus, see al-Khwarizmi; Sacrobosco, Johannes de; Villa Dei, Alexander de

al-Khwarizmi, 410–413, 414; translations of his book, 411, 417, 445

Alphabet, history of, 121, 262–267 passim; and the Ionian-Milesian alphabet, 264

Alphabetical numerals, 196, 275; Arabic, 265, 276, 410; Gothic, 259–261, 263; Greek, 262–263, 264–265, 270–274 passim, 278; Hebrew, 265–266; Iranian, 278; Katapaya system of, 275; and word calculations, 266–267. *See also* Letters, and numbers

Altaic, subfamily of languages, 114; Japanese belongs to, 454

Anglo-Saxon numerals, 75, 83, 94, 152, 155–156

Annolied, Das, 154

Apices, 324–331 passim, 341, 439, 445

Arabic numerals: alphabetical, 265, 276, 410; *gubar* (West Arabic), 325, 327, 413, 415–417 passim; *siyaq* script, 74, 276–277; Indian-East Arabic, 114, 115, 116, 411, 412–417 passim, 418–421 passim; formation of number words, 73, 74, 80–81; numbers as nouns, adjectives and the dual, 12, 15; number words written out, 262, 276; zero in, 325, 401, 413

Arabs: alphabet, 263, 276, 410; finger counting, 212–214; history of, 319, 320–322 passim, 406–410 passim; and Sicilian coin, 439

Aramaic, 114, 263, 394; Jesus spoke, 115

Archimedes, "sand-reckoner" of, 139–142, 270, 316

Aristotle, 104, 180, 408–409

Arithmetic: pay for teachers of, 316; personification of, 339, 350, 365, 431; various operations of, 351

Arithmetic textbooks: al-Khwarizmi, 410–413, 414, 417, 445; Bamberg block-printed book, 335–336, 420; Bamberger Rechenbuch (Wagner), 335, 356–357, 419, 420, 441, 442; Chinese, 416, 462; on counters (French texts), 338, 362, 367; on the counting board (Albrecht), 349–350; in Denmark, 338; Egyptian, 414–415; English (Recorde and St. Albans), 337, 432; French text for merchants, 401; Grammateus, 360; Italian, 335, 428, 441; Japanese, on Chinese stick digits, 368–371 passim; Japanese, on the *soroban*, 307–308, 309; in Netherlands, 338; on pebble-placing (Planudes), 301; Reisch, 337, 339, 350, 431; entirely in Roman numerals (Köbel), 286; Sacrobosco, 402–403; Salem Codex (Algorism), 20, 206, 411–412, 423; Spanish, 144, 215, 338; Widmann, 429–430, 437

on finger counting: Artavazdos, 214, 272; Aventin, 216, Bede, 201–207, 217, 322; Leonardo of Pisa, 216, 218, 425–426, 445; Leupold, 204–205, 207–208, 418; Pacioli, 203, 204, 205, 215; Spanish codex, 215

See also Köbel, Jacob; Riese, Adam

Armenian language and numerals, 14, 91, 103

Article, and number one, 19

Articulus, 45

Aryans, and Aryan groups of languages, 91, 102, 103, 105–106, 396. *See also* Iranian language and numerals; Sanskrit

Asian hand abacus, 277, 306–315, 435

Astronomy, 168; Babylonian writings on, 398–399, Brahmagupta, 410; Ptolemy, 168, 399, 405–406, 409

A CATALOG OF SELECTED

DOVER BOOKS

IN ALL FIELDS OF INTEREST

A CATALOG OF SELECTED DOVER
BOOKS IN ALL FIELDS OF INTEREST

CONCERNING THE SPIRITUAL IN ART, Wassily Kandinsky. Pioneering work by father of abstract art. Thoughts on color theory, nature of art. Analysis of earlier masters. 12 illustrations. 80pp. of text. 5⅜ x 8½.   23411-8 Pa. $4.95

ANIMALS: 1,419 Copyright-Free Illustrations of Mammals, Birds, Fish, Insects, etc., Jim Harter (ed.). Clear wood engravings present, in extremely lifelike poses, over 1,000 species of animals. One of the most extensive pictorial sourcebooks of its kind. Captions. Index. 284pp. 9 x 12.   23766-4 Pa. $14.95

CELTIC ART: The Methods of Construction, George Bain. Simple geometric techniques for making Celtic interlacements, spirals, Kells-type initials, animals, humans, etc. Over 500 illustrations. 160pp. 9 x 12. (USO)   22923-8 Pa. $9.95

AN ATLAS OF ANATOMY FOR ARTISTS, Fritz Schider. Most thorough reference work on art anatomy in the world. Hundreds of illustrations, including selections from works by Vesalius, Leonardo, Goya, Ingres, Michelangelo, others. 593 illustrations. 192pp. 7⅛ x 10¼.   20241-0 Pa. $9.95

CELTIC HAND STROKE-BY-STROKE (Irish Half-Uncial from "The Book of Kells"): An Arthur Baker Calligraphy Manual, Arthur Baker. Complete guide to creating each letter of the alphabet in distinctive Celtic manner. Covers hand position, strokes, pens, inks, paper, more. Illustrated. 48pp. 8¼ x 11.   24336-2 Pa. $3.95

EASY ORIGAMI, John Montroll. Charming collection of 32 projects (hat, cup, pelican, piano, swan, many more) specially designed for the novice origami hobbyist. Clearly illustrated easy-to-follow instructions insure that even beginning papercrafters will achieve successful results. 48pp. 8¼ x 11.   27298-2 Pa. $3.50

THE COMPLETE BOOK OF BIRDHOUSE CONSTRUCTION FOR WOOD-WORKERS, Scott D. Campbell. Detailed instructions, illustrations, tables. Also data on bird habitat and instinct patterns. Bibliography. 3 tables. 63 illustrations in 15 figures. 48pp. 5¼ x 8½.   24407-5 Pa. $2.50

BLOOMINGDALE'S ILLUSTRATED 1886 CATALOG: Fashions, Dry Goods and Housewares, Bloomingdale Brothers. Famed merchants' extremely rare catalog depicting about 1,700 products: clothing, housewares, firearms, dry goods, jewelry, more. Invaluable for dating, identifying vintage items. Also, copyright-free graphics for artists, designers. Co-published with Henry Ford Museum & Greenfield Village. 160pp. 8¼ x 11.   25780-0 Pa. $10.95

HISTORIC COSTUME IN PICTURES, Braun & Schneider. Over 1,450 costumed figures in clearly detailed engravings–from dawn of civilization to end of 19th century. Captions. Many folk costumes. 256pp. 8⅜ x 11¾.   23150-X Pa. $12.95

STICKLEY CRAFTSMAN FURNITURE CATALOGS, Gustav Stickley and L. & J. G. Stickley. Beautiful, functional furniture in two authentic catalogs from 1910. 594 illustrations, including 277 photos, show settles, rockers, armchairs, reclining chairs, bookcases, desks, tables. 183pp. 6½ x 9¼. 23838-5 Pa. $11.95

AMERICAN LOCOMOTIVES IN HISTORIC PHOTOGRAPHS: 1858 to 1949, Ron Ziel (ed.). A rare collection of 126 meticulously detailed official photographs, called "builder portraits," of American locomotives that majestically chronicle the rise of steam locomotive power in America. Introduction. Detailed captions. xi + 129pp. 9 x 12. 27393-8 Pa. $13.95

AMERICA'S LIGHTHOUSES: An Illustrated History, Francis Ross Holland, Jr. Delightfully written, profusely illustrated fact-filled survey of over 200 American lighthouses since 1716. History, anecdotes, technological advances, more. 240pp. 8 x 10¾. 25576-X Pa. $12.95

TOWARDS A NEW ARCHITECTURE, Le Corbusier. Pioneering manifesto by founder of "International School." Technical and aesthetic theories, views of industry, economics, relation of form to function, "mass-production split" and much more. Profusely illustrated. 320pp. 6⅛ x 9¼. (USO) 25023-7 Pa. $9.95

HOW THE OTHER HALF LIVES, Jacob Riis. Famous journalistic record, exposing poverty and degradation of New York slums around 1900, by major social reformer. 100 striking and influential photographs. 233pp. 10 x 7⅞. 22012-5 Pa. $11.95

FRUIT KEY AND TWIG KEY TO TREES AND SHRUBS, William M. Harlow. One of the handiest and most widely used identification aids. Fruit key covers 120 deciduous and evergreen species; twig key 160 deciduous species. Easily used. Over 300 photographs. 126pp. 5⅜ x 8½. 20511-8 Pa. $3.95

COMMON BIRD SONGS, Dr. Donald J. Borror. Songs of 60 most common U.S. birds: robins, sparrows, cardinals, bluejays, finches, more–arranged in order of increasing complexity. Up to 9 variations of songs of each species. Cassette and manual 99911-4 $8.95

ORCHIDS AS HOUSE PLANTS, Rebecca Tyson Northen. Grow cattleyas and many other kinds of orchids–in a window, in a case, or under artificial light. 63 illustrations. 148pp. 5⅜ x 8½. 23261-1 Pa. $5.95

MONSTER MAZES, Dave Phillips. Masterful mazes at four levels of difficulty. Avoid deadly perils and evil creatures to find magical treasures. Solutions for all 32 exciting illustrated puzzles. 48pp. 8¼ x 11. 26005-4 Pa. $2.95

MOZART'S DON GIOVANNI (DOVER OPERA LIBRETTO SERIES), Wolfgang Amadeus Mozart. Introduced and translated by Ellen H. Bleiler. Standard Italian libretto, with complete English translation. Convenient and thoroughly portable–an ideal companion for reading along with a recording or the performance itself. Introduction. List of characters. Plot summary. 121pp. 5¼ x 8½. 24944-1 Pa. $3.95

TECHNICAL MANUAL AND DICTIONARY OF CLASSICAL BALLET, Gail Grant. Defines, explains, comments on steps, movements, poses and concepts. 15-page pictorial section. Basic book for student, viewer. 127pp. 5⅜ x 8½. 21843-0 Pa. $4.95

BRASS INSTRUMENTS: Their History and Development, Anthony Baines. Authoritative, updated survey of the evolution of trumpets, trombones, bugles, cornets, French horns, tubas and other brass wind instruments. Over 140 illustrations and 48 music examples. Corrected and updated by author. New preface. Bibliography. 320pp. 5⅜ x 8½. 27574-4 Pa. $9.95

HOLLYWOOD GLAMOR PORTRAITS, John Kobal (ed.). 145 photos from 1926-49. Harlow, Gable, Bogart, Bacall; 94 stars in all. Full background on photographers, technical aspects. 160pp. 8⅜ x 11¼. 23352-9 Pa. $12.95

MAX AND MORITZ, Wilhelm Busch. Great humor classic in both German and English. Also 10 other works: "Cat and Mouse," "Plisch and Plumm," etc. 216pp. 5⅜ x 8½. 20181-3 Pa. $6.95

THE RAVEN AND OTHER FAVORITE POEMS, Edgar Allan Poe. Over 40 of the author's most memorable poems: "The Bells," "Ulalume," "Israfel," "To Helen," "The Conqueror Worm," "Eldorado," "Annabel Lee," many more. Alphabetic lists of titles and first lines. 64pp. 5³⁄₁₆ x 8¼. 26685-0 Pa. $1.00

PERSONAL MEMOIRS OF U. S. GRANT, Ulysses Simpson Grant. Intelligent, deeply moving firsthand account of Civil War campaigns, considered by many the finest military memoirs ever written. Includes letters, historic photographs, maps and more. 528pp. 6⅛ x 9¼. 28587-1 Pa. $12.95

AMULETS AND SUPERSTITIONS, E. A. Wallis Budge. Comprehensive discourse on origin, powers of amulets in many ancient cultures: Arab, Persian Babylonian, Assyrian, Egyptian, Gnostic, Hebrew, Phoenician, Syriac, etc. Covers cross, swastika, crucifix, seals, rings, stones, etc. 584pp. 5⅜ x 8½. 23573-4 Pa. $15.95

RUSSIAN STORIES/PYCCKNE PACCKA3bI: A Dual-Language Book, edited by Gleb Struve. Twelve tales by such masters as Chekhov, Tolstoy, Dostoevsky, Pushkin, others. Excellent word-for-word English translations on facing pages, plus teaching and study aids, Russian/English vocabulary, biographical/critical introductions, more. 416pp. 5⅜ x 8½. 26244-8 Pa. $9.95

PHILADELPHIA THEN AND NOW: 60 Sites Photographed in the Past and Present, Kenneth Finkel and Susan Oyama. Rare photographs of City Hall, Logan Square, Independence Hall, Betsy Ross House, other landmarks juxtaposed with contemporary views. Captures changing face of historic city. Introduction. Captions. 128pp. 8¼ x 11. 25790-8 Pa. $9.95

AIA ARCHITECTURAL GUIDE TO NASSAU AND SUFFOLK COUNTIES, LONG ISLAND, The American Institute of Architects, Long Island Chapter, and the Society for the Preservation of Long Island Antiquities. Comprehensive, well-researched and generously illustrated volume brings to life over three centuries of Long Island's great architectural heritage. More than 240 photographs with authoritative, extensively detailed captions. 176pp. 8¼ x 11. 26946-9 Pa. $14.95

NORTH AMERICAN INDIAN LIFE: Customs and Traditions of 23 Tribes, Elsie Clews Parsons (ed.). 27 fictionalized essays by noted anthropologists examine religion, customs, government, additional facets of life among the Winnebago, Crow, Zuni, Eskimo, other tribes. 480pp. 6⅛ x 9¼. 27377-6 Pa. $10.95

FRANK LLOYD WRIGHT'S HOLLYHOCK HOUSE, Donald Hoffmann. Lavishly illustrated, carefully documented study of one of Wright's most controversial residential designs. Over 120 photographs, floor plans, elevations, etc. Detailed perceptive text by noted Wright scholar. Index. 128pp. 9¼ x 10¾. 27133-1 Pa. $11.95

THE MALE AND FEMALE FIGURE IN MOTION: 60 Classic Photographic Sequences, Eadweard Muybridge. 60 true-action photographs of men and women walking, running, climbing, bending, turning, etc., reproduced from rare 19th-century masterpiece. vi + 121pp. 9 x 12. 24745-7 Pa. $10.95

1001 QUESTIONS ANSWERED ABOUT THE SEASHORE, N. J. Berrill and Jacquelyn Berrill. Queries answered about dolphins, sea snails, sponges, starfish, fishes, shore birds, many others. Covers appearance, breeding, growth, feeding, much more. 305pp. 5¼ x 8¼. 23366-9 Pa. $9.95

GUIDE TO OWL WATCHING IN NORTH AMERICA, Donald S. Heintzelman. Superb guide offers complete data and descriptions of 19 species: barn owl, screech owl, snowy owl, many more. Expert coverage of owl-watching equipment, conservation, migrations and invasions, etc. Guide to observing sites. 84 illustrations. xiii + 193pp. 5⅜ x 8½. 27344-X Pa. $8.95

MEDICINAL AND OTHER USES OF NORTH AMERICAN PLANTS: A Historical Survey with Special Reference to the Eastern Indian Tribes, Charlotte Erichsen-Brown. Chronological historical citations document 500 years of usage of plants, trees, shrubs native to eastern Canada, northeastern U.S. Also complete identifying information. 343 illustrations. 544pp. 6½ x 9¼. 25951-X Pa. $12.95

STORYBOOK MAZES, Dave Phillips. 23 stories and mazes on two-page spreads: Wizard of Oz, Treasure Island, Robin Hood, etc. Solutions. 64pp. 8¼ x 11. 23628-5 Pa. $2.95

NEGRO FOLK MUSIC, U.S.A., Harold Courlander. Noted folklorist's scholarly yet readable analysis of rich and varied musical tradition. Includes authentic versions of over 40 folk songs. Valuable bibliography and discography. xi + 324pp. 5⅜ x 8½. 27350-4 Pa. $9.95

MOVIE-STAR PORTRAITS OF THE FORTIES, John Kobal (ed.). 163 glamor, studio photos of 106 stars of the 1940s: Rita Hayworth, Ava Gardner, Marlon Brando, Clark Gable, many more. 176pp. 8⅜ x 11¼. 23546-7 Pa. $14.95

BENCHLEY LOST AND FOUND, Robert Benchley. Finest humor from early 30s, about pet peeves, child psychologists, post office and others. Mostly unavailable elsewhere. 73 illustrations by Peter Arno and others. 183pp. 5⅜ x 8½. 22410-4 Pa. $6.95

YEKL and THE IMPORTED BRIDEGROOM AND OTHER STORIES OF YIDDISH NEW YORK, Abraham Cahan. Film Hester Street based on Yekl (1896). Novel, other stories among first about Jewish immigrants on N.Y.'s East Side. 240pp. 5⅜ x 8½. 22427-9 Pa. $6.95

SELECTED POEMS, Walt Whitman. Generous sampling from *Leaves of Grass*. Twenty-four poems include "I Hear America Singing," "Song of the Open Road," "I Sing the Body Electric," "When Lilacs Last in the Dooryard Bloom'd," "O Captain! My Captain!"–all reprinted from an authoritative edition. Lists of titles and first lines. 128pp. 5¹⁵⁄₁₆ x 8¼. 26878-0 Pa. $1.00

THE BEST TALES OF HOFFMANN, E. T. A. Hoffmann. 10 of Hoffmann's most important stories: "Nutcracker and the King of Mice," "The Golden Flowerpot," etc. 458pp. 5⅜ x 8½.   21793-0 Pa. $9.95

FROM FETISH TO GOD IN ANCIENT EGYPT, E. A. Wallis Budge. Rich detailed survey of Egyptian conception of "God" and gods, magic, cult of animals, Osiris, more. Also, superb English translations of hymns and legends. 240 illustrations. 545pp. 5⅜ x 8½.   25803-3 Pa. $13.95

FRENCH STORIES/CONTES FRANÇAIS: A Dual-Language Book, Wallace Fowlie. Ten stories by French masters, Voltaire to Camus: "Micromegas" by Voltaire; "The Atheist's Mass" by Balzac; "Minuet" by de Maupassant; "The Guest" by Camus, six more. Excellent English translations on facing pages. Also French-English vocabulary list, exercises, more. 352pp. 5⅜ x 8½.   26443-2 Pa. $9.95

CHICAGO AT THE TURN OF THE CENTURY IN PHOTOGRAPHS: 122 Historic Views from the Collections of the Chicago Historical Society, Larry A. Viskochil. Rare large-format prints offer detailed views of City Hall, State Street, the Loop, Hull House, Union Station, many other landmarks, circa 1904-1913. Introduction. Captions. Maps. 144pp. 9⅜ x 12¼.   24656-6 Pa. $12.95

OLD BROOKLYN IN EARLY PHOTOGRAPHS, 1865-1929, William Lee Younger. Luna Park, Gravesend race track, construction of Grand Army Plaza, moving of Hotel Brighton, etc. 157 previously unpublished photographs. 165pp. 8⅞ x 11¾.
                    23587-4 Pa. $13.95

THE MYTHS OF THE NORTH AMERICAN INDIANS, Lewis Spence. Rich anthology of the myths and legends of the Algonquins, Iroquois, Pawnees and Sioux, prefaced by an extensive historical and ethnological commentary. 36 illustrations. 480pp. 5⅜ x 8½.   25967-6 Pa. $10.95

AN ENCYCLOPEDIA OF BATTLES: Accounts of Over 1,560 Battles from 1479 B.C. to the Present, David Eggenberger. Essential details of every major battle in recorded history from the first battle of Megiddo in 1479 B.C. to Grenada in 1984. List of Battle Maps. New Appendix covering the years 1967-1984. Index. 99 illustrations. 544pp. 6½ x 9¼.   24913-1 Pa. $16.95

SAILING ALONE AROUND THE WORLD, Captain Joshua Slocum. First man to sail around the world, alone, in small boat. One of great feats of seamanship told in delightful manner. 67 illustrations. 294pp. 5⅜ x 8½.   20326-3 Pa. $6.95

ANARCHISM AND OTHER ESSAYS, Emma Goldman. Powerful, penetrating, prophetic essays on direct action, role of minorities, prison reform, puritan hypocrisy, violence, etc. 271pp. 5⅜ x 8½.   22484-8 Pa. $7.95

MYTHS OF THE HINDUS AND BUDDHISTS, Ananda K. Coomaraswamy and Sister Nivedita. Great stories of the epics; deeds of Krishna, Shiva, taken from puranas, Vedas, folk tales; etc. 32 illustrations. 400pp. 5⅜ x 8½.   21759-0 Pa. $12.95

BEYOND PSYCHOLOGY, Otto Rank. Fear of death, desire of immortality, nature of sexuality, social organization, creativity, according to Rankian system. 291pp. 5⅜ x 8½.
                    20485-5 Pa. $8.95

A THEOLOGICO-POLITICAL TREATISE, Benedict Spinoza. Also contains unfinished Political Treatise. Great classic on religious liberty, theory of government on common consent. R. Elwes translation. Total of 421pp. 5⅜ x 8½. 20249-6 Pa. $9.95

MY BONDAGE AND MY FREEDOM, Frederick Douglass. Born a slave, Douglass became outspoken force in antislavery movement. The best of Douglass' autobiographies. Graphic description of slave life. 464pp. 5⅜ x 8½. 22457-0 Pa. $8.95

FOLLOWING THE EQUATOR: A Journey Around the World, Mark Twain. Fascinating humorous account of 1897 voyage to Hawaii, Australia, India, New Zealand, etc. Ironic, bemused reports on peoples, customs, climate, flora and fauna, politics, much more. 197 illustrations. 720pp. 5⅜ x 8½. 26113-1 Pa. $15.95

THE PEOPLE CALLED SHAKERS, Edward D. Andrews. Definitive study of Shakers: origins, beliefs, practices, dances, social organization, furniture and crafts, etc. 33 illustrations. 351pp. 5⅜ x 8½. 21081-2 Pa. $8.95

THE MYTHS OF GREECE AND ROME, H. A. Guerber. A classic of mythology, generously illustrated, long prized for its simple, graphic, accurate retelling of the principal myths of Greece and Rome, and for its commentary on their origins and significance. With 64 illustrations by Michelangelo, Raphael, Titian, Rubens, Canova, Bernini and others. 480pp. 5⅜ x 8½. 27584-1 Pa. $9.95

PSYCHOLOGY OF MUSIC, Carl E. Seashore. Classic work discusses music as a medium from psychological viewpoint. Clear treatment of physical acoustics, auditory apparatus, sound perception, development of musical skills, nature of musical feeling, host of other topics. 88 figures. 408pp. 5⅜ x 8½. 21851-1 Pa. $11.95

THE PHILOSOPHY OF HISTORY, Georg W. Hegel. Great classic of Western thought develops concept that history is not chance but rational process, the evolution of freedom. 457pp. 5⅜ x 8½. 20112-0 Pa. $9.95

THE BOOK OF TEA, Kakuzo Okakura. Minor classic of the Orient: entertaining, charming explanation, interpretation of traditional Japanese culture in terms of tea ceremony. 94pp. 5⅜ x 8½. 20070-1 Pa. $3.95

LIFE IN ANCIENT EGYPT, Adolf Erman. Fullest, most thorough, detailed older account with much not in more recent books, domestic life, religion, magic, medicine, commerce, much more. Many illustrations reproduce tomb paintings, carvings, hieroglyphs, etc. 597pp. 5⅜ x 8½. 22632-8 Pa. $12.95

SUNDIALS, Their Theory and Construction, Albert Waugh. Far and away the best, most thorough coverage of ideas, mathematics concerned, types, construction, adjusting anywhere. Simple, nontechnical treatment allows even children to build several of these dials. Over 100 illustrations. 230pp. 5⅜ x 8½. 22947-5 Pa. $8.95

DYNAMICS OF FLUIDS IN POROUS MEDIA, Jacob Bear. For advanced students of ground water hydrology, soil mechanics and physics, drainage and irrigation engineering, and more. 335 illustrations. Exercises, with answers. 784pp. 6⅛ x 9¼. 65675-6 Pa. $19.95

SONGS OF EXPERIENCE: Facsimile Reproduction with 26 Plates in Full Color, William Blake. 26 full-color plates from a rare 1826 edition. Includes "The Tyger," "London," "Holy Thursday," and other poems. Printed text of poems. 48pp. 5¼ x 7. 24636-1 Pa. $4.95

OLD-TIME VIGNETTES IN FULL COLOR, Carol Belanger Grafton (ed.). Over 390 charming, often sentimental illustrations, selected from archives of Victorian graphics—pretty women posing, children playing, food, flowers, kittens and puppies, smiling cherubs, birds and butterflies, much more. All copyright-free. 48pp. 9¼ x 12¼. 27269-9 Pa. $7.95

PERSPECTIVE FOR ARTISTS, Rex Vicat Cole. Depth, perspective of sky and sea, shadows, much more, not usually covered. 391 diagrams, 81 reproductions of drawings and paintings. 279pp. 5⅜ x 8½. 22487-2 Pa. $7.95

DRAWING THE LIVING FIGURE, Joseph Sheppard. Innovative approach to artistic anatomy focuses on specifics of surface anatomy, rather than muscles and bones. Over 170 drawings of live models in front, back and side views, and in widely varying poses. Accompanying diagrams. 177 illustrations. Introduction. Index. 144pp. 8⅜ x11¼. 26723-7 Pa. $8.95

GOTHIC AND OLD ENGLISH ALPHABETS: 100 Complete Fonts, Dan X. Solo. Add power, elegance to posters, signs, other graphics with 100 stunning copyright-free alphabets: Blackstone, Dolbey, Germania, 97 more–including many lower-case, numerals, punctuation marks. 104pp. 8⅛ x 11. 24695-7 Pa. $8.95

HOW TO DO BEADWORK, Mary White. Fundamental book on craft from simple projects to five-bead chains and woven works. 106 illustrations. 142pp. 5⅜ x 8. 20697-1 Pa. $5.95

THE BOOK OF WOOD CARVING, Charles Marshall Sayers. Finest book for beginners discusses fundamentals and offers 34 designs. "Absolutely first rate . . . well thought out and well executed."–E. J. Tangerman. 118pp. 7¾ x 10⅝. 23654-4 Pa. $7.95

ILLUSTRATED CATALOG OF CIVIL WAR MILITARY GOODS: Union Army Weapons, Insignia, Uniform Accessories, and Other Equipment, Schuyler, Hartley, and Graham. Rare, profusely illustrated 1846 catalog includes Union Army uniform and dress regulations, arms and ammunition, coats, insignia, flags, swords, rifles, etc. 226 illustrations. 160pp. 9 x 12. 24939-5 Pa. $10.95

WOMEN'S FASHIONS OF THE EARLY 1900s: An Unabridged Republication of "New York Fashions, 1909," National Cloak & Suit Co. Rare catalog of mail-order fashions documents women's and children's clothing styles shortly after the turn of the century. Captions offer full descriptions, prices. Invaluable resource for fashion, costume historians. Approximately 725 illustrations. 128pp. 8⅜ x 11¼. 27276-1 Pa. $11.95

THE 1912 AND 1915 GUSTAV STICKLEY FURNITURE CATALOGS, Gustav Stickley. With over 200 detailed illustrations and descriptions, these two catalogs are essential reading and reference materials and identification guides for Stickley furniture. Captions cite materials, dimensions and prices. 112pp. 6½ x 9¼. 26676-1 Pa. $9.95

EARLY AMERICAN LOCOMOTIVES, John H. White, Jr. Finest locomotive engravings from early 19th century: historical (1804–74), main-line (after 1870), special, foreign, etc. 147 plates. 142pp. 11⅜ x 8¼. 22772-3 Pa. $10.95

THE TALL SHIPS OF TODAY IN PHOTOGRAPHS, Frank O. Braynard. Lavishly illustrated tribute to nearly 100 majestic contemporary sailing vessels: Amerigo Vespucci, Clearwater, Constitution, Eagle, Mayflower, Sea Cloud, Victory, many more. Authoritative captions provide statistics, background on each ship. 190 black-and-white photographs and illustrations. Introduction. 128pp. 8⅞ x 11¾. 27163-3 Pa. $14.95

EARLY NINETEENTH-CENTURY CRAFTS AND TRADES, Peter Stockham (ed.). Extremely rare 1807 volume describes to youngsters the crafts and trades of the day: brickmaker, weaver, dressmaker, bookbinder, ropemaker, saddler, many more. Quaint prose, charming illustrations for each craft. 20 black-and-white line illustrations. 192pp. 4⅝ x 6. 27293-1 Pa. $4.95

VICTORIAN FASHIONS AND COSTUMES FROM HARPER'S BAZAR, 1867–1898, Stella Blum (ed.). Day costumes, evening wear, sports clothes, shoes, hats, other accessories in over 1,000 detailed engravings. 320pp. 9⅜ x 12¼. 22990-4 Pa. $15.95

GUSTAV STICKLEY, THE CRAFTSMAN, Mary Ann Smith. Superb study surveys broad scope of Stickley's achievement, especially in architecture. Design philosophy, rise and fall of the Craftsman empire, descriptions and floor plans for many Craftsman houses, more. 86 black-and-white halftones. 31 line illustrations. Introduction 208pp. 6½ x 9¼. 27210-9 Pa. $9.95

THE LONG ISLAND RAIL ROAD IN EARLY PHOTOGRAPHS, Ron Ziel. Over 220 rare photos, informative text document origin (1844) and development of rail service on Long Island. Vintage views of early trains, locomotives, stations, passengers, crews, much more. Captions. 8⅞ x 11¾. 26301-0 Pa. $13.95

THE BOOK OF OLD SHIPS: From Egyptian Galleys to Clipper Ships, Henry B. Culver. Superb, authoritative history of sailing vessels, with 80 magnificent line illustrations. Galley, bark, caravel, longship, whaler, many more. Detailed, informative text on each vessel by noted naval historian. Introduction. 256pp. 5⅜ x 8½. 27332-6 Pa. $7.95

TEN BOOKS ON ARCHITECTURE, Vitruvius. The most important book ever written on architecture. Early Roman aesthetics, technology, classical orders, site selection, all other aspects. Morgan translation. 331pp. 5⅜ x 8½. 20645-9 Pa. $8.95

THE HUMAN FIGURE IN MOTION, Eadweard Muybridge. More than 4,500 stopped-action photos, in action series, showing undraped men, women, children jumping, lying down, throwing, sitting, wrestling, carrying, etc. 390pp. 7⅞ x 10⅝. 20204-6 Clothbd. $27.95

TREES OF THE EASTERN AND CENTRAL UNITED STATES AND CANADA, William M. Harlow. Best one-volume guide to 140 trees. Full descriptions, woodlore, range, etc. Over 600 illustrations. Handy size. 288pp. 4½ x 6⅜. 20395-6 Pa. $6.95

SONGS OF WESTERN BIRDS, Dr. Donald J. Borror. Complete song and call repertoire of 60 western species, including flycatchers, juncoes, cactus wrens, many more–includes fully illustrated booklet. Cassette and manual 99913-0 $8.95

GROWING AND USING HERBS AND SPICES, Milo Miloradovich. Versatile handbook provides all the information needed for cultivation and use of all the herbs and spices available in North America. 4 illustrations. Index. Glossary. 236pp. 5⅜ x 8½. 25058-X Pa. $7.95

BIG BOOK OF MAZES AND LABYRINTHS, Walter Shepherd. 50 mazes and labyrinths in all–classical, solid, ripple, and more–in one great volume. Perfect inexpensive puzzler for clever youngsters. Full solutions. 112pp. 8⅛ x 11. 22951-3 Pa. $4.95

PIANO TUNING, J. Cree Fischer. Clearest, best book for beginner, amateur. Simple repairs, raising dropped notes, tuning by easy method of flattened fifths. No previous skills needed. 4 illustrations. 201pp. 5⅜ x 8½.   23267-0 Pa. $6.95

A SOURCE BOOK IN THEATRICAL HISTORY, A. M. Nagler. Contemporary observers on acting, directing, make-up, costuming, stage props, machinery, scene design, from Ancient Greece to Chekhov. 611pp. 5⅜ x 8½.   20515-0 Pa. $12.95

THE COMPLETE NONSENSE OF EDWARD LEAR, Edward Lear. All nonsense limericks, zany alphabets, Owl and Pussycat, songs, nonsense botany, etc., illustrated by Lear. Total of 320pp. 5⅜ x 8½. (USO)   20167-8 Pa. $7.95

VICTORIAN PARLOUR POETRY: An Annotated Anthology, Michael R. Turner. 117 gems by Longfellow, Tennyson, Browning, many lesser-known poets. "The Village Blacksmith," "Curfew Must Not Ring Tonight," "Only a Baby Small," dozens more, often difficult to find elsewhere. Index of poets, titles, first lines. xxiii + 325pp. 5⅜ x 8¼.   27044-0 Pa. $8.95

DUBLINERS, James Joyce. Fifteen stories offer vivid, tightly focused observations of the lives of Dublin's poorer classes. At least one, "The Dead," is considered a masterpiece. Reprinted complete and unabridged from standard edition. 160pp. 5³⁄₁₆ x 8¼.   26870-5 Pa. $1.00

THE HAUNTED MONASTERY and THE CHINESE MAZE MURDERS, Robert van Gulik. Two full novels by van Gulik, set in 7th-century China, continue adventures of Judge Dee and his companions. An evil Taoist monastery, seemingly supernatural events; overgrown topiary maze hides strange crimes. 27 illustrations. 328pp. 5⅜ x 8½.   23502-5 Pa. $8.95

THE BOOK OF THE SACRED MAGIC OF ABRAMELIN THE MAGE, translated by S. MacGregor Mathers. Medieval manuscript of ceremonial magic. Basic document in Aleister Crowley, Golden Dawn groups. 268pp. 5⅜ x 8½.   23211-5 Pa. $9.95

NEW RUSSIAN-ENGLISH AND ENGLISH-RUSSIAN DICTIONARY, M. A. O'Brien. This is a remarkably handy Russian dictionary, containing a surprising amount of information, including over 70,000 entries. 366pp. 4½ x 6⅛.   20208-9 Pa. $10.95

HISTORIC HOMES OF THE AMERICAN PRESIDENTS, Second, Revised Edition, Irvin Haas. A traveler's guide to American Presidential homes, most open to the public, depicting and describing homes occupied by every American President from George Washington to George Bush. With visiting hours, admission charges, travel routes. 175 photographs. Index. 160pp. 8¼ x 11.   26751-2 Pa. $11.95

NEW YORK IN THE FORTIES, Andreas Feininger. 162 brilliant photographs by the well-known photographer, formerly with *Life* magazine. Commuters, shoppers, Times Square at night, much else from city at its peak. Captions by John von Hartz. 181pp. 9¼ x 10¾.   23585-8 Pa. $13.95

INDIAN SIGN LANGUAGE, William Tomkins. Over 525 signs developed by Sioux and other tribes. Written instructions and diagrams. Also 290 pictographs. 111pp. 6⅛ x 9¼.   22029-X Pa. $3.95

ANATOMY: A Complete Guide for Artists, Joseph Sheppard. A master of figure drawing shows artists how to render human anatomy convincingly. Over 460 illustrations. 224pp. 8⅜ x 11¼. 27279-6 Pa. $11.95

MEDIEVAL CALLIGRAPHY: Its History and Technique, Marc Drogin. Spirited history, comprehensive instruction manual covers 13 styles (ca. 4th century thru 15th). Excellent photographs; directions for duplicating medieval techniques with modern tools. 224pp. 8⅜ x 11¼. 26142-5 Pa. $12.95

DRIED FLOWERS: How to Prepare Them, Sarah Whitlock and Martha Rankin. Complete instructions on how to use silica gel, meal and borax, perlite aggregate, sand and borax, glycerine and water to create attractive permanent flower arrangements. 12 illustrations. 32pp. 5⅜ x 8½. 21802-3 Pa. $1.00

EASY-TO-MAKE BIRD FEEDERS FOR WOODWORKERS, Scott D. Campbell. Detailed, simple-to-use guide for designing, constructing, caring for and using feeders. Text, illustrations for 12 classic and contemporary designs. 96pp. 5⅜ x 8½. 25847-5 Pa. $3.95

SCOTTISH WONDER TALES FROM MYTH AND LEGEND, Donald A. Mackenzie. 16 lively tales tell of giants rumbling down mountainsides, of a magic wand that turns stone pillars into warriors, of gods and goddesses, evil hags, powerful forces and more. 240pp. 5⅜ x 8½. 29677-6 Pa. $6.95

THE HISTORY OF UNDERCLOTHES, C. Willett Cunnington and Phyllis Cunnington. Fascinating, well-documented survey covering six centuries of English undergarments, enhanced with over 100 illustrations: 12th-century laced-up bodice, footed long drawers (1795), 19th-century bustles, 19th-century corsets for men, Victorian "bust improvers," much more. 272pp. 5⅜ x 8¼. 27124-2 Pa. $9.95

ARTS AND CRAFTS FURNITURE: The Complete Brooks Catalog of 1912, Brooks Manufacturing Co. Photos and detailed descriptions of more than 150 now very collectible furniture designs from the Arts and Crafts movement depict davenports, settees, buffets, desks, tables, chairs, bedsteads, dressers and more, all built of solid, quarter-sawed oak. Invaluable for students and enthusiasts of antiques, Americana and the decorative arts. 80pp. 6½ x 9¼. 27471-3 Pa. $8.95

HOW WE INVENTED THE AIRPLANE: An Illustrated History, Orville Wright. Fascinating firsthand account covers early experiments, construction of planes and motors, first flights, much more. Introduction and commentary by Fred C. Kelly. 76 photographs. 96pp. 8¼ x 11. 25662-6 Pa. $8.95

THE ARTS OF THE SAILOR: Knotting, Splicing and Ropework, Hervey Garrett Smith. Indispensable shipboard reference covers tools, basic knots and useful hitches; handsewing and canvas work, more. Over 100 illustrations. Delightful reading for sea lovers. 256pp. 5⅜ x 8½. 26440-8 Pa. $8.95

FRANK LLOYD WRIGHT'S FALLINGWATER: The House and Its History, Second, Revised Edition, Donald Hoffmann. A total revision—both in text and illustrations—of the standard document on Fallingwater, the boldest, most personal architectural statement of Wright's mature years, updated with valuable new material from the recently opened Frank Lloyd Wright Archives. "Fascinating"—*The New York Times*. 116 illustrations. 128pp. 9¼ x 10¾. 27430-6 Pa. $12.95

PHOTOGRAPHIC SKETCHBOOK OF THE CIVIL WAR, Alexander Gardner. 100 photos taken on field during the Civil War. Famous shots of Manassas Harper's Ferry, Lincoln, Richmond, slave pens, etc. 244pp. 10⅝ x 8¼. 22731-6 Pa. $10.95

FIVE ACRES AND INDEPENDENCE, Maurice G. Kains. Great back-to-the-land classic explains basics of self-sufficient farming. The one book to get. 95 illustrations. 397pp. 5⅜ x 8½. 20974-1 Pa. $7.95

SONGS OF EASTERN BIRDS, Dr. Donald J. Borror. Songs and calls of 60 species most common to eastern U.S.: warblers, woodpeckers, flycatchers, thrushes, larks, many more in high-quality recording. Cassette and manual 99912-2 $9.95

A MODERN HERBAL, Margaret Grieve. Much the fullest, most exact, most useful compilation of herbal material. Gigantic alphabetical encyclopedia, from aconite to zedoary, gives botanical information, medical properties, folklore, economic uses, much else. Indispensable to serious reader. 161 illustrations. 888pp. 6½ x 9¼. 2-vol. set. (USO) Vol. I: 22798-7 Pa. $9.95
Vol. II: 22799-5 Pa. $9.95

HIDDEN TREASURE MAZE BOOK, Dave Phillips. Solve 34 challenging mazes accompanied by heroic tales of adventure. Evil dragons, people-eating plants, blood-thirsty giants, many more dangerous adversaries lurk at every twist and turn. 34 mazes, stories, solutions. 48pp. 8¼ x 11. 24566-7 Pa. $2.95

LETTERS OF W. A. MOZART, Wolfgang A. Mozart. Remarkable letters show bawdy wit, humor, imagination, musical insights, contemporary musical world; includes some letters from Leopold Mozart. 276pp. 5⅜ x 8½. 22859-2 Pa. $7.95

BASIC PRINCIPLES OF CLASSICAL BALLET, Agrippina Vaganova. Great Russian theoretician, teacher explains methods for teaching classical ballet. 118 illustrations. 175pp. 5⅜ x 8½. 22036-2 Pa. $5.95

THE JUMPING FROG, Mark Twain. Revenge edition. The original story of The Celebrated Jumping Frog of Calaveras County, a hapless French translation, and Twain's hilarious "retranslation" from the French. 12 illustrations. 66pp. 5⅜ x 8½. 22686-7 Pa. $3.95

BEST REMEMBERED POEMS, Martin Gardner (ed.). The 126 poems in this superb collection of 19th- and 20th-century British and American verse range from Shelley's "To a Skylark" to the impassioned "Renascence" of Edna St. Vincent Millay and to Edward Lear's whimsical "The Owl and the Pussycat." 224pp. 5⅜ x 8½. 27165-X Pa. $5.95

COMPLETE SONNETS, William Shakespeare. Over 150 exquisite poems deal with love, friendship, the tyranny of time, beauty's evanescence, death and other themes in language of remarkable power, precision and beauty. Glossary of archaic terms. 80pp. 5³⁄₁₆ x 8¼. 26686-9 Pa. $1.00

BODIES IN A BOOKSHOP, R. T. Campbell. Challenging mystery of blackmail and murder with ingenious plot and superbly drawn characters. In the best tradition of British suspense fiction. 192pp. 5⅜ x 8½. 24720-1 Pa. $6.95

CATALOG OF DOVER BOOKS

THE WIT AND HUMOR OF OSCAR WILDE, Alvin Redman (ed.). More than 1,000 ripostes, paradoxes, wisecracks: Work is the curse of the drinking classes; I can resist everything except temptation; etc. 258pp. 5⅜ x 8½. 20602-5 Pa. $6.95

SHAKESPEARE LEXICON AND QUOTATION DICTIONARY, Alexander Schmidt. Full definitions, locations, shades of meaning in every word in plays and poems. More than 50,000 exact quotations. 1,485pp. 6½ x 9¼. 2-vol. set.
Vol. 1: 22726-X Pa. $17.95
Vol. 2: 22727-8 Pa. $17.95

SELECTED POEMS, Emily Dickinson. Over 100 best-known, best-loved poems by one of America's foremost poets, reprinted from authoritative early editions. No comparable edition at this price. Index of first lines. 64pp. 5$\frac{3}{16}$ x 8¼.
26466-1 Pa. $1.00

CELEBRATED CASES OF JUDGE DEE (DEE GOONG AN), translated by Robert van Gulik. Authentic 18th-century Chinese detective novel; Dee and associates solve three interlocked cases. Led to van Gulik's own stories with same characters. Extensive introduction. 9 illustrations. 237pp. 5⅜ x 8½. 23337-5 Pa. $7.95

THE MALLEUS MALEFICARUM OF KRAMER AND SPRENGER, translated by Montague Summers. Full text of most important witchhunter's "bible," used by both Catholics and Protestants. 278pp. 6⅝ x 10. 22802-9 Pa. $12.95

SPANISH STORIES/CUENTOS ESPAÑOLES: A Dual-Language Book, Angel Flores (ed.). Unique format offers 13 great stories in Spanish by Cervantes, Borges, others. Faithful English translations on facing pages. 352pp. 5⅜ x 8½.
25399-6 Pa. $8.95

THE CHICAGO WORLD'S FAIR OF 1893: A Photographic Record, Stanley Appelbaum (ed.). 128 rare photos show 200 buildings, Beaux-Arts architecture, Midway, original Ferris Wheel, Edison's kinetoscope, more. Architectural emphasis; full text. 116pp. 8¼ x 11. 23990-X Pa. $9.95

OLD QUEENS, N.Y., IN EARLY PHOTOGRAPHS, Vincent F. Seyfried and William Asadorian. Over 160 rare photographs of Maspeth, Jamaica, Jackson Heights, and other areas. Vintage views of DeWitt Clinton mansion, 1939 World's Fair and more. Captions. 192pp. 8⅞ x 11. 26358-4 Pa. $12.95

CAPTURED BY THE INDIANS: 15 Firsthand Accounts, 1750-1870, Frederick Drimmer. Astounding true historical accounts of grisly torture, bloody conflicts, relentless pursuits, miraculous escapes and more, by people who lived to tell the tale. 384pp. 5⅜ x 8½. 24901-8 Pa. $8.95

THE WORLD'S GREAT SPEECHES, Lewis Copeland and Lawrence W. Lamm (eds.). Vast collection of 278 speeches of Greeks to 1970. Powerful and effective models; unique look at history. 842pp. 5⅜ x 8½. 20468-5 Pa. $14.95

THE BOOK OF THE SWORD, Sir Richard F. Burton. Great Victorian scholar/adventurer's eloquent, erudite history of the "queen of weapons"–from prehistory to early Roman Empire. Evolution and development of early swords, variations (sabre, broadsword, cutlass, scimitar, etc.), much more. 336pp. 6⅛ x 9¼.
25434-8 Pa. $9.95

AUTOBIOGRAPHY: The Story of My Experiments with Truth, Mohandas K. Gandhi. Boyhood, legal studies, purification, the growth of the Satyagraha (nonviolent protest) movement. Critical, inspiring work of the man responsible for the freedom of India. 480pp. 5⅜ x 8½. (USO)   24593-4 Pa. $8.95

CELTIC MYTHS AND LEGENDS, T. W. Rolleston. Masterful retelling of Irish and Welsh stories and tales. Cuchulain, King Arthur, Deirdre, the Grail, many more. First paperback edition. 58 full-page illustrations. 512pp. 5⅜ x 8½.   26507-2 Pa. $9.95

THE PRINCIPLES OF PSYCHOLOGY, William James. Famous long course complete, unabridged. Stream of thought, time perception, memory, experimental methods; great work decades ahead of its time. 94 figures. 1,391pp. 5⅜ x 8½. 2-vol. set.
Vol. I: 20381-6 Pa. $13.95
Vol. II: 20382-4 Pa. $14.95

THE WORLD AS WILL AND REPRESENTATION, Arthur Schopenhauer. Definitive English translation of Schopenhauer's life work, correcting more than 1,000 errors, omissions in earlier translations. Translated by E. F. J. Payne. Total of 1,269pp. 5⅜ x 8½. 2-vol. set.
Vol. 1: 21761-2 Pa. $12.95
Vol. 2: 21762-0 Pa. $12.95

MAGIC AND MYSTERY IN TIBET, Madame Alexandra David-Neel. Experiences among lamas, magicians, sages, sorcerers, Bonpa wizards. A true psychic discovery. 32 illustrations. 321pp. 5⅜ x 8½. (USO)   22682-4 Pa. $9.95

THE EGYPTIAN BOOK OF THE DEAD, E. A. Wallis Budge. Complete reproduction of Ani's papyrus, finest ever found. Full hieroglyphic text, interlinear transliteration, word-for-word translation, smooth translation. 533pp. 6½ x 9¼.
21866-X Pa. $11.95

MATHEMATICS FOR THE NONMATHEMATICIAN, Morris Kline. Detailed, college-level treatment of mathematics in cultural and historical context, with numerous exercises. Recommended Reading Lists. Tables. Numerous figures. 641pp. 5⅜ x 8½.
24823-2 Pa. $11.95

THEORY OF WING SECTIONS: Including a Summary of Airfoil Data, Ira H. Abbott and A. E. von Doenhoff. Concise compilation of subsonic aerodynamic characteristics of NACA wing sections, plus description of theory. 350pp. of tables. 693pp. 5⅜ x 8½.   60586-8 Pa. $14.95

THE RIME OF THE ANCIENT MARINER, Gustave Doré, S. T. Coleridge. Doré's finest work; 34 plates capture moods, subtleties of poem. Flawless full-size reproductions printed on facing pages with authoritative text of poem. "Beautiful. Simply beautiful."—*Publisher's Weekly.* 77pp. 9¼ x 12.   22305-1 Pa. $7.95

NORTH AMERICAN INDIAN DESIGNS FOR ARTISTS AND CRAFTSPEOPLE, Eva Wilson. Over 360 authentic copyright-free designs adapted from Navajo blankets, Hopi pottery, Sioux buffalo hides, more. Geometrics, symbolic figures, plant and animal motifs, etc. 128pp. 8⅜ x 11. (EUK)   25341-4 Pa. $8.95

SCULPTURE: Principles and Practice, Louis Slobodkin. Step-by-step approach to clay, plaster, metals, stone; classical and modern. 253 drawings, photos. 255pp. 8⅛ x 11.
22960-2 Pa. $11.95

THE INFLUENCE OF SEA POWER UPON HISTORY, 1660–1783, A. T. Mahan. Influential classic of naval history and tactics still used as text in war colleges. First paperback edition. 4 maps. 24 battle plans. 640pp. 5⅜ x 8½. 25509-3 Pa. $14.95

THE STORY OF THE TITANIC AS TOLD BY ITS SURVIVORS, Jack Winocour (ed.). What it was really like. Panic, despair, shocking inefficiency, and a little heroism. More thrilling than any fictional account. 26 illustrations. 320pp. 5⅜ x 8½. 20610-6 Pa. $8.95

FAIRY AND FOLK TALES OF THE IRISH PEASANTRY, William Butler Yeats (ed.). Treasury of 64 tales from the twilight world of Celtic myth and legend: "The Soul Cages," "The Kildare Pooka," "King O'Toole and his Goose," many more. Introduction and Notes by W. B. Yeats. 352pp. 5⅜ x 8½. 26941-8 Pa. $8.95

BUDDHIST MAHAYANA TEXTS, E. B. Cowell and Others (eds.). Superb, accurate translations of basic documents in Mahayana Buddhism, highly important in history of religions. The Buddha-karita of Asvaghosha, Larger Sukhavativyuha, more. 448pp. 5⅜ x 8½. 25552-2 Pa. $12.95

ONE TWO THREE . . . INFINITY: Facts and Speculations of Science, George Gamow. Great physicist's fascinating, readable overview of contemporary science: number theory, relativity, fourth dimension, entropy, genes, atomic structure, much more. 128 illustrations. Index. 352pp. 5⅜ x 8½. 25664-2 Pa. $8.95

ENGINEERING IN HISTORY, Richard Shelton Kirby, et al. Broad, nontechnical survey of history's major technological advances: birth of Greek science, industrial revolution, electricity and applied science, 20th-century automation, much more. 181 illustrations. ". . . excellent . . ."–*Isis*. Bibliography. vii + 530pp. 5⅜ x 8¼. 26412-2 Pa. $14.95

DALÍ ON MODERN ART: The Cuckolds of Antiquated Modern Art, Salvador Dalí. Influential painter skewers modern art and its practitioners. Outrageous evaluations of Picasso, Cézanne, Turner, more. 15 renderings of paintings discussed. 44 calligraphic decorations by Dalí. 96pp. 5⅜ x 8½. (USO) 29220-7 Pa. $4.95

ANTIQUE PLAYING CARDS: A Pictorial History, Henry René D'Allemagne. Over 900 elaborate, decorative images from rare playing cards (14th–20th centuries): Bacchus, death, dancing dogs, hunting scenes, royal coats of arms, players cheating, much more. 96pp. 9¼ x 12¼. 29265-7 Pa. $12.95

MAKING FURNITURE MASTERPIECES: 30 Projects with Measured Drawings, Franklin H. Gottshall. Step-by-step instructions, illustrations for constructing handsome, useful pieces, among them a Sheraton desk, Chippendale chair, Spanish desk, Queen Anne table and a William and Mary dressing mirror. 224pp. 8⅛ x 11¼. 29338-6 Pa. $13.95

THE FOSSIL BOOK: A Record of Prehistoric Life, Patricia V. Rich et al. Profusely illustrated definitive guide covers everything from single-celled organisms and dinosaurs to birds and mammals and the interplay between climate and man. Over 1,500 illustrations. 760pp. 7½ x 10⅛. 29371-8 Pa. $29.95

Prices subject to change without notice.

Available at your book dealer or write for free catalog to Dept. GI, Dover Publications, Inc., 31 East 2nd St., Mineola, N.Y. 11501. Dover publishes more than 500 books each year on science, elementary and advanced mathematics, biology, music, art, literary history, social sciences and other areas.